Mathematical
Biology

Ω IAS/PARK CITY
MATHEMATICS SERIES
Volume 14

Mathematical Biology

Mark A. Lewis
Mark A. J. Chaplain
James P. Keener
Philip K. Maini
Editors

American Mathematical Society
Providence, Rhode Island

Institute for Advanced Study
Princeton, New Jersey

John C. Polking, Series Editor
Mark A. Lewis, Volume Editor
Mark A. J. Chaplain, Volume Editor
James P. Keener, Volume Editor
Philip K. Maini, Volume Editor

IAS/Park City Mathematics Institute runs mathematics education programs that bring together high school mathematics teachers, researchers in mathematics and mathematics education, undergraduate mathematics faculty, graduate students, and undergraduates to participate in distinct but overlapping programs of research and education. This volume contains the lecture notes from the Graduate Summer School program.

2000 *Mathematics Subject Classification.* Primary 34-02, 35-02, 37-02, 92-02.

Library of Congress Cataloging-in-Publication Data

Mathematical biology / Mark A. Lewis... [et al.], editors.
p. cm. — (IAS/Park City mathematics series ; v. 14)
ISBN 978-0-8218-4765-7 (alk. paper)
1. Biology–Mathematical models. I. Lewis, M. (Mark), 1962–

QH323.5.M36369 2009
570.1′5118–dc22 200847401

∞ The paper used in this book is acid-free and falls within the guidelines
established to ensure permanence and durability.
Visit the AMS home page at http://www.ams.org/

10 9 8 7 6 5 4 3 2 1 14 13 12 11 10 09

Contents

Preface

The IAS/Park City Mathematics Institute (PCMI) was founded in 1991 as part of the "Regional Geometry Institute" initiative of the National Science Foundation. In mid 1993 the program found an institutional home at the Institute for Advanced Study (IAS) in Princeton, New Jersey.

The IAS/Park City Mathematics Institute encourages both research and education in mathematics and fosters interaction between the two. The three-week summer institute offers programs for researchers and postdoctoral scholars, graduate students, undergraduate students, high school teachers, undergraduate faculty, and researchers in mathematics education. One of PCMI's main goals is to make all of the participants aware of the total spectrum of activities that occur in mathematics education and research: we wish to involve professional mathematicians in education and to bring modern concepts in mathematics to the attention of educators. To that end the summer institute features general sessions designed to encourage interaction among the various groups. In-year activities at the sites around the country form an integral part of the High School Teachers Program.

Each summer a different topic is chosen as the focus of the Research Program and Graduate Summer School. Activities in the Undergraduate Summer School deal with this topic as well. Lecture notes from the Graduate Summer School are being published each year in this series. The first fourteen volumes are:

- Volume 1: *Geometry and Quantum Field Theory* (1991)
- Volume 2: *Nonlinear Partial Differential Equations in Differential Geometry* (1992)
- Volume 3: *Complex Algebraic Geometry* (1993)
- Volume 4: *Gauge Theory and the Topology of Four-Manifolds* (1994)
- Volume 5: *Hyperbolic Equations and Frequency Interactions* (1995)
- Volume 6: *Probability Theory and Applications* (1996)
- Volume 7: *Symplectic Geometry and Topology* (1997)
- Volume 8: *Representation Theory of Lie Groups* (1998)
- Volume 9: *Arithmetic Algebraic Geometry* (1999)
- Volume 10: *Computational Complexity Theory* (2000)
- Volume 11: *Quantum Field Theory, Supersymmetry, and Enumerative Geometry* (2001)
- Volume 12: *Automorphic Forms and their Applications* (2002)
- Volume 13: *Geometric Combinatorics* (2004)
- Volume 14: *Mathematical Biology* (2005)

Volumes are in preparation for subsequent years.

Some material from the Undergraduate Summer School is published as part of the Student Mathematical Library series of the American Mathematical Society. We hope to publish material from other parts of the IAS/PCMI in the future. This will include material from the High School Teachers Program and publications documenting the interactive activities which are a primary focus of the PCMI. At the summer institute late afternoons are devoted to seminars of common interest to all participants. Many deal with current issues in education, while others treat mathematical topics at a level which encourages broad participation. The PCMI has also spawned interactions between universities and high schools at a local level. We hope to share these activities with a wider audience in future volumes.

John C. Polking
Series Editor
October 2008

Introduction

M. A. Lewis and J. Keener

IAS/Park City Mathematics Series
Volume 14, 2005

Introduction

M. A. Lewis and J. Keener

Mathematical Biology is a rapidly growing field that defies definition. In simplest terms it is the use of mathematics to study and understand problems of biology. In fact, there is no area of mathematics that is precluded from use and no area of biology that is beyond reach. Obviously, then, an introduction to Mathematical Biology can only begin to scratch the surface of this field. However, if there is a unifying theme to these lecture notes it is that the study of biology is the study of how things change, that is, it is the study of dynamics. So in the introductory lectures of Jim Keener and Mark Lewis, we are given a brief overview of dynamical systems analysis that has been proven to be widely useful in the study of biology.

Chemical reactions, whether in the test tube or in the cell, are a natural place to begin when modeling biological dynamics. Jim Keener builds up the tools and theory of chemical reaction dynamics from first principles, and shows how bistability, thresholds and excitability can arise from simple chemical reactions, and how the resulting dynamics can give rise to biologically significant behavior, such as switches, oscillations and quorum sensing.

Spatial dynamics, whether in an ecosystem, a growing tumor, or in developing tissue patterns, is an area central to mathematical biology. The mathematical problems are challenging, as they involve both time and space components. Mark Lewis and coauthors introduce the theory pattern formation and traveling waves in biology using the so-called stream paradox of spatial ecology as a motivating example. How do species persist in streams while being constantly washed downriver by the current?

The lectures of the main speakers each emphasize a specific biological problem. The lectures of Jim Cushing show how discrete dynamical systems have been used to model the population dynamics of flour beetles; those of David Earn show how the SARS epidemic can be studied using SIR models; Leon Glass uses topological arguments to study the behavior of oscillatory biological dynamics; Helen Byrne shows how populations of cancer cells can be modeled and studied; Paul Bressloff describes the dynamics of neural systems;

Jim Cushing's chapter applies nonlinear matrix models to the study of population dynamics. Matrix population models have a long and distinguished history in mathematical biology — their application to ecological systems goes back more than half a century and they are widely applied by biologists to understand dynamics of structured populations. However, Cushing's work brings new excitement to this well-established subject. In his study of insect populations he includes specific nonlinearities that describe overcompensation and cannibalism in structured populations. These give rise to very complex dynamics, which are influenced by chaotic attractors. Part of Cushing's chapter focuses on understanding the bifurcation structure and qualitative dynamics using mathematical tools. Another part focuses on analyzing data from experimental case studies from real laboratory populations of flour beetles, and on relating these data to the nonlinear dynamical systems using maximum likelihood methods. This coupling of complex nonlinear dynamics and large biological data sets brings a level of biological realism that moves the mathematics from "understanding the theory" to "testing the hypotheses."

David Earn's chapter introduces the reader to the science and art of modeling diseases using dynamical systems. He introduces the dynamical systems from a refreshingly data-oriented perspective. Starting with time series data for real epidemics, Earn asks how we can tease out patterns for infection levels, using statistics. Which diseases persist? Which oscillate? Are there characteristic scales for outbreaks? He then introduces mechanistic models for disease, based on ordinary differential equations, which can connect the dynamical outbreak behaviors to underlying mechanisms. The mechanisms, in turn are depicted as terms in the equations. He adds increasing layers of realism to the models, with the inclusion of stochasticity and environmental forcing, and shows that these factors can have a major influence on observed dynamics.

Leon Glass explores topological approaches to biological dynamics. His approach exploits the topologies common to many physiological systems: a heart cell beats, forming a time-periodic signal; travel far enough around a developing limb bud and you may return to the same spot on the limb bud; a wave of excitation on a spherical surface, such as a whole heart, can travel over the entire surface, returning to the spot where it started; and a network of genes can repeatedly cycle through a variety of different states. What happens when the heart cell is over-stimulated, when a right limb bud is transplanted onto a left stump, when dead region on the excitable surface of the heart is formed, or when the gene network is perturbed or damaged? The result, in a very general sense, is a perturbation of the dynamical system, described on a particular topology (such as a circle or a sphere). Glass develops mathematical tools for studying fixed points, limit cycles, entrainment, and phase resetting in a topological setting, and applies these to understanding the nonlinear dynamical properties to a broad spectrum of biological systems.

Our understanding of cancer tumor growth is rapidly expanding through the application of mathematics and computer modeling to study tumor dynamics. Helen Byrne weaves a framework of dynamical systems models for the various stages of cancer solid tumor dynamics. These stages progress from avascular (little blood supply) through angiogenesis (formation of new blood vessels) to vascular (supplied with blood). Progression from avascular to vascular mirrors an increase in the size and seriousness of tumors, and a detailed understanding of the processes governing the progression may eventually be key to medical treatments, including optimal radiotherapy and chemotherapy. Byrne's dynamical models address the growth, size, shape and the complex movement patterns of tumors and their cells. A mixture of analysis and numerical simulation is used to understand the processes from first principles.

The brain and its neurons remains a source of fascination for many modelers. How can we understand the overwhelming complexity exhibited by the brain, based on the chemical and cellular processes that describe individual brain cells, or neurons? Paul Bressloff's chapter takes on this daunting challenge, starting at the level of a single neuron, and building up to neural pattern formation in the visual cortex of the brain. One of the mathematical challenges lies in the nonlocal nature of interactions between nerves: excitation or inhibition can occur over distant spatial scales as one part of the brain signals to another. The resulting models, coupling nonlinear dynamics to nonlocal interactions, provide insight as to the form and function of the brain.

While the first two chapters are designed as introductory material, each of the remaining chapters of the book stand alone, as snapshots of in-depth research within sub-areas of mathematical biology. In this respect, the volume sacrifices breadth for depth; many other important sub-areas, ranging from cell structure to evolution to immunology are not covered. While these could fill several more volumes, we trust that the excitement and richness of research topics shown here will encourage the reader to explore these other areas on their own.

Introduction to Dynamics of Biological Systems

James P. Keener

IAS/Park City Mathematics Series
Volume 14, 2005

Introduction to Dynamics of Biological Systems

James P. Keener

1. Rates of reaction

Biology is all about change. To keep track of this change, one might take frequent censuses and then using this data with curve fitting, make predictions about the future size of the population under consideration. However, one might also want to understand the mechanisms underlying this change, and use this understanding to build predictive models.

There are only two ways that the amount of some object can change. It can either be created/destroyed, or it can be added to/removed from the region in which the census is being conducted. Thus, for example, to keep track of the number of people in a stadium, one need only count the number of people who go in or out through the turnstiles. In a hospital, keeping a correct census requires keeping track of the births and deaths as well. The measured quantity could be a population of chemicals, cells, or organisms, or it could be a physical entity like heat or momentum. Regardless, the principles involved in mathematizing these are the same, and can be stated in mathematical terms as

$$\frac{d}{dt}\int_{\Omega} u dV = \int_{\partial\Omega} J \cdot n ds + \int_{\Omega} f dV, \tag{1}$$

where u is the density of the measured quantity, J is the flux of u across the boundary of Ω, f is the production rate density, and Ω is the domain under consideration (a cell, a room, a city, etc.)

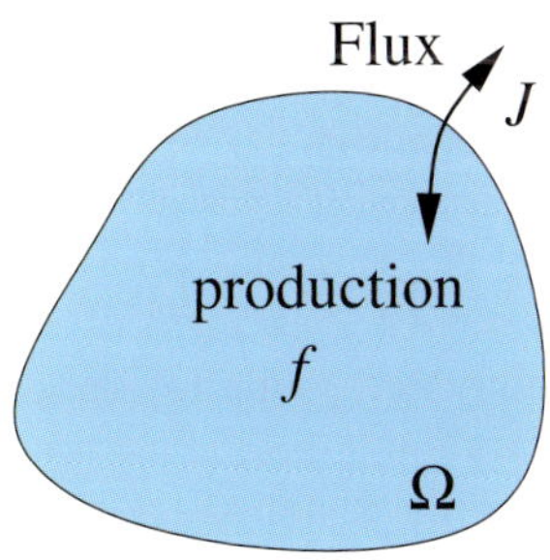

For chemical reactions, (as well as for many other interacting populations; see for example the lectures by David Earn), the most common model for the production rate f is derived using the law of mass action. For example, if species

Department of Mathematics, University of Utah, Salt Lake City, UT 84112
E-mail address: keener@math.utah.edu

A is converted to species B at rate k, denoted

$$A \xrightarrow{k} B, \tag{2}$$

the differential equations describing the populations of A and B are

$$\frac{da}{dt} = -\frac{db}{dt} = -ka, \tag{3}$$

where lower case letters denote concentrations of upper case quantities. Typical chemical reactions can proceed in both directions, denoted

$$A \underset{k_-}{\overset{k_+}{\rightleftarrows}} B, \tag{4}$$

in which case the differential equations for the populations of A and B are

$$\frac{da}{dt} = -\frac{db}{dt} = -k_+ a + k_- b. \tag{5}$$

Bimolecular reactions such as

$$A + B \underset{k_-}{\overset{k_+}{\rightleftarrows}} C \tag{6}$$

are also modeled using the law of mass action with

$$\frac{da}{dt} = \frac{db}{dt} = -\frac{dc}{dt} = k_- c - k_+ ab. \tag{7}$$

It should be pointed out that this "law" is actually not a law in a strict sense, but is an approximation that works reasonably well in some ranges of the variables, however, it fails if either of the concentrations is quite large or quite small.

For the reaction (6), there are two conservation laws, one of which is $a + c = A_0$. Using this, it follows that at steady state,

$$a = \frac{K_d A_0}{K_d + b}, \qquad c = \frac{bA_0}{K_d + b}, \tag{8}$$

where $K_d = \frac{k_-}{k_+}$ is called the equilibrium constant. The equilibrium constant has two important interpretations. Notice that B can be viewed as controlling the way A_0 is distributed between A and C. That is, as B is increased so also is the percentage of A_0 that is in the form of C. Thus, at $b = K_d$, half of A_0 is in the form A and half is in the form C. If K_d is small, then B is said to have a high affinity for A, since B readily combines with A to form C. Conversely, if K_d is large, then B has a low affinity for A, since only a small percentage of A combines with B to form C.

As a specific example consider a simple polymerization reaction in which a monomer, denoted A_1 can polymerize by adding one monomer onto one end of a polymer of length n, denoted A_n, via

$$A_{n-1} + A_1 \underset{k_-}{\overset{k_+}{\rightleftarrows}} A_n. \tag{9}$$

Using the law of mass action, the differential equations for the population of n-polymers A_n for $n > 1$ is

$$\frac{da_n}{dt} = k_+ a_{n-1} a_1 - k_+ a_n a_1 + k_- a_{n+1} - k_- a_n. \tag{10}$$

There are four terms on the right hand side of this equation because there are two ways a molecule of A_n can be created and two ways it can be destroyed. It can be created when a polymer of length $n-1$ combines with a monomer, or when a polymer of length $n+1$ loses a monomer. It is destroyed when it combines with a monomer to form an $n+1$-mer, and when it loses a monomer to become an $n-1$-mer. For a_1, there are an infinite number of terms,

$$\frac{da_1}{dt} = 2k_-a_2 - 2k_+a_1^2 - \sum_{n=2}^{\infty} k_+a_ka_1 + \sum_{k=3}^{\infty} k_-a_k. \tag{11}$$

The factor 2 in two terms of this equation comes from the fact that two molecules of A_1 are involved in the reaction. Since the total number of component molecules cannot change it must be that

$$\sum_{n=1}^{\infty} n\frac{da_n}{dt} = 0. \tag{12}$$

Reactions that are describable using the simple law of mass action are called elementary reactions, because presumably they proceed directly from interaction of the reactants. Reactions that do not follow mass action kinetics usually proceed by a complex mechanism consisting of two or more elementary reaction steps.

One of the most fundamental examples of a reaction that is not an elementary reaction is the enzymatic reaction

$$S + E \underset{k_-}{\overset{k_+}{\rightleftharpoons}} C \overset{k_2}{\rightarrow} P + E, \tag{13}$$

where E represents an enzyme whose task is to convert the substrate S into the product P. This it does by first binding the substrate to form the complex C, and then releasing it as the product.

One can easily write down the law of mass action for these reactants. These are

$$\frac{ds}{dt} = k_-(e_0 - e) - k_+se, \tag{14}$$

$$\frac{de}{dt} = -k_+se + (k_- + k_2)(e_0 - e), \tag{15}$$

$$\frac{dp}{dt} = k_2c, \tag{16}$$

where we have used the fact that total enzyme is conserved, $e + c = e_0$.

Now, an important observation is that quite often in chemical reaction networks, different reactions proceed at vastly different rates. If one can identify reactions that proceed much faster than others, then it may be possible to simplify the reaction system by assuming that the fast reactions equilibrate quickly. For example, in the above example, if the total amount of enzyme is much smaller than the total amount of substrate, then the reaction (15) is a fast reaction that goes quickly to its quasi-equilibrium,

$$(k_- + k_2)(e_0 - e) - k_+se = 0, \tag{17}$$

so that

$$e = e_0\frac{K_m}{s + K_m}, \tag{18}$$

where $K_m = \frac{k_- + k_2}{k_+}$. This approximation, called the quasi-steady state approximation, is valid only after some brief initial time period during which this equilibrium is reached. However, if it is valid, equations for p and s reduce to

$$\frac{dp}{dt} = -\frac{ds}{dt} = k_2 c = k_2 e_0 \frac{s}{K_m + s}. \tag{19}$$

This expression for the rate at which a substrate is enzymatically converted to product is called the *Michaelis-Menten reaction rate*, (and K_m is called the Michaelis-Menten constant,) and is used routinely (often without checking the underlying hypotheses).

Enzymatic reactions can be modified in a number of important ways. For example, enzyme activity can be inhibited (or poisoned) by another substrate of the enzyme, via,

$$S + E \rightleftarrows C \overset{k_2}{\to} P + E, \qquad I + E \rightleftarrows C_2.$$

In this case the Michaelis-Menten reaction rate is modified to

$$\frac{dp}{dt} = -\frac{ds}{dt} = k_2 e_0 \frac{s}{s + K_m(1 + \frac{i}{K_i})}. \tag{20}$$

Notice that the effect of the inhibitor is to increase the effective Michaelis-Menten constant, i.e., slowing the reaction by shifting the rate of reaction to the right.

Enzymes can have more than one active binding site, and these can "cooperate" as in

$$S + E \underset{k_{-1}}{\overset{k_1}{\rightleftarrows}} C_1 \overset{k_2}{\to} P + E, \qquad S + C_1 \underset{k_{-3}}{\overset{k_3}{\rightleftarrows}} C_2 \overset{k_4}{\to} P + E. \tag{21}$$

The Michaelis-Menten rate for this reaction involves a number of additional terms, however one can show that in the limit $k_3 \to \infty$ and $k_1 \to 0$ with $k_1 k_3$ held fixed, the rate of reaction is

$$\frac{dp}{dt} = -\frac{ds}{dt} = V_{max} \frac{s^2}{K_m^2 + s^2}. \tag{22}$$

This reaction rate is also called a "Hill function" with Hill coefficient 2.

The significance of these observations can be illustrated in a simple but important example of oxygen and carbon dioxide transport and exchange by the blood. As you know, most oxygen is transported in the blood by binding with hemoglobin, while most CO_2 (87 %) is transported in the form of bicarbonate ions.

Hemoglobin has four highly cooperative binding sites for oxygen, so the binding of oxygen to hemoglobin can be reasonably represented with a Hill function with Hill constant four. However, there are a number of other factors that affect the binding of oxygen to hemoglobin, the most important of which is the hydrogen ion, which is an inhibitor of oxygen binding. A simple model of this reaction is depicted in Fig. 1. We assume that the hemoglobin molecule can exist in four different states: with H^+ bound (concentration Z), with O_2 bound (concentration Y), with neither bound (concentration X), or with both bound (concentration W).

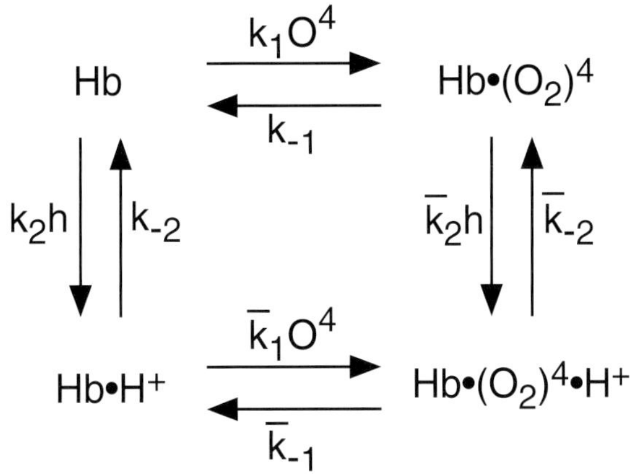

FIGURE 1. Binding diagram for the binding of hydrogen ions and oxygen to hemoglobin. We assume a single hydrogen ion binding site, and a simplified mechanism for oxygen binding. Hb denotes hemoglobin.

At equilibrium it must be that

$$O^4X = K_1Y, \tag{23}$$
$$hX = K_2Z, \tag{24}$$
$$O^4Z = \bar{K}_1W, \tag{25}$$
$$X+Y+Z+W = T_{\mathrm{Hb}}, \tag{26}$$

where h denotes $[\mathrm{H}^+]$, O denotes unbound $[\mathrm{O}_2]$, T_{Hb} denotes the total concentration of hemoglobin, $K_1 = k_{-1}/k_1$ and similarly for $K_2, \bar{K}_1$ and $\bar{K}_2$ (which is used below). Solving these four equations we find the total concentration of hemoglobin with oxygen bound to be

$$Y+W = \frac{O^4T_{\mathrm{Hb}}}{\phi(h)+O^4}, \tag{27}$$

where

$$\phi(h) = \frac{K_1\bar{K}_1(K_2+h)}{K_2\bar{K}_1+hK_1}. \tag{28}$$

The function (27) is called a *saturation curve*, as it represents the saturation of hemoglobin-oxygen binding sites as a function of free oxygen. Since hydrogen ions inhibit oxygen binding, we assume that $\bar{K}_1 > K_1$. Clearly, h does not change the maximal saturation, although it shifts the mid-point of the curve. In fact, since $\phi(h)$ is an increasing function of h, increasing h shifts the saturation curve to the right, decreasing the amount of oxygen that is bound to hemoglobin.

The second important reaction is the conversion of carbon dioxide to bicarbonate. In both plasma and red blood cells CO_2 combines with water to form carbonic acid (H_2CO_3), which then dissociates quickly into hydrogen ions and bicarbonate ions. Thus,

$$\text{(29)} \qquad CO_2 + H_2O \underset{r_{-1}}{\overset{r_1}{\rightleftharpoons}} H_2CO_3 \underset{r_{-2}}{\overset{r_2}{\rightleftharpoons}} H^+ + HCO_3^- .$$

This reaction proceeds slowly in plasma but much more rapidly in red blood cells because of the presence there of the enzyme *carbonic anhydrase*, which increases the speed of CO_2 hydration by more than a thousand times.

At equilibrium,

$$\text{(30)} \qquad c_T = (1 + \frac{1}{K_r h})c,$$

where c_T is the total carbon dioxide (free carbon dioxide and bicarbonate), c is the free carbon dioxide concentration, and $K_r = \frac{r_{-1}}{r_1}$. Here we see that hydrogen regulates the amount of carbon dioxide that is in the form of bicarbonate, with more bicarbonate when there is less hydrogen, and vice versa.

Now we can describe the full carbon dioxide-oxygen exchange cycle. In the tissues, carbon dioxide is produced and is therefore high, so it diffuses down its concentration gradient into the plasma and into the red blood cells, where much of it is converted to bicarbonate. In the lungs the reaction is reversed, as CO_2 diffuses down its concentration gradient to be excreted in the alveolar air and then the expired air, enabling bicarbonate to be converted to carbon dioxide.

In a similar way, oxygen is high in the lungs and flows down its concentration gradient into the blood, where it is bound to hemoglobin. At the tissues, where oxygen is low because it is metabolized, oxygen again flows down its concentration gradient, allowing more oxygen to be released by hemoglobin.

The importance of the regulatory effect of H^+ on these reactions can now be understood. In the tissues, because of the high local CO_2 concentration, its conversion to bicarbonate causes an increase in the local concentration of H^+ which in turn results in a decreased affinity of hemoglobin for oxygen, thus increasing oxygen release to the tissues. In the lungs, the reverse occurs; the low local CO_2 concentration and the ensuing conversion of bicarbonate to carbon dioxide causes a decrease in H^+ concentration which results in an increase in hemoglobin oxygen affinity, and thus increased oxygen uptake.

2. Thresholds and bistability

An important feature of biochemical networks is their use as switches. We illustrate biochemical switches here with the example of quorum sensing. Quorum sensing denotes the ability of an organism to change its behavior on the basis of the size of the colony in which it resides, and is an example of decision making by an organism based on features of its external environment.

An interesting example of quorum sensing is by the bacteria *P. aeruginosa*. These bacteria are a major cause of infection in hospitals and a primary infector of people with cystic fibrosis. *P. aeruginosa* live in two forms, as planktonic, or freely moving, or in biofilm. In biofilm, these bacteria are remarkably resistant to antibiotics, and are essentially impossible to eliminate.

The decision to be planktonic or biofilm secreting involves a number of factors, one of which is the size of the colony, i.e., quorum sensing. Quorum sensing is mediated by a molecule called an autoinducer (homoserine lactone or HSL). Two features of autoinducer are that it regulates its own production via positive feedback and it readily diffuses across the cell membrane.

A simple model for the dynamics of autoinducer is as follows:

$$\frac{dA}{dt} = \kappa + V_m \frac{A^2}{K_a + A^2} - d(A - E). \tag{31}$$

Here the three terms for the intracellular production of autoinducer are a slow background rate of production κ, a Hill function for its self-induced production, and a loss term to represent loss by diffusion to the outside of the cell. Here A and E represent intracellular and extracellular autoinducer, respectively. Outside the cell, autoinducer is simply degraded, so we write

$$(1 - \rho)(\frac{dE}{dt} + k_d E) = \rho d(A - E). \tag{32}$$

Here there are two terms that specify the rate of change of E, namely a term due to degradation, and a term because of diffusion of autoinducer from the cell. The extra factors ρ and $1 - \rho$ in (32) are necessary to account for the fact that the size of extracellular space into which the cellular autoinducer is secreted is related to the density of cells, ρ.

The system of equations (31)-(32) is easy to analyze because it is a two variable system and is therefore amenable to phase plane analysis. The nullclines for these dynamics are the two curves

$$E = A - \frac{\kappa}{d} - \frac{V_m}{d}\frac{A^2}{K_a + A^2}, \qquad (\frac{dA}{dt} = 0), \tag{33}$$

and

$$A = E + (1 - \rho)\frac{k_d}{\rho d}E, \qquad (\frac{dE}{dt} = 0). \tag{34}$$

These nullclines are shown plotted in Fig. 2, with curve (33) shown solid and (34) shown dashed. A decreases when above and increases when below the nullcline $\frac{dA}{dt} = 0$, while E increases when to the left and decreases when to the right of the nullcline $\frac{dE}{dt} = 0$. The slope of the nullcline $\frac{dE}{dt} = 0$ is large when ρ is small, and is small when ρ is close to one.

For intermediate values of ρ, as depicted in Fig. 2, there are three critical points. It is easy to verify that the two critical points with smallest and largest values of A and E are both stable nodes, while the intermediate critical point is a saddle point. Hence, in this case, this is a bistable system, having two stable critical points.

Depending on the value of ρ there can be either one or three (counting multiplicity) critical points. For small values of ρ, there is only one critical point, with both A and E quite small, whereas for ρ close to one the one critical point has both A and E large.

The relationship between the steady solution A and the parameter ρ is given by

$$\rho = \frac{k_d(A - f(A))}{k_d(A - f(A)) + df(A)}, \qquad f(A) = \frac{\kappa}{d} + \frac{V_m}{d}\frac{A^2}{K_a + A^2}. \tag{35}$$

It is clearly easier to view this expression as specifying ρ as a function of A, rather than as A specified as a function of ρ, since A is not a single-valued function of ρ. In fact, the behavior of the steady solution A, shown plotted against the cell density ρ in Fig. 3, reveals "hard switch" or hysteretic behavior.

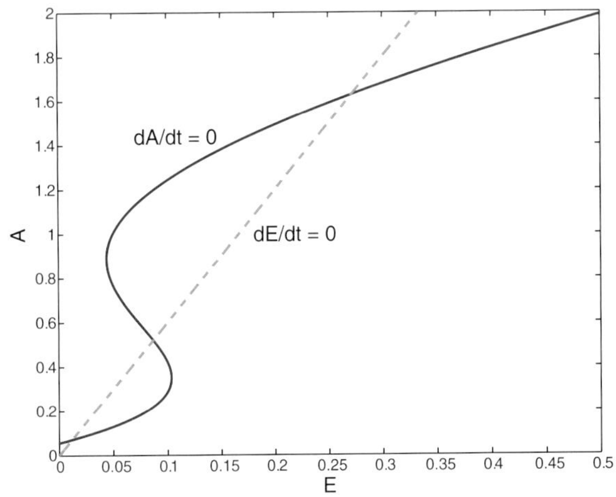

FIGURE 2. Nullclines for the system of equations (31)-(32) with parameter values $\frac{\kappa}{d} = 0.05$, $\frac{V_{max}}{d} = 1.8$, $K_a = 1.0$, $\frac{k_d}{d} = 5$, and $\rho = 0.5$.

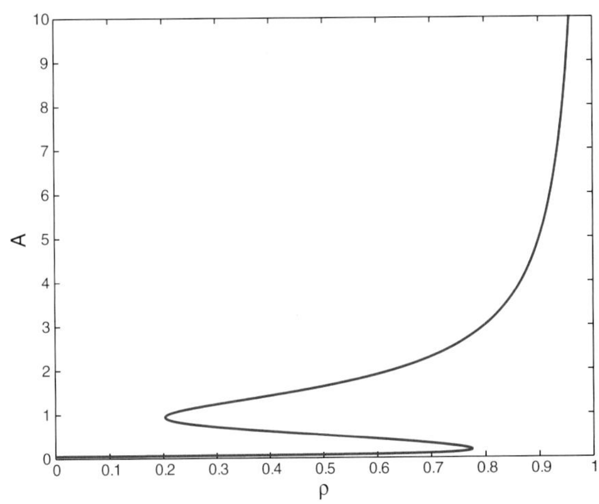

FIGURE 3. Steady state solution A for the system of equations (31)-(32) plotted against ρ with parameter values $\frac{\kappa}{d} = 0.05$, $\frac{V_{max}}{d} = 1.8$, $K_a = 1.0$, and $\frac{k_d}{d} = 5$.

From Fig. 3, we can understand how quorum sensing works. At low densities, the production of extracellular autoinducer is quite slow, so it remains at low concentration. However, at larger densities, because there are many more cells, extracellular autoinducer concentration builds up and impedes its diffusion out of the cells. As intracellular autoinducer increases, the positive feedback kicks in and causes the rapid amplification of autoinducer production, leading to large amounts of autoinducer. Thus, the small steady solution ceases to exist and there is only the large steady solution. This large increase in the amount of autoinducer causes the cell to differentiate and become biofilm secreting.

Biofilm secretion has the effect of decreasing the density of the colony. However, because of the hysteresis, a decrease in density created by biofilm secretion does not lead to down-regulation of autoinducer.

(Remark: The thermostat that controls the heat in your home is hysteretic. What is the purpose of this hysteresis?)

3. Excitability

Positive feedback and bistability illustrated in the last section are fundamental building blocks for the decision making and signaling networks used by biological organisms. It is easy to understand how an oscillatory signaling mechanism might be constructed from the ingredients of the previous example.

Suppose that autoinducer A has an effect on density ρ such that ρ increases when A is small, but decreases when A is large. Then, one could imagine that there would be regular oscillations which repetitively move around the hysteresis loop.

While no such relationship between autoinducer and cell density is known, there are numerous biochemical systems which have exactly the same mathematical structure as that described here.

One such system is the calcium handling system in cardiac cells. Contraction of cardiac cells is controlled by intracellular calcium. Normally, intracellular (cytosolic) calcium concentration is quite low, however, when it is high, it causes contraction of the cell. Thus, contraction of the cell is regulated by the shuttling of calcium into and out of the cytosol.

Calcium is quite high in the extracellular space as well as in intracellular stores called the sarcoplasmic reticulum (SR). Calcium can flow into the cytosol through channels call L-type calcium channels, which are regulated by the membrane voltage. Calcium can also enter the cytosol from the SR through channels called Ryanodine receptors (RYR), which are regulated by calcium. Calcium is removed from the cytosol into the extracellular space primarily through sodium calcium exchangers (NCX) and into the SR through ATP-powered calcium pumps (called Sarcoplasmic or Endoplasmic Reticulum Calcium ATP-ase, SERCA for short). It follows that cytosolic calcium c is governed by the equation

$$\frac{dc}{dt} = J_L + J_{RYR} - J_{NCX} - J_{SERCA}. \tag{36}$$

We now need to determine models for these four different fluxes.

The most significant observation is that the RYR's are regulated by calcium, so here there is an opportunity for positive feedback, which as we saw in the previous example, was a crucial ingredient for bistability. A basic model for the regulation of a ryanodine receptor is that there are two calcium binding sites, one of which activates the RYR when calcium is bound and the other which inactivates the RYR when calcium is bound. A diagram for this reaction scheme, which looks quite similar to the binding diagram for hemoglobin, is shown in Fig. 4. Here the different binding states are denoted as R_{ij} where i (the activating site) and j (the inactivating site) can be 0 or 1, to denote if the site is unbound (0) or bound (1). An important point is that the activating calcium binds to the RYR much faster than does the inactivating calcium.

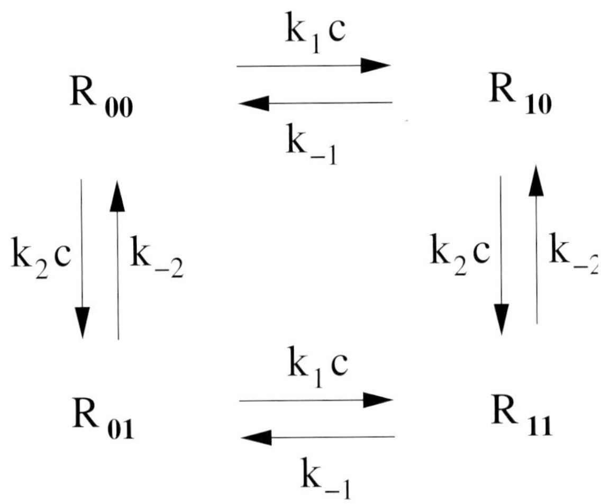

FIGURE 4. Calcium binding diagram for a ryanodine receptor.

Flux through the RYR is assumed to occur (mostly) when in the activated state R_{10}. Thus, we take

$$J_{RYR} = (g_R r_{10} + g_l)(c_s - c), \tag{37}$$

where g_R is the conductance of the activated state, g_l is a "leak" conductance, and c_s is the concentration of SR calcium, assumed for this model to be constant. Here r_{ij} is the percentage of a large population of ryanodine receptors that is in the ij-state.

To quantify this model, we must determine how the states of the RYR evolve in time. Using standard mass action kinetics, we write

$$\frac{dr_{00}}{dt} = k_{-1}r_{10} + k_{-2}r_{10} - (k_1 + k_2)cr_{00} \tag{38}$$

$$\frac{dr_{10}}{dt} = (k_1 r_{00} - k_2 r_{10})c - k_{-1}r_{10} + k_{-2}r_{11} \tag{39}$$

$$\frac{dr_{11}}{dt} = (k_2 r_{10} + k_1 r_{01})c - k_{-1}r_{11} - pk_{-2}r_{11} \tag{40}$$

with $r_{00} + r_{10} + r_{01} + r_{11} = 1$. Now we quantify the assumption that the activating binding site binds calcium much more rapidly than does the inactivating site, by assuming that $k_{-1} >> k_{-2}$ and that $k_1 >> k_2$. Following from these assumptions, we take r_{00} to be in quasi-equilibrium with r_{10} and we take r_{01} to be in quasi-equilibrium with r_{11}, that is

$$k_1 cr_{00} = k_{-1}r_{10}, \qquad k_1 cr_{01} = k_{-1}r_{11}. \tag{41}$$

Now we set $y = r_{00} + r_{10}$ and observe that

$$\frac{dy}{dt} = k_{-2}(1 - y) - k_2 cy. \tag{42}$$

Thus,

$$r_{10} = \frac{k_1 cy}{k_1 c + k_{-1}}. \tag{43}$$

To complete the specification of the model, we take

$$J_{SERCA} + J_{NCX} = \frac{p_1 c^2}{p_2^2 + c^2}, \tag{44}$$

and we take J_L to be a specified function of time. Thus our full model is

$$\frac{dc}{dt} = (g_R \frac{k_1 c y}{k_1 c + k_{-1}} + g_l)(c_s - c) - \frac{p_1 c^2}{p_2^2 + c^2} + J_L(t), \tag{45}$$

$$\frac{dy}{dt} = k_{-2}(1 - y) - k_2 c y. \tag{46}$$

Typical parameter values are given in Table 1.

k_1	=	$15\ \mu M^{-1}s^{-1}$	k_{-1}	=	$7.6\ s^{-1}$
k_2	=	$0.8\ \mu M^{-1}s^{-1}$	k_{-2}	=	$0.84\ s^{-1}$
p_1	=	$192\ \mu M^{-1}s^{-1}$	p_2	=	$0.12\ \mu M$
g_R	=	$14.8 s^{-1}$	c_s	=	$25 \mu M$

TABLE 1. Parameter values for the model of calcium dynamics (45)-(46).

The solutions of this system of equations can be understood by examining its phase portrait, an example of which is shown in Fig. 5, upper panel. The first things to notice about the phase portrait are the nullclines (shown dashed in Fig. 5). The nullcline on which $\frac{dc}{dt} = 0$ is "N"-shaped, reflecting the fact that for y between about 0.6 and 1, the equation for c, (45), is bistable. The nullcline on which $\frac{dy}{dt} = 0$,

$$y = \frac{k_{-2}}{k_{-2} + k_2 c} \equiv y_\infty(c), \tag{47}$$

is a monotone decreasing function of c, having one or three intersections (counting multiplicity) with the nullcline $\frac{dc}{dt} = 0$. For the parameter values in Table 1 and with $g_l = 0$, the nullclines have a single intersection, and it is a globally stable critical point.

Excitability is characterized by a threshold response to stimuli. If a small stimulus is applied, then the solution returns promptly to the unique steady solution. However, if the applied stimulus is sufficiently large, then there is a large amplitude excursion or transient before the solution returns to the stable steady state. An example of such a super-threshold response is shown by the solid curve in the upper and lower panels of Fig. 5. The dashed curve in the lower panel shows the applied stimulus $J_L(t)$.

The superthreshold response is characterized by two phases. In the first phase there is a rapid calcium transient which also quickly reduces the value of y. In the second phase, y returns gradually to its rest value. Notice that there are actually two sub-phases of this second phase. At first, when y is quite small (smaller than about 0.6), the right hand side of (45) is monostable. In this phase the system is said to be *absolutely refractory* and this phase is called the *absolute refractory period*. Later, as y increases, the system regains its bistability, although the threshold for

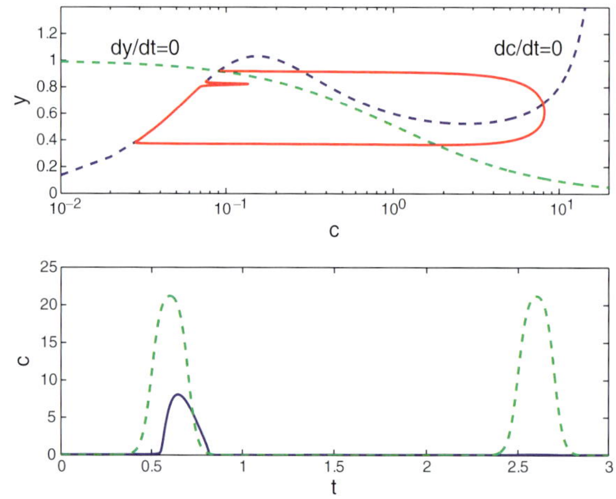

FIGURE 5. Solution of equations (45)-(46) with $g_l = 0$. Upper panel shows the solution in the phase portrait, with c plotted on a logarithmic scale, and the lower panel shows the time dependent calcium concentration c (solid curve), with the applied stimulus shown as a dashed curve.

a response decreases with time, i.e. the calcium subsystem gradually regains its sensitivity or excitability. This phase is called the *relatively refractory period.*

The solution in Fig. 5 actually shows the response to two identical stimuli applied 2s apart. The response to the first of these is a large amplitude calcium transient, while the response to the second is so small it is only visible in the upper panel, where it c plotted on a logarithmic scale. This failure to respond is because the stimulus was applied while the system was in its relatively refractory phase and its excitability was not fully recovered.

The second important feature of an excitable system is that in certain parameter ranges it is self-oscillatory. An example of this behavior is shown in Fig. 6 where $g_l = 1.4s^{-1}$. In this case, the two nullclines again have a single intersection, but now the critical point is unstable, and the globally attracting stable behavior is a limit cycle oscillation, as shown in the lower panel.

Excitable systems can exhibit a large range of interesting dynamical behaviors in response to periodic stimuli. One such behavior is shown in Fig. 7. Here the applied stimulus has period 2.63s and there is a large amplitude calcium transient for three of every four stimuli. This response is called a 3:4 Wenckebach pattern. These patterns have been studied extensively in the context of cardiac cells. The key feature of these patterns is that the amplitude of the response is steadily decreasing until a stimulus fails to evoke a response. Then, with the new found additional time to recover, the subsequent transient has a large amplitude, and the pattern repeats.

It is easy to observe a wide variety of interesting behaviors as parameters are varied. However, rather than describing these for you, my suggestion is that you write a simple MATLAB code to integrate this system of equations and explore the parameter space for yourself.

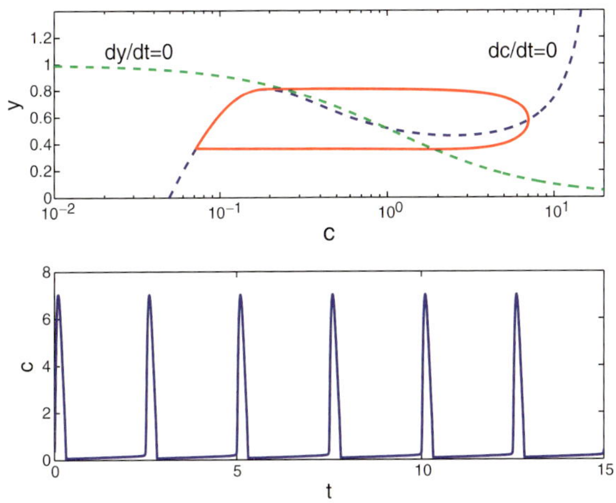

FIGURE 6. Solution of (45)-(46) with $g_l = 1.4s^{-1}$. Upper panel shows the solution in the phase portrait, and the lower panel shows the time dependent calcium concentration c.

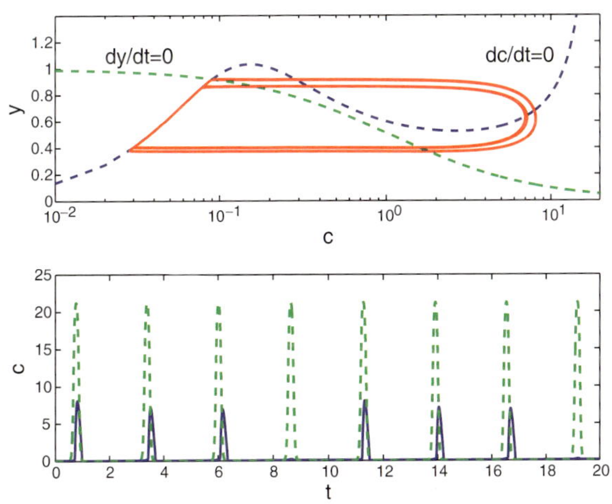

FIGURE 7. Solution of (45)-(46) with a periodic stimulus, $T = 2.63s$. Upper panel shows the solution in the phase portrait, and the lower panel shows the time dependent calcium concentration c (solid curve), with the applied stimulus shown as a dashed curve.

4. Summary

This brief introduction to biological dynamics was intended to introduce and illustrate three of the fundamental tools or ideas of the trade.

- Changing quantities are tracked by following the production/destruction rates and their influx/efflux rates;

- Because interactions and reactions can occur on many different time scales, quasi-steady state approximations are often quite useful;
- For two variable systems, much can be learned from the "phase portrait".

If these ideas are unfamiliar to you, then this introduction was far from adequate. Much more complete treatments of these ideas are found in other books, my personal favorite being [**4**]. There are also helpful discussions of these topics in [**2**].

The model of quorum sensing described here was a condensation of the model introduced in [**1**]. In particular, there a more detailed description of the autoinducer biochemistry is given as well as an analysis of a more realistic model of autoinducer dynamics.

There are a vast number of models of excitable systems, the most famous of which are the Hodgkin-Huxley equations and the FitzHugh-Nagumo caricature thereof. The lectures by Leon Glass and Paul Bressloff make reference to several of these. The model of calcium dynamics described here is a simplification of a model originally introduced in [**6**]. The illustrative power of this simple system is hard to beat. However, nowadays there are much more sophisticated and accurate models of these phenomena, an up-to-date discussion of which can be found in [**5**].

There is much more to say about dynamical systems in biology, some of which is said in the lectures to follow. Here no mention whatever was made of discrete dynamical systems, although these play an important role in the lectures of Leon Glass and Jim Cushing. Another nice reference on these and related topics is [**3**].

Bibliography

[1] J. D. Dockery and J. P. Keener. A mathematical model for quorum sensing in *Pseudomonas Aeruginosa*. *Bull. Math. Biol.*, 63:95–116, 2001.

[2] C. Fall, E. Marland, J. Wagner, and J. Tyson. *Computational Cell Biology*. Springer-Verlag, New York, 2002.

[3] L. Glass and M. C. Mackey. *From Clocks to Chaos*. Princeton University Press, Princeton, 1988.

[4] J. P. Keener and J. Sneyd. *Mathematical Physiology*. Springer-Verlag, New York, 1998.

[5] J. Sneyd. *Tutorials in Mathematical Biosciences II*, volume 187 of *Lecture Notes in Mathematics*, chapter Mathematical modeling of calcium dynamics and signal transduction. Springer, 2005.

[6] Y. Tang and H. G. Othmer. A model of calcium dynamics in cardiac myocytes based on the kinetics of ryanodine-sensitive calcium channels. *Biophys. J.*, 67:2223–2235, 1994.

Spatial Dynamics in Ecology

Mark A. Lewis, Thomas Hillen,
and Frithjof Lutscher

IAS/Park City Mathematics Series
Volume 14, 2005

Spatial Dynamics in Ecology

Mark A. Lewis, Thomas Hillen, and Frithjof Lutscher

1. Introduction

The purpose of this chapter is to introduce the student to the methods of spatial dynamics in biology. They are to form the building blocks for later lectures in the PCMI series. The approach taken here is to introduce some well known ideas and methods in spatial dynamics in the context of a new problem in mathematical biology. Many of the general results here can be found in the excellent texts by Fife [**7**], Kot [**10**], and Okubo [**13**], as well as in the new text by de Vries et al [**6**].

The problem that provides the context for this chapter is called the *drift paradox* in stream ecology [**12**]. The paradox can be summarized by a question. Why can populations persist in streams when they are being constantly washed downstream? While the problem has been recognized for half a century, the application of methods of spatial dynamics to the problem has occurred in the last five years, starting with the work of Speirs and Gurney [**15**]. Speirs and Gurney modeled the drift paradox using partial differential equations. The analysis of the equations was then used, for example, to show why invasive zebra mussels *Dreissena polymorpha* cannot persist in certain US rivers, or why stone flies can persist, but fish cannot persist in certain creeks in southeast UK. Here the mathematical analysis allows us to link persistence of species to environmental conditions in streams and rivers.

Even though the models discussed here have been used to understand specific stream ecosystems, the models themselves are simple, and leave out many biologically relevant factors. The primary motivation is to use the model to understand the essential processes governing species persistence and spread. Extensions of Speirs and Gurney's work to more realistic stream ecosystems have been made by Pachepsky et al [**14**] and Lutscher et al [**11**]. However, we will stick to the simplest versions of the models for the purposes of this chapter.

Mark A. Lewis, Department of Mathematical & Statistical Sciences, University of Alberta, Edmonton, Alberta, Canada
The research of Mark Lewis was supported by a discovery grant from NSERC and a Canada Research Chair.
Thomas Hillen, Department of Mathematical & Statistical Sciences, University of Alberta, Edmonton, Alberta, Canada
The research of Thomas Hillen was supported by a discovery grant from NSERC
Frithjof Lutscher, Dept. of Mathematics and Statistics, University of Ottawa, Ottawa, Ontario, Canada
The research of Frithjof Lutscher was supported by a discovery grant from NSERC
E-mail address: mlewis@ualberta.ca, thillen@ualberta.ca, flutsche@uOttawa.ca
Thanks to Caspar Williams for help in preparing some of the figures. The notes are partially based on Chapters 4 and 5 of de Vries, Hillen, Lewis, Müller, and Schönfisch [**6**].

2. Deriving the model

We first focus on deriving a model that includes the relevant spatial and growth dynamics. This model takes the form of a reaction-advection-diffusion model. The 'reaction' describes birth and death dynamics, the 'advection' describes drift in the stream, and the 'diffusion' describes random movement in the stream.

There are two fundamental ways to derive the model. The first is via a *conservation law*. This approach comes from mathematical physics. The conservation equation tracks the density of individuals over time and space by specifying forms for the 'reaction', which describes population growth, and for the flux of individuals, which describes drift and random motion.

The second approach comes from stochastic processes. If we ignore birth and death dynamics for now, the location of an individual is a random variable, indexed by time, (i.e., a stochastic process). This random variable has an associated probability density function. The probability density function changes with time, as the individual executes steps in a biased random walk. The resulting model is an advection-diffusion equation referred to variously as the Fokker-Planck equation, or the forward Kolmogorov equation.

2.1. Conservation law derivation

Assume a population with density $u(\mathbf{x}, t)$ is living and moving in a given region. To describe movement, we introduce another dependent quantity, the population flux, $\mathbf{J}(\mathbf{x}, t) \in \mathbb{R}^n$, where n is the space dimension. At each location $\mathbf{x}$ and at each time t, the flux $\mathbf{J}(\mathbf{x}, t)$ is a vector which points in the direction of movement at that location. Its units are density times velocity, and its magnitude, $|\mathbf{J}(\mathbf{x}, t)|$, is proportional to the amount of particles which flow in that direction per unit time.

We assume that the population density and the flux are smooth functions of space and time. We consider a test volume Ω with boundary Γ and we balance the fluxes inward and outward on Ω through Γ (see Figure 1). In words,

$$\text{Change of } u \text{ in } \Omega = \text{flux into } \Omega + \text{change due to birth, death, interactions.}$$

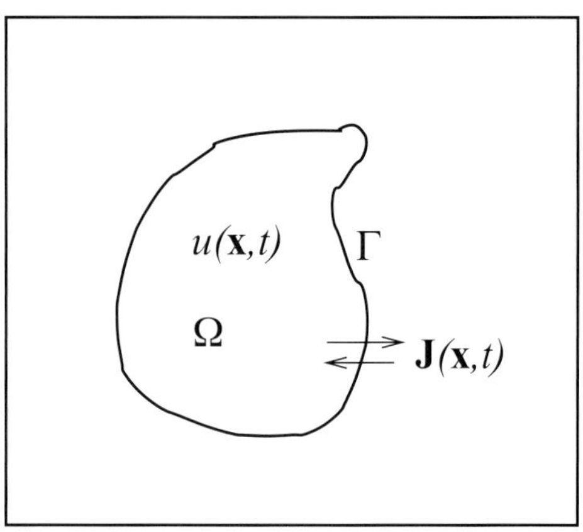

FIGURE 1. Sketch of a test volume Ω with boundary Γ, population density $u(\mathbf{x}, t)$, and flux $\mathbf{J}(\mathbf{x}, t)$ through the boundary.

Written in mathematical terms, this means

$$\frac{d}{dt}\int_\Omega u(\mathbf{x},t)dV = -\int_\Gamma \mathbf{J}(\mathbf{x},t)\cdot\mathbf{n}\,dS + \int_\Omega f(u(x,t))dV,$$

where $\mathbf{n}$ is the outwardly oriented unit vector normal to Γ.

We use the Divergence Theorem

$$\int_\Gamma \mathbf{J}(\mathbf{x},t)\cdot\mathbf{n}\,dS = \int_\Omega \nabla\cdot\mathbf{J}(\mathbf{x},t)dV,$$

and we get

$$\int_\Omega \left(\frac{\partial}{\partial t}u - f(u) + \nabla\cdot\mathbf{J}\right)dV = 0.$$

The above relation is satisfied in each test volume Ω. Then it follows that

$$\frac{\partial}{\partial t}u - f(u) + \nabla\cdot\mathbf{J} = 0. \tag{1}$$

Next, we need an expression of the flux in terms of the population density. Simple diffusive flux arises from Fick's second law[1]

$$\mathbf{J} = -D\nabla u. \tag{2}$$

This law assumes that the flux $\mathbf{J}$ is proportional to the negative gradient of the density of individuals. Note that, for the units to match in (2), the diffusion coefficient D must have units space^{2}time^{-1}.

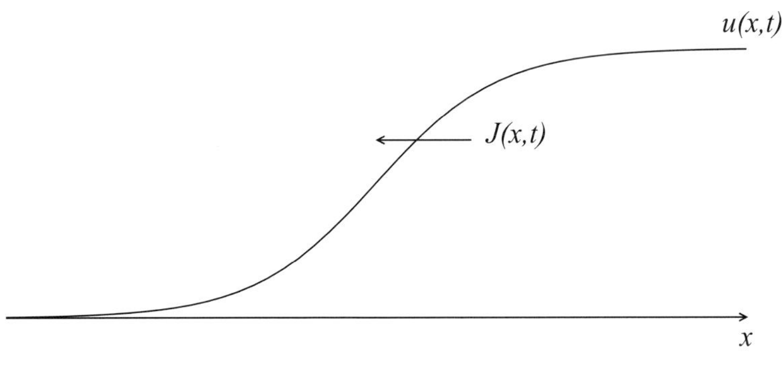

FIGURE 2. Schematic of Fick's second law. A positive gradient of u gives rise to a negative flux J.

In Figure 2, we show a positive gradient of u ($\frac{\partial}{\partial x}u(x,t) > 0$). The flux points to the left, tending to equilibrate high and low levels of u. If there is also advection (drift) with velocity $\mathbf{v}$ then Fick's law is modified to

$$\mathbf{J} = -D\nabla u + \mathbf{v}u. \tag{3}$$

If we combine the balance law (1) with the general version of Fick's law (3), we get a *reaction-advection-diffusion equation*. When the diffusion coefficient D is constant this equation is

$$\frac{\partial u}{\partial t} + \nabla\cdot(\mathbf{v}u) = D\Delta u + f(u), \tag{4}$$

[1]In the interpretation of heat transport, this law is known as *Fourier's Law*.

where the Laplacian Δu is defined as

$$\Delta u(x,t) = \frac{\partial^2}{\partial x_1^2} u(x,t) + \cdots + \frac{\partial^2}{\partial x_n^2} u(x,t), \quad x = (x_1, \ldots, x_n) \in \mathbb{R}^n.$$

If there are no growth dynamics ($f = 0$), and there is no advection ($\mathbf{v} = 0$) then equation (4) is simply the *diffusion equation* or *heat equation.*

2.2. The Fokker-Planck equation

Most living organisms move in space. Given that we have some information about how an organism moves over short time scales, can we determine where it is likely to be over long time scales? If movement rules are simple, mathematical models can be used to translate the movement rules into equations. As we will show in this section, analysis of the resulting equations yields an equation for a probability density function that can be used to track the changing location of the organism over time.

We consider an individual executing a random walk in one-dimensional space. At each time-step, the individual either jumps to the right or left, and its new position is determined by its current position plus a random increment to the left or right. This is called a one-step *Markov process*, because it is a stochastic process which requires only the knowledge of its current location, plus the random increment, to determine the next position. The precise path taken to get to the current location plays no role in determining future positions. A very readable introduction to stochastic processes is Allen [**2**].

We now consider the probability density function (PDF) for the location of the individual, released at location $x = 0$ at time $t = 0$, for times $t > 0$. Rather than calculating the probability density directly, we show that the PDF satisfies a diffusion-advection equation called the Fokker-Planck equation. As we will show in the subsequent section, when the Fokker-Planck equation comes from a simple movement model, it can be directly solved.

We start with a *master equation* describing possible locations of the individual from one time step to the next. Let $X(t)$ be a stochastic process describing the location of an individual at time t, which is released at location $x = 0$ at time $= 0$ ($X(0) = 0$). We define a time-dependent probability density function $p(x,t)$ and small space interval λ, such that $p(x,t)\lambda =$ probability an individual released at $x = 0$ and time $= 0$ is between x and $x + \lambda$ at time t.

We start by assuming that the random walk is unbiased, with individuals moving at every time step. This means that the probability of jumping to the right, R, or left, L in a given time step is one half. The *master equation* which describes movement between points situated distance λ apart is

$$p(x, t+\tau) = \frac{1}{2}\, p(x-\lambda, t) + \frac{1}{2}\, p(x+\lambda, t) \tag{5}$$

(see Figure 3).

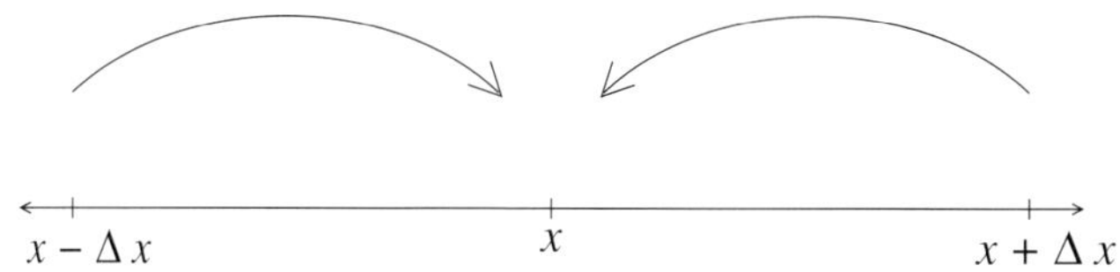

FIGURE 3. Movement on the lattice giving rise to the diffusion equation. Here an individual arriving at x can come from locations $x - \Delta x$ and $x + \Delta x$, where $\Delta x = \lambda$ in equation (5).

Expanding the Taylor series gives:

$$\begin{aligned} p(x,t) &+ \tau\frac{\partial p}{\partial t}(x,t) + \frac{(\tau)^2}{2}\frac{\partial^2 p}{\partial t^2}(x,t) + \text{h.o.t.} \\ &= \tfrac{1}{2}\ \Big\{ p(x,t) - \lambda\frac{\partial p}{\partial x}(x,t) + \frac{(\lambda)^2}{2}\frac{\partial^2 p}{\partial x^2}(x,t) + \text{h.o.t.} \\ &+ \ p(x,t) + \lambda\frac{\partial p}{\partial x}(x,t) + \frac{(\lambda)^2}{2}\frac{\partial^2 p}{\partial x^2}(x,t) + \text{h.o.t.}\Big\}, \end{aligned}$$

where h.o.t. indicates higher order terms in the Taylor series. The above equation can be approximated by dropping the higher order terms and reorganizing to yield

$$\frac{\partial p}{\partial t} + \frac{\tau}{2}\frac{\partial^2 p}{\partial t^2} = \frac{(\lambda)^2}{2\tau}\frac{\partial^2 p}{\partial x^2}.$$

Taking the limit as $\lambda, \tau \to 0$ so that $\frac{(\lambda)^2}{2\tau} \to D$ yields the diffusion equation:

$$\frac{\partial p}{\partial t} = D\frac{\partial^2 p}{\partial x^2}. \tag{6}$$

A slight bias in the movement direction ($R = 0.5 + \gamma\lambda$, $L = 0.5 - \gamma\lambda$) yields the advection-diffusion equation

$$\frac{\partial p}{\partial t} + v\frac{\partial p}{\partial x} = D\frac{\partial^2 p}{\partial x^2}, \tag{7}$$

where $\frac{\gamma(\lambda)^2}{2\tau} \to v$. The biased random walk model can be extended to higher spatial dimensions to yield the n-dimensional advection diffusion equation ((4) with $f = 0$).

Exercise (Random walk derivation of a diffusion-advection equation)
Describe, by means of a master equation, a random walk with movement to nearest neighbors on a 2D lattice with spacing λ and time steps of size τ. Denote L, R, U and V to be the probability of moving to the left, right up or down, respectively. Assume that $L = 0.25 - \lambda\gamma_1$, $R = 0.25 + \lambda\gamma_1$, $U = 0.25 + \lambda\gamma_2$ and $V = 0.25 - \lambda\gamma_2$. Derive a diffusion-advection equation

$$\frac{\partial p}{\partial t} + \mathbf{v}\cdot\nabla p = \nabla\cdot D\nabla p$$

by taking the appropriate diffusion limit of the random walk. Here $\mathbf{v}$ is an advection vector which you should describe in terms of γ_1 and γ_2.

When the probabilities of jumping to right and left depend on spatial location a more complex form of the Fokker Planck equation arises.

Exercise (Spatially varying diffusion model)

Using Taylor Series and the diffusion limit, derive an equation for $\partial p/\partial t$ where

(a) $p(x,t+\tau) = \alpha(x-\lambda)p(x-\lambda,t) + N(x)p(x,t) + \alpha(x+\lambda)p(x+\lambda,t)$ and
(b) $p(x,t+\tau) = \alpha(x-\lambda/2)p(x-\lambda,t) + N(x)p(x,t) + \alpha(x+\lambda/2)p(x+\lambda,t)$.

In the first model, $2\alpha(x)$ and $N(x)$ describe the probabilities of leaving the site or staying at the site. In the second model $\alpha(x-\lambda/2)+\alpha(x+\lambda/2)$ and $N(x)$ describe the probabilities of leaving a site or staying at a site, respectively. Note that $N(x)+2\alpha(x) = 1$ in the first model and $N(x)+\alpha(x-\lambda/2)+\alpha(x+\lambda/2) = 1$ in the second model. Show that the first problem can be written in the form

$$\frac{\partial p}{\partial t} = \frac{\partial^2}{\partial x^2}\left\{A(x)p\right\},$$

where $A(x)$ is a function you should determine. (HINT: when deriving Taylor expansions it may be easiest to expand the product of two functions.) Show that the second problem can be written in the form

$$\frac{\partial p}{\partial t} = \frac{\partial}{\partial x}\left\{B(x)\frac{\partial p}{\partial x}\right\},$$

where $B(x)$ is a function you should determine. Compare and contrast the assumptions for the two models and the resulting equations for p_t. Show that the difference between the two models can be expressed as an advection term that appears in one model but not the other. For further reading on this see Aronson [**3**].

So far we have derived the diffusion-advection equation two different ways, first from a conservation law, and second from random walk. If we are to calculate the density of individuals (or, alternately, the probability density function for a single individual) we must solve this equation. This is what we consider in the next section.

2.3. Fundamental solution to the diffusion equation

The fundamental solution is a particular solution of the diffusion equation (equation (6), or equivalently, equation (4) with $f = 0$ and $\mathbf{v} = 0$), that can be used to find other solutions by convolution (see, for example, Britton [**5**]).

The *fundamental solution* can be interpreted as the solution to the Fokker-Planck equation (6) for an individual which starts at the location $x = 0$ at time $t = 0$.

To denote this single individual whose initial location is known precisely we use a *δ-distribution* $\delta_0(x)$. The δ-distribution denotes a unit impulse, centered at $x = 0$. Mathematically, it is defined by its action on smooth functions. If $h(x)$ is a smooth function, then $\delta_0(x)$ is the one and only object which satisfies

$$\int_R \delta_0(x)h(x)dx = h(0)$$

and

$$\int_R \delta_0(x)dx = 1.$$

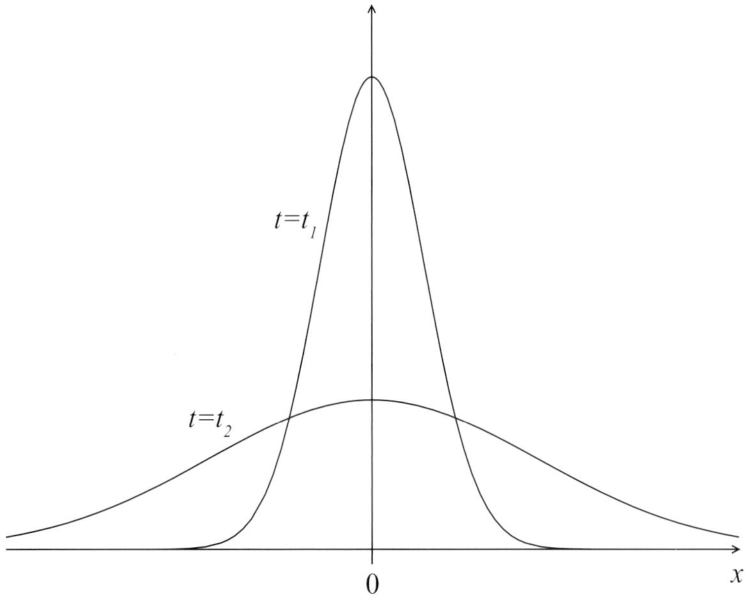

FIGURE 4. Solutions of the diffusion equation (6) are given for three time values, $t = 0$, $t = t_1 > 0$, and $t = t_2 > t_1$.

Consider the *initial-value problem* for a particle which diffuses in one dimension and starts with certainty at 0:

$$w_t = D w_{xx}, \quad w(x,0) = \delta_0(x). \tag{8}$$

The *fundamental solution* (in one dimension) is

$$w(x,t) = \frac{1}{2\sqrt{\pi D t}} e^{-\frac{x^2}{4Dt}}. \tag{9}$$

This can be found by using Fourier Transform methods [**9**].

Exercise (Fundamental solution)

(a) Verify that the function

$$w(x,t) = \frac{1}{2\sqrt{\pi D t}} e^{-\frac{x^2}{4Dt}}$$

satisfies the diffusion equation $w_t = D w_{xx}$.

(b) Derive the fundamental solution to the diffusion-advection equation $u_t + v u_x = D u_{xx}$.

In Figure 4, we show this solution for time steps $t = 0$, $t = t_1 > 0$, $t = t_2 > t_1$, and $D = 1$. As time increases, the variance of the Gaussian increases, and hence the location of the individual becomes less certain.

Although the initial condition is not continuous, the solution (9) is continuous for all $t > 0$. In fact, it is infinitely often continuously differentiable, a property which is known as the *regularizing* property of the diffusion equation.

If we study the diffusion equation with a general initial condition,

$$u_t = Du_{xx}, \quad u(x,0) = h(x), \tag{10}$$

then the solution can be found by convolution with w:

$$u(x,t) = (h * w(\cdot,t))(x),$$

where the convolution integral is given by

$$(h * w(\cdot,t))(x) = \int_{-\infty}^{\infty} h(y)\, w(x-y,t)\, dy \tag{11}$$

$$= \frac{1}{2\sqrt{\pi D t}} \int_{-\infty}^{\infty} h(y) e^{-\frac{(x-y)^2}{4Dt}}\, dy. \tag{12}$$

This convolution of the fundamental solution with initial conditions works also for general linear operators, including the diffusion advection equation (7).

3. Population spread

One measure of whether a population can persist in a stream is whether it will invade spatially when introduced into localized region of the stream. We initially analyze a very long (in fact, infinitely long) stream, so there are no losses at upstream and downstream boundaries. These kinds of boundary losses are considered in the subsequent section. We also initially choose the drift term (advection velocity) to be zero. While this is clearly an unreasonable assumption, we make the assumption only for clarity of exposition. The analysis of the zero advection case can be easily extended to include advection and we do this later in the section.

We write our reaction-diffusion model for growth and dispersal in the stream in one spatial dimension. Here x is the distance down river and $n(x,t)$ is the density of individuals in the stream. Random motion, coupled to growth, yields

$$\frac{\partial n}{\partial t} = D\frac{\partial^2 n}{\partial x^2} + f(n), \tag{13}$$

Here typical growth dynamics are logistic growth

$$f(n) = rn\left(1 - \frac{n}{K}\right). \tag{14}$$

However, some populations have a reduced per capita growth rate at low densities, so on average they cannot replace themselves when densities are low. This is referred to as an 'Allee effect' in honor of Allee, a biologist from the first part of the 20th century who deduced that these effects may be important in the dynamics of populations (Allee [**1**]). The bistable growth function

$$f(n) = rn\left(1 - \frac{n}{K}\right)\left(\frac{n-C}{K}\right) \tag{15}$$

is an appropriate model for the Allee effect.

We can simplify our model by rescaling population density by the carrying capacity, and time and space by characteristic time and length scales.

$$u = \frac{n}{K}, \quad t^* = rt, \quad x^* = \sqrt{\frac{r}{D}}x.$$

This rescaling renders our new quantities u, t^* and x^* dimensionless. We now drop the asterisks on t and x for notational simplicity. Our new equation is

$$\frac{\partial u}{\partial t} = \frac{\partial^2 u}{\partial x^2} + g(u), \tag{16}$$

where $g(u) = u(1-u)$ for logistic growth and $g(u) = u(1-u)(u-a)$ for bistable growth.

Our first step in analyzing population spread is to consider travelling wave solutions of (16). Travelling wave solutions are translationally invariant solutions which move with a fixed profile and constant speed.

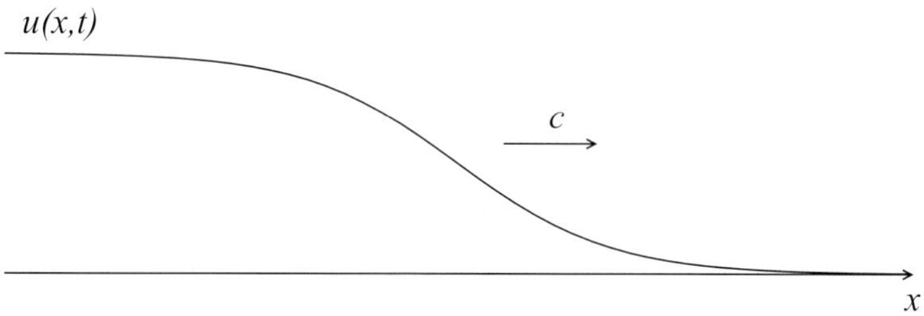

FIGURE 5. A typical invasion travelling wave.

In particular, we seek solutions $u(x,t)$ that have the form shown in Figure 5, joining the extinction steady state $u = 0$ to the carrying capacity steady state $u = 1$, and moving with constant speed c. A solution of this type can be expressed as

$$u(x,t) = U(x - ct).$$

The function $U(x-ct)$ is the function $U(x)$ shifted to the right by ct, see Figure 6. The parameter c is the *wave speed*, the new variable $z = x - ct$ is called the *wave variable*, and the function $U(z)$ is called the *wave profile*.

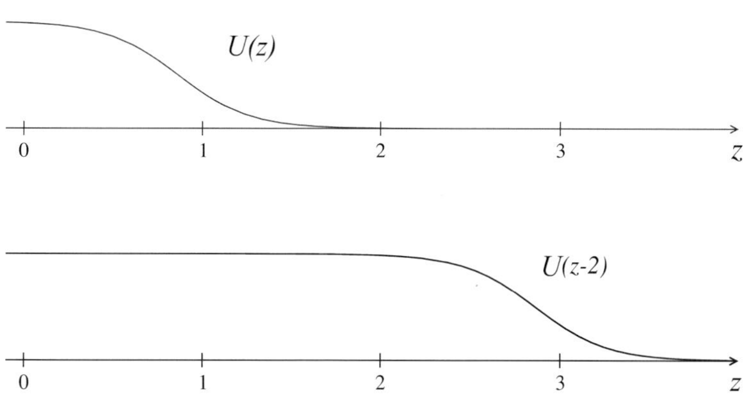

FIGURE 6. The profile $U(x)$ from the top figure is shifted by 2 to the right (bottom).

The travelling wave ansatz, with boundary conditions is

$$u(x,t) = U(x-ct), \quad U(-\infty) = 1, \quad U(+\infty) = 0, \tag{17}$$

For $x \to -\infty$, the population has already grown to its carrying capacity (1 in this case), and for $x \to +\infty$, the population has not yet arrived.

From (17), we obtain

$$\frac{\partial}{\partial t}u(x,t) = -cU', \quad \frac{\partial^2}{\partial x^2}u(x,t) = U'',$$

and (16) reduces to the following ordinary differential equation f or $U(z)$

$$-cU' = U'' + g(U) \tag{18}$$

To allow analysis in the phase plane, we introduce a new variable, $V = U'$, and write (18) as a 2×2 system

$$\begin{aligned} U' &= V, \\ V' &= -cV - g(U) \end{aligned} \tag{19}$$

Before doing further analysis, we need to specify the dynamics $g(u)$ in equation (16) precisely. We consider the case with logistic growth so that $g(U) = U(1-U)$. The model (16) with logistic growth is referred to as Fisher's equation in honor of R.A. Fisher, a quantitative geneticist who used this equation to study the spatial spread of an advantageous gene into a new environment (Fisher [**8**]). The analysis of the bistable case is left as an exercise (below).

The equilibria of (19) are $P_1 = (0,0)$ and $P_2 = (1,0)$. Using the linearization, we find that the point $P_1 = (0,0)$ is stable for $c > 0$. It is a stable spiral for $c < 2$, and a stable node for $c \geq 2$. The point $P_2 = (1,0)$ is always a saddle.

Recall that the boundary conditions for the wave profile are $U(-\infty) = 1$ and $U(+\infty) = 0$. Moreover, from the form of U as shown in Figure 7, it is clear that $V(-\infty) = V(+\infty) = 0$. In the phase portrait of system (19), we must find a connection from the saddle $(1,0)$ to the stable point $(0,0)$. We show these connections for $c < 2$ in Figure 8 (saddle-focus), and for $c \geq 2$ in Figure 9 (saddle-node).

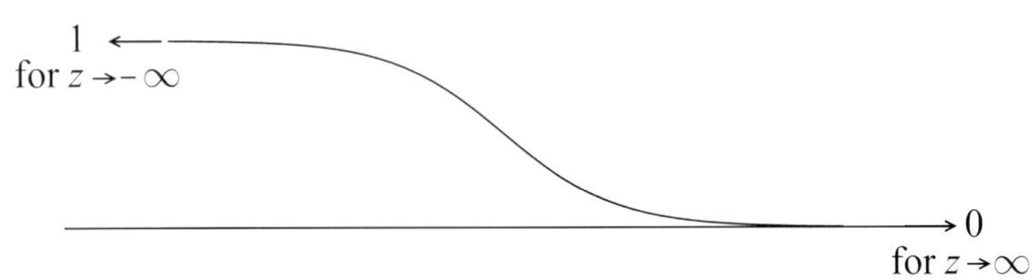

FIGURE 7. The travelling wave as a function of the wave variable z.

Exercise (Phase plane analysis for the bistable travelling wave)

Analyze the bistable travelling wave problem in the phase plane. Show that a heteroclinic connection from $U = 1$ to $U = 0$ yields a saddle-saddle connection in the phase plane.

The function U is the profile of the population density; hence it must be nonnegative. Thus solutions for $c < 2$ are not biologically relevant. (Nor are they mathematically possible as solutions to (13), provided initial conditions are nonnegative (Fisher [**7**]).) They correspond to an oscillating front (see Figure 10).

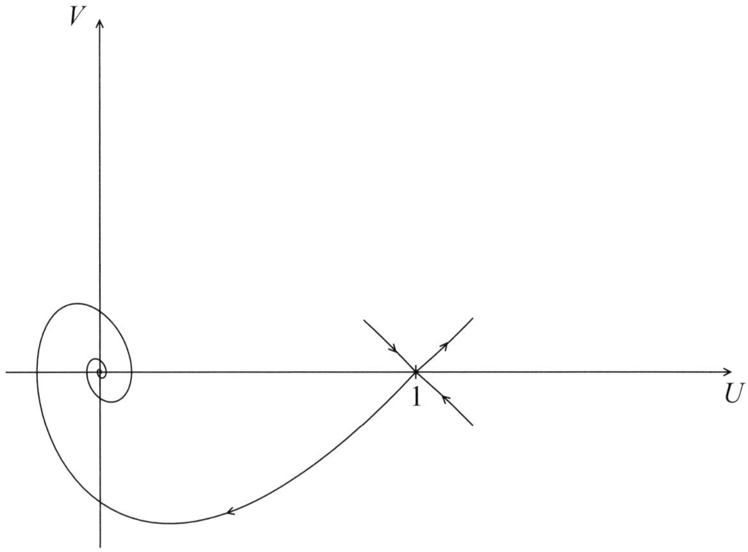

FIGURE 8. Heteroclinic connection from the saddle at $(1, 0)$ to the stable spiral at $(0, 0)$. Here $\mu = D = 1$ and $c < 2$. There is no non-negative travelling wave.

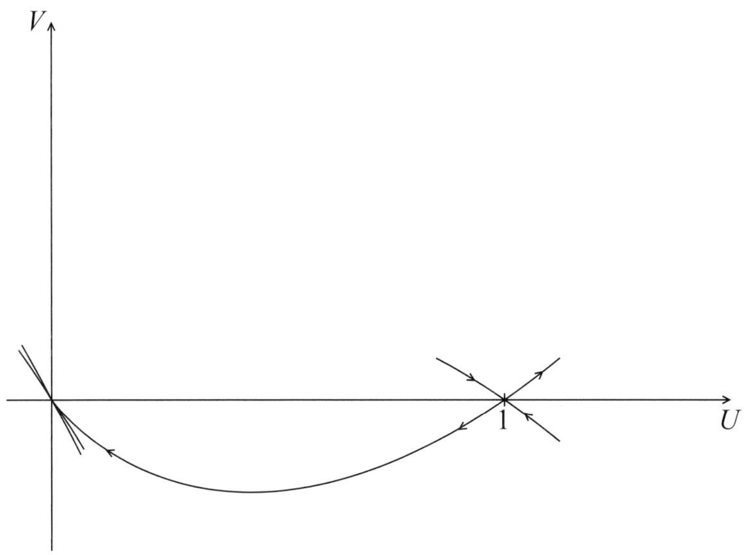

FIGURE 9. Heteroclinic connection from the saddle at $(1, 0)$ to the stable node at $(0, 0)$. Here $\mu = D = 1$, $c > 2$. There exists a non-negative travelling wave.

Hence $c \geq 2$ is a necessary condition for a heteroclinic orbit. Is it a sufficient condition? Because the stable manifold at $P_1 = (0, 0)$ is of dimension 2, it is reasonable to expect that, when $c \geq 2$, a trajectory leaving the saddle at $P_2 = (1, 0)$, as shown in Figure 9, will approach the origin with $U \geq 0$. You are asked to prove this result in the exercise below.

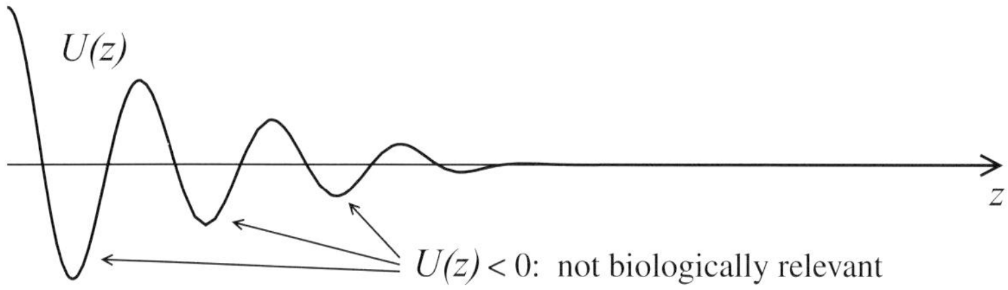

FIGURE 10. Oscillations at the leading edge of the wave from Figure 8.

Exercise (Trapping region)
Show that $c \geq 2$ is a sufficient condition for the heteroclinic orbit to exist as shown in Figure 9). A proof can be constructed as follows:

(a) Create a triangular 'trapping region' whose edges are $V = 0$, $U = 1$ and $V = -\alpha U$, $\alpha > 0$.
(b) The unstable manifold at $P_2 = (1, 0)$ emanates out of the equilibrium and into the triangular region. Furthermore, within the triangle, the flow must move to the left. Why?
(c) Start to show that the 'trapping region' is invariant with respect to the flow (the flow starting at any point (U, V) inside the triangle can never leave the triangle). Do this by showing that on the edge $V = 0$, $V' < 0$ and on the edge $U = 1$, $U' < 0$.
(d) Finally show that α can be chosen so that dot product of the inward normal to the line $V = -\alpha U$ and the flow on the line $V = -\alpha U$ is positive. In other words show that α can be chosen so that $(\alpha, 1)\cdot(U, -cU - V(1 - V)) > 0$ whenever $V = -\alpha U$. HINT: show that an α value that satisfies $\alpha^2 - c\alpha + 1 \leq 0$ will do this. Hence a value of α lying between $(c - \sqrt{c^2 - 4})/2$ and $(c + \sqrt{c^2 - 4})/2$ will suffice.
(e) Put the parts of the argument together to conclude that there must be a heteroclinic orbit that stays within the trapping region when $c \geq 2$.

Thus a non-negative travelling wave solution exists for each $c \geq 2$ and no such solution exists for $c < 2$. The minimum wave speed for which there is a positive solution is $c = 2$. In dimensional terms this minimum wave speed is $c = 2\sqrt{rD}$.

The travelling wave problem, expressed in travelling wave coordinates (18) has an alternative 'physical' interpretation if we take U to be displacement of a nonlinear spring and z to be 'time'. Then $U' = V$ is velocity and $V^2/2$ is kinetic energy. We assume that the potential energy in the spring is a nonlinear function of displacement

$$G(U) = \int_0^U g(w)\, dw. \tag{20}$$

Defining the energy to be the sum of kinetic and potential energies $E = V^2/2 + G(U)$ we calculate the rate of change of energy as

$$\begin{aligned} E' &= \frac{dE}{dU}U' + \frac{dE}{dV}V' \\ &= g(U)V + V(-cV - g(U)) = -cV^2. \end{aligned} \tag{21}$$

The last term on the right hand side of the equation can be interpreted as a damping term, with c the friction coefficient. In the case of the Fisher equation, when the spring is stretched close to $U = 1$ it moves towards $U = 0$. When there is little or no friction the spring will oscillate. The critical wave speed $c = 2$ is the critical friction coefficient that prevents oscillation about the well in potential energy that is found at $U = 0$. Any friction coefficient that exceeds this value will also prevent oscillations, and will allow the spring to come to rest without overshooting.

Exercise (Energy methods for waves)
Plot $G(U)$ for Fisher's equation and for the bistable equation. Use the above kind of heuristic reasoning to argue that while, Fisher's equation allows travelling waves for all speeds exceeding the minimum speed $c = 2$, a heteroclinic connection between $U = 1$ and $U = 0$ for the bistable equation must have a unique wave speed. (HINT by plotting $G(U)$ show that $U = 0$ is not a well in potential energy, but a local peak.)

Although our initial interest was in spread rate of an introduced population, our analysis, so far, has been restricted to travelling waves and their speeds. The relationship between minimum travelling wave speed and spread rate was derived by Aronson and Weinberger [**4**]. Here the spread rate refers to the rate at which a locally introduced population (zero outside a compact set) will spread outwards. Mathematically c^* is the spread rate if, providing the population is introduced locally on a sufficiently large region, a moving reference frame that expands at a speed slower than c^* eventually sees the carrying capacity $u = 1$, and a moving reference frame that expands at speed faster than c^* eventually sees the uninvaded steady state. Mathematically these statements are written as for all $0 < \epsilon \ll c^*$

(22) $$\lim_{t\to\infty} \sup_{|x|>(c^*+\epsilon)t} u(x,t) = 0$$

and

(23) $$\lim_{t\to\infty} \sup_{|x|<(c^*-\epsilon)t} |u(x,t) - 1| = 0$$

The result from Aronson and Weinberger [**4**] shows that the spread rate for Fisher's equation is exactly the minimum wave speed $c^* = 2\sqrt{rD}$.

Returning to the problem of invading a stream system with unidirectional flow, we now include the drift (advection) back into our model (13-14) so that

(24) $$\frac{\partial n}{\partial t} + \frac{\partial}{\partial x}(vn) = D\frac{\partial^2 n}{\partial x^2} + rn\left(1 - \frac{n}{K}\right).$$

However, this drift term is equivalent to simply using a moving reference frame in (13-14). In other words moving from (x, t) to $(x - vt, t)$ transforms (24) to (13-14). Hence, the above analysis can be interpreted as follows. A locally introduced population will spread upstream at speed $c^* - v$ and downstream at speed $c^* + v$. In other words, the population will be successful at spreading upstream when the rate $c^* = 2\sqrt{rD}$ exceeds the drift rate v, and is washed downstream when $v > 2\sqrt{rD}$ (Figure 11).

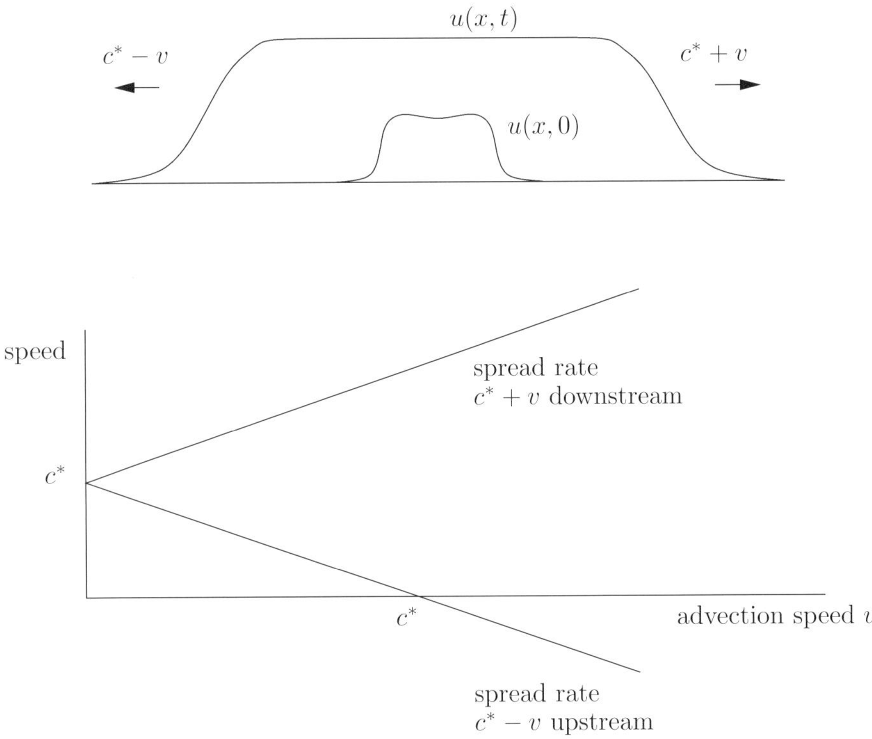

FIGURE 11. A locally introduced population spreads upstream at rate $c^* - v$ and downstream at rate $c^* + v$. Upstream and downstream spread rates are shown as a function of drift velocity.

4. Critical domain size problem

Another measure of whether a population will persist in a stream is whether it will grow when rare. Consider a short stream where the population can grow, outside of which it dies quickly. We start by asking whether the length of stream L is sufficiently large to allow the species to grow when rare. If the length of stream is too small, then individuals will leave the region, move into the hostile surroundings and die. When the stream is long, most individuals can grow and reproduce without leaving the region and moving into the hostile surroundings.

4.1. Classical problem

As we did earlier, we first evaluate the case with zero flow ($v = 0$), and then return to the case $v > 0$ later. As the species is assumed to be rare, it is reasonable to assume that it is at low density. Hence the linearization of (13-14) about $n = 0$ will approximate the dynamics of (13-14). This linearized system, along with boundary conditions indicating a hostile exterior, and an initial condition describing the initial spatial profile of the population is given as

$$\frac{\partial n}{\partial t} = rn + D\frac{\partial^2 n}{\partial x^2}, \quad n(0,t) = n(L,t) = 0, \quad n(x,0) = n_0(x). \tag{25}$$

Because our system is linear, it can be solved by separation of variables. If we consider separable solutions of the form $n \propto \exp(\lambda t)f(x)$, then substitution into

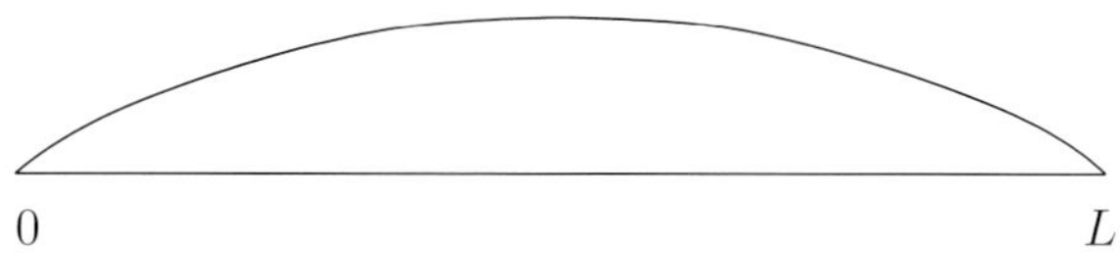

FIGURE 12. A typical solution of (29).

(25) yields

$$f'' + \frac{r-\lambda}{D} f = 0, \quad f(0) = f(L) = 0. \tag{26}$$

Solutions $f(x)$ that satisfy the boundary conditions take the form $f(x) = A\cos(\mu x) + B\sin(\mu x)$, where $\mu^2 = (r-\lambda)/D$. The application of boundary condition $f(0) = 0$ implies $A = 0$ and the application of boundary condition $f(L) = 0$ implies $\mu = k\pi/L$. Hence $(k\pi/L)^2 = (r-\lambda)/D$ so $\lambda = r - D(k\pi/L)^2$. Using the principle of superpositioning the solution to (25) is given as

$$n(x,t) = \sum_{k=1}^{\infty} B_k e^{\left(r - D\left(\frac{k\pi}{L}\right)^2\right)t} \sin\left(\frac{k\pi}{L}x\right), \tag{27}$$

where the constants B_k are determined by the initial conditions $n_0(x)$ in equation (25).

Hence the population will grow if $\lambda > 0$ for some k. The fastest growing mode is associated with $k = 1$. In other words, the population will grow if L exceeds

$$L_c = \pi\sqrt{D/r}. \tag{28}$$

Suppose the critical domain size is exceeded and the population grows. How far it will grow? We can answer this question by considering steady state solutions to time-independent version of the nonlinear problem (13-14). Rescaling n by K to yield $u = n/K$, steady state solutions satisfy

$$\frac{r}{D}u(1-u) + u'' = 0, \quad u(0) = u(L) = 0 \tag{29}$$

(Figure 12). Multiplying (29) through by u' and integrating with respect to x yields

$$\frac{w^2}{2} + \frac{r}{D}F(u) = \frac{r}{D}F(\omega), \tag{30}$$

where $F(u) = u^2/2 - u^3/3$ and $w = u'$. This is a Hamiltonian system. In other words it has a first integral (30) that is conserved. Here ω is the maximum value of u which is found at the interior critical point $w = 0$. Symmetry of the diffusion operator and the boundary conditions imply that the maximum occurs at $x = L/2$.

Hence, on the interval $0 < x < L/2$

$$w = \frac{du}{dx} = \sqrt{\frac{2r}{D}}(F(\omega) - F(u)). \tag{31}$$

Separating variables for the previous equation and integrating yields

$$\begin{aligned} L(\omega) &= \sqrt{\frac{2D}{r}} \int_0^{\omega} \frac{du}{\sqrt{(F(\omega) - F(u))}} \\ &= \sqrt{\frac{2D}{r}} \int_0^1 \frac{\omega dz}{\sqrt{(F(\omega) - F(\omega z))}}, \end{aligned} \tag{32}$$

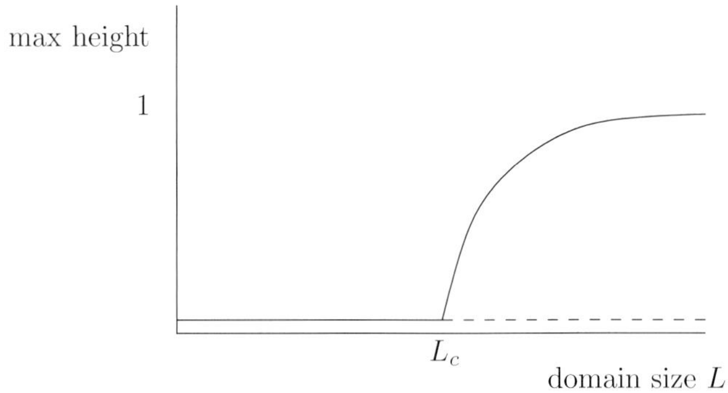

FIGURE 13. A bifurcation curve describing the maximum density of the population μ as a function of the domain size L. Solid lines indicate a stable solution, and dashed lines an unstable.

where $z = u/\omega$. This equation yields a bifurcation diagram for ω versus L. A branch of nontrivial steady state solutions arises from the bifurcation point $(L, \omega) = (\pi\sqrt{D/r}, 0)$ (Figure 13).

Using (32) it is possible to show that, as the maximum value ω approaches zero, L approaches $L_c = \pi\sqrt{D/r}$.

Exercise (Bifurcation point in critical domain problem) Using (32), and carefully taking the limit, show that $\lim_{\omega\to 0} L(\omega) = L_c = \pi\sqrt{D/r}$.

4.2. Critical domain size problem in a stream

We now undertake the analysis of the critical domain size problem in the stream ecosystem. Our analysis will conceptually tie together the classical critical domain size problem with travelling wave theory.

When we include drift, the linearized equation (25) becomes

$$\frac{\partial n}{\partial t} + v\frac{\partial n}{\partial x} = rn + D\frac{\partial^2 n}{\partial x^2}, \quad n(0,t) = n(L,t) = 0, \quad n(x,0) = n_0(x). \tag{33}$$

Typical streams have an upstream source, and then empty into a larger body of water. Hence reasonable boundary conditions are zero-flux at the upstream end, where the population cannot leave, and hostile at the downstream end, where environmental conditions may change

$$vn - D\frac{\partial n}{\partial x} = 0 \quad \text{at} \quad x = 0, n = 0 \quad \text{at} \quad x = 0. \tag{34}$$

Proceeding as we did before, the spatial eigenfunction problem becomes

$$\lambda f + vf' = rf + Df'', \quad vf(0) - Df'(0) = 0, \quad f(L) = 0. \tag{35}$$

This kind of problem is referred to as a Sturm-Liouville boundary value problem [**9**]. Solutions that satisfy the boundary conditions take the form

$$f = A\exp(\xi x)\cos(\rho x) + B\exp(\xi x)\sin(\rho x), \tag{36}$$

where $\xi = v/(2D)$ and $\rho = \sqrt{(r-\lambda)/D - (v/(2D))^2}$. Application of the boundary conditions (after some algebra) yields

$$\tan(\rho L) = -2D\rho/v \tag{37}$$

and hence, the critical domain size is given as $\lambda \to 0$ by

$$L = L_c = \frac{\sqrt{D/r}}{\sqrt{1 - v^2/(4rD)}} \arctan\left(-\frac{2\sqrt{rD}}{v}\left(1 - v^2/(4rD)\right)\right). \tag{38}$$

Exercise (Critical stream size problem) Verify equations (36–38). See also the electronic Appendix A of Speirs and Gurney [**15**].

A plot of critical domain size L_c and the up- and down-stream spread rates is given in Figure 14. As the stream velocity reaches the critical speed $c^* = 2\sqrt{rD}$, the critical domain size approaches infinity while the upstream spread rate $c^* - v$

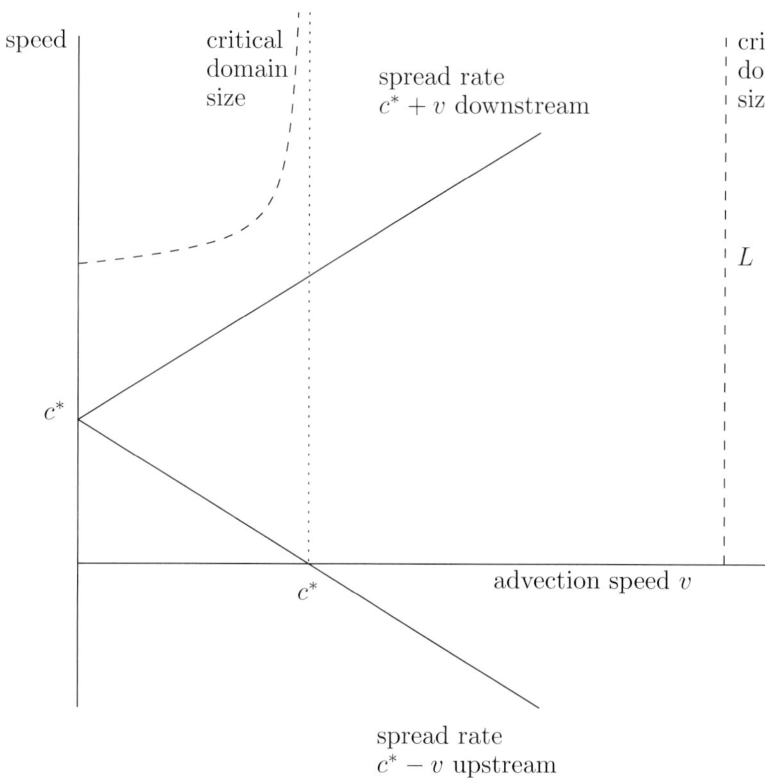

FIGURE 14. A plot of the critical domain size L_c 38 and the up- and down-stream spread rates $c^* - v$ and $c^* + v$ for a range of drift velocities. Note that, for $v = 0$ the critical domain size $\pi\sqrt{D/r}/2$ is exactly half that of the classical problem $\pi\sqrt{D/r}$. This is because of the differing boundary conditions.

simultaneously approaches zero. That these two events occur simultaneously is intuitively appealing. This also provides a link to two seemingly disjoint areas of analysis in mathematical biology (critical domain size analysis/bifurcation theory for one result and travelling wave/spread rate theory for the other).

Further extensions of this work to include more biological scenarios, such as complex mobile and stationary life-history states (Pachepsky et al [**14**]) and non-diffusive movement via turbulent flow (Lutscher et al [**11**]) have provided a similar connection between the critical domain size and upstream spread rate.

In summary, the mathematical analysis shows that having a positive spread rate is a necessary, but not sufficient condition for persistence when there are hostile downstream boundary conditions. The deleterious effects of these boundary conditions become more pronounced as the stream shortens. These ideas are used in the final section of the Speirs and Gurney paper [**15**], where the authors apply their mathematical results to explaining persistence and extinction in a variety of river ecosystems.

Bibliography

[1] W. C. Allee. *The Social Life of Animals.* W. W. Norton and Co., New York, New York, 1938.

[2] L.J.A. Allen. *Stochastic Processes with Applications to Biology.* Prentice Hall, 1st edition, 2003.

[3] D. G. Aronson. The role of diffusion in mathematical population biology: Skellam revisited. In V. Capaso, E. Grosso, and S. L. Paveri-Fontana, editors, *Mathematics in Biology and Medicine*, pages 2–6. Springer-Verlag, Berlin, 1985.

[4] D. G. Aronson and H. F. Weinberger. Nonlinear diffusion in population genetics, combustion, and nerve pulse propagation. In J. A. Goldstein, editor, *Lecture Notes in Mathematics*, volume 446, pages 5–49. Springer-Verlag, Berlin, 1975.

[5] N.F. Britton. *Reaction–Diffusion Equations and Their Applications to Biology.* Academic Press, London, 1986.

[6] G. de Vries, T. Hillen, M.A. Lewis, J. Müller, and B. Schönfisch. *A Course in Mathematical Biology.* SIAM, Boston, 2006.

[7] P. C. Fife. *Mathematical Aspects of Reacting and Diffusing Systems*, volume 28 of *Springer Lecture Notes in Biomathematics.* Springer Verlag, Berlin, 1979.

[8] R.A. Fisher. The advance of advantageous genes. *Ann. Eugenics*, 7:355–369, 1937.

[9] R. Haberman. *Applied Partial Differential Equations with Fourier Series and Boundary Value Problems.* Pearson, 4th edition, 2003.

[10] Mark Kot. *Elements of Mathematical Ecology.* Cambridge University Press, Cambridge, 2001.

[11] F. Lutscher, E. Pachepsky, and M.A. Lewis. The effect of dispersal patterns on stream populations. *SIAM J. Appl. Math.*, 65:1305–1327, 2005.

[12] K. Müller. The colonization of freshwater insects. *Oecologica*, 53:202–207, 1982.

[13] A. Okubo. *Diffusion and Ecological Problems: Modern Perspectives*, volume 14 of *Biomathematics.* Springer-Verlag, Berlin, Germany / Heidelberg, Germany / London, UK / etc., 2001.

[14] E. Pachepsky, F. Lutscher, R.M. Nisbet, and M.A. Lewis. Persistence, spread and the drift paradox. *Theor. Pop. Biol.*, 67:61–73, 2005.

[15] D.C. Speirs and W.S.C Gurney. Population persistence in rivers and estuaries. *Ecology*, 82:1219–1237, 2001.

Matrix Models and Population Dynamics

J. M. Cushing

IAS/Park City Mathematics Series
Volume 14, 2005

Matrix Models and Population Dynamics

J. M. Cushing

Introduction

These lectures are intended to serve as an introduction to the application of nonlinear matrix models in the study of population dynamics. Matrix models define discrete time (semi) dynamical systems that can be used to project population state variables from time t to $t+1$. They have a long tradition of use in describing the dynamics of so-called structured populations. By a structured population is meant one in which individual members are assigned to classification categories and thereby are not considered identical (as they are in many classical models in population dynamics and theoretical ecology). These categories are most often based on physiological characteristics regarded as important to vital processes driving the dynamics of the population. In their seminal papers, Lewis [**49**] and Leslie [**47, 48**] classified individuals by means of (discrete) chronological age classes, but their methodology is straightforwardly applicable to other kinds of classifications such as body size, gender, life cycle stage, location in a spatial habitat, and so on [**46**]. Individual characteristics can correlate closely with vital rates that effect and determine the dynamics of the population as a whole, including rates of reproduction, mortality, immigration/emigration and dispersal, resource utilization and consumption, and so on.

In these lectures I am interested in discrete time models that describe the dynamics of (a finite number of) discrete classes based on a specified categorization of individuals making up a population. Such matrix models, like all models, have their advantages and disadvantages. They are appropriate under some circumstances and not under others (for example when continuous time and/or continuous structuring variables are more appropriate). Caswell gives a nice discussion of these modeling issues in his book [**1**]. The books [**55**] and [**4**] contain treatments of continuous models for structured populations.

Lecture 1 develops a general framework for the derivation of matrix models that describe the dynamics of a population structured by a finite number of distinct classes. A fundamental question is the extinction or the persistence of the population. In Lecture 1 this question is related to two fundamental quantities, the inherent growth rate r and the inherent net reproductive number n. This is done by means of general theorems implying the loss of stability of the extinction

Department of Mathematics & Interdisciplinary Program in Applied Mathematics, University of Arizona, Tucson, AZ 85721
E-mail address: cushing@math.arizona.edu
The author was supported in part by NSF grant DMS-0414212 .

equilibrium at $r = 1$ (equivalently $n = 1$) and population persistence for $r > 1$ (equivalently $n > 1$).

The bifurcation of viable equilibrium states that occurs at $n = r = 1$ is a main theme of Lecture 2. This primary bifurcation can lead to stable or unstable equilibrium states for the population, depending on the relative roles of negative and positive feedbacks from increased (but low level) population density. (Negative feedback is called density regulation and positive feedback is referred to as an Allee effect.) Stable bifurcating equilibria can undergo further bifurcations, including those that give rise to periodic cycles or quasi-periodic motion on invariant loops. Bifurcation cascades to chaos are a well known property of recursive formulas such as those defined by matrix equations. The most famous is the period doubling route to chaos that commonly occurs for the one dimensional case of unstructured population models. For structured models many other bifurcation sequences are also possible, such as the sequences that arise in the case studies presented in Lecture 3.

The major themes in Lecture 3 include the connection of matrix models to time series data, the evaluation of models (which necessitates stochastic versions of the deterministic matrix models in Lectures 2 and 3), the use of models to study nonlinear phenomena (such as bifurcation routes to chaos) by means of controlled experiments, and a discussion of some issues that arose from these studies (e.g., the importance of habitat size and the effects of discrete state space variables).

Lecture 4 extends the basic theory concerning time autonomous models presented in Lectures 1 and 2 to models with periodically forced model parameters. Experimental observations of a nonlinear resonance phenomenon motivate this extended theory.

Lecture 5 returns to autonomous models, but considers interacting (structured) species. Motivated by the use of discrete models in the early history of competition theory, this lecture focuses on the (interference) competition between two species.

I would like to thank Robert F. Costantino, Brian Dennis, Robert A. Desharnais, Shandelle M. Henson, and Aaron King. My collaborations with this interdisciplinary group of biologists, statisticians, and mathematicians (some of which is reported in these lectures) has been a wonderful adventure into the marvelous world of nonlinear population dynamics.

LECTURE 1

Matrix models

One of the most important first steps in deriving a mathematical model of a physical or biological system is the choice of the state variables that will be used to describe the system. In the classical logistic differential equation

$$x' = r\left(1 - \frac{x}{K}\right)x, \tag{1.1}$$

as well as many other classic models in population dynamics and theoretical ecology, the state variable is the total population size $x = x(t)$ (e.g., the number, density, or biomass of all individuals), which is assumed to vary with time t according to this equation. Dynamic equations for the time evolution of the state variable(s) derive from specified assumptions, hypotheses, and laws. The resulting models are characterized as much by phenomena they ignore as they are by phenomena they include. In the logistic equation, for example, since the state variable is total population size, all individuals in the population are in effect treated as identical to each other. Thus, the population is assumed homogeneous in its composition, as if the individual organisms were identical billiard balls or molecules. And yet biologists will often point out that there is often more variability among individuals within a biological population or species than there is among populations or species. Even for micro-organisms, for which this homogeneous population structure assumption might be considered appropriate, a close examination usually reveals significant differences among individuals (size, shape, weight, age, gender, etc.). Indeed, the currency of Darwinian evolution is the reproducing individual, among all of whom there are heritable variations on which natural selection can work.

There are other significant homogeneity assumptions in models such as the logistic differential equation (1.1) and many other similar classical differential equation models (e.g., the Lotka-Volterra competition and predator-prey models). The model parameter K is the equilibrium to which all solutions tend asymptotically. If we view this "carrying capacity" as a descriptor of the population's environment, then the assumption that K is a constant means that we ignore all environmental fluctuations. Similarly, the inherent growth rate r (at which the population would exponentially grow if unrestrained by negative effects of large population sizes) is assumed unchanging in time. Certainly some populations live in relatively constant environments, but a great many do not. For many there are variations in innumerable physical and biological environmental factors (temperature, humidity, food resources, etc.), some of which are irregular (random or stochastic) and some of which are approximately periodic (e.g., seasonal or daily or monthly cycles). Autonomous differential equations are convenient for mathematicians (since there are many tools available to analyze them), but they do entail these simplifying homogeneity assumptions which in many cases are biologically unwarranted.

One of the goals of these lectures is to provide an introduction to a theory for the dynamics of structured populations in which individuals are not assumed identical. The modeling methodology uses classes of individuals as state variables (based, for example, on age, body size, life cycle stage, etc.). This is a compromise between, at one extreme, treating all individuals as identical and, at the other extreme, treating each individual as a state variable. I will develop some basic dynamic properties and features of these models in a mathematically general context from the point of view of bifurcation theory.

Another goal of the lectures is to apply the modeling methodology and analysis in some specific biological and ecological contexts. Cases studies involving applications to some controlled and replicated laboratory experiments not only illustrate the effectiveness of matrix models, but provide an investigation into the occurrence of a variety of model predicted nonlinear phenomena in a real biological population. I will address several issues in these case studies. One issue concerns the connection of theoretical models to data for the purpose of calibrating (parameterizing) the model and evaluating its descriptive accuracy. Another aim is to provide examples in which theoretical models are quantitatively predictive of the dynamics of a real biological population. That is to say, the parameterized model is used to create protocols for experiments designed to test (observe) and study model predicted dynamics. This combined modeling and experimental system provides a powerful tool for the study of nonlinear population dynamics. One can use it to corroborate the theoretical predictions of nonlinear theory as they can occur in a real population (e.g., equilibrium destabilization, bifurcation to non-equilibrium dynamics, chaos, etc.). Moreover, the modeling/experimental system often provides explanations for observations that previously had none and in several instances predicts new and sometimes unexpected phenomena that are subsequently corroborated by experiments. It can also provide a means to investigate the extent to which tenets and principles based on classical ecological models can be validated (or contradicted) in a controlled experimental setting.

This and the next lecture develop an autonomous theory for a single population (appropriate for the case of a constant environment and constant vital rates). Lectures 3 and 4 will extend the theory to include both stochastic and regular periodic fluctuations in both environmental and individual characteristics. Lecture 5 contains a brief look at competition between two structured populations.

Broadly speaking, population dynamic models fall into two classes: those in which time t is continuous and those in which time is discrete. Time is continuous in the logistic differential equation (1.1) and the model is based, as is typical in this case, on the consideration of the rate of change of the state variable x. For discrete time models the approach is typically to build a model by prescribing instead the state variable from one time step to the next. The point of view is that given a sequence of population counts

$$x(0),\ x(1),\ x(2),\ \cdots$$

we want a rule that predicts the next census count $x(t+1)$ from the current census count $x(t)$. Here $x(t)$ can be a vector of state variables observable at the census times $t = 0, 1, 2, \cdots$.

For example, if the solution

$$x(t) = Kx(0)\frac{e^{rt}}{K - x(0) + x(0)e^{rt}}$$

of the logistic differential equation (1.1) is sampled at times $t = 0, 1, 2, \cdots$, the resulting sequence of population sizes satisfies the *discrete logistic* difference (or recursive) equation [**47, 48, 59, 60**]

$$x(t+1) = b\frac{1}{1+cx(t)}x(t) \tag{1.2}$$

where $b \triangleq e^r$ and $c \triangleq (b-1)/K$. We can use this difference equation to predict the population size from one census to the next.

Derivations of discrete time models are not best made from continuous time models. The reason is that approximation errors are likely to be introduced in such derivations and the resulting discrete time models can have dynamic properties that are attributable to these errors and not to any biological mechanisms. This is not the case for the discrete logistic (1.2), whose census counts exactly match those obtained from the solution of the differential equation. However, rarely is a solution formula for a differential equation model available. If instead we use an approximating discretization procedure to obtain the prediction $x(t+1)$ from $x(t)$, then the resulting difference equation is no longer exact. For example, the Euler method applied to the logistic differential equation (1.1) yields the approximation scheme $x(t+1) = x(t) + b(1 - x(t)/K)x(t)$ or $x(t+1) = (1 + b - bx(t)/K)\,x(t)$. This quadratic difference equation (often also called the discrete logistic equation) makes dynamic predictions, such as periodic cycles and chaos, that differ drastically from those of the differential equation.

Discrete time models are best derived from first principles as discrete time models. As a warm up example related to models I will use below, consider a population that reproduces at discrete times (which we census just before reproduction). Suppose there are $x(t)$ adult individuals at census time t who then produce $bx(t)$ newborns, where b is the per capita birth rate. If all these newborns survive one unit of time, then they contribute $bx(t)$ individuals to the adult census count at time $t+1$. However, suppose the survival of a newborn depends on the number of adults $x(t)$ present. Suppose that a contact between a newborn and an adult increases the probability that the newborn will die. (One example, relevant to the case studies in Lecture 2, is cannibalism.) Specifically, suppose the probability that a newborn encounters one specific adult, and as a result dies (with a certain probability), is approximately proportional to the length of time that passes. If we divide the time unit t to $t+1$ into $1/\Delta t$ steps and consider encounters during each sub-step as independent probabilities, then the probability that a newborn survives exposure to one adult is approximately $(1 - c\Delta t)^{1/\Delta t}$ where $c > 0$ is the constant of proportionality. If encounters with adults are independent events, then the probability that the newborn will survive in the presence of $x(t)$ adults is approximately $(1 - c\Delta t)^{x(t)/\Delta t}$. Letting $\Delta t \to 0$, we find the probability that a newborn will survive and join the adult population one time unit later is equal to $\exp(-cx(t))$. Thus, the contribution of the newborns to the next adult census is $bx(t)\exp(-cx(t))$. This contribution is added to the surviving adults. The resulting difference equation is

$$x(t+1) = bx(t)e^{-cx(t)} + (1-\mu)x(t)$$

where μ is the adult death rate ($0 \le \mu \le 1$). If $\mu < 1$ the population is *iteroparous* (adults reproduce more than once). If $\mu = 1$ it is *semelparous* (adults reproduce only once) and the equation reduces to the *Ricker equation* [**62**]

$$x(t+1) = be^{-cx(t)}x(t). \tag{1.3}$$

Over the entire range of the parameter values $b > 0$ and $c > 0$, the Ricker equation makes quite different predictions from those of the discrete logistic (1.2). Both predict population extinction ($x(0) > 0 \implies \lim_{t\to+\infty} x(t) = 0$) if $b < 1$ and the existence of a unique positive equilibrium (fixed point) if $b > 1$, namely, $x = c^{-1}(b-1)$ for the discrete logistic and $x = c^{-1}\ln b$ for the Ricker equation. The discrete logistic predicts that when $b > 1$ all populations $x(0) > 0$ tend to the equilibrium as $t \to +\infty$. (We can see this from the way in which the equation was derived from the logistic equation, or we can prove it directly without reference to this derivation.) This is (not necessarily) true for the Ricker equation (1.3). When b is sufficiently large, populations can tend asymptotically to non-equilibrium states, such as periodic cycles or chaotic attractors. This story (which is not easy to analyze rigorously from a mathematical point of view) is summarized in Figure 1 which shows the attractor plotted against the parameter b. This bifurcation diagram, which is perhaps familiar to the reader, has become a virtual icon of chaos theory and illustrates the extraordinary complexity that discrete models of populations dynamics can exhibit. This complexity was first noted by Lord May in several influential papers in the 1970's that served to popularize chaos theory; e.g., see [**28, 52, 53**] (also see [**51**]).

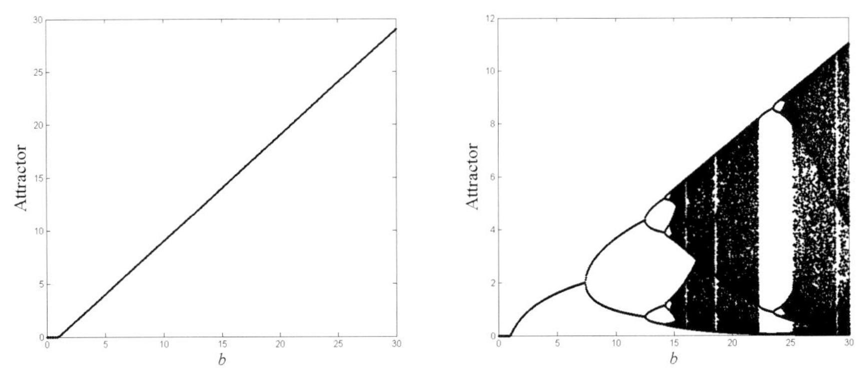

FIGURE 1. (a) The final state (attractor) of all solutions $x(0) > 0$ of the discrete logistic equation (1.2) with $c = 1$ is the equilibrium $x = 0$ for $b < 1$ and the unique positive equilibrium $x = c^{-1}(b-1)$ for $b > 1$. (b) The attractor of solutions $x(0) > 0$ of the Ricker equation (1.3) with $c = 1$ is shown plotted against $b > 0$. For $b < 0$ the attractor is the equilibrium $x = 0$ and for $b > 1$ but less than (approximately) 7.4 the attractor is the unique positive equilibrium $x = c^{-1}\ln(b-1)$. For larger values of b the attractor consists of more than one point. For example, for b between (approximately) 7.4 and 12.5 the attractor consists of the two points of a periodic 2-cycle. For some values of b the attractor apparently consists of a large number of points. The plot was constructed by iterating the Ricker equation 1000 times from the initial condition $x(0) = 1$ and plotting the last 100 points.

The state variable in the discrete time logistic and Ricker difference equations is a single class of homogeneous individuals. To motivate discrete time models that structure the population into more than one class, I consider the classic example of such models in which the structuring classes are based on chronological age. The *Leslie model* age structured model classifies individuals into m age classes x_i of equal length l and tracks the changes in these classes at discrete time intervals of duration l which we take, for notational simplicity, to be $l = 1$. Then $x_{i+1}(t+1) = \tau_{i+1,i}x_i(t)$ where $\tau_{i+1,i}$ is the fraction of individuals of age i that survive one unit of time (and hence move from class i to class $i+1$). The survivors at time $t+1$ are the components of the vector $Tx(t)$ where $x(t) = \mathrm{col}(x_i(t))$ is an m-dimensional column vector (and hence lies in m-dimensional Euclidean space $R^m \triangleq R \times \cdots \times R$) and where

$$T = \begin{pmatrix} 0 & 0 & \cdots & 0 & 0 \\ \tau_{21} & 0 & \cdots & 0 & 0 \\ 0 & \tau_{32} & \cdots & 0 & 0 \\ \vdots & \vdots & & \vdots & \vdots \\ 0 & 0 & \cdots & \tau_{m,m-1} & 0 \end{pmatrix}$$

is the matrix of age class survivorships (or class survival probabilities). This assumes no individual lives more than m time units. Sometimes this assumption is not made and the final class x_m consists of all individuals of age greater than or equal to m, in which case a survival fraction τ_{mm} appears in the lower right hand corner of T. This model also assumes no individuals migrate in or out of the population.

Since by definition newborn individuals are in the first age class x_1, the total number of newborns at time $t+1$ is $x_1(t+1) = \sum_{i=1}^{m} f_{1i}x_i(t)$ where f_{1i} is the number of newborns (i.e., class $i = 1$ individuals), contributed by individuals of age i, who survive to $t+1$. For a non-reproductive age class j, we have $f_{1j} = 0$. The matrix equation

$$\begin{pmatrix} x_1(t+1) \\ x_2(t+1) \\ \vdots \\ x_m(t+1) \end{pmatrix} = \begin{pmatrix} f_{11} & f_{12} & \cdots & f_{1,m-1} & f_{1m} \\ 0 & 0 & \cdots & 0 & 0 \\ 0 & 0 & \cdots & 0 & 0 \\ \vdots & \vdots & & \vdots & \vdots \\ 0 & 0 & \cdots & 0 & 0 \end{pmatrix} \begin{pmatrix} x_1(t) \\ x_2(t) \\ \vdots \\ x_m(t) \end{pmatrix}$$
$$+ \begin{pmatrix} 0 & 0 & \cdots & 0 & 0 \\ \tau_{21} & 0 & \cdots & 0 & 0 \\ 0 & \tau_{32} & \cdots & 0 & 0 \\ \vdots & \vdots & & \vdots & \vdots \\ 0 & 0 & \cdots & \tau_{m,m-1} & 0 \end{pmatrix} \begin{pmatrix} x_1(t) \\ x_2(t) \\ \vdots \\ x_m(t) \end{pmatrix}$$

summarizes the bookkeeping of the census counts in the age categories from one census time to the next. The matrix

$$F = \begin{pmatrix} f_{11} & f_{12} & \cdots & f_{1,m-1} & f_{1m} \\ 0 & 0 & \cdots & 0 & 0 \\ 0 & 0 & \cdots & 0 & 0 \\ \vdots & \vdots & & \vdots & \vdots \\ 0 & 0 & \cdots & 0 & 0 \end{pmatrix}.$$

contains the per capita fertilities of each age class in the first row. The remaining rows of zeros indicate that all newborns belong to the first age class. The *projection matrix* $P = F + T$ in this example is called a *Leslie matrix.*

Other classification schemes for the individuals in a population result in transition and fertility matrices T and F of different forms. (See [**1**] for a detailed treatment of matrix models in population dynamics.) In general, we obtain a recursive matrix (or difference) equation of the form

$$x(t+1) = (F+T)\,x(t). \tag{1.4}$$

The entry τ_{ij} in the *transition matrix* $T = (\tau_{ij})$ is the fraction of j class individuals that survive and become i class individuals in one unit of time. The entry f_{ij} in the *fertility matrix* $F = (f_{ij})$ is the (surviving) number of i class offspring born to a j class individual in one unit of time. Thus,

$$f_{ij} \geq 0, \quad 0 \leq \tau_{ij} \leq 1, \quad \sum_{i=1}^{m} \tau_{ij} \leq 1. \tag{1.5}$$

The reason for the last inequality on the column sums of the transition matrix T is that the number of individuals from class i that get distributed to all classes (including the class i itself) by time $t+1$ cannot exceed the number available in class i at time t. Notice that if newborns cannot arise in class i, then the entire i^{th} row of F consists of zeros. For example if, as in the Leslie age class model, newborns arrive in only one class (usually taken to be the class $i = 1$), then all rows except the first consists entirely of 0's.

A matrix is called *positive (non-negative)* if all its entries are positive (non-negative). In applications to population dynamics positive (non-negative) class distribution vectors $x > 0$ (and $x \geq 0$) are of interest. We call the set of positive (m-dimensional) column vectors $x > 0$ the *positive cone* in R^m and denote this set by R^m_+. The set of non-negative vectors $x \geq 0$ is the non-negative cone $\bar{R}^m_+$.

The entries in the fertility and transition matrices, and hence the projection matrix $P = F + T$, is non-negative. This implies the solution $x(t)$ of (1.4) starting from a non-negative initial condition $x(0) \geq 0$ remains non-negative for all $t = 1, 2, 3 \cdots$. The non-negative cone $\bar{R}^m_+$ is said to be *forward invariant* with respect to the matrix equation (1.4).

A matrix is *reducible* if we can, by reordering the classes, put it in block triangular form

$$\begin{pmatrix} A & 0 \\ C & D \end{pmatrix}$$

where A and D are square matrices. It is *irreducible* if it is not reducible. The projection matrix P is irreducible means that individuals from the i-class contribute to the j-class in a finite number of steps (by transition or by births), for every i and j. (It also means that the directed graph associated with P is strongly connected.) As an example, a Leslie matrix is irreducible if all $\tau_i > 0$ and $f_m > 0$ (the oldest class is fertile).

The eigenvalue of P with the largest absolute value is the *dominant eigenvalue.* The absolute value of the dominant eigenvalue is the *spectral radius* of P, which we denote by r or sometimes by $\rho(P)$.

By iterating the matrix equation (1.4), we obtain the formula $x(t) = P^t x(0)$ for the solution of (1.4) with initial condition $x(0)$.

Theorem 1.1 ([**37**], Theorem 5.612). *Consider the matrix equation (1.4) under the assumptions (1.5). If* $r = \rho(P) < 1$*, then all solutions satisfy* $\lim_{t\to+\infty} x(t) = 0$.

The famous Perron-Frobenius Theorem implies that if the non-negative matrix P is irreducible, then r is an algebraically simple eigenvalue of P. Moreover, r has a positive eigenvector $v > 0$ and no other eigenvector associated with any other eigenvalue is non-negative. An irreducible matrix P is said to be *primitive* if $r = \rho(P)$ is *strictly* dominant (the magnitude of all other eigenvalues are *strictly* less than r). It turns out that P is primitive (hence irreducible by definition) if and only if some integer power of P is positive, i.e., there exists a positive integer k such that $P^k > 0$ ([**26**], p. 80).

Theorem 1.2. *Consider the matrix equation (1.4) under the assumptions (1.5). Assume* P *is primitive. If* $r = \rho(P) > 1$*, then all solutions with initial conditions* $x(0) \geq 0$, $x(0) \neq 0$ *satisfy* $\lim_{t\to+\infty} |x(t)| = \infty$.

The case $r = \rho(P) = 1$ is not covered by Theorems 1.1 and 1.2. In this case, solutions with an initial condition $x(0)$ that is a multiple of an eigenvector of P remain constant: $x(t) = v$ for all $t = 0, 1, 2, \cdots$. Thus, at this (and only at this) one value of $r = \rho(P) = 1$ the population dynamics do not necessarily imply extinction or unlimited growth.

Suppose we view the dynamics of the linear model (1.4) as a function of the spectral radius $r = \rho(P)$. Theorems 1.1 and 1.2 imply that a *bifurcation* occurs at $r = 1$ in the sense that there is a drastic change in the asymptotic dynamics for $r < 1$ and $r > 1$, a change from extinction to unbounded growth. At the *bifurcation point* $r = 1$ there is a branch of equilibria (all scalar multiplies of v) that intersects the equilibrium $x = 0$; that is to say, two equilibrium branches intersect at $r = 1$. Therefore, with regard to the existence of nontrivial equilibria, the linear model (1.4) has a "point spectrum". Feasible asymptotic population dynamics – that is, non-extinction states that are not unbounded – occur only at an isolated value of the bifurcation parameter.

The lowest dimensional case, when the number of classes $m = 1$, is a one dimensional map $x(t+1) = (f+\tau)x(t)$ for which $r = f + \tau$. The conditions $r < 1$, $r = 1$, and $r > 1$ are equivalent to $n < 1$, $n = 1$, $n > 1$ where

$$n \triangleq \frac{f}{1-\tau}.$$

This number, which we can rewrite as

$$n = f + \tau f + \tau^2 f + \cdots + \tau^t f + \cdots,$$

has an important biological interpretation. If a newborn survives for t time steps, then f offspring result. Since the probability of surviving t time steps is τ^t, the expected number of offspring from the newborn at time t is $\tau^t f$. As a result, n is the expected number of offspring (for each newborn) over the course of its entire lifetime. The number n is called the *net reproductive number* (often denoted R_0).

In the $m > 1$ dimensional case we can define the net reproductive number n as follows. Assume that $I - T$ is invertible (i.e., 1 is not an eigenvalue of the transition matrix T). The net reproductive number n is the spectral radius of $F(I-T)^{-1}$, i.e., $n = \rho\left(F(I-T)^{-1}\right)$.

A theorem relating the spectral radius r to n appears in [**50**]. A slightly different version of this result appears in [**15**].

Theorem 1.3 ([**50**], Theorem 3.3). *Suppose the projection matrix* $P = F + T$, *where* F *and* T *satisfy (1.5), is primitive. Let* $r = \rho(P)$ *and* $n = \rho\left(F(I-T)^{-1}\right)$. *If* $T \neq 0$ *and* $\rho(T) < 1$, *then one and only one of the following holds:*

$$r = n = 1 \qquad \text{or} \qquad 1 < r < n \qquad \text{or} \qquad 0 < n < r < 1$$

The assumption that $\rho(T) < 1$ means that no individual has an infinite life expectancy. This follows from the fact that iterations of the transition component of the model $T^t x(0)$ tend to 0 if and only if $\rho(T) < 1$ ([**37**], Theorem 5.16). The condition on T appearing in (1.5) implies $\rho(T) \leq 1$, but not necessarily that $\rho(T) < 1$. ($T = I$ is an example.)

It follows from Theorem 1.3 that we can use either quantity r or n to determine the asymptotic fate of a population modeled by (1.4). However, whereas formulas for r (in terms of the matrix entries) are not available except for models of low dimension (m small), formulas for n often are available for models of arbitrary dimension [**4**].

For example, for a Leslie matrix we have

$$n = \textstyle\sum_{i=1}^{m} f_{1i} \prod_{j=1}^{i} \tau_{j,j-1} \tag{1.6}$$

where, for notational convenience, $\tau_{10} = 1$. The biological interpretation of this formula for n when P is a Leslie matrix is straightforward: $\prod_{j=1}^{i} \tau_{j,j-1}$ is the probability of surviving to age i, $f_{1i} \prod_{j=1}^{i} \tau_{j,j-1}$ is the number of offspring produced when at age i, given survival to age i. Therefore, n is the sum total of all offspring produced during all ages, i.e., is the expected number of offspring per newborn per lifetime.

The general biological interpretation of n is more complicated. The i,j entry in $(I-T)^{-1}$ is the expected amount of time a j class newborn will spend in the i class during its lifetime. The i,j entry in $F(I-T)^{-1}$ is the expected number of i class newborns that will be produced by a j class newborn during its lifetime. If (without loss in generality) the classes into which offspring can be born are listed first, then the last $m-k$ rows of the fertility matrix F will consist 0's and, as a result, so will the last $m-k$ rows of $F(I-T)^{-1}$. The $k \times k$ matrix Q in

$$F(I-T)^{-1} = \begin{pmatrix} Q & S \\ 0 & 0 \end{pmatrix}$$

is nontrivial, non-negative and irreducible ([**50**], Proposition 4.1) and $n = \rho(Q)$ is its dominant eigenvalue. If we denote the first k entries in x by z, a well known formula for the dominant eigenvalue of a non-negative irreducible matrix is

$$n = \max_{z \geq 0} \min_{1 \leq i \leq m, z_i \neq 0} \frac{(Qz)_i}{z_i}$$

where z_i is the ith component of z and $(Qz)_i$ is the ith component of Qz ([**26**], p. 65). Consider an initial cohort z of newborns. The quantity $(Qz)_i / z_i$ is the expected number of i class offspring produced by the cohort per member of the i class in the cohort. According to the formula above, the net reproductive number n is found by selecting the smallest of these expected numbers, from among all classes represented in the cohort z ($z_i \neq 0$), and then maximizing this minimum over all possible initial cohorts z of newborns. If there is only one newborn class ($k = 1$), then the biological interpretation of n is the same as that the Leslie matrix case given above.

Having a formula for the net reproductive number n, in terms of parameters appearing in F and T, means we can study how n depends on any of these model parameters. While Theorem 1.3 provides some comparisons between n and r, it does not provide a functional relationship between r and n. Thus, for example, it is not clear whether an increase in a selected parameter in one of the model matrices F and/or T which results in an increase in n (presumably interpreted as favorable for the population) will result in an increase or decrease in r. Such questions arise in the "sensitivity analysis" of models [**1**].

A linear matrix model (1.4) implies extinction or unbounded growth, except when $n = 1$. Long term sustainable population growth is only possible at the point spectrum $n = 1$. This situation represents a "vertical" bifurcation, by which is meant that a branch of nontrivial equilibria intersects the branch of trivial (extinction) equilibria $x = 0$ at and only at $n = 1$. (By Theorem 1.3 we could also use r.)

Many parameters appear in the projection matrix P (f_{ij} and τ_{ij}) but the asymptotic dynamics of the matrix model (1.4) depends on only one composite parameter, namely, n or equivalently r. These important quantities do not explicitly appear in the matrix model. However, since formulas relating n to model parameters are often available, one can often introduce n explicitly into the model. For example, we can in principle scale the entries in the fertility matrix by n so that $F = n\Phi$ where the dominant eigenvalue of $\Phi\,(I - T)^{-1}$ equals 1. Then (1.4) becomes

$$x(t+1) = (n\Phi + T)\, x(t).$$

To have long term sustainable population growth, for other than the isolated value $n = 1$, the matrices Φ and T cannot remain constant in time. One important way in which the projection matrix can change in time is through a dependence on $x = x(t)$. In this case, the model is called *density dependent*. As will be seen below, in general such nonlinear models have a branch of nontrivial equilibria that bifurcates from the extinction equilibrium, just as linear models do. However, unlike the case for linear models, the spectrum of n for which unbounded but sustainable growth will, in general, no longer be an isolated point, but instead be an interval. The vertical branch of nontrivial equilibria of a linear model gets "bent" into a non-vertical curve for a density dependent (nonlinear) model, producing a continuous spectrum.

Consider a nonlinear (density dependent) matrix model

$$x(t+1) = \left(F(x(t)) + T(x(t))\right) x(t) \tag{1.7}$$

where now the entries in the fertility and transition matrices $F(x)$ and $T(x)$ are (continuously differentiable) functions of x on some open domain $D \subset R^m$ that contains the non-negative cone $\bar{R}^m_+ \subset D$:

$$f_{ij} \in C^1\left(D, [0, +\infty)\right), \quad \tau_{ij} \in C^1\left(D, [0, 1]\right). \tag{1.8}$$

Assume $F(0)$ and $T(0)$ satisfy (1.5), that $F(0)+T(0)$ is primitive, and that $T(0) \neq 0$ satisfies $\rho\,(T(0)) < 1$ so that $n > 0$ is defined. Normalize the fertility matrix so that $F(x) = n\Phi(x)$ where $\rho\left(\Phi(0)\,(I - T(0))^{-1}\right) = 1$.

In summary, consider the matrix equation

$$x(t+1) = \left(n\Phi(x(t)) + T(x(t))\right) x(t), \quad x(0) \in \bar{R}_+^m \tag{1.9}$$

under the assumptions

$$\varphi_{ij} \in C^1\left(D, [0, +\infty)\right), \quad \tau_{ij} \in C^1\left(D, [0,1]\right) \tag{1.10}$$

and

$$\begin{gathered} n\Phi(0) + T(0) \text{ is primitive for } n > 0 \\ \sum_{i=1}^{m} \tau_{ij}(0) \leq 1, \quad T(0) \neq 0, \quad \rho\left(T(0)\right) < 1 \\ \rho\left(\Phi(0)\left(I - T(0)\right)^{-1}\right) = 1 \end{gathered} \tag{1.11}$$

on the normalized fertility matrix $\Phi = (\varphi_{ij})$ and the transition matrix $T = (\tau_{ij})$. The quantity n is called the *inherent* net reproductive number associated with the nonlinear model (1.9). That is to say, n is the reproductive number when population density is low (mathematically is zero).

Example 1.4. *The $m = 3$ dimensional projection matrix $P(x) = F(x) + T(x)$ with*

$$F(x) = \begin{pmatrix} 0 & 0 & f_{13}\exp\left(-c_{11}x_1 - c_{13}x_3\right) \\ 0 & 0 & 0 \\ 0 & 0 & 0 \end{pmatrix}$$

$$T(x) = \begin{pmatrix} 0 & 0 & 0 \\ \tau_{21} & 0 & 0 \\ 0 & \tau_{32}\exp\left(-c_{33}x_3\right) & \tau_{33} \end{pmatrix}$$

$$f_{13}, c_{ij} > 0, \quad 0 < \tau_{21}, \tau_{32} \leq 1, \quad 0 < \tau_{33} < 1$$

is a nonlinear Leslie matrix that describes a common life cycle that occurs in insect populations. Class $i = 1$ is the (only) newborn class and class $i = 3$ is the only reproducing (adult) class. Class $i = 2$ is an intermediate non-reproductive developmental stage. We can think of class $i = 1$ as a larval stage, after one unit of time in which surviving individuals become pupae. After another unit of time surviving pupae become adults. The component model equations are

$$\begin{aligned} L(t+1) &= bA(t)\exp\left(-\frac{c_{el}}{V}L(t) - \frac{c_{ea}}{V}A(t)\right) \\ P(t+1) &= \left(1 - \mu_l\right)L(t) \\ A(t+1) &= \left(1 - \mu_p\right)P(t)\exp\left(-\frac{c_{pa}}{V}A(t)\right) + \left(1 - \mu_a\right)A(t) \end{aligned}$$

where, in order to conform with the notation used in the literature [**5, 11**], *some parameters are re-labeled as*

$$\begin{aligned} x &= \operatorname{col}\left(x_1, x_2, x_3\right) = \operatorname{col}\left(L, P, A\right) \\ f_{13} &= b, \quad \tau_{21} = 1 - \mu_l, \quad \tau_{32} = 1 - \mu_p, \quad \tau_{33} = 1 - \mu_a \\ c_{11} &= c_{el}/V, \quad c_{13} = c_{ea}/V, \quad c_{33} = c_{pa}/V. \end{aligned}$$

When $x = 0$, the projection matrix $P(0)$ is a Leslie matrix and by (1.6) we find that

$$n = b\frac{\left(1 - \mu_l\right)\left(1 - \mu_p\right)}{\mu_a}$$

provided $\mu_a > 0$. *This model has the form (1.9) with normalized fertility matrix*

$$\Phi(x) = \begin{pmatrix} 0 & 0 & \frac{\mu_a}{(1-\mu_l)(1-\mu_p)} \exp\left(-\frac{c_{el}}{V}L - \frac{c_{ea}}{V}A\right) \\ 0 & 0 & 0 \\ 0 & 0 & 0 \end{pmatrix}.$$

Calculations show

$$T(0) = \begin{pmatrix} 0 & 0 & 0 \\ 1-\mu_l & 0 & 0 \\ 0 & 1-\mu_p & 1-\mu_a \end{pmatrix}$$

$$n\Phi(0) + T(0) = \begin{pmatrix} 0 & 0 & n\frac{\mu_a}{(1-\mu_l)(1-\mu_p)} \\ 1-\mu_l & 0 & 0 \\ 0 & 1-\mu_p & 1-\mu_a \end{pmatrix}$$

$$\Phi(0)\left(I - T(0)\right)^{-1} = \begin{pmatrix} 1 & (1-\mu_l)^{-1} & (1-\mu_l)^{-1}(1-\mu_p)^{-1} \\ 0 & 0 & 0 \\ 0 & 0 & 0 \end{pmatrix}$$

and $\rho(T(0)) = 1 - \mu_a < 1$, $\rho\left(\Phi(0)\left(I - T(0)\right)^{-1}\right) = 1$. *Conditions (1.10) and (1.11) are met.*

The nonlinear matrix equation (1.9) has the trivial (extinction) equilibrium $x = 0$ for all values of n. An important question is whether or not the solution of (1.9) tends to $x = \mathrm{col}(0,0,0)$ as $t \to +\infty$ or not. For the linear case, according to Theorem 1.3, the solution does so when $n < 1$ and does not do so when $n > 1$. The nonlinear case requires some definitions and theorems.

Let O be an open subset of R^m and $f : O \to R^m$. An equilibrium $x_e \in O$ of the (difference) equation $x(t+1) = f(x(t))$ is a fixed point of f, i.e., $x_e = f(x_e)$. The following definitions are standard in stability theory for dynamical systems.

Definition 1.5. An equilibrium x_e of the equation $x(t+1) = f(x(t))$ is **stable** if $\forall \varepsilon > 0 \; \exists \; \delta = \delta(\varepsilon) > 0$ such that $|x(0) - x_e| < \delta \Longrightarrow |x(t) - x_e| < \varepsilon$ for $t = 0, 1, 2, \cdots$. An equilibrium x_e is a **local attractor** if $\exists \; \delta_0 > 0$ such that $|x(0) - x_e| < \delta_0 \Longrightarrow \lim_{t\to+\infty} |x(t) - x_e| = 0$ and it is **locally asymptotically stable** if it is a stable local attractor. The set B of all $x(0) \in O$ for which $\lim_{t\to+\infty} |x(t) - x_e| = 0$ is the **basin of attraction** of x_e. An equilibrium is a **global attractor** with respect to a set G if $G \subseteq B$. An equilibrium is **globally asymptotically stable** with respect to a set G if it is both asymptotically stable and a global attractor with respect to G.

For the linear matrix model (1.4) the extinction equilibrium $x_e = 0$ is globally asymptotically stable (with respect to R^m) if $\rho(P) < 1$. If $\rho(P) > 1$ then $x_e = 0$ is unstable (not stable). For nonlinear models local stability of equilibria can be determined by the linearization principle. See [**24**] (chapter 4).

Theorem 1.6. *Assume x_e is a fixed point of $f \in C^1(O, R^m)$ and J is the Jacobian matrix*[1] *evaluated at x_e. If $\rho(J) < 1$, then x_e is locally asymptotically stable. If $\rho(J) > 1$ then x_e is unstable.*

[1]The Jacobian of f is the matrix $(\partial_i f_j)$ of first order partial derivatives of f, where $\partial_i f_j$ is the derivative of the jth component f_j of f with respect to the ith component x_i of x.

Consider the matrix model (1.9). The Jacobian of $f(x) = (n\Phi(x) + T(x))\,x$ evaluated at the equilibrium $x_e = 0$ is $n\Phi(0) + T(0)$. It follows from Theorem 1.3 and 1.6 that, just as does for the linear model (1.4), the extinction equilibrium of the nonlinear model (1.9) loses stability as n (or r) increases through 1.

Theorem 1.7. *The extinction equilibrium $x_e = 0$ of the matrix model*

$$x(t+1) = (n\Phi(x(t)) + T(x(t)))\,x(t),$$

under the assumptions (1.10)-(1.11), is locally asymptotically stable if $n < 1$ and is unstable if $n > 1$.

This theorem applies, for example, to the three life cycle stage model in Example 1.4.

We will see examples below for which $x_e = 0$ is *not* globally asymptotically stable for $n < 1$. Further restrictions are required on the dependence of Φ and T on x in order to have global stability.

Suppose, for example, that all entries $\varphi_{ij}(x)$ and $\tau_{ij}(x)$ in the matrices $\Phi(x)$ and $T(x)$ satisfy the inequalities

$$\varphi_{ij}(x) \leq \varphi_{ij}(0), \qquad \tau_{ij}(x) \leq \tau_{ij}(0), \qquad x \in \bar{R}_+^m. \tag{1.12}$$

Then for $x(0) \geq 0$ we have

$$0 \leq x(t+1) \leq (n\Phi(0) + T(0))\,x(t).$$

By induction $0 \leq x(t) \leq y(t)$ where $y(t)$ solves the linear initial value problem

$$y(t+1) = (n\Phi(0) + T(0))\,y(t), \qquad y(0) = x(0).$$

If $n < 1$ then, using Theorems 1.1 and 1.3 with $P = n\Phi(0) + T(0)$, we conclude that $\lim_{t\to+\infty} y(t) = 0$ and hence $\lim_{t\to+\infty} x(t) = 0$.

Theorem 1.8. *If $n < 1$, the extinction equilibrium $x_e = 0$ of the matrix model*

$$x(t+1) = (n\Phi(x(t)) + T(x(t)))\,x(t),$$

under the assumptions (1.10)-(1.11) and (1.12), is globally asymptotically stable on the non-negative cone $\bar{R}_+^m$.

The conditions (1.12) describe a strong case of deleterious density dependence (or negative feedback) in the sense that the fertility and transition/survivorship rates of all age classes are decreased in the presence of (stage specific) population densities. These conditions are satisfied by the LPA model in Example 1.4.

What are the dynamics of a nonlinear matrix model (1.9) when $x = 0$ is unstable (for example when $n > 1$)? Two basic questions are: when are solutions bounded (i.e., when is unlimited population growth avoided) and when do solutions not tend to 0 (i.e., when do no populations go extinct)?

Define the vector norm $|x| = \sum_{i=1}^m |x_i|$, which for population models is the *total population size*. For this norm the induced matrix norm[2] is $||M|| = \sup_j \Sigma_i |m_{ij}|$, i.e., the largest vector norm from among the columns of M. Assume

$$|F(x)x| \leq f_\infty < +\infty, \quad ||T(x)|| \leq \tau_\infty < 1 \text{ for all } x \in \bar{R}_+^m. \tag{1.13}$$

[2]Let $|x|$ be a vector norm on R^m. The *induced (or operator) norm* of a matrix M is $||M|| = \sup_{|x|=1} |Mx| = \sup_{|x|\neq 0} |Mx| \,/\, |x|$. In addition to the defining properties of a norm, the induced norm has the property $||MN|| \leq ||M||\,||N||$. By definition, $|Mx| \leq ||M||\,|x|$ for all $x \in R^m$.

This assumption on F implies that the total number of newborns (per unit time) in all classes, produced by adults from all classes, is bounded. This is a reasonable biological assumption. The assumption on T implies that there is always some loss in the transitions among classes during a unit of time (due, for example, to mortality). This is also a reasonable biological assumption, although occasionally a model will violate this condition because its transition matrix has a column sum equal to 1.

The matrix model (1.9) is *dissipative* if there exists a constant $\beta > 0$ (that does not depend on the initial state $x(0)$) and a time t^* such that $t \geq t^*$ (which might depend on $x(0)$) implies $|x(t)| \leq \beta$. This means populations are of total size no greater than β after a finite number of time steps.

Theorem 1.9. *Assume (1.8). The conditions (1.13) imply the matrix model (1.7) is dissipative.*

Proof. From the inequalities

$$0 \leq |x(t+1)| \leq f_\infty + ||T(x(t)|| \, |x(t)| \leq f_\infty + \tau_\infty \, |x(t)|$$

and an induction we find that $|x(t)| \leq y(t)$ where $y(t) \geq 0$ is the solution of the scalar linear difference equation $y(t+1) = f_\infty + \tau_\infty y(t)$ that satisfies $y(0) = |x(0)|$. The formula $y(t) = \tau_\infty^t \, |x(0)| + f_\infty \sum_{i=0}^{t-1} \tau_\infty^i$ for the solution of this linear recursion equation implies $\lim_{t\to+\infty} y(t) = f_\infty / (1 - \tau_\infty)$. Thus, we can take, for example, $\beta = 1 + f_\infty / (1 - \tau_\infty)$ in the definition of dissipativity. ■

The second question, when $n > 1$ and $x = 0$ is unstable, concerns the possibility that some solutions with $x(0) \geq 0$ tend to 0 (i.e., that some populations will still go extinct). The matrix model (1.9) is *uniformly persistent* (with respect to $x_e = 0$) if there exists an $\alpha > 0$ (that does not depend on $x(0)$) such that $0 \leq x(0) \neq 0$ implies $\liminf_{t\to+\infty} |x(t)| \geq \alpha$. For a uniformly persistent matrix model, the total size of all populations (starting from a nonzero initial class distribution) eventually is no less that α.

The following theorem follows from Theorem 3 in [**45**]. Note: $\bar{R}_+^m/\{0\}$ consists of all nonzero, non-negative $x \in R^m$.

Theorem 1.10. *Assume (1.10), (1.11) and $n > 1$. Assume the matrix model (1.9) is dissipative and that $\bar{R}_+^m/\{0\}$ is forward invariant. Then the matrix model (1.9) is uniformly persistent.*

That $\bar{R}_+^m/\{0\}$ is forward invariant means $x(t) \in \bar{R}_+^m/\{0\}$ implies $x(t+1) \in \bar{R}_+^m/\{0\}$. This is the same thing as saying that no population goes extinct in a finite number of steps. This assumption is necessary in Theorem 1.10, as the $m = 1$ scalar example $x(t+1) = nf(x(t))x(t)$, $f(x) \triangleq (1 - x + |1 - x|)/2$, shows. In this example, if $n > 4$ and $x(0) = 1/2$, then $x(2) = 0$.

Example 1.11. *For the LPA model*

$$\begin{aligned}
L(t+1) &= bA(t) \exp\left(-\frac{c_{el}}{V} L(t) - \frac{c_{ea}}{V} A(t)\right) \\
P(t+1) &= (1 - \mu_l) L(t) \\
A(t+1) &= (1 - \mu_p) P(t) \exp\left(-\frac{c_{pa}}{V} A(t)\right) + (1 - \mu_a) A(t)
\end{aligned}$$

we have

$$F(x)x = \begin{pmatrix} bA\exp\left(-\frac{c_{el}}{V}L - \frac{c_{ea}}{V}A\right) \\ 0 \\ 0 \end{pmatrix}, \quad T(x) = \begin{pmatrix} 0 & 0 & 0 \\ 1-\mu_l & 0 & 0 \\ 0 & 1-\mu_p & 1-\mu_a \end{pmatrix}$$

($x = \operatorname{col}(L, P, A)$*) and hence*

$$\begin{aligned} |F(x)x| &= bA\exp\left(-\frac{c_{el}}{V}L - \frac{c_{ea}}{V}A\right) \\ &\leq bA\exp\left(-\frac{c_{ea}}{V}A\right) \\ &\leq b\frac{V}{c_{ea}}e^{-1} \triangleq f_\infty < +\infty \\ ||T(x)|| &= \sup\{1-\mu_l, 1-\mu_p, 1-\mu_a\}. \end{aligned}$$

The conditions (1.13) are satisfied if μ_l*,* μ_p*,* $\mu_a > 0$ *and as a result the model is dissipative by Theorem 1.9 for all values of* b *(and* V*).*

Suppose $x = \operatorname{col}(L(t), P(t), A(t)) > 0$*. Then it is easy to see from the model equations that* $\operatorname{col}(L(t+1), P(t+1), A(t+1)) > 0$ *and as a result* $R^m_+/\{0\}$ *is forward invariant.*

It follows from Theorem 1.10 that if μ_l*,* μ_p*,* $\mu_a > 0$*, then the model is uniformly persistent whenever*

$$n \triangleq b\frac{(1-\mu_l)(1-\mu_p)}{\mu_a} > 1.$$

EXERCISES

Exercise 1. (a) The discrete logistic (1.2) is called a monotone map because $bx/(1+cx)$ is a monotone function of $x > 0$. Use this fact to prove that $x(0) > 0$ implies $\lim_{t\to+\infty} x(t) = (b-1)/c$. (b) Show that the change of variables $y = 1/x$ transforms (1.2) to a linear difference equation. Use this fact to find a solution formula for $x(t)$ and use the solution formula to prove that $x(0) > 0$ implies $\lim_{t\to+\infty} x(t) = (b-1)/c$.

Exercise 2. Suppose the non-negative matrix P is primitive and diagonalizable in the matrix equation $x(t+1) = Px(t)$. For $x(0) \geq 0$, $x(0) \neq 0$ prove

$$\lim_{t\to+\infty} \frac{x(t)}{|x(t)|} = \frac{v}{|v|}$$

where $v > 0$ is the positive eigenvector associated with the dominant eigenvalue of P. This result (which is also true when P is not diagonalizable) is known as the Strong Ergodic Theorem or the Fundamental Theorem of Demography.

Exercise 3. Prove Theorem 1.2 when P is diagonalizable (i.e., R^m has a basis of eigenvectors of P).

Exercise 4. Consider the fertility and transition matrices

$$F = \begin{pmatrix} f_{11} & f_{12} & \cdots & f_{1,m-1} & f_{1m} \\ 0 & 0 & \cdots & 0 & 0 \\ 0 & 0 & \cdots & 0 & 0 \\ \vdots & \vdots & & \vdots & \vdots \\ 0 & 0 & \cdots & 0 & 0 \end{pmatrix}, \text{ and}$$

$$T = \begin{pmatrix} \tau_{11} & 0 & \cdots & 0 & 0 \\ \tau_{21} & \tau_{22} & \cdots & 0 & 0 \\ 0 & \tau_{32} & \cdots & 0 & 0 \\ \vdots & \vdots & & \vdots & \vdots \\ 0 & 0 & \cdots & \tau_{m,m-1} & \tau_{mm} \end{pmatrix}.$$

The matrix $P = F + T$ is sometimes called an Usher matrix and the associated matrix model the "standard size structured model". Here the categories are size classes (weight, volume, length, biomass, etc.) and an individual either remains in the same class or grows into the next class in one unit of time. An individual cannot skip a size class nor shrink in size. All newborns lie in the smallest size class. Derive a formula for the net reproductive number n. Use the formula to obtain a biological interpretation of n.

Exercise 5. Consider a size structured (Usher) matrix with two newborn and two reproducing classes, large and small. (See Exercise 4.) Assume larger size individuals produce larger size offspring while smaller size individuals produce smaller size offspring. A model for this population has projection matrix $P = F + T$ where

$$F = \begin{pmatrix} 0 & 0 & f_{13} & 0 \\ 0 & 0 & 0 & f_{14} \\ 0 & 0 & 0 & 0 \\ 0 & 0 & 0 & 0 \end{pmatrix}, \qquad T = \begin{pmatrix} \tau_{11} & 0 & 0 & 0 \\ \tau_{21} & \tau_{22} & 0 & 0 \\ 0 & \tau_{32} & \tau_{33} & 0 \\ 0 & 0 & \tau_{43} & \tau_{44} \end{pmatrix}.$$

Derive a formula for the net reproductive number n. Use the formula to obtain a biological interpretation of n.

Exercise 6. The matrix model

$$F = \begin{pmatrix} 0 & f_{13} \\ 0 & 0 \end{pmatrix}, \quad T = \begin{pmatrix} \tau_{11} & 0 \\ \tau_{21} & \tau_{22} \end{pmatrix}, \quad 0 \leq \tau_{ii} < 1, \ 0 < \tau_{21} \leq 1$$

describes a population with a juvenile stage and an adult stable. Find formulas for n and r. Find a functional relationship $r = \phi(n)$ that relates n to r. Show that $\phi(n)$ is an increasing, concave (down) function of n.

Exercise 7. (a) Show all eigenvalues of a 2×2 matrix

$$A = \begin{pmatrix} a & b \\ c & d \end{pmatrix}$$

lie inside the unit complex circle if and only if the "Jury Conditions"

$$|\det A| < 1, \quad |\mathrm{tr}\ A| < |1 + \det A|$$

hold. (b) Show $\mathrm{tr}A = 1 + \det A$ implies $\lambda = 1$ is an eigenvalue and $\mathrm{tr}A = -(1+\det A)$ implies $\lambda = -1$ is an eigenvalue. (c) Show $|\det A| = 1$, $|\mathrm{tr}\ A| < |1 + \det A|$ imply the eigenvalues are complex conjugates with absolute value equal to 1.

Bifurcations

This lecture concerns the bifurcation of positive equilibrium solutions of the nonlinear matrix equation

$$x(t+1) = (n\Phi(x(t)) + T(x(t)))\, x(t). \tag{2.1}$$

In addition to the smoothness assumption (1.10) on the fertility and transition matrices Φ and T, it is assumes for non-negative $x \geq 0$ and $n > 0$ that

$$\text{the matrix } n\Phi(x) + T(x) \text{ is primitive,}$$

$$\sum_{i=1}^{m} \tau_{ij}(x) \leq 1, \quad T(x) \neq 0, \quad \rho\,(T(x)) < 1 \text{ and} \tag{2.2}$$

$$\rho\left(\Phi(0)\,(I - T(0))^{-1}\right) = 1.$$

Assumptions (2.2) implies (1.11).

Nontrivial equilibria $x \in R^m$, and in particular non-negative or positive equilibria, of the matrix equation satisfy the equilibrium equation

$$x = (n\Phi(x) + T(x))\, x.$$

Using the Taylor expansion

$$n\Phi(x) + T(x) = n\Phi(0) + T(0) + H(n, x),$$

where $\|H(n,x)\| = O\,(|x|)$,[1] we can rewrite the equilibrium equation as

$$x = nLx + g(n, x) \tag{2.3}$$

where[2]

$$L \triangleq (I - T(0))^{-1}\, \Phi(0), \quad g(n,x) \triangleq (I - T(0))^{-1}\, H(n,x)x, \quad |g(n,x)| = O\left(|x|^2\right).$$

Those values of n are sought for which there is a non-negative, nontrivial solution of (2.3) (i.e., an equilibrium of (2.1) lying in $\bar{R}^m_+/\{0\}$). We call such a pair (n, x) a *nontrivial, non-negative equilibrium pair* of (2.1). A nontrivial equilibrium pair is a *positive equilibrium pair* when $x > 0$ (i.e., when $x \in R^m_+/\{0\}$).

Define $\mathcal{S}$ to be the set of all nontrivial solution pairs of (2.1). If $(n_c, 0) \in \bar{\mathcal{S}}$ (the closure of $\mathcal{S}$) we say $(n_c, 0)$ is a bifurcation point. A necessary condition that $(n_c, 0)$ is a bifurcation point is that n_c is a characteristic value of L, that is to say, $v = n_c L v$ for some vector $v \neq 0$ in R^m, or in other words n_c is the reciprocal of a nonzero eigenvalue of L. (See Exercise 8.) Note that a characteristic value

[1] This means for any interval $n_1 \leq n \leq n_2$ there are constants $a, b > 0$ such that $\|H(n,x)\| \leq b\,|x|$ for all $|x| \leq a$ and $n_1 \leq n \leq n_2$.

[2] This means for any interval $n_1 \leq n \leq n_2$ there are constants $a, b > 0$ such that $|g(n,x)| \leq b\,|x|^2$ for all $|x| \leq a$ and $n_1 \leq n \leq n_2$.

of $L = (I - T(0))^{-1} \Phi(0)$ is a characteristic value of the matrix $\Phi(0)(I - T(0))^{-1}$ (and vice versa).

The following theorem states that for quite general nonlinear matrix models, $n_c = 1$ is a bifurcation value (just as it is for linear matrix models).

Theorem 2.1. *Consider the nonlinear matrix model (2.1) under assumptions (1.10) and (2.2). There exists a continuum*[3] $\mathcal{C}^+$ *in* $\bar{S}$ *that contains the bifurcation point* $(n_{cr}, 0) = (1, 0)$ *and such that* $\mathcal{C}^+ / \{(1,0)\}$ *contains only positive equilibrium pairs. The range of the continuum* $\mathcal{C}^+$ *(that is to say, the set* $\{x : (n, x) \in \mathcal{C}^+ / \{(1,0)\}\}$*), is unbounded in* R_+^m *and the spectrum* $\sigma(\mathcal{C}^+) \triangleq \{n : (n, x) \in \mathcal{C}^+ / \{(1,0)\}\}$ *associated with* $\mathcal{C}^+$ *contains only positive* $n > 0$.

This theorem follows from Theorem 1.2.7 in [**4**] (which is proved using the global bifurcation theorems of Rabinowitz [**39, 61**]).

Example 2.2. *Consider the matrix model (1.7) with*

$$F(x_1, x_2) = \begin{pmatrix} cbx_2e^{-x_2} & be^{-x_2} \\ 0 & 0 \end{pmatrix}, \qquad T(x_1, x_2) = \begin{pmatrix} 0 & 0 \\ \tau_{21} & \tau_{22} \end{pmatrix}$$

$$b, c > 0, \qquad 0 < \tau_{21} \leq 1, \qquad 0 \leq \tau_{22} < 1.$$

This describes a population in which newborns lie in class 1. If $c = 0$ *then the two classes* $x_1 = J$ *and* $x_2 = A$ *could properly be referred to as juveniles and adults (reproductively immature and mature). With* $c > 0$ *both classes reproduce, with fertility dependent on the population density* x_2 *of class 2. In this case we can refer to a population with a younger class and an older class.*

In this model, fertility of the older individuals is negatively affected by increased density of the older class, x_2*. Younger individuals reproduce very little when few older individuals are present (*x_2 *is small), but their reproduction increases as* x_2 *increases, until a maximum is reached (at* $x_2 = 1$*) after which negative density effects reduce their reproduction. This kind of effect, in which increased fertility occurs with increased population density (in this case, adult density) is called an Allee effect. Allee effects have been attributed to a variety of biological mechanisms (e.g., enhanced mating opportunities in denser populations, the care and nurture of young, social behavior, group defense, and many others).*

The model equations are

$$J(t+1) = cbA(t)e^{-A(t)}J(t) + be^{-A(t)}A(t) \tag{2.4}$$
$$A(t+1) = \tau_{21}J(t) + \tau_{22}A(t).$$

The calculation

$$F(0,0)(I - T(0,0))^{-1} = \begin{pmatrix} b\frac{\tau_{21}}{1-\tau_{22}} & b\frac{1}{1-\tau_{22}} \\ 0 & 0 \end{pmatrix}$$

yields the formula $n = b\tau_{21}/(1 - \tau_{22})$ *for the inherent net reproductive number. The solution of the equilibrium equations*

$$J = cbAe^{-A}J + be^{-A}A$$
$$A = \tau_{21}J + \tau_{22}A$$

[3] A continuum is a closed, connected set.

for nontrivial solutions $(J, A) \neq (0,0)$ *reduces to solving the scalar equation*

$$e^A = \left(c \frac{1-\tau_{22}}{\tau_{21}} A + 1 \right) n$$

for $A \neq 0$ *and letting* $J = (1-\tau_{22}) A/\tau_{21}$. *This equation for* A *leads to two kinds of bifurcation graphs in the* (n, A)*-plane, as shown in Figure 2. When* $c \leq \tau_{21}(1-\tau_{22})^{-1}$ *the spectrum* $\sigma(\mathcal{C}^+) = [1, +\infty)$ *and positive equilibria exist for and only for* $n > 1$. *However, if* $c > \tau_{21}(1-\tau_{22})^{-1}$ *then the spectrum is a larger interval and includes the bifurcation value* $n = 1$ *on its interior. In this case, positive equilibria (two of them) also exist for some values of* $n < 1$.

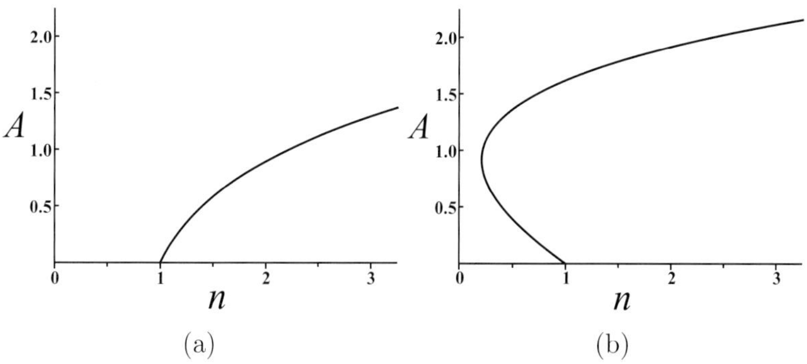

FIGURE 2. The A components of the positive equilibria of (2.4) are plotted against n. (a) For $c \leq \tau_{21}/(1-\tau_{22})$ the bifurcation is supercritical. (b) For $c > \tau_{21}/(1-\tau_{22})$ it is subcritical.

The two distinct cases that occur in Example 2.2 are distinguished by the *direction of bifurcation* at $n = 1$. When, near the bifurcation point $(1,0)$, positive equilibrium pairs (n, x) exist for $n > 1$, then the bifurcation is called *supercritical* (or forward). On the other hand, when near the bifurcation point $(1,0)$ the positive equilibrium pairs (n, x) exist for $n < 1$, then the bifurcation is *subcritical* (or backward). Allee effects typically can lead to a subcritical bifurcation. Models in which fertility and mortality never increase with increases in population density lead to supercritical bifurcations.

One technique that can determine properties of the bifurcating continuum of positive equilibria is based on the *net reproductive number* $n(x)$ at equilibrium x defined as the dominant eigenvalue of $F(x)(I - T(x))^{-1}$. This quantity should not be confused with the *inherent* net reproductive number n. We write $n(x) = n\nu(x)$ where $n = n(0)$ is the inherent net reproductive number and where the dominant eigenvalue $\nu(x)$ of $\Phi(x)(I - T(x))^{-1}$ satisfies $\nu(x) > 0$, $\nu(0) = 1$. If x is an equilibrium associated with inherent net reproductive number $n(x)$, then

$$n\nu(x) = 1. \tag{2.5}$$

Thus, $n(x) = n\nu(x)$ is an invariant (2.5) along the bifurcating branch $\mathcal{C}^+$ in Theorem 2.1. Properties of $\nu(x)$ can often give information about the direction of bifurcation at $n = 1$ and the spectrum $\sigma(\mathcal{C}^+)$. For example, if $\nu(x) < \nu(0)$ for x near 0 satisfying $0 \leq x \neq 0$, then to maintain the invariance of $n\nu(x)$ the local bifurcation must be supercritical (i.e., near the bifurcation point n would have to satisfy $n > 1$). For more on this approach see [**4**].

As an example, for the model in Example 2.2 it turns out that

$$\nu(J, A) = \frac{1 - \tau_{22}}{\tau_{21}} cAe^{-A} + e^{-A}.$$

If $c < \tau_{21} (1 - \tau_{22})^{-1}$ then $\partial_A \nu(0,0) < 0$. It follows in this case that, near $(J, A) = (0,0)$, the inequality $\nu(J, A) < 0$ for positive equilibria. Thus, near bifurcation the bifurcation point $n = 1$, it must be the case that $n > 1$ and hence the bifurcation is supercritical. Similarly, we deduce that if $c > \tau_{21} (1 - \tau_{22})^{-1}$ then the bifurcation is subcritical (since $\partial_A \nu(0,0) < 0$). We can also deduce from $\nu(J, A)$ that the spectrum is unbounded. This is because the continuum $\mathcal{C}^+$ is unbounded by Theorem 2.1 and consequently either the spectrum is unbounded or the equilibria on the branch are unbounded. However, even in the latter case the spectrum must be unbounded (as a consequence of the invariant (2.5)) because in this case $\nu(x) = \nu(J, A)$ tends to 0 as A tends to $+\infty$.

An important problem is to determine when the positive equilibria guaranteed by Theorem 2.1 are stable or unstable. It turns out that the stability of the positive equilibria near the bifurcation point usually depends on the direction of bifurcation. The bifurcation $n = 1$ described in Theorem 2.1 is called *stable* if the equilibria from the positive equilibrium pairs near the bifurcation point $(n, x) = (1, 0)$ are (locally asymptotically) stable. If these positive equilibria are unstable then the bifurcation is *unstable*.

Under some additional conditions, the bifurcation at $(n, x) = (1, 0)$ in Theorem 2.1 is stable when it is supercritical and unstable when it is subcritical. This fact is a consequence of the Theorem 1.2.6 in [**4**].

Define γ_{ij} to be the gradient of $\varphi_{ij}(x) + \tau_{ij}(x)$ evaluated at $(n, x) = (1, 0)$:

$$\gamma_{ij} \triangleq \nabla_x \left(\varphi_{ij}(x) + \tau_{ij}(x) \right)\big|_{(n,x)=(1,0)} \in R^m.$$

Let

$$d_{ij} \triangleq \gamma_{ij}^* v$$

(the prime " * " denotes transpose) and form the matrix $D = (d_{ij})$. Let $w > 0$ and $v > 0$ denote the positive left and right eigenvalues of $\Phi(0) + T(0)$ and define

$$\kappa \triangleq -wDv.$$

Theorem 2.3. *Assume*

$$\varphi_{ij} \in C^2 \left(D, [0, +\infty) \right), \quad \tau_{ij} \in C^2 \left(D, [0, 1] \right)$$

(1.11), and $\kappa \neq 0$. Then the bifurcation of positive equilibrium pairs in Theorem 2.1 is stable if it is supercritical ($\kappa > 0$) and unstable if it is subcritical ($\kappa < 0$).

Since $w > 0$ and $v > 0$, some of the d_{ij} must be positive in order for a subcritical bifurcation to occur, i.e., at least one gradient γ_{ij} must have some positive components. This means Allee effects of some kind are necessary for a subcritical to occur (although they are not necessarily sufficient).

Example 2.4. *For the juvenile/adult model in Example 2.2*

$$\Phi(x_1, x_2) = \begin{pmatrix} c\frac{1-\tau_{22}}{\tau_{21}} x_2 e^{-x_2} & \frac{1-\tau_{22}}{\tau_{21}} e^{-x_2} \\ 0 & 0 \end{pmatrix}, \quad T(x_1, x_2) = \begin{pmatrix} 0 & 0 \\ \tau_{21} & \tau_{22} \end{pmatrix}$$

and $n = b\tau_{21}/(1-\tau_{22})$. *Positive right and left eigenvalues of*

$$\Phi(0,0)+T(0,0)=\begin{pmatrix} 0 & \frac{1-\tau_{22}}{\tau_{21}} \\ \tau_{21} & \tau_{22} \end{pmatrix}$$

associated with eigenvalue 1 *are*

$$w=\begin{pmatrix} \tau_{21} & 1 \end{pmatrix}, \qquad v=\begin{pmatrix} 1-\tau_{22} \\ \tau_{21} \end{pmatrix}.$$

The matrix D *evaluated at* $(n,x)=(1,0)$ *is*

$$D=\begin{pmatrix} c(1-\tau_{22}) & -(1-\tau_{22}) \\ 0 & 0 \end{pmatrix}$$

and

$$\kappa=\tau_{21}(1-\tau_{22})^2\left(\frac{\tau_{21}}{1-\tau_{22}}-c\right).$$

Thus, the bifurcation at $(n,x)=(1,0)$ *is stable (supercritical) if* $c<\tau_{21}/(1-\tau_{22})$ *and unstable (subcritical) if* $c>\tau_{21}/(1-\tau_{22})$.

Even when they contain Allee effects at low population densities[4], most models assume negative density effects for large population densities. The latter assumption is responsible for the "turning around" of the bifurcating branch, at a point (n_{sn},x_{sn}), as illustrated in Figure 2(b). This creates a sub-interval in the spectrum $\sigma(\mathcal{C}^+)$ on which multiple positive equilibria exist.

Figure 3 shows the basins of attraction[5] of the two stable equilibria (the extinction equilibrium and one of the positive equilibria) when a subcritical bifurcation occurs in Example 2.2. This plot illustrates the commonly occurring situation in which the "smaller" positive equilibria from the "lower" portion of the bifurcating branch of positive equilibria is unstable (consistent with Theorem 2.3), while the "larger" positive equilibrium from the "upper" portion as stable. For those n values less than 1 there are two stable equilibria, the unstable positive equilibrium lies of the boundary of the basins of attraction.

In the typical subcritical bifurcation scenario illustrated in Figure 4, a population can survive for $n<1$ provided its initial state $x(0)$ is sufficiently far from the extinction equilibrium $x_e=0$; otherwise it will go extinct. Notice the sudden collapse and extinction of the population as $n<1$ decreases (and the population follows the stable equilibrium) below the critical value $n_{sn}<1$ where the bifurcation curve "turns around". If we increase $n<n_{sn}$ (in an attempt to recover the population) the population follows the extinction equilibrium (unless there is also a perturbation so as to move the population sufficiently far from $x_e=0$) until $n=1$ is exceeded, at which point the population recovers. This phenomenon where the collapse point n_{sn} is less than the recovery point $n=1$ is called *hysteresis*.

As we move along the bifurcating branch from the bifurcation point $(n,x)=(1,0)$ in the subcritical case illustrated in Figure 4, equilibrium stability is gained as we move "around" the turn the branch at the point (n_{sn},x_{sn}) where the lower and upper branches meet. This occurs because the spectral radius of the Jacobian of

[4]Some argue that accurate population models should contain Allee effects [**16**], especially if one is concerned with the possibility of extinction.

[5]The basin of attraction of an equilibrium is the set of initial conditions $x(0)$ that tend to the equilibrium.

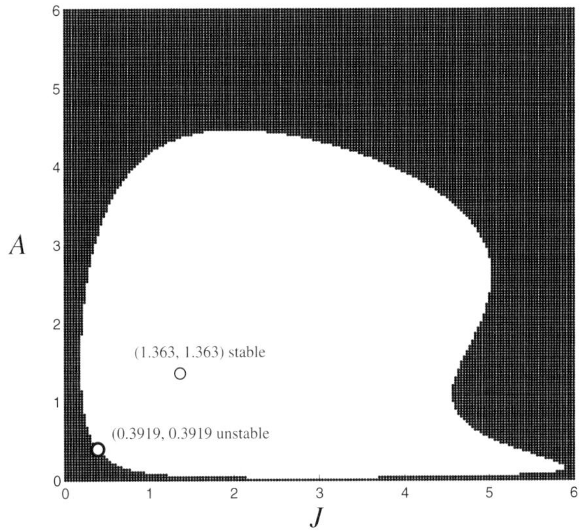

FIGURE 3. The set of initial conditions whose orbits tend to an equilibrium is the equilibrium's basin of attraction. For $c = 2.5$, $\tau_{21} = 0.75$, $\tau_{22} = 0.25$ and $n = 0.5$ ($b = 0.5$) in the juvenile-adult model in Example 2.2 there are two stable equilibria: the extinction equilibrium $(J, A) = (0, 0)$, whose basin of attraction is the dark region, and $(J, A) = (1.363, 1.363)$ whose basin of attraction is the light region. The unstable equilibrium $(J, A) = (0.3919, 0.3919)$ lies on the basin boundary on which orbits tend to this unstable equilibrium. The unstable equilibrium is a *saddle* and the basin boundary is its *stable manifold.*

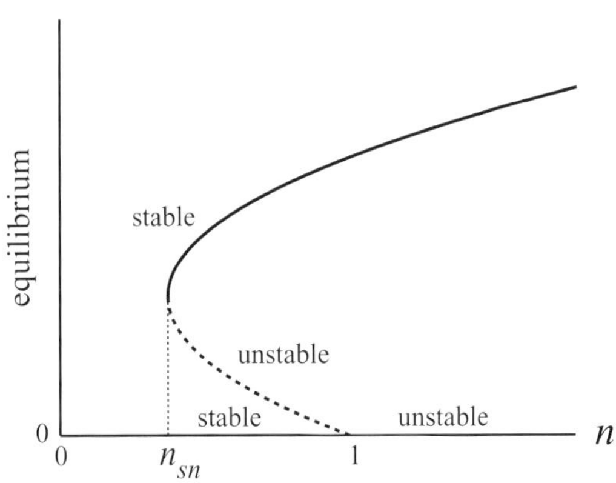

FIGURE 4. A subcritical (unstable) bifurcation caused by Allee effects at low population densities and negative density regulated at high population densities leads to a hysteresis effect.

the model system, evaluated at the equilibrium, changes from greater to less than 1. Thus, an eigenvalue of the Jacobian moves into the unit circle in the complex plane as we move around the turning point, from the lower to the upper branch, and, as a result, the Jacobian at the point (n_{sn}, x_{sn}) has an eigenvalue of magnitude equal to 1.

We can distinguish three ways in which an eigenvalue λ of the Jacobian can migrate out of (or into) the complex unit circle as a model parameter changes: either λ passes through $+1$ or -1 or $e^{i\theta}$ for some $\theta \neq 0, \pi$. Different kinds of bifurcations result according to which of these three cases occurs.

If λ passes through $+1$ as a parameter in the model is changed, the bifurcation generally involves equilibria. Theorem 2.1 shows that such a bifurcation occurs quite generally in matrix models. When n (or r) increases through 1 two branches of equilibria intersect in what is called a *transcritical bifurcation*[6]. It is characteristic of transcritical bifurcations that one branch losses stability and the other branch gains it, a phenomenon called an *exchange of stability*. Another common bifurcation that can occur when λ leaves the unit circle at $+1$ is that exemplified by the multiple equilibrium case in Figure 4 (as can occur in the subcritical case caused by Allee effects). The bifurcation at $n = n_{sn}$, which occurs as two nontrivial equilibria are created when n increases through n_{sn}, is called a *saddle-node* or (*tangent* or *blue sky*) bifurcation. Other bifurcation scenarios involving equilibria can occur when destabilization results from λ passing through $+1$ (for example, the *pitch fork* bifurcation), but the saddle-node and transcritical are the most common ones to occur in population dynamic models.

When, as a model parameter is changed, an eigenvalue of the Jacobian leaves the unit circle through -1 and an equilibrium thereby loses stability, typically there exists a branch of 2-cycles (periodic solutions of the matrix equation of period 2) that intersects the equilibrium branch at the critical parameter value where $\lambda = -1$. The 2-cycles collapse into an equilibrium at the bifurcation point and so we say small amplitude cycles "bifurcate out of" the equilibrium as it loses stability. Such a *2-cycle* (or *period doubling bifurcation*[7]) can involve stable or unstable cycles and can occur as the model parameter increases or decreases, depending on the circumstances. This famous bifurcation is familiar in maps of dimension $m = 1$ (such as the Ricker map (1.3)).

There are formal theorems to be found in many textbooks that provide conditions under which period doubling and transcritical bifurcations are guaranteed to occur in the cases $\lambda = \pm 1$, and conditions that determine the direction of bifurcation and the stability properties of the equilibria and cycles involved. See for example [**24, 67**].

Transcritical and saddle-node bifurcations commonly occur in population matrix models (1.4) (or 2.1)). Period doubling bifurcations occur commonly in $m = 1$ dimensional models. After all, in $m = 1$ models only two possibilities arise for an eigenvalue λ of the Jacobian (derivative) of the equation, which is necessarily real, to leave the unit circle, namely ± 1. In higher dimensional models, in which eigenvalues can be complex, λ can leave the unit circle at any of infinitely many points $e^{i\theta}$ other than ± 1. (Of course, in the latter case a complex conjugate pair of eigenvalues λ crosses the unit circle.)

---

[6]Although not considered here, the general bifurcation theory for matrix models presented here also implies a bifurcating continuum of negative equilibrium pairs from the point $(n, x) = (1, 0)$. Thus, the set $\mathcal{S}$ contains a continuum of nontrivial equilibrium pairs that transversely crosses the extinction equilibrium branch.

[7]An equilibrium is 1-cycle.

When an equilibrium destabilizes because an eigenvalue λ of the Jacobian crosses the unit circle at a point $e^{i\theta} \neq \pm 1$, a more complicated bifurcation typically occurs. As a warm up for this case, consider the $m = 2$ dimensional linear matrix model $x(t+1) = Px(t)$ with coefficient matrix

$$P = \begin{pmatrix} a\cos\theta & -a\sin\theta \\ a\sin\theta & a\cos\theta \end{pmatrix}$$

where $a > 0$ and $0 < \theta < \pi$ are model parameters. The spectral radius is $\rho(P) = a$ and for $a < 1$ the origin $x_e = 0$ is globally asymptotically stable while for $a > 1$ nontrivial solutions grow exponentially in magnitude. For $a = 1$ orbits remain on the circle $x_1^2 + x_2^2 = x_1^2(0) + x_2^2(0)$ in the x_1, x_2-plane as each successive point is rotated θ radians counter-clockwise. Each circle, centered at the origin, is an invariant loop and the characteristics of an orbit lying on the circle depend on θ. If θ is rationally related to 2π, then the orbits are periodic cycles. If θ is irrationally related to 2π, then the orbits are neither periodic, nor asymptotically periodic, but are dense on the circle and are termed quasi-periodic ([**63**], p. 428). A bifurcation diagram that represents this situation is vertical in that the invariant loops occur only at $a = 1$, where the equilibrium $x_e = 0$ is destabilized.

Branches of invariant loops can also bifurcate from a destabilized equilibrium in nonlinear matrix equations (1.4). In this case, the branch of invariant loops is generally non-vertical and the spectrum is no longer a single point, but instead is an interval of parameter values. This is the subject of the *invariant loop* (or *Sacker/Neimark* or *discrete Hopf*) *bifurcation theorem*. This theorem (see [**24, 67**]) gives rigorous conditions under which a branch of invariant loops bifurcates out of an equilibrium that destabilizes because an eigenvalue λ of the Jacobian passes through the unit circle at a point $e^{i\theta} \neq \pm 1$. The theorem also gives conditions that determine the direction of bifurcation and stability properties of the invariant loops. (One of these conditions, rather unexpectedly, requires that $\theta \neq 2\pi/3$ or $2\pi/4$ or in other words that λ not equal a cube or quartic root of 1. What happens in these exceptional "resonance" cases is apparently still not fully understood.) The motion of orbits on the bifurcating invariant loops can, as in the linear example above, be periodic or tend to a periodic cycle or be quasi-periodic and dense in the loop. As the model (bifurcation) parameter changes, the dynamics on the loop alternates between periodicity (in which case the system is said to be *period locked*) and quasi-periodicity.

Example 2.5. *Figure 5 shows a sample bifurcation diagram for the matrix model*

$$\begin{pmatrix} J(t+1) \\ A(t+1) \end{pmatrix} = \begin{pmatrix} cbA(t)e^{-A(t)} & be^{-A(t)} \\ \tau_{21} & \tau_{22} \end{pmatrix} \begin{pmatrix} J(t) \\ A(t) \end{pmatrix} \tag{2.6}$$

in Example 2.2 using the bifurcation parameter $n = b\tau_{21}/(1-\tau_{22})$. Plotted against each n value are 200 points of the A component of the attractor found after first iterating this matrix model 500 times (in order to arrive at an attractor). When the attractor is quasi-periodic 200 different points are plotted, which appear as a nearly solid vertical line segment above the corresponding n value. When the attractor is a cycle of period p, then only p points appear above the value of n. For the parameter values indicated in the figure caption, we see an invariant loop bifurcation occurs at a critical value of $n \approx 8.1$. A plot of one such invariant loop, for $n = 9$, appears in Figure 5(b) and the time series of the J component of a dense orbit on the loop in Figure 5(c). It is often difficult to tell the difference between a cycle with a long *period and a quasi-periodic orbit. Figures 5(d) and (e) show plots of what appear to be a 33-cycle when* $n = 8.9624$.

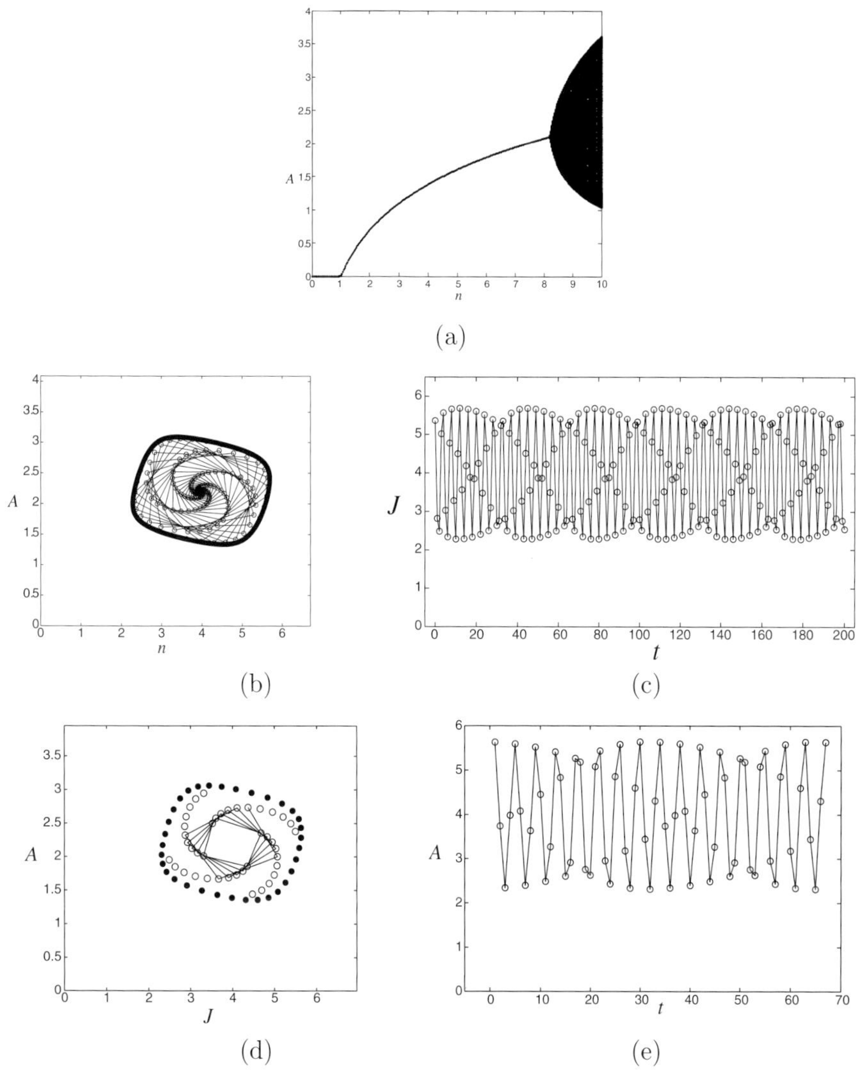

FIGURE 5. Choose $\tau_{21} = 0.5$, $\tau_{22} = 0.1$ and $c = 0$ (no Allee effect) in the juvenile-adult matrix model (2.6). (a) The bifurcation diagram shows the transcritical bifurcation of positive equilibria at $n = 1$ and the destabilization of the positive equilibria with a resulting an invariant loop bifurcation at $n \approx 8.1$. (b) An orbit starting near the unstable equilibrium spirals out to the invariant loop when $n = 9$. (c) The time series for the J component of a quasi-periodic orbit on the invariant loop in (b). A time series plot of the A component exhibits similar oscillations. (d) At $n = 8.9624$ an orbit spirally approaches a 33-cycle (residing on an invariant loop). (e) The J component of the 33-cycle in (d).

While the *primary bifurcation* of positive equilibria from the extinction equilibrium $x_e = 0$ at $n = 1$ is a quite general phenomenon in population models, the occurrence of further bifurcations through the loss of stability of the positive equilibria (or other positive attractors such as cycles or invariant loops) is dependent on the details of specific models. Period doubling and invariant loop bifurcations might or might not occur in a specific model.

In many cases numerous bifurcations occur as n increases, such as the famous period doubling cascade of the Ricker equation. In principle the study of cycles and cycle bifurcations is reducible to the study of equilibrium bifurcations of a composition of the model equations. (A 2-cycle is a fixed point of the first composite, and so forth.) For $m = 1$ dimensional maps, the common occurrence of a $\lambda = -1$ bifurcation can lead to repeated bifurcations of period doublings for higher and higher composites. Therefore, period doubling cascades are common for the $m = 1$ dimensional case. For the higher dimensional cases which arise in the study of structured population dynamics, invariant loop bifurcations often play a significant role.

Cycles also can lose stability through invariant loop bifurcations. For example, a 2-cycle can bifurcate into two disjoint invariant loops (surrounding each point of the cycle). Bifurcations involving invariant loops are more complicated and difficult to study analytically. As n increases invariant loops often suddenly disappear from the bifurcation diagram, yielding to other attractors (e.g. to periodic cycles in what are called period locking windows), or they can become twisted and convoluted or even break into pieces, as the dynamics become more complex and chaotic. Given the complexity of the dynamics in such regions of parameter space, it is not surprising that there are many technical definitions of chaos in the literature and a growing list of exotic bifurcation phenomena identified and classified by mathematicians (usually with the aid of computer simulations). Definitions of a chaotic attractors generally assume (or imply) that such attractors contain infinitely many unstable periodic cycles, bounded orbits that are not periodic and do not tend to a periodic cycle, orbits that are dense in the attractor, phase space directions in which orbits expand and other directions in which orbits contract, and sensitivity to initial conditions. The latter property is perhaps the most famous property of chaotic attractors. It means that orbits initially infinitesimally close diverge exponentially from one another over time, a property that has serious consequences with regard to the ability to predict the future state of a system, even though its dynamics are deterministic.

Needless to say, the study of the complicated dynamics that can occur in some models (e.g., the bifurcation diagram in Figure 6) is usually analytically intractable. Not only can the attractors at selected values of the bifurcation parameter n be chaotic, but the mix of chaotic and tame dynamics throughout in an interval of n values can be complicated. Chaotic and non-chaotic attractors can be densely packed throughout a parameter interval. For this reason, as well as the fact that parameter estimates are not exact and come at best with confidence intervals, in these regions of parameter space the best point of view to take would not be that a particular times series of data has a specific type of model attractor type – chaotic, for example. It might be better to say in such cases that the population's dynamics are "influenced by chaos" [**20**].

Example 2.6. *Figure 6(a) shows the same bifurcation diagram for the juvenile/adult model in Example 2.2 shown in Figure 5, but extended over a larger interval of n values. After the invariant loop bifurcation at $n \approx 8.1$, there occurs a complicated sequence of bifurcations in which periodic cycles play a role as well as other complicated chaotic and strange attractors. In Figure 6(b) we see the break up of an invariant loop into a strange attractor consisting of several distinct pieces.*

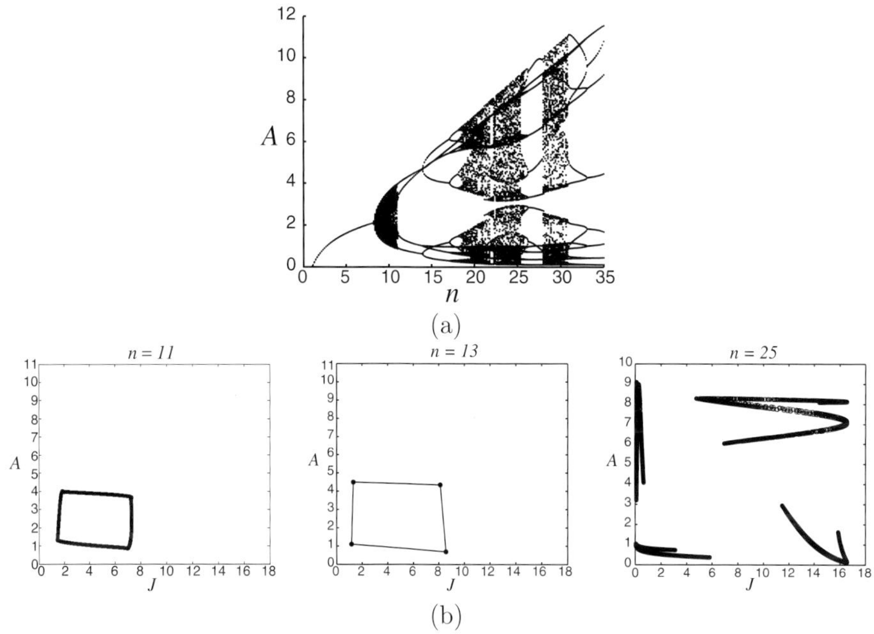

FIGURE 6. Let $\tau_{21} = 0.5$, $\tau_{22} = 0.1$ and $c = 0$ (no Allee effect) in the juvenile-adult matrix model (2.6). (a) The bifurcation diagram shows complicated bifurcations after the invariant loop bifurcation at $n \approx 8.1$. (b) The attractor evolves from an invariant loop at $n = 11$ to a 4-cycle at $n = 13$ to a strange chaotic attractor at $n = 25$.

The parameter n is a convenient parameter to use in the general theory and discussion above. It is well defined in a general setting, there often exists an explicit formula relating it to model parameters, and it has an important biological interpretation (the inherent net reproductive number). In some applications, however, we might want to study the dynamics of a matrix model as a function of a different parameter, one that appears explicitly in the model equations. Theorems generalizing Theorems 2.1 and 2.3 appear in [**4**] (Theorems 1.2.3 - 1.2.6) for matrix models in which $F(0)$ is linear in a parameter μ, that is, for matrix models of the form

$$x(t+1) = (A + \mu B + H(\mu, x))\, x \tag{2.7}$$

where $|H(\mu, x)| = O(|x|)$ near $x = 0$. These theorems provide more alternatives for the bifurcating continuum $\mathcal{C}^+$ than are given in Theorem 2.1 (e.g., see Theorem 4.1 in Lecture 4). Nonetheless, the direction of bifurcation of the positive equilibrium branch still determines the stability of the equilibria, although the relationship between direction and stability can be reversed. An example using the death rate μ_a in the LPA model of Example 1.4 as the bifurcation parameter appears in [**4**].

Example 2.7. *Figures 7 and 8 show two sample bifurcation diagrams for the LPA model in Example 1.4 using* μ_a *and* c_{pa} *as bifurcation parameters. In Figure 7 the* $x_1 = A$ *component of the attractor is plotted against the bifurcation parameter* μ_a. *In Figure 7(a) we see a period doubling bifurcation at* $\mu_a \approx 0.045$ *and an invariant loop bifurcation at* $\mu_a \approx 0.83$. *There is a re-equilibration that occurs because of a subcritical 2-cycle bifurcation of the equilibrium at* $\mu_a \approx 0.565$ *and a 2-cycle saddle-node bifurcation at* $\mu_a \approx 0.59$*; this sequence of bifurcations is seen more clearly in Figure 7(b).*

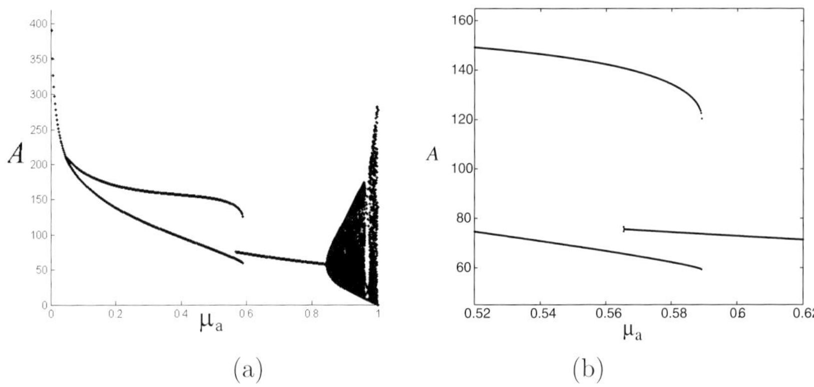

FIGURE 7. (a) and (b) show bifurcation diagrams for the LPA model in Example 1.4 using the adult stage death rate μ_a as the bifurcation parameter. The $x_1 = A$ component of the attractor is plotted against μ_a. Other parameter values are $b = 10$, $\mu_l = 0.2$ and $c_{ea} = c_{ea} = c_{pa} = 0.01$.

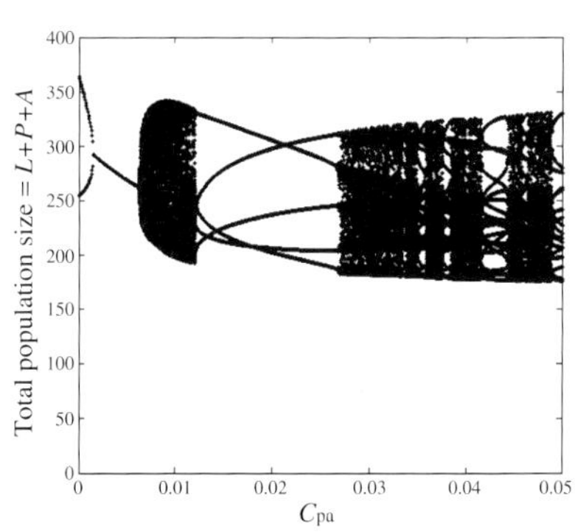

FIGURE 8. In this bifurcation diagram for the LPA model in Example 1.4, the total population size $x_1 + x_2 + x_3 = L + P + A$ of the attractor is plotted against the bifurcation parameter c_{pa}. Other parameter values are $b = 10$, $\mu_l = 0.2$, $\mu_a = 0.95$ and $c_{ea} = c_{ea} = 0.01$.

In Figure 8 the total population size $x_1+x_2+x_3 = L+P+A$ *of the attractor is plotted against* c_{pa}. *A 2-cycle at* $c_{pa} = 0$ *undergoes a reverse period doubling bifurcation to an equilibrium at* $c_{pa} \approx 0.0025$. *The equilibrium undergoes an invariant loop bifurcation at* $c_{pa} \approx 0.00625$. *Further bifurcations occur for large* c_{pa} *values, bifurcations that lead to period locking windows (such as the large window on the interval* $0.012 < \mu_a < 0.03$*) interspersed with strange, chaotic attractors.*

EXERCISES

Exercise 8. If $(n^*, 0)$ is a bifurcation point of $x = nLx + h(n, x)$, prove that n^* must be a characteristic value of L

Exercise 9. (a) Calculate the quantity κ for the LPA model in Example 1.4. (b) Show that for $n > 1$ there exists exactly one nontrivial equilibrium and it is positive.

Exercise 10. (a) Consider the linear scalar ($m = 1$) difference equation $y(t+1) = ay(t) + b(t)$ where $0 \leq a$ is a constant satisfying $0 \leq a < 1$ and where $b(t)$ is a bounded sequence: $0 \leq b(t) \leq b_\infty < \infty$, $t = 0, 1, 2, \cdots$. Show that solutions $y(t)$ with non-negative initial conditions $y(0) = c \geq 0$ are bounded. Specifically, show there exists a constant $\beta > 0$ (independent of c) and an integer $t^*(c)$ such that $0 \leq y(t) \leq \beta$ for $t \geq t^*(c)$. (b) Consider the linear scalar equation $z(t+1) = \alpha(t)z(t) + \beta(t)$ for which the coefficients $\alpha(t)$, $\beta(t)$ satisfy $|\alpha(t)| \leq a < 1$ and $|\beta(t)| \leq b(t) \leq b_\infty < \infty$, $t = 0, 1, 2, \cdots$. Show that solutions $z(t)$ are bounded. Specifically, show there exists a constant $\beta > 0$ (independent of c) and an integer $t^*(c)$ such that $0 \leq |z(t)| \leq \beta$ for $t \geq t^*(c)$. (c) Consider the nonlinear scalar ($m = 1$) difference equation $w(t+1) = \alpha(t, w(t)w(t) + \beta(t, w(t))$ for which the coefficients satisfy $|\alpha(t, w)| \leq a < 1$ and $|\beta(t, w)| \leq b(t) \leq b_\infty < \infty$ for $t = 0, 1, 2, \cdots$ and all $w \in R^1$. Show that solutions $w(t)$ with are bounded. Specifically, show there exists a constant $\beta > 0$ (independent of c) and an integer $t^*(c)$ such that $0 \leq |w(t)| \leq \beta$ for $t \geq t^*(c)$.

Exercise 11. (a) Consider the juvenile/adult model in Example 2.2

$$\begin{aligned} J(t+1) &= cbA(t)e^{-A(t)}J(t) + be^{-A(t)}A(t) \\ A(t+1) &= \tau_{21}J(t) + \tau_{22}A(t) \end{aligned}$$

under the assumption that $c > 0$ satisfies the inequality

$$c < e\frac{\tau_{21}}{1 - \tau_{22}}.$$

Recall $n \triangleq b\tau_{21}/(1 - \tau_{22})$. Show the model is dissipative when

$$1 < n < \frac{e}{c}\frac{\tau_{21}}{1 - \tau_{22}}.$$

(Hint: Use the results in the preceding Exercise 10.) (b) Show the model is uniformly persistent if, in addition, $0 < \tau_{21}, \tau_{22} < 1$.

Exercise 12. Analyze the transcritical bifurcation at $(J, A) = (0, 0)$, $n = 1$ of juvenile/adult model

$$\begin{aligned} J(t+1) &= b_1 A(t)\frac{1}{1 + A(t)}J(t) + b_2\frac{1}{1 + A(t)}A(t) \\ A(t+1) &= \tau_{21}J(t) + \tau_{22}A(t). \end{aligned}$$

Here $n \triangleq b_2\tau_{21}/\left(1 - \tau_{22}\right)$ is the inherent net reproductive number.

Exercise 13. The $m = 2$ dimensional matrix model with fertility and transition matrices

$$F = \begin{pmatrix} 0 & b \\ 0 & 0 \end{pmatrix}, \quad T = \begin{pmatrix} 0 & 0 \\ \tau_{21} & \tau_{22} \end{pmatrix}$$

describes a population classified by juveniles and adults: $x_1 = J$ and $x_2 = A$ respectively. If all three parameters b, τ_{21}, and τ_{22} are constants, the model is linear. Consider the nonlinear model in which τ_{21} is replaced by $\tau_{21} f(J)$ where f models a fractional decrease in juvenile maturation due to the density J of juveniles. Thus, assume $b > 0$, $0 < \tau_{21}, \tau_{22} < 1$ and $f(0) = 1$, $0 < f(J) \leq 1$ for $J \geq 0$. (a) Let $f(J) = 1/(1 + J)$. Show positive equilibria exist for, and only for, $n > 1$ and that they are (locally asymptotically) stable. (b) Let $f(J) = e^{-J}$. Show positive equilibria exist for, and only for, $n > 1$. Show that there exists an $n_{cr} > 1$ such that the positive equilibria are (locally asymptotically) stable for $1 < n < n_{cr}$. What happens at $n = n_{cr}$?

LECTURE 3

Experimental case studies

The influential papers of Li and Yorke [**51**] and Lord May [**52, 28, 53**] ignited interest in complex (chaotic) dynamics that can result from simple rules described by difference equations. Ecologists (and other scientists) found this possibility fascinating for several reasons. From a philosophical point of view, there is a paradigm shift away from the long held notion that complex phenomena require complex causes (and, conversely, that only simple phenomena result from simple causes) to one in which extraordinarily complex phenomena can result from simple rules. Indeed, dynamics virtually indistinguishable from randomness or stochasticity can result from simple deterministic rules. From a practical perspective, there arose the possibility that some of the apparent randomness or "noise" that is so common in population and ecosystem data might have a deterministic – indeed a simple (i.e., low dimensional) – component. This opened up new avenues to explore for possible explanations of observed patterns in data and, thereby, to deeper insights that might lead to improved predictability.

Complex sequences of bifurcations and complex/chaotic dynamics are readily found in nonlinear matrix models for the dynamics of a structured population. This is also true for more complicated situations such as multi-species interactions, models with spatial structure, temporally forced models for non-constant environments, and so on. This being the case, we are led to wonder about the role that such bifurcations and chaotic dynamics might play in the ecological world. During the last few decades there have been many studies that examine ecological data for the possibility of chaotic dynamics. Besides the inherent similarity between chaos and stochasticity and the difficulty in distinguishing between the two[1], there are numerous obstacles to carrying out such investigations, including the lack of sufficiently long time series of data, the measurement accuracy (sampling error) of data, and the general lack of quantitatively accurate, mechanistic mathematical models on which to base analyses [**58**].

A study of how a population responds to perturbations in vital parameters is an alternative to the approach of focusing on time series of observations from a population in a specific set of circumstances and trying to identify and classify the dynamic as a particular type or another. This approach is directly based on what a model bifurcation diagram is all about. It is a common approach in science and engineering to disturb, perturb, change, and tweak a system in order to gain an understanding from its responses into how it works. Furthermore, from a practical point of view, the question of how a population or ecosystem responds to certain

[1]Chaos fascinated scientists and mathematics because of the "noise-like" complexity that can arise from deterministic models (especially simple low dimensional models). Of course complicated fluctuations arising from noisy systems are not unexpected or surprising. So, mathematical chaos is of interest because it is deterministic phenomenon.

perturbations is an important one in a world in which hardly any biological system is unperturbed by natural and/or human causes. This is important not only for the purposes of basic scientific understanding, but for the purposes of managing biological resources, preserves, and parks; controlling of invasive species and diseases; insuring the survival of species endangered by resource loss, environmental change, habitat pollution, habitat fragmentation, and so on. Nonlinearity can imply unexpected responses to perturbations. Fluctuations of increasing complexity caused by dynamic bifurcations, for example, can be surprising.

Of course, one usually cannot study an ecosystem by subjecting it to a scheme of perturbations any more that an astrophysicist can subject the solar system or a galaxy to an experimental plan of perturbations to see how it will respond. One can however, like a physicist, study smaller and isolated components of a system in order to gain insight into its workings. Similarly, there can be a role for focused experimental study of population systems in manageable and controllable situations in promoting an understanding how larger communities and ecosystems are put together. To quote E. O. Wilson ([**68**], p. 111)

> When observation and theory collide, scientists turn to carefully designed experiments for resolution. Their motivation is especially high in the case of biological systems, which are typically far too complex to be grasped by observation and theory alone. The best procedure, as in the rest of science is first to simplify the system, then to hold it more or less constant while varying the important parameters one or two at a time to see what happens.

The main goal of experimental studies is not always limited to a better understanding of the particular biological species involved (although that results too). Such studies help to develop methods for the study the dynamics of populations in general, to test and corroborate various principles and dogma, to find explanations of observed phenomena that previously had none, and (as inevitably occurs in such studies) discover new phenomena and develop methods to deal with them. Thus, even if a study is specialized to a particular biological organism (as is, of course, necessarily the case in experimental studies) a broader goal is to demonstrate, elucidate, refine, extend, and discover new principles, methods and hypotheses for the dynamics of ecological populations which apply to and aid in the study of other systems.

In order to carry out a study of nonlinear dynamics along the lines discussed above it is necessary to connect mathematical models with data. There is a plethora of mathematical models in theoretical ecology and population dynamics. Precious few, however, have been tied to experimental or observed data to a standard of description and predictability found in the so-called "hard" sciences. This is not to say that mathematical models have not contributed to ecological theory. They have in fact helped establish most of the fundamental principles found today in ecology (exponential growth, logistic growth, carrying capacity, r and K selectors, competitive exclusion, ecological niche, limiting similarity, predator-prey oscillations, and so on). Nonetheless, if mathematical models are to rise above only qualitative accuracy and verbal metaphors, they must be more closely tied to data. This is far easier said than done; ecosystems are notoriously complex.

One approach to the study of biological systems is that advocated by Wilson in the quote above. In that spirit, the focus here will be on some successful case studies that utilized a manipulatable experimental system. The goal in looking at these selected research projects is to touch (briefly) on several general topics: some methods for connecting models to data; the use of a dynamic model to design experiments for the purpose of testing theoretical predictions and tenets; the ability of models to provide new explanations for dynamic patterns observed in data; how models can predict unexpected and counter-intuitive phenomena that subsequent experiments corroborate; and how the interdisciplinary wedding of mathematical models with controlled experimental studies can lead to the discovery of new phenomena and new general principles.

Any mathematical model is based on assumptions concerning the biological and physical mechanisms to be included in the model and those to be left out. These assumptions might, of course, need re-evaluation when the model predictions are confronted with data. Suppose we have available a time sequence of observations $y(0), y(1), \cdots, y(q)$ of the stage distributions. How well does the model describe this data? How are the model parameters chosen to best describe the data (i.e., how is the model parameterized or calibrated)?

Of course, the model might be so fundamentally inappropriate that it simply cannot describe the data well no matter what parameter values are used. This eventuality should ultimately be revealed in an evaluation of the parameterized model, in which case we take the model back to the drawing board for an overhaul. If the parameterized model is deemed an adequate descriptor of the data, then it can be further evaluated by analyzing how well it describes other data not used in the parametrization procedure. (Perhaps some of the original data is left aside for this purpose, or other data sets are obtained by further experiments.) If the model passes these evaluations, then we gain some confidence in its ability to describe – and more importantly predict – the dynamics of the population under various (possibly changed) circumstances. Experiments based on the model's predictions, whose data bear out the model's predictions, provide further validation of the model's accuracy.

Some model parameters might be known from experiments designed specifically to measure them. Other parameters we can estimate from the data $y(i)$. Given numerical values for its parameters, the model makes a prediction $x(t+1)$ from each data point $y(t)$. We do not expect this prediction to be 100% accurate, of course, and so there is a residual, i.e., there is a nonzero difference between the model prediction $x(t+1)$ and the actual observation $y(t+1)$. Most model parametrization and evaluation procedures are based, in one way or another, on the study of these residuals.

Nonzero residuals arise for many reasons. The model could be perfectly accurate and the data contain measurement errors. Of course, no model perfectly describes a population's dynamics and residuals can also arise because some processes are inaccurately described in the model (or even excluded from the model). Such processes can be deterministic and/or stochastic. Stochastic events (noise) is ubiquitous in ecological data. This point will forcibly be brought home in the case studies below in which experimental results contain significant noise even though they were obtained from highly controlled laboratory experiments with virtually no measurement error.

One way to incorporate stochasticity into a model is include both the deterministic and the stochastic aspects of the population's dynamics into the model's components. We can use the resulting stochastic model to study residuals, to parametrize and validate the model, to obtain descriptions of data from the validated model, and ultimately to make model predictions.

There are many ways to build a stochastic version of a matrix model (1.7). One class of stochastic models (called nonlinear autoregression models or NLAR models) adds a random term to the deterministic model (the skeleton) on an appropriate scale, i.e., for an appropriate transformation of the data y (say $w = g(y)$) and the model predictions x ($n = g(x)$). "Appropriate" here means that the model is transformed so that additive noise has a stabilized variance. This means in the $m = 1$ dimensional case, for example, that the stochastic model has the form $n(t+1) = f(n(t)) + E(t)$ where for each t the random variable $E(t)$ has mean 0 and a constant variance $v = \sigma^2$ that is independent of t. (It is also assume that $E(0)$, $E(1)$, $E(2), \cdots$ are independent random variables.) For the multi-variate case $m > 1$, $E(t)$ is a vector of random variables with means 0 and a (symmetric) variance-covariance matrix Σ.

Ecologists distinguish two fundamental types of noise in biological systems: environmental noise and demographic noise [**53**]. Roughly speaking, environmental noise arises from random disturbances that effect all individuals in the same way whereas demographic noise is due to random differences among individuals (differences in birth rates, survival rates, etc.).

In [**11**] an instructive example is given that illustrates one way to model environmental and demographic stochasticity. (Also see [**17**].) Consider a simple survival process and let μ denote the fraction of individuals who die during a unit of time. In the case of environmental stochasticity μ is a random variable. The total number of survivors at time $t + 1$

$$x(t+1) = (1-\mu)\,x(t) \tag{3.1}$$

is then a random variable that depends on the number of individuals $x(t)$ present at time t. The mean of this random variable is $Mean\,[1-\mu]\,x(t)$ where $Mean\,[1-\mu]$ is the mean of the random variable $1-\mu$ and the variance is

$$Var\,[x(t+1)] = Var\,[1-\mu]\,x^2(t) = Var\,[\mu]\,x^2(t) \tag{3.2}$$

where $Var\,[\mu]$ is the variance of the random variable μ.

In the case of demographic stochasticity each individual in the population has an independent random chance of dying with probability μ. In this case the number of survivors $x(t+1)$ at time $t+1$ is a binomial random variable of $x(t)$ trials with success probability $1-\mu$. The mean of this random variable is $(1-\mu)\,x(t)$ and the variance is

$$Var\,[x(t+1)] = (1-\mu)\,\mu x(t)\ . \tag{3.3}$$

Many statistical techniques are available for stochastic processes of the (nonlinear autoregressive or NLAR) form

$$n(t+1) = f(n(t)) + E(t) \tag{3.4}$$

where $E(t)$ is a normal random variable with mean 0 and a constant variance σ^2 and $E(0)$, $E(1)$, $E(2)$, ... are uncorrelated. The nonlinear function f defines the deterministic skeleton $n(t+1) = f(n(t))$, i.e., the the nonlinear model that describes the model dynamics in the absence of noise [**65**]. The stochastic survival processes derived in the motivational illustrations above for environmental and demographic stochasticity do not have the additive form (3.4).

One approach to take is to transform the state variable $x(t)$ in such a way that the stochastic process for the resulting transformed random variable $n(t)$ does have (at least approximately) the NLAR form (3.4). Toward this end, define a transformation $n(t) = g(x(t))$ by means of a function g satisfying

$$n = g(x), \qquad \frac{dg(x)}{dx} > 0.$$

Consider the first order Taylor polynomial approximation

$$n(t+1) = g(x(t+1)) \approx g(x(t)) + g'(x(t))(x(t+1) - x(t))$$

in which $x(t+1)$ is a random variable conditioned on a given value of $x(t)$. Then, from basic facts about the variance of a random variable, we find that

$$\begin{aligned} Var(n(t+1)) &\approx Var[g(x(t)) + g'(x(t))(x(t+1) - x(t))] \\ &= Var[g'(x(t))(x(t+1) - x(t))] \\ &= Var[x(t+1) - x(t)](g'(x(t)))^2 \\ &= Var[x(t+1)](g'(x(t)))^2 . \end{aligned}$$

If the (conditional) variance of $x(t+1)$ is a function of $x(t)$, so that

$$Var[x(t+1)] = v(x(t)), \tag{3.5}$$

then

$$Var[x(t+1)] \approx v(x(t))(g'(x(t)))^2 .$$

The goal is to find a transformation $g(x)$ so that $Var[x(t+1)]$ is approximately constant. That is to say, we need to solve the differential equation

$$v(x)(g'(x))^2 = c_0$$

for $g(x)$, where c_0 is a (yet to be specified) constant. The general solution of this equation is

$$g(x) = \int \left(\frac{c_0}{v(x)}\right)^{1/2} dx + c_1 \tag{3.6}$$

where c_1 is another arbitrary constant.

For a stochastic model for which the variance of $x(t+1)$ is a function $v(x(t))$ of $x(t)$ as in (3.5), the function $g(x)$ defined by (3.6) transforms the model (approximately) to an NLAR of the form (3.4) for the transformed state variable $n(t) = g(x(t))$. The arbitrary constants c_0, c_1 of integration in (3.6) are free to be used to simplify $g(x)$ in any convenient way. Here are two fundamental examples.

For environmental stochasticity, from (3.2) we found above that

$$v(x) = Var[\mu]\, x^2$$

which by (3.6) leads to the transformation

$$g(x) = \left(\frac{c_0}{Var[\mu]}\right)^{1/2} \ln x + c_1.$$

For demographic stochasticity, from (3.3) we have

$$v(x) = \mu(1-\mu)\, x.$$

which by (3.6) leads to the transformation

$$g(x) = \left(\frac{c_0}{Var\,[\mu]}\right)^{1/2} \ln x + c_1.$$

In these expressions c_0 and c_1 are arbitrary constants. If $c_0 = Var\,[\mu]$ and $c_1 = 0$ in the environmental case and $c_0 = (1-\mu)\mu/4$ and $c_1 = 0$ in the demographic case, then transformations simplify to log and square root transformations $g(x) = \ln x$ and $\sqrt{x}$ respectively.

Thus, one way to model environmental stochasticity for a $m = 1$ dimensional model with deterministic skeleton

$$x(t+1) = f(x(t))$$

is to add noise on the logarithm scale or, equivalently, by means of the stochastic equation

$$x(t+1) = f(x(t))e^{E(t)}. \tag{3.7}$$

Here we proceed by analogy with the survivorship model and assumes that the variance of $x(t+1)$ conditioned on $x(t)$ is approximately constant on the logarithmic scale. (In a higher dimensional case $m > 1$, random exponential factors appear in each component of the skeleton model.)

Similarly, a model for demographic stochasticity utilizes a square root transformation of the state variables. Thus, in the $m = 1$ case, a demographic stochastic version of the model is $x(t+1) = \left(\sqrt{f(x(t)} + E(t)\right)^2$, although $\sqrt{f(x(t)} + E(t)$ should be replaced by 0 if it turns out to be negative:

$$x(t+1) = \left(\frac{\phi(t) + |\phi(t)|}{2}\right)^2 \text{ where } \phi(t) \triangleq \sqrt{f(x(t))} + E(t).$$

Example 3.1. *An environmental stochastic version of the Ricker equation (1.3) adds a random variable $E(t)$ (of mean 0) to the equation on the log scale:*

$$x(t) = be^{-cx(t)}x(t)e^{E(t)}.$$

A demographic stochastic version of the discrete logistic equation (1.2) adds a random variable $E(t)$ on the square root scale:

$$x(t+1) = \left(\frac{\phi(t) + |\phi(t)|}{2}\right)^2, \quad \phi(t) \triangleq \sqrt{b\frac{1}{1+cx(t)}x(t)} + E(t).$$

A stochastic version of the juvenile/adult model in Example (2.2) in which both life cycle stages are subject to environmental noise is

$$J(t+1) = \left[cbA(t)e^{-A(t)}J(t) + be^{-A(t)}A(t)\right]e^{E_1(t)}$$

$$A(t+1) = \left[\tau_{21}J(t) + \tau_{22}A(t)\right]e^{E_2(t)}$$

where $E(t) = (E_1(t), E_2(t))$ is a vector of random variables $E_i(t)$.

Given $q+1$ observations $y(0), y(1), \cdots, y(q)$ what is the probability that these stage vectors could be produced by a stochastic version of a matrix model (1.7), for some choice of the model parameters? A formula that relates this probability to model parameters is the *likelihood function*, and parameters that maximize this function are *maximum likelihood parameter (ML) estimates*.

A maximum likelihood function is basic to statistical inference and analysis, e.g., for estimation of parameters, the calculation of confidence intervals, and hypothesis testing. (For more about maximum likelihood methods and comparisons with other methods see [**18**].) A requirement for deriving an explicit formula for a ML function is a distributional assumption about the random variable $E(t)$. A mathematical formula for the *pdf* (probability distribution function) of $E(t)$ permits a formula for the ML function and to maximize it as a function of the model parameters. This maximization calculation can rarely be done analytically and usually must be done numerically with the aid of a computer.

Consider the $m = 1$ dimensional case. Suppose $w(0), w(1), \cdots, w(q)$ are transformed observations of a population's transformed state variable. Let $n(t+1) = f(n(t), \theta) + E(t)$ be the model equations for the transformed variable, where θ is the vector of model parameters appearing in the deterministic skeleton f. If the random variables $E(t)$ are normally distributed with mean 0 and variance $v = \sigma^2$ (and uncorrelated in time), then the probability of observing $w(t)$ given the observation $w(t-1)$ is

$$\frac{1}{\sqrt{2\pi v}} \exp\left(-\frac{(w(t) - f(w(t-1), \theta)^2}{2v}\right)$$

and the probability of observing the q data points $w(1), \cdots, w(q)$ is the product

$$\Lambda \triangleq \prod_{t=1}^{q} \frac{1}{\sqrt{2\pi v}} \exp\left(-\frac{(w(t) - f(w(t-1), \theta)^2}{2v}\right).$$

$\Lambda = \Lambda(\theta, v)$ is a function of the parameters θ and v. The maximum likelihood parameter estimates for θ and v are those that maximize $\Lambda(\theta, v)$.

In practice one often maximizes instead the log likelihood function

$$l(\theta, v) \triangleq \ln \Lambda(\theta, v) = -\frac{1}{2} q \ln 2\pi - \frac{1}{2} q \ln v - \frac{1}{2v} S(\theta)$$

where

$$S(\theta) \triangleq \sum_{t=1}^{q} \left(w(t) - f(w(t-1), \theta)\right)^2 \tag{3.8}$$

is the *sum of squared (transformed) one-step residuals*. It is possible to show that the maximum occurs for

$$v = \frac{1}{q} S(\theta) \tag{3.9}$$

where θ maximizes

$$l(\theta, S(\theta)/q) = -\frac{1}{2} q \ln(2\pi) - \frac{1}{2} q \ln\left(\frac{1}{q} S(\theta)\right) - \frac{q}{2}.$$

Thus, in the $m = 1$ case, the maximum likelihood estimates of the parameters in the vector θ are those that minimize the sum of squared residuals $S(\theta)$ together with v defined by (3.9).[2] (For $m > 1$ the ML estimates for the parameters in θ do not necessarily minimize $S(\theta)$.)

Example 3.2. *Consider the time sequence of* 101 ($q = 100$) *data points plotted in Figure 9. To calculate the ML parameter estimates of the environmental stochastic version (3.7) of the discrete logistic (1.2) we transform the data to the log scale and minimizes* $S(\theta)$, *given by (3.8) with* $f(w, \theta) = \ln b + w - \ln(1 + ce^w)$, *with respect to* b *and* c *(with the help of a computer program). The results are (recall (3.9))*

$$b \approx 13.14, \quad c \approx 0.06059, \quad v \approx 0.08163.$$

For the environmental stochastic version of the Ricker model (1.3) the ML estimates are, obtained , using the same minimization procedure with $f(w, \theta) = \ln b + w - ce^w$, *are*

$$b \approx 7.591, \quad c \approx 0.009751, \quad v \approx 0.009702.$$

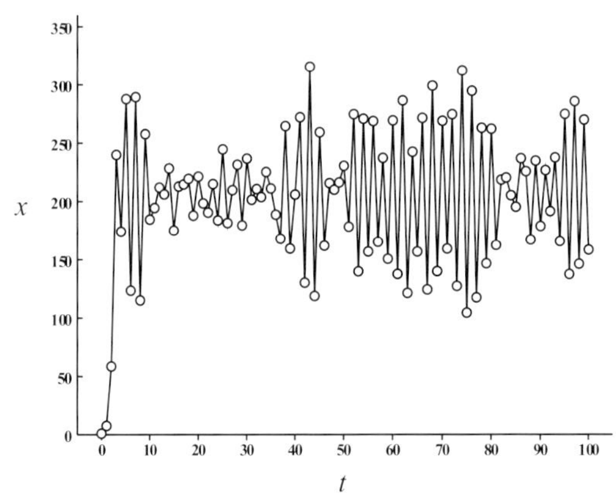

FIGURE 9. This plot shows the data points used to calculate ML parameter estimates for the environmental stochastic versions of the logistic and the Ricker models in Example 3.2.

All solutions of the discrete logistic (1.2) model equilibrate for $b > 1$. If we were to accept the parameterized version of this model in Example 3.2, then we would conclude that the data in Figure 9 is a noisy equilibrium $x_e = (b-1)/c \approx 200.4$. On the other hand, if we were to accept the parameterized Ricker model in Example 3.2 then, because the Ricker model (1.3) predicts a 2-cycle (that oscillates between $x_1 \approx 166.6$ and $x_2 \approx 249.2$) when $b = 7.591$, we would conclude that the data in Figure 9 is a noisy 2-cycle. Which of these different conclusions should we accept?

One of the fundamental issues with regard to connecting models to data is the assessment or validation of a model. This is not usually an easy problem. It can involve, among other things, a considerable amount of sophisticated statistical analyses and careful considerations of the model's assumptions (both the deterministic and the stochastic portions) with regard to the biological mechanisms thought to most predominant in the application. In the end we reject a model as inadequate or

[2]Note that the one-step residuals in $S(\theta)$ are not those calculated from points on an orbit of the model. This method does not fit an orbit to the data.

accept it (at least tentatively) as reasonably accurate. There is no clear cut, algorithmic path to arrive at such decisions and there will always be room for criticism and improvement.

One way to confront a model with data is to think in terms of two aspects: an analysis of fit and an analysis of prediction. So, in Example 3.2 we ask "how well does the environmental stochastic logistic" fit the data in Figure 9? We can also ask that question of the environmental stochastic Ricker, and then ask "which does a better job"? Answers help us choose between these two models, but it must be remembered that there are other (infinitely many other) models. Having made a choice, we then ask how well the chosen model predicts, in the sense that it also adequately describes other data not used in the parametrization estimation procedure. This other data might be subset of a given data set deliberately set aside for this purpose. Or it might be another data set collected as a replicate or from a different experiment or set of observations. A strong test of the model is to test its ability to make accurate predictions (perhaps even unusual or unexpected predictions) that are subsequently corroborated by controlled experiments or observations. This is perhaps the ultimate goal in modeling building and validation. Accurate and trustworthy predictions not only serve scientific and engineering purposes, but serve to further validate the model and add confidence in its accuracy. This prediction aspect of a model will be a an important component of the case studies described below.

Returning to Example 3.2, we consider only a couple of basic ways to quantify an assessment of the two models. We want more than a plot of a model orbit superimposed on the data that visually looks good. For example, if we plot the solution of the parameterized logistic starting from the initial data point $x(0) = 1$ in Example 3.2, we obtain an orbit that rather convincingly "goes through the middle of the data". Similarly, however, the orbit of the parameterized Ricker also looks visually reasonable when superimposed on the original data time series. See Figure 10. We need quantifiable ways to assess the results of these two parameterized models.

One way to proceed is to recall that the stochastic version of the model claims that the one-step residuals are normally distributed with a mean 0 and with the estimated variance v. The residuals constitute a data set that we can test for normality using a number of normality tests from statistics. We will not dwell on these statistical issues here, but simply be content with a visual inspection histogram plots the residuals shown in Figure 11. The residuals from the ML parameterized Ricker model appear more normally distributed than those of the ML parameterized logistic model.

Moreover, the residuals of the parameterized Ricker model have a smaller variance (by an order of magnitude) than those of the parameterized logistic model. That is to say, a one-step prediction of the deterministic Ricker model, made from the data point at time t, is more often closer to the next data point at time $t+1$ than is the one-step prediction of the deterministic logistic model. One way researchers quantify this observation is by means of an "R-squared" value:

$$R^2 \triangleq 1 - \frac{v}{v_0}$$

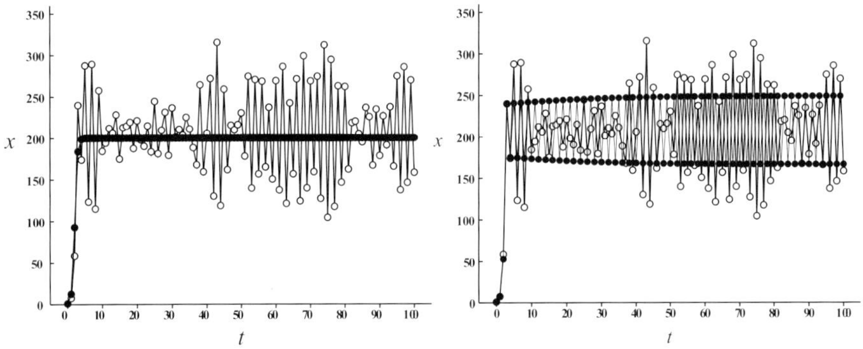

(a) (b)

FIGURE 10. (a) and (b) show plots of the orbits of the discrete logistic and the Ricker models (1.2) and (1.3), respectively, superimposed on the data from Figure 9 used to parametrize the models. The initial condition is the first data point $x(0) = 1$ and the parameters are the ML parameters estimated in Example 3.2. It should be pointed out that the ML parameter estimates in, for example (a), were not obtained by minimizing the residuals between the data and this orbit (as in a regression analysis). This common procedure does not calculate the maximum likelihood estimates of any obvious model and almost certainly not the one being parameterized. The ML parameter estimates, which minimize the residuals from the conditional one-step predictions (i.e., the residual at $t + 1$ conditioned on starting from the data point at t), are not explicitly designed to "optimally fit an orbit through the data".

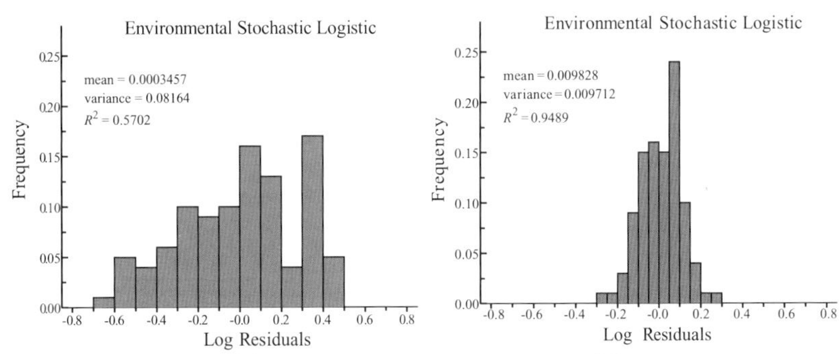

(a) (b)

FIGURE 11. These plots show frequency histograms of the one-step conditional log residuals of the ML parameterized logistic and Ricker, using the data in Figure 9. For the data in Figure 9, $v_0 = 0.1899$ and $w_0 = 5.258179$.

where $v = S/q$ (see (3.8)) and v_0 is the variance of the transformed data $w(t)$:

$$v_0 \overset{\circ}{=} \frac{1}{q}\sum_{t=1}^{q}\left(w(t) - w_0\right)^2, \qquad w_0 \triangleq \frac{1}{q}\sum_{t=1}^{q} w(t).$$

One says that R^2 is "the fraction of the variability in the data that is explained by the model". So, if $R^2 = 0.75$, then one says that the model "explains 75%" of the variability in the data. R^2 values reported in Figure 11 show that the ML parameterized logistic model explains 57% of the variability in the data in Figure 9, while the ML parameterized Ricker model explains 95%. On this basis, it is reasonable to conclude that the Ricker model is the more accurate model for this data set.[3]

Figure 12 shows histograms of the log residuals of both parameterized models applied to another data set from the same source – a replicate data set. The

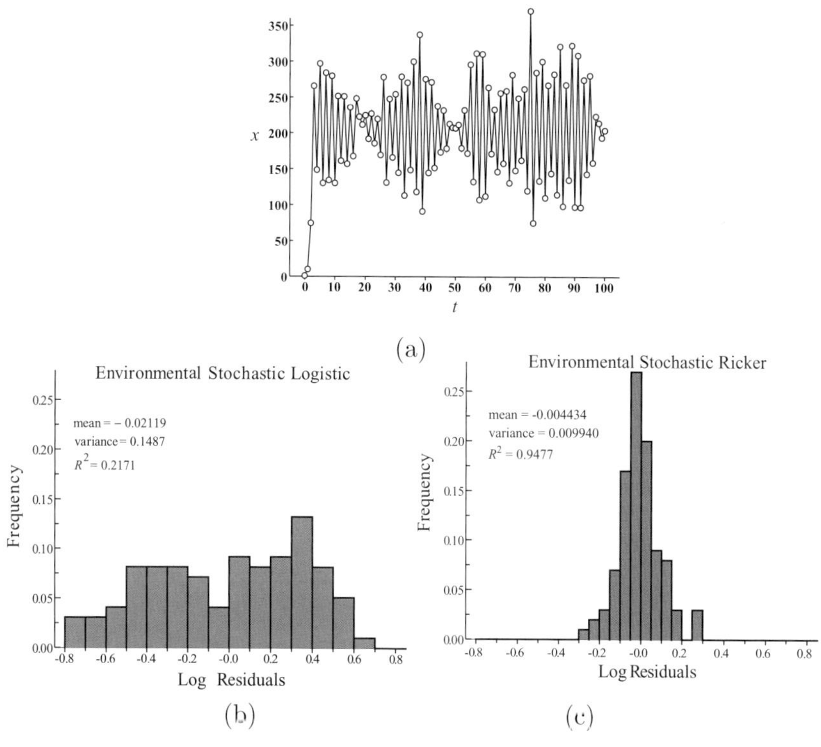

FIGURE 12. The plot in (a) shows data from a replicate of the experiment that produced the data in Figure 9 used for the ML parametrization of the logistic and the Ricker models in Example 3.2. Frequency histograms of the one-step log residuals of the parameterized logistic and Ricker appear in (b) and (c). Note that the data in (a) was not used to obtain the ML parameter estimates. The R^2 value of the logistic model drops over 50% while that of the logistic remains nearly unchanged.

[3]To compare models one shouldn't always simply compare their variances or R^2 values. Such a comparison is reasonable if both models have the same number of estimated parameters. However, if one model has more parameters than the other, then its smaller R^2 value might be due more to its extra parameters than to its superior structure as a model. When two models have a different number of parameters, one should use statistics that compensate, or penalize, for the number of parameters when making comparisons. One example is the *Akaike information criterion* (AIC) which quantifies the relative goodness-of-fit of statistical models, given a sample of data. The model is preferred that has the smallest quantity $A \triangleq 2k + q\ln(S/q)$ where k is the number of estimated parameters.

predictive accuracy of the Ricker model is reflected in a virtually unchanged R^2 value from that of the Ricker model fit shown in Figure 11,even though these data were not used to estimate the ML parameters for the model. In contrast, the R^2 value of the parameterized logistic drops over 50% in going from the fitted data to the predicted data sets.

Figure 13(a) shows a data set taken from a laboratory experiment. The data show the adult numbers in an insect population (the beetle *Tribolium castaneum*) obtained from census counts taken at two week intervals. The ML parameter estimates for the environmental stochastic Ricker model turn out to be

$$b \approx 3.5733, \qquad c \approx 0.03026, \qquad v \approx 7.756. \tag{3.10}$$

The histogram of one step log residuals in Figure 13(b) shows a marked departure from a normal distribution, a fact that leads us to reject this model as a description of this data set. Indeed, if we were to accept the model as accurate then we would conclude that the data in Figure 13(a) is a (very) noisy equilibrium. (The Ricker map as an equilibrium $x_e = c^{-1} \ln b \approx 42.09$ for the parameter estimates of b and c above.) Yet the data in Figure 13(a) seems to have distinctive non-equilibrium patterns that we might reasonably expect to be amenable to model description and explanation. We can critique the rejected environmental Ricker in two ways. Its deterministic skeleton (the Ricker equation) could be ill formulated for the data in Figure 13(a) and/or the environmental stochastic version might be a poor description of the noise present in the system. We can, of course, modify either of these model components in any of innumerable ways.

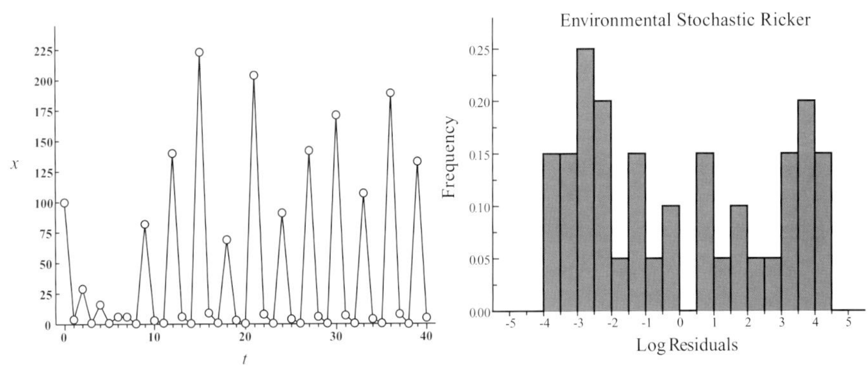

(a) (b)

FIGURE 13. If we use the data time series in (a) to estimate the parameters in the environmental stochastic Ricker model (using the maximum likelihood parameter estimation procedure), we obtain the estimates (3.10). However, the histogram of one-step log residuals in (b) show a significant departure from a normal distribution.

To obtain guidance in the construction of an improved model, one should consider the important features of the population's biology and the physical features of its environment. A good model is based on mechanisms that are known, or thought to be, significantly important in determining the population's dynamics. Biologists who study a particular species generally have a very good idea what such mechanisms are. Indeed, biologists (by the nature of their training) generally know a formidable number of details about a species, and it is often a challenge for a modeler, when collaborating with a biologist, is to determine which of a plethora of facts and details known to the biologist are to be incorporated into the model. Mathematicians (by their training) tend to be generalists and the contrast with the focused attention to a complexity of details that is characteristic of a biologist's understanding of a biological system can make for lively (and frustrating) conversations in this regard. Keeping in mind that a model with enough parameters can fit virtually any data set, whether it is a reasonable model or not (after all, one can find a high degree polynomial that passes through any number of given data points, with all residuals equal to 0!), we try to find a few mechanisms that dominate the dynamics of the population and build from them a model with as few parameters as possible.

The data in Figure 13 are adult population numbers of an insect species, the beetle *T. castaneum.* This beetle is a significant agricultural pest, and as a result researchers have extensively studied this species (and others from the genus *Tribolium*) since the early 20th century. Grown in a finite volume of medium[4], which is regularly renewed, sustainable populations must regulate their numbers in some manner. The species *T. castaneum* regulates its population growth by means of cannibalism.

T. castaneum has four main life cycle stages, egg, larval, pupal, and adult, and the moving stages cannibalize the non-moving stages (although larvae cannibalism of pupae is rare). This suggests a model with at least four state variables. However, the egg stage is relatively short (2 to 3 days) compared to the census time (which is taken to be the length of the larval stage, namely 2 week). We can utilize this fact to construct a model with three state variables: larvae, pupae, and adults. In this way, by using a stage structured model we are able to account for inter-stage cannibalistic interactions, the mechanisms known to drive the dynamics of this species.

It turns out that the pupal stage in *T. castaneum* is very nearly of the same duration as the larval stage (two weeks). As a result a matrix model for the population density distribution vector $x = \operatorname{col}(L, P, A)$ has the form of a Leslie age-structured model with only the third stage fertile, i.e., a projection matrix $P = F + T$ with

$$F = \begin{pmatrix} 0 & 0 & b \\ 0 & 0 & 0 \\ 0 & 0 & 0 \end{pmatrix}, \quad T = \begin{pmatrix} 0 & 0 & 0 \\ 1-\mu_l & 0 & 0 \\ 0 & 1-\mu_p & 1-\mu_a \end{pmatrix}.$$

Here the μ_i are the stage mortality fractions and b is the larval recruitment per adult per unit time.

The cannibalism of eggs by larvae and adults decrease larval recruitment b by a fraction dependent on the density of larvae and adults L and A. Cannibalism of pupae by adults decreases adult recruitment fraction $1 - \mu_p$ by an additional fraction dependent on A (but not on L since observation has shown that for *T. castaneum* pupal cannibalism by larvae is negligible). In this model Ricker type exponential nonlinearities account for the cannibalistic interactions. The rationale for this is as follows.

[4]For many species of *Tribolium* growing in a finite container of medium is not far removed from their "natural" habitat, since they have grown in containers of grain products stored by humans for centuries. For example, evidence of *Tribolium* has been found in ancient Egyptian urns.

Cannibalism in *T. castaneum* occurs, as far as careful observation has revealed, because of random encounters among individual adult or larvae (as they continually move and feed in the medium) with eggs or pupae (which are stationary). If we begin with the assumption that the probability of an encounter between one larva, say, with an egg during Δt time units is proportional to Δt and inversely proportional to the volume V of the culture medium [**2**], then the probability that the egg is not encountered by the larva is $1 - c_{el}\Delta t/V$ where $c_{el} > 0$ is a constant of proportionality (the cannibalism coefficient). The probability that the egg will not encounter the larva during one full unit of time is $(1 - c_{el}\Delta t/V)^{1/\Delta t}$. If L larvae are present, then the egg's survivorship probability is the product

$$\Pi_{i=1}^{L}\left(1 - c_{el}\Delta t/V\right)^{1/\Delta t} = \left(1 - c_{el}\Delta t/V\right)^{L/\Delta t}.$$

As $\Delta t \to 0$, the probability approaches $\exp\left(-c_{el}L/V\right)$. A similar argument holds for adult/egg and hence the probability that an egg survives cannibalism by both larvae and adults is $\exp\left(-c_{el}L/V\right)\exp\left(-c_{ea}A/V\right)$. As a result larval recruitment is reduced from bA in the absence of cannibalism to $bA\exp\left(-c_{el}L - c_{ea}A\right)$ when cannibalism is taken into account. A similar argument introduces a fractional reduction term $\exp(-c_{pa}A)$ in the survivorship of pupae to the adult stage.

The result of introducing exponential nonlinearities into the transition matrix T is the LPA model in Example 1.4, namely

$$\begin{aligned} L(t+1) &= bA(t)\exp\left(-\frac{c_{el}}{V}L(t) - \frac{c_{ea}}{V}A(t)\right) \\ P(t+1) &= \left(1-\mu_l\right)L(t) \\ A(t+1) &= \left(1-\mu_p\right)P(t)\exp\left(-\frac{c_{pa}}{V}A(t)\right) + \left(1-\mu_a\right)A(t). \end{aligned} \tag{3.11}$$

This $m = 3$ dimensional matrix model has eight parameters. The medium volume V is known in experimental situations. Moreover, extensive observations have shown that pupal survivorship of *T. castaneum* (and for the species in the genus *Tribolium* in general) in the absence of cannibalism is virtually 100% and therefore we can safely assume $\mu_p = 0$. This leaves six parameters to be estimated, unless of course some are controlled as part of an experimental protocol.

The three life cycle stages of *Tribolium* are easily recognized and can be routinely counted. (The example data in Figure 13(a) is, in fact, the $x = A$ stage component from a time series of all three life cycles L, P and A.) We can estimate the model parameters from a time series of L, P, A data by using the maximum likelihood procedure described above extended to the multivariate case when x is a vector. (Sometimes we can instead estimate a parameter another ways. For example, if dead adults are counted then we can estimate μ_a from the results.)

Suppose $w(0), w(1), \cdots, w(q)$ are transformed observations of a population's transformed ($m > 1$ dimensional vector) state variable. Let $n(t+1) = f(n(t), \theta) + E(t)$ be the model equations for the transformed (vector) variable, where θ is the vector of model parameters appearing in the deterministic skeleton f. Here we assume that $E(t)$ is a vector of normal distributions (a multivariate normal distribution) with mean 0 and (symmetric) variance-covariance matrix Σ. The variances of the components of $E(t)$ appear on the diagonal and covariances among the components appear in the off diagonal elements. In situations in which a disturbance that occurs in one stage is unlikely to effect other stages then the covariances can be assumed equation to 0 and Σ is diagonal. The likelihood function is

$$\Lambda \triangleq \prod_{t=1}^{q} \frac{1}{\sqrt{(2\pi)^m \det \Sigma}} \exp\left(-\frac{(w(t)-f(w(t-1),\theta))^* \Sigma^{-1} (w(t)-f(w(t-1),\theta))}{2}\right)$$

and the log likelihood function is

$$l \triangleq \ln L = -\frac{m}{2} q \ln(2\pi) - \frac{1}{2} q \ln(\det \Sigma) - \frac{1}{2} \sum_{t=1}^{q} (w(t)-f(w(t-1),\theta))^* \Sigma^{-1} (w(t)-f(w(t-1),\theta)).$$

(The asterisk denotes the transpose of a vector or matrix.) The log likelihood function l is a function of the skeleton's parameters in the vector θ and of the entries in the (symmetric) variance-covariance matrix Σ. The ML parameter estimates are those values of θ and the entries in Σ that maximize l.

It turns out that the maximum occurs when $\Sigma = RR^*/q$ where R is the matrix with the residuals $w(t) - f(w(t-1), \theta)$ as columns. If this expression for Σ is substituted into l, then the problem becomes to maximize the resulting formula for l as a function of θ only (and calculate Σ from R, in which the residuals are calculated using the calculated ML estimate for θ).

Stochastic versions of the LPA model (3.11), and various modifications of the model, constitute the basis for over fifteen years of experimental and theoretical investigations of a long list of nonlinear phenomena in population dynamics[5]. Reports on these studies appear in [**1, 3, 11, 56**]. I will give here only a summary report of some of these studies.

There are several levels on which we can view the results of the studies. First of all, we can consider them to be illustrative examples of the modeling methodology discussed above. Second, the studies will demonstrate that a mathematical model can "work" in population dynamics; that is to say, a mathematical model can make accurate quantitative descriptions of the dynamics of a biological organism and, importantly, can make predictions that are later born out by controlled experiments. Thus, we can go beyond the statement that a model is only "qualitatively correct", as is found so often in the literature, and make a small contribution towards raising population dynamics and theoretical ecology above verbal metaphors. Third, these studies often provide explanations for observations that had no previous explanation. Fourth, an accurately validated model can make surprising predictions, sometimes counter-intuitive, that are subsequently corroborated by experimental observations. This leads to a deeper understanding of the population's dynamics and provides an ability to predict the outcome of disturbances and altered circumstances (environmental perturbations, generic changes, application of control policies, etc.). Fifth, new phenomena are discovered and new mathematical methodologies are developed to handle them. And finally, going beyond the particular experimental system used, the studies often elucidate, confirm, and sometimes

[5]The senior researchers involved are R. F. Costantino (University of Arizona), R. A. Desharnais (California State University at Los Angeles), Brian Dennis (University of Idaho), Shandelle M. Henson (Andrews University), Aaron A. King (University of Michigan) and J. M. Cushing (University of Arizona). Many undergraduate and graduate students have also made significant contributions over the years. A list of contributors can be found in [**11**].

challenge accepted ecological principles and tenets. This provides some guidance to the study of nonlinear dynamics of other systems, both laboratory and field systems. Each of these aspects have played, and continue to play, a role in the application of the LPA model to experimental cultures of *Tribolium* [**3**].

In the experiments described below, all three life cycle stages are counted every two weeks, and the cultures returned to their habitats with refreshed medium[6]. In some experiments manipulations are performed at this time in order to force certain values for selected parameters in the model. Many studies, particularly those in which demographic stochasticity predominate, have shown that the covariances (off diagonal entries in Σ) are quite small compared to the variances (diagonal entries) and as a result these covariances are assumed equal to 0 in the stochastic model. These details, which can change from one experimental protocol to another, reflect on the number of parameters in the model that need to be estimated.

Details of a preliminary parametrization of the LPA model using historical data for *T. castaneum* and an environmental stochastic version of the LPA model appear in [**18**]. This initial study used an environmental stochastic model because population sizes were large and demographic stochasticity is generally less of a factor in large populations [**53**]. In subsequent studies and experiments, in which large oscillations resulted in numbers in individual life cycle stages being small (but not total population size), it was found that a demographic stochastic version of the LPA model was a more accurate model [**19, 20**]. Details of the statistical analyses that validates the model are found in [**18**] (also see [**11, 20**]).

The goal of one study was to demonstrate that a biological population would traverse a model predicted bifurcation route-to-chaos. A preliminary experiment tested the predictive capability of the parameterized LPA model for cultures of *T. castaneum* on the basis of a relatively simple bifurcation diagram. In that experiment the bifurcation sequence results by varying the adult death rate μ_a from its parameterized value (holding all other parameters fixed). Varying μ_a between its allowable limits of 0 and 1, we obtain a bifurcation diagram similar to the that in Figure 7(a). An experiment placed replicated cultures of *T. castaneum* at selected positions in the bifurcation diagram, in order to corroborate that the predicted bifurcations would indeed occur in the dynamics of beetle cultures. (There were also unmanipulated and replicated control cultures.) The historical data used to obtain the preliminary parameter estimates for the LPA model came from a strain of *T. castaneum* that was no longer available for experiments. As a result, the preliminary bifurcation experiment used different strains. The experiment was conducted twice, once for two *T. castaneum* strains called SS and RR. For this reason, the data obtained from the experimental treatments were used to re-parametrize the LPA model and re-calculate the bifurcation diagram (which, as it turned out, did not significantly change from the original bifurcation diagram). The result for the SS strain appears in Figure 14 (taken from [**11**]).

The arrows in Figure 14 indicate where experimental cultures of *T. castaneum* (strain SS) were placed by experimental manipulation of the adult survival fraction μ_a. The model predictions for these treatments, for increasing μ_a, were: an equilibrium at $\mu_a = 0.04$, a 2-cycle at both $\mu_a = 0.27$ and 0.50, another equilibrium at

[6]R. F. Costantino and R. A. Desharnais carried out the experiments in laboratories at the University of Rhode Island and University of Arizona. Details of the experimental protocols can be found in [**11**] and the research papers cited therein.

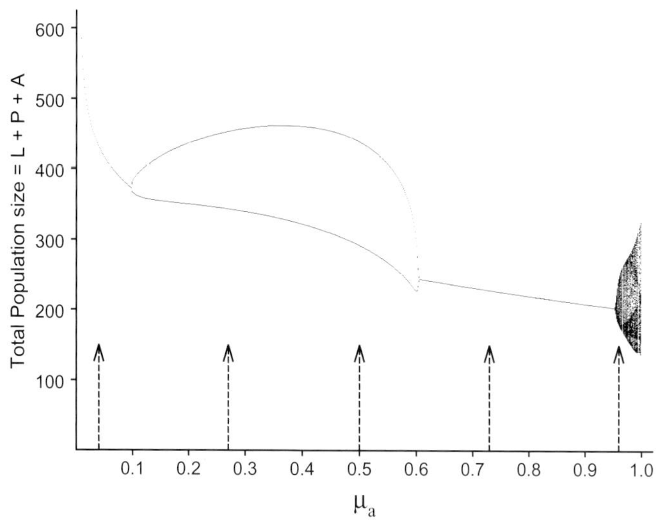

FIGURE 14. The bifurcation diagram for the LPA model (3.11) using the adult death rate μ_a as a bifurcation parameter and plotting total population size $L+P+A$ for the attractor. Other ML parameter values, obtained from the environmental stochastic LPA model and data for the SS strain, are $b = 7.483$, $\mu_p = 0$, $\mu_l = 0.2670$, $c_{ea} = 0.009170$, $c_{el} = 0.01200$, $c_{pa} = 0.004139$. (Confidence intervals for these parameters and the estimates for Σ appear in [**11, 20**].) [Reprinted from [**11**], with the permission of Academic Press.]

$\mu_a = 0.7$, and quasi-periodic fluctuations at $\mu_a = 0.96$. The state space plots of the data shown in Figure 15 visually show how these predictions were born out by the experimental data. Model evaluations using a variety of tests and analyses provide statistical support for the accuracy of the model predictions [**19**]. (The experiment was duplicated with similar results using the RR strain of *T. castaneum*. See [**11**].)

It is interesting to note that the bifurcation diagram in Figure 14 makes predictions that are perhaps counter-intuitive to what one might expect if the adults in an insect population are subjected to increasing mortality (as part of, say, a pest control procedure). It is probably not intuitive that increasing adult mortality can lead to the crash boom outbreaks of a 2-cycle, followed by a re-equilibration, with very little drop in infestation level, and ultimately at very high mortality rates to quasi-periodic (and hence difficult to predict) outbreaks.

The success in the predictive capability of the LPA model parameterized for the flour beetle *T. Castaneum* in the preliminary bifurcation experiment reported in Figures 14 and 15 was sufficient to encourage an investment into a longer term experiment (which ultimately lasted over 8 years or over 100 generations of beetles). A considerably more complicated LPA model predicted bifurcation diagram – one that involved a route-to-chaos – formed the basis of a more elaborate experiment using *T. Castaneum*. The ML parameter estimates for the RR strain produce the bifurcation diagram in Figure 16 when two parameters are manipulated. Specifically, the adult death rate is held fixed at $\mu_a = 0.96$ and the adult-on-pupae

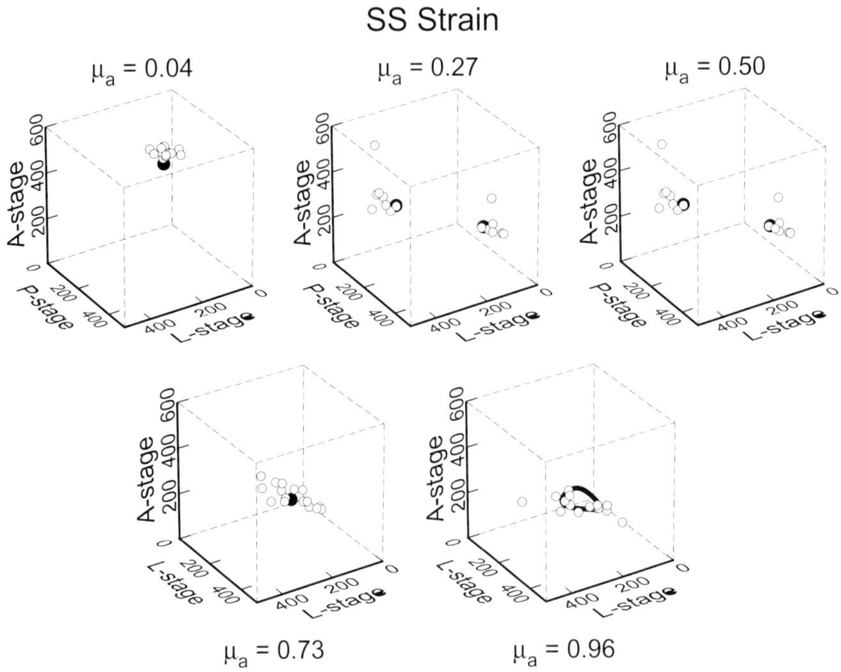

FIGURE 15. Open circles are experimental data points plotted in $m = 3$ dimensional (L, P, A) phase space. Transients are removed in order to emphasize the predicted attractors, which appear as solid points and lines. These results are for the *SS* strain of *T. Castaneum*. A duplicate experiment, with similar results, was carried out for the *RR* strain. [Reprinted from [**11**], with the permission of Academic Press.]

cannibalism rate c_{pa} is varied as a bifurcation parameter[7]. The arrows in Figure 16 show the locations in the bifurcation diagram where replicated treatments were placed in the experiment. These locations were chosen in order to determine if the beetle populations would exhibit the distinctively different attractors predicted by the LPA model for these parameter values. Figure 17 shows four examples of the attractors that are possible: equilibria, cycles, quasi-periodic oscillations (invariant loops), and chaotic strange attractors.

The state space plots of the experimental data displayed in Figures 18 and 19 show how the data bear out the model predictions. Model evaluations using a variety of tests and analyses provide statistical support for the accuracy of the model predictions [**20**]. Also note in Figures 18 and 19 how closely simulations from the demographic stochastic LPA model resembles the data.

Of particular interest in Figures 18 and 19 is the treatment at $c_{pa} = 0.35$ which lies in a parameter interval that contains chaotic dynamics. Indeed, for the point estimates of the parameters in this treatment, the model predicted attractor is chaotic. (This has not been proved rigorously in a mathematical sense, but is indicated by computer simulations and the calculation of Liapunov exponents, which measure sensitivity to initial conditions.) As is typical in models with chaotic

[7] See [**11, 20**] for a description of how c_{pa} was experimentally manipulated.

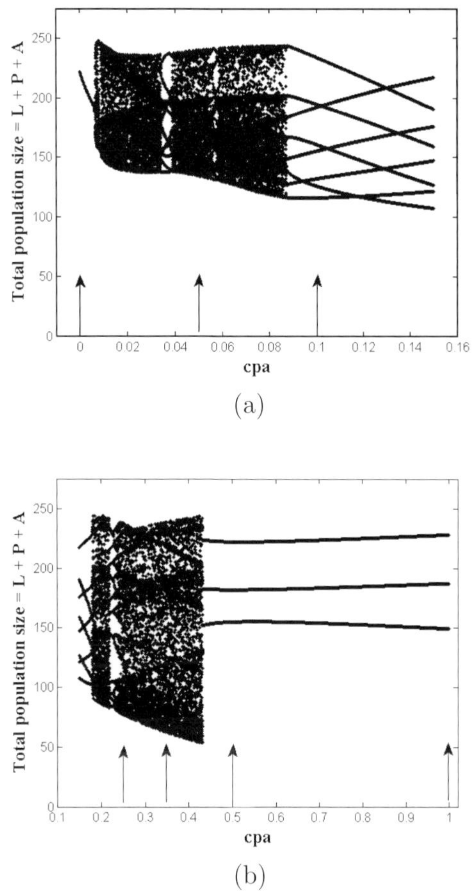

FIGURE 16. Two plots show bifurcation diagrams for the LPA model (3.11) using the adult-on-pupa cannibalism rate c_{pa} as a bifurcation parameter and plotting total population size $L + P + A$ for the attractor. Other ML parameter values for the RR strain (calculated with the environmental stochastic version of the LPA model) are $b = 7.876$, $\mu_p = 0$, $\mu_l = 0.1613$, $c_{el} = 0.01385$, $c_{ea} = 0.01114$, $\mu_a = 0.96$. (Confidence intervals for these parameters and the estimates for Σ appear in [**11, 20**].) In (a) c_{pa} ranges from 0 to 0.15 and in (b) c_{pa} ranges from 0.15 to 1. A post-experiment re-parametrization using the demographic stochastic LPA model forms the basis of detailed statistical analyses of the data, which show slightly better validation results. The bifurcation diagram changes little, however. For a robustness study of the bifurcation diagram see [**11, 20**]. [Reprinted from [**11**], with the permission of Academic Press.]

dynamics, non-chaotic dynamics can be found for nearby parameter values lying within the confidence intervals for the estimated parameters. Nonetheless, for such parameter values it is typical that transients still exhibit chaotic-like dynamics in their approach to the attractor. Moreover, the bifurcation diagram in Figure 20 is quite robust throughout the confidence intervals [**11, 20**]. We could reasonably

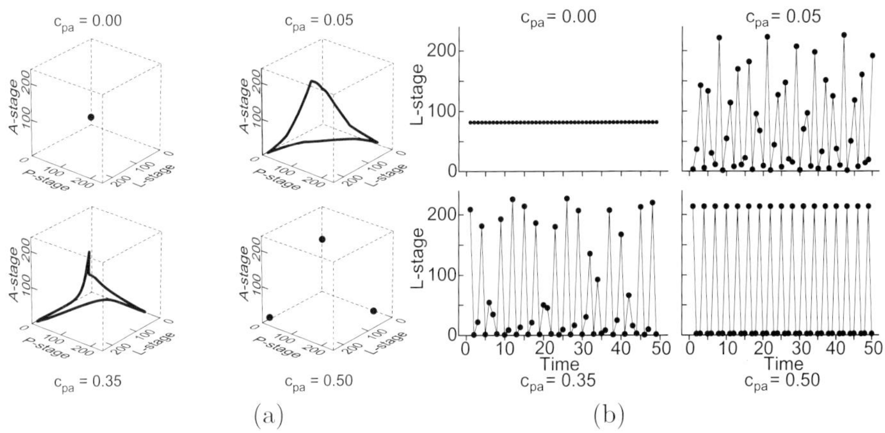

FIGURE 17. (a) These plots show the attractors in phase space at four locations in the bifurcation diagram of Figure 16. At $c_{pa} = 0$ the attractor is an equilibrium. At $c_{pa} = 0.05$ the attractor is an invariant loop. At $c_{pa} = 0.25$ the attractor is an chaotic strange attractor. At $c_{pa} = 1$ the attractor is a 3-cycle. (b) The L stage components of the attractors in (a) appear plotted against time t. [Reprinted from [**11**], with the permission of Academic Press.]

refer to the data from this treatment as chaotic, even though it is perhaps better to say that the data (i.e., the beetle population's dynamics) are "influenced" by the chaotic dynamics present through the confidence intervals for the parameters.

A study of chaotic dynamics requires a long time series of data and for that reason the three replicates of the $c_{pa} = 0.35$ treatment were continued for over eight years (100 generations). The resulting data provide an opportunity to study various issues concerning chaos as realized in a biological population. I will briefly discuss only three such topics: transients, state space lattice effects, and habitat size.

One way to view the effects of demographic or environmental noise is that such disturbances continually produce transient dynamics. Even should the trajectory of a population reach the vicinity of a deterministic attractor such perturbations will move it away and cause it to follow a transient path back to the attractor. Such perturbations might even be sufficient to place the population far from an attractor in which case the influence of other deterministic entities might come into play, such as the stable and unstable manifolds of non-attracting invariant sets. Such events are frequently observed in the experimental data shown in Figures 18 and 19.

An example appears in Figure 21 which shows the time series from one replicate taken from the treatment at $c_{pa} = 0.05$ in the bifurcation diagram Figure 16. The deterministic model predicted attractor for this treatment is an invariant loop. At the sixteenth week of the experiment a random perturbation placed the data orbit, which until then (for four generations) had been reasonably close to the predicted invariant loop, near the model predicted equilibrium. After residing near this unstable equilibrium for two months (two generations) the data orbit moves away from the equilibrium. This transitory pattern, which lasts 10 weeks until the

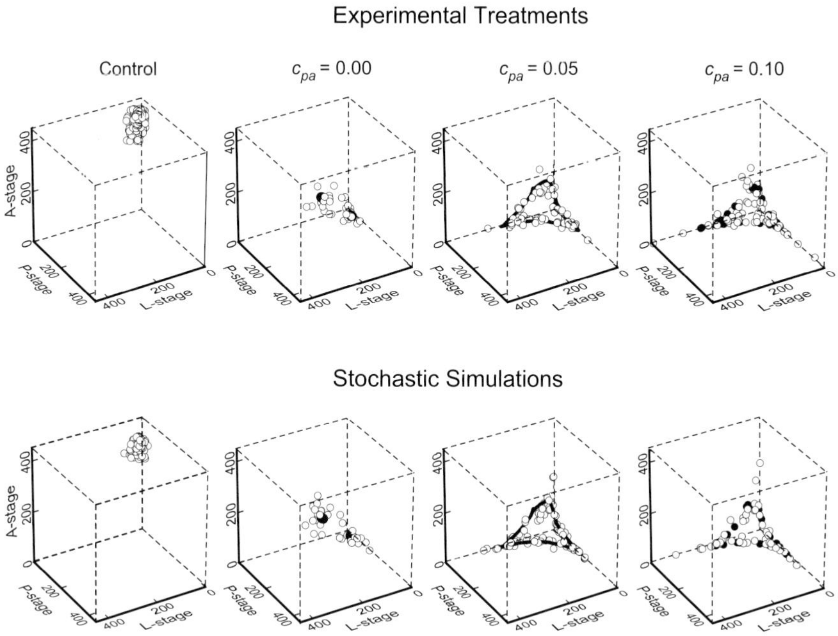

FIGURE 18. The top row of plots shows the experimental data obtained from cultures placed at the selected values of c_{pa} indicated in the bifurcation diagram in Figure 16(a) and from a control culture (which the model predicts should equilibrate). The bottom row of plots shows simulations from the demographic (square root scale) LPA model. [Reprinted from [**11**], with the permission of Academic Press.]

data orbit returns to the attractor, has a rotational pattern that is remarkably well predicted by the linearization of the model at the equilibrium. Details are explained in the caption of Figure 21.

See [**11**] for other examples of "saddle flybys" and of random visitations near other kinds of unstable invariant sets of the LPA model observed in the data from the route-to-chaos experiment. Of course, if the initial condition is not close to the attractor, and especially if it is near an unstable saddle or its stable manifold, then the model would predict such a transient saddle flyby. However, as Figure 21 shows, occasionally a data orbit will exhibit a flyby later in an experiment, after the vicinity of the attractor has been reached. A dramatic case of this is seen in Figure 22 where a saddle equilibrium flyby is observed in one replicate of the chaotic $c_{pa} = 0.35$ treatment during weeks 358 to 380 of the 424 week (106 generation long) experiment. Moreover, a data orbit will occasionally experience several flybys during the course of the experiment. Stochastically caused flybys of unstable invariant sets of other kinds of attractors, such as cycles, are also observed. See [**11**] for examples.

Another effect of random perturbations in populations with oscillatory dynamics can be a change in the phase of the oscillation. For example, the 3-cycle attractor in the $c_{pa} = 1.0$ treatment has three different phases. The three replicates in the treatment, by the end of the experiment, were all out of phase because each had suffered a perturbation that adjusted the phase and, as it happened, all three

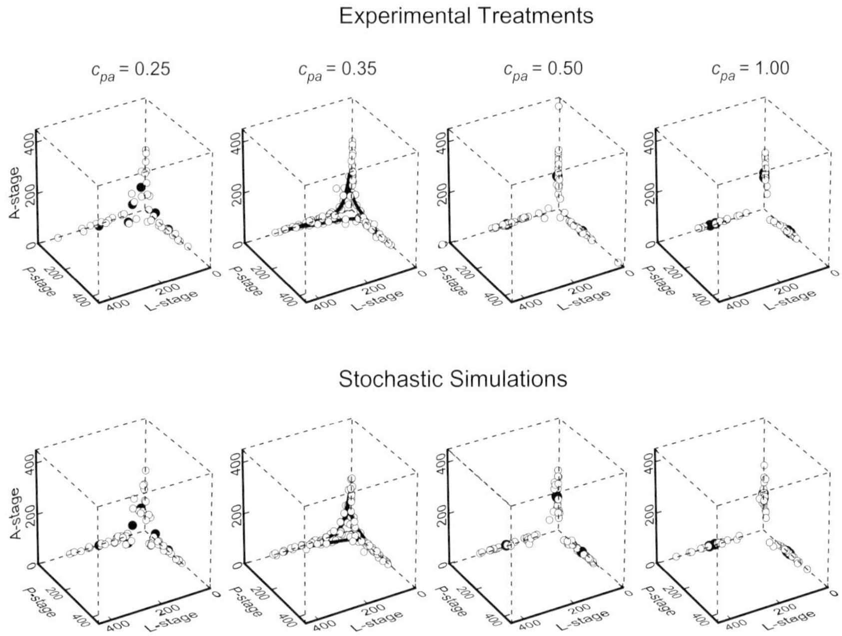

FIGURE 19. The top row of plots shows the experimental data obtained from cultures placed at the selected values of c_{pa} indicated in the bifurcation diagram in Figure 16(b). The bottom row of plots shows simulations from the demographic (square root scale) LPA model. The solid circles and lines are the model predicted attractors using a re-parametrization based on the route-to-chaos data itself (rather than the parameter estimates used Figure 16 for the a priori bifurcation sequence obtained from historical data): $b = 10.45$, $\mu_p = 0$, $\mu l = 0.2000$, $c_{el} = 0.01731$, $c_{ea} = 0.01310$, $\mu_a = 0.96$. [Reprinted from [**11**], with the permission of Academic Press.]

phases were represented in the three replicates at the end of the experiment. See Figure 23. Other examples of phase shifts are given in [**11**]. One way of explaining these shifts is to view the dynamics in composite phase space, where each point of a cycle is a different equilibrium, and study the stochastic hops from one basin of attraction to another. See [**32**] for several applications of this approach. One take home message here is the erroneous view of a population's dynamics that might result from the common practice of averaging over replicates. For example, out of phase cycles, when averaged, will result in a time series that will likely look like an equilibrium.

A close look at the data from the chaotic treatment $c_{pa} = 0.35$ reveals several distinctive cyclic patterns. Figure 24 shows one cyclic pattern, with period 11 (22 weeks, or 5.5 generations). Chaotic attractors typically contain infinitely many unstable (saddle) cycles and it turns out that a saddle 11-cycle is located on the chaotic attractor predicted for this treatment. Moreover, this 11-cycle has a strong influence on the dynamics of the chaotic attractor (as, for example, a spectral analysis shows) [**11**]. Thus, the attractor of the deterministic LPA model offers an explanation of this particular pattern observed in the data.

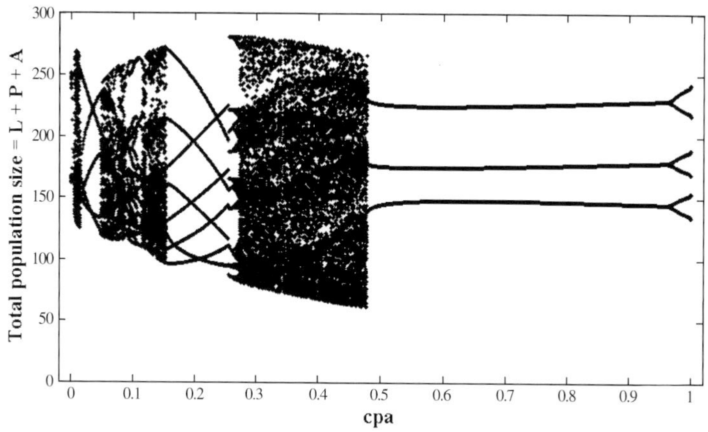

FIGURE 20. The re-calibrated bifurcation diagram the attractors are: an invariant loop at $c_{pa} = 0.00$ (two disjoint loops that appear nearly as a 2-cycle); chaos at $c_{pa} = 0.05$; a 26-cycle at $c_{pa} = 0.10$; an 8-cycle at $c_{pa} = 0.25$; chaos at $c_{pa} = 0.35$; a 3-cycle at $c_{pa} = 0.50$; a 6-cycle at $c_{pa} = 1.00$. [Reprinted from [**11**], with the permission of Academic Press.]

Another cyclic pattern – a near 6-cycle pattern – also appears prominently in data from the chaotic treatment $c_{pa} = 0.35$. Examples appear in Figure 25(a). The LPA model does not, however, provide an explanation for this period 6 pattern. Although there is yet no rigorous proof, extensive computer searchers suggest that a 6-cycle solution (stable or unstable) of the LPA model does not exist for the parameter values associated with this experimental treatment. We can, however, find a model explanation for these 6-cycle patterns in an unexpected way.

The census counts of individuals in each life cycle state are, of course, whole integers. The predictions of the LPA model (3.11), on the other hand, are not whole integers; the state space for the LPA model is continuous. A simple mathematical way to get integer predictions from the LPA model is to round the right hand sides of the three model equations. The "integerized" LPA model obtained in this manner, namely

$$\begin{aligned}
L(t+1) &= \text{round}\left[bA(t)\exp\left(-\frac{c_{el}}{V}L(t) - \frac{c_{ea}}{V}A(t)\right)\right] \\
P(t+1) &= \text{round}\left[(1-\mu_l)\,L(t)\right] \\
A(t+1) &= \text{round}\left[(1-\mu_p)\,P(t)\exp\left(-\frac{c_{pa}}{V}A(t)\right)\right] \\
&\quad + \text{round}\left[(1-\mu_a)\,A(t)\right],
\end{aligned} \tag{3.12}$$

predicts the 6-cycle shown in Figure 25(b), when the model parameters and the initial conditions of the $c_{pa} = 0.35$ treatment are used[8]. The 6-cycle in Figure 25(b) is strikingly similar to the cyclic patterns often observed in the data (such as the examples in Figure 25(a)). Therefore, we can attribute the observed period 6-cycle pattern to the fact that the data reside on a integer lattice in state space,

[8]The same is true if in the third equation the sum is rounded instead of the individual terms.

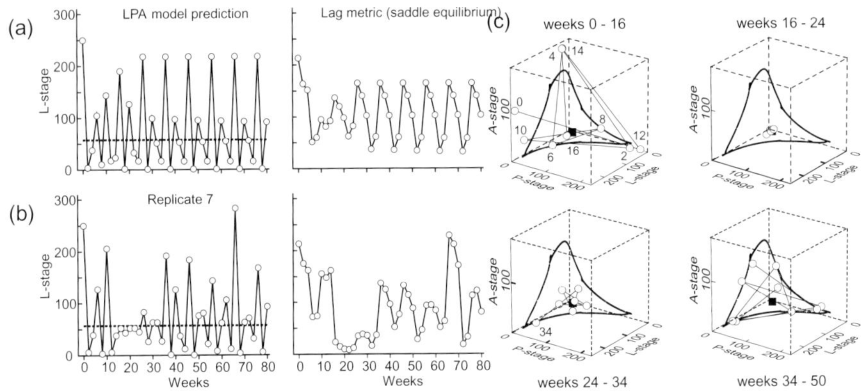

FIGURE 21. [**11, 10**] (a) The LPA model (3.11) does not predict a flyby of the saddle equilibrium for the treatment $c_{pa} = 0.05$ in the bifurcation diagram of Figure 16 when the orbit is started at the experimental initial conditions $(L(0), P(0), A(0)) = (250, 5, 100)$. The lag metric plotted here is the Euclidean distance in $m = 3$ dimensional phase space between data points and the model predicted equilibrium $(L, P, A) = (22.41, 4.625, 334.77)$, as a function of time. (b) Data from one from of the replicates for the $c_{pa} = 0.05$ treatment shows a remarkable visitation near the model predicted equilibrium (which is an unstable saddle). A random perturbation occurs at week sixteen that places the population near this equilibrium. (c) The flyby of the saddle equilibrium by the data orbit is strikingly apparent in state space. Note the star-like rotational motion in state space that occurs during weeks twenty-four through thirty-four when the data orbit leaves the vicinity of the saddle equilibrium. This geometrically distinctive path is in fact predicted by the deterministic LPA model. The linearization at the equilibrium has a conjugate pair of complex eigenvalues $re^{\pm i\theta}$ of magnitude $r \approx 1.265 > 1$ and polar angle $\theta \approx 2.576$ (and a third real positive eigenvalue $\lambda \approx 0.3945 < 1$). This complex eigenvalue implies a rotational motion away from the saddle of approximately $2\pi/\theta \approx 2.439$ radians (139.8 degrees) degrees per step, the motion occurring approximately in a plane parallel to that spanned by the eigenvectors $(L, P, A) \approx (1, -1.166, 0.4860)$ and $(1, -0.3526, -0.2817)$. These characteristics are seen in the orbit data during weeks 24 to 34. [Reprinted from [**10**] with the permission of Elsevier]

not the continuous state space of the LPA model (3.11). A comparison of Figures 24 and 25 shows intermittent episodes when the data from a portion of one replicate were close to the 11-cycle (the chaotic attractor) and the lattice model predicted 6-cycle.

Lattice effects (i.e., patterns in data, or in stochastic model simulations, attributed to the fact that data lies on a finite lattice in phase space) in the chaotic treatment $c_{pa} = 0.35$ are the object of study in [**41**]. The take home message from that study is that in order to explain the observed patterns in the dynamics of a biological population it might be necessary to use both a continuous state space model and a discrete state space model (based on whole integer numbers or on fractions as densities obtained from integers divided by an area or volume). What we expect

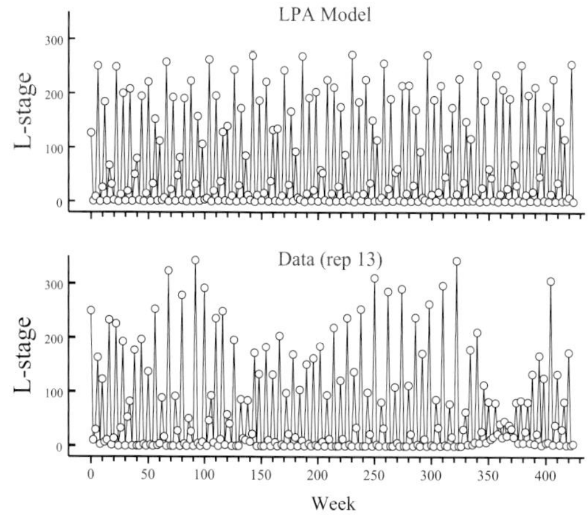

(a)

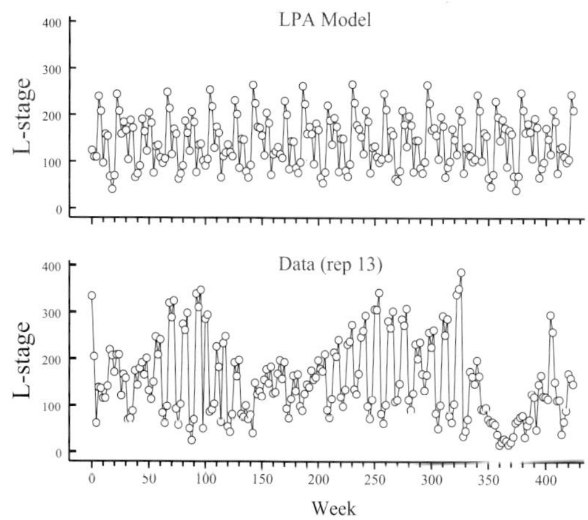

(b)

FIGURE 22. (a) A comparison of the L-stage components of one replicate of the chaotic $c_{pa} = 0.35$ treatment to those of the chaotic attractor. Notice the damped oscillations in the replicate's L-stage during weeks 358 to 380. The lag metric plots (distance to model predicted equilibrium in phase space) shown in (b) indicate that this damping is caused by a flyby of the equilibrium.

to observe are episodes of lattice cycle attractors (there are generally more than one) randomly mixed by stochastic events with transient dynamics that resemble the continuous state space attractors (and unstable invariant sets, such as saddle equilibria and cycles). Figure 26 shows an example of lattice effect phenomena in a stochastic, integerized version of the Ricker model [**11, 34, 40**].

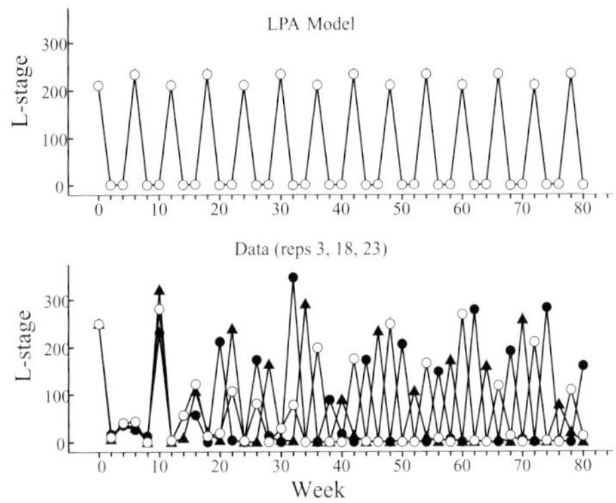

FIGURE 23. The three replicate cultures of the treatment $c_{pa} = 0.1$ whose predicted attractor is a 3-cycle are all out of phase by the end of the experiment.

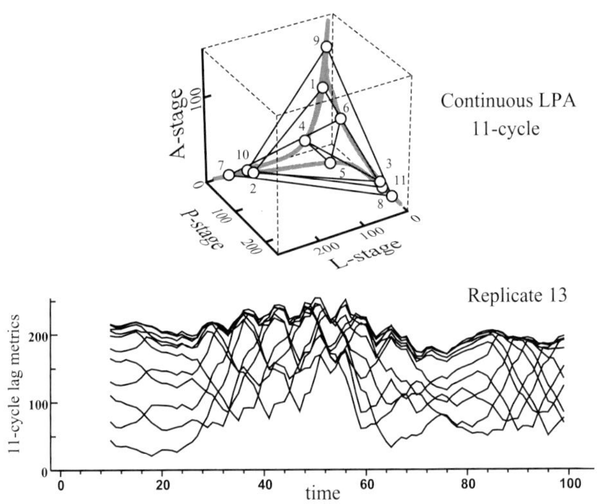

FIGURE 24. An unstable (saddle) 11-cycle is located on the chaotic attractor predicted for the treatment $c_{pa} = 0.35$ in Figure 16(b). This saddle cycle has a significant influence on the dynamics of orbits on and near the chaotic attractor [**11**]. We can view this 11-cycle as a signature of the chaotic attractor. The lag metric for the 11-cycle is the (average) distance that 11 consecutive data points are from the 11-cycle. Since there are 11 phases of the 11-cycle, there are 11 lag metrics associated with the 11-cycle, as shown in the lower plot for a portion of one replicate of data. Low lag metric values and an "unbraiding" of the lag metric plots indicate episodes when the data are near a phase of 11-cycle. Note that 11 time steps in the LPA model span nearly 3 generations of *T. castaneum*.

Henson et al. [**35**] give one way to provide a unifying point of view for this conclusion about discrete state space lattice effects and continuous state space attractors. From an probabilistic model that describes the prediction of state variables on a lattice from time t to time $t + 1$ as random variables, we can derive a

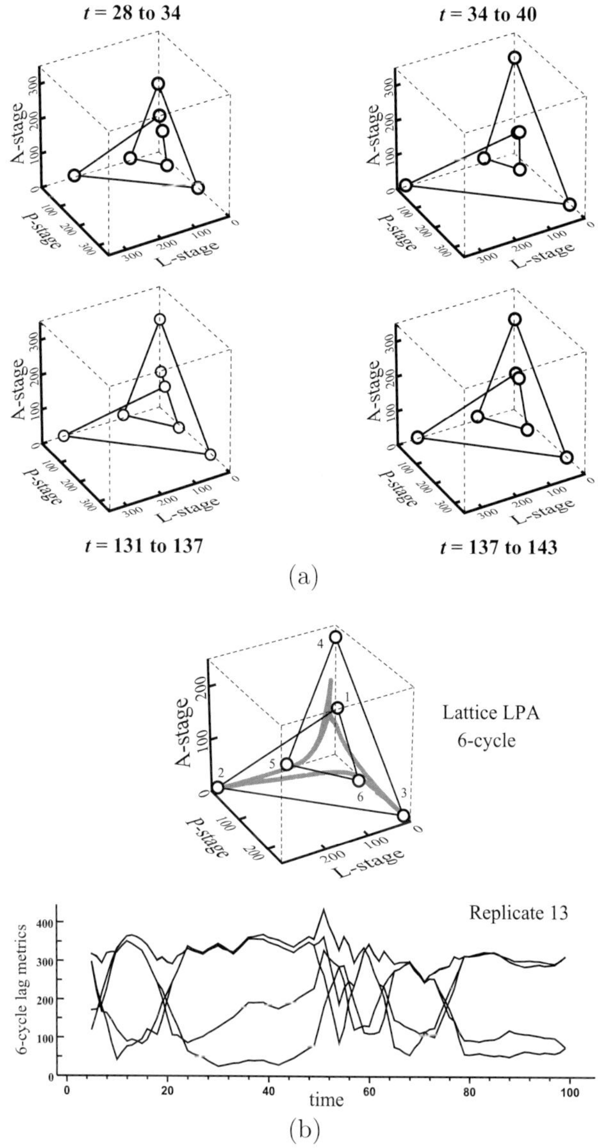

FIGURE 25. (a) Sample temporal 6-cycle patterns appearing in the data from the chaotic treatment $c_{pa} = 0.35$. (b) The upper plot shows the lattice 6-cycle predicted by the integerized LPA model (3.12) for the chaos treatment $c_{pa} = 0.35$ in Figure 16(b). The lower plot shows the lag metrics for the phases of this 6-cycle for a portion of one replicate of data. Low lag metric values and an "unbraiding" of the lag metric plots indicate episodes when the data are near a phase of the 6-cycle.

deterministic prediction in different ways. If the expectation (mean) of the random variable is taken as the prediction at $t + 1$, then the resulting deterministic model has a continuous state space. If instead we take the mode as the prediction at $t+1$, the resulting deterministic model still lies on the lattice.

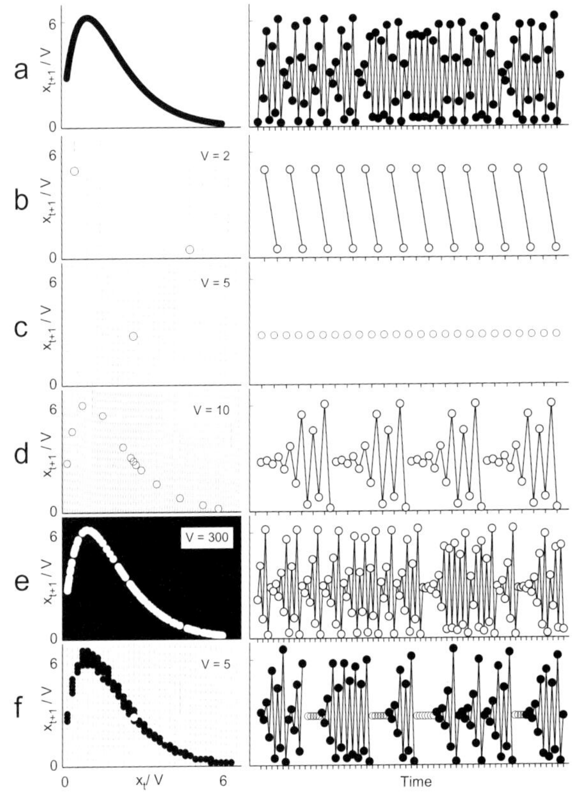

FIGURE 26. Lattice effects are illustrated in these plots using the Ricker model and its integerized version $x(t+1) = \text{round}\left[bx(t)e^{-cx(t)/V}\right]$. As V (habitat volume) increases, the mesh size of the lattice of densities x/V is refined. The plot (a) shows that continuous state space chaotic attractor for the Ricker model with $b = 17$ and $c = 1$. Plots (b)-(d) show the (density) lattice attractor of the integerized Ricker model obtained (in finite time) from the initial condition $x(0) = 5V$ for a sequence of habitat volumes V. The attractor in each case is a periodic cycle (case (c) is a 1-cycle or an equilibrium). On the lattice in plot (e) the 117-cycle lattice cycle resembles the chaotic attractor of the continuous state space in (a). Plot (f) shows a typical simulation of the environmental stochastic version of the lattice Ricker model in volume $V = 5$ when the lattice attractor is an equilibrium. In that plot we see episodes that resemble the equilibrium randomly interspersed with episodes that resemble the chaotic attractor. [From [**34**]. Reprinted with permission from AAAS]

For example, consider the following probabilistic model for the dynamics of a population with three life cycle stages (such as *Tribolium*) subject to demographic stochasticity [**20, 35**]:

$$\begin{array}{llll} & \text{(a)} & L_{t+1} & \sim \text{Poisson}\left(ba_t \exp\left(-\frac{c_{el}}{V}l_t - \frac{c_{ea}}{V}a_t\right)\right) \\ (3.13) & \text{(b)} & P_{t+1} & \sim \text{binomial}\left(l_t, 1-\mu_l\right) \\ & \text{(c)} & A_{t+1} & \sim \text{binomial}\left(p_t, \exp\left(-\frac{c_{pa}}{V}a_t\right)\right) + \text{binomial}\left(a_t, 1-\mu_a\right). \end{array}$$

Here "$\sim$" means "is a random variable distributed as". This model assumes mortality is described by a binomial random variable and fertility is described by a Poisson random variable. (The first equation involving larval recruitment involves both adult fertility and survivorship. The composition of a Poisson and a binomial random variable is a Poisson random variable.) If we extract a deterministic model from the Poisson-binomial LPA model (3.13) by choosing the means of each life cycle stage as the "most likely" observations at time $t+1$, then we obtain the deterministic, continuous state space LPA model (3.11). If instead we choose the modes as the most likely observations, then we obtain a deterministic, integer state space model

$$\begin{aligned} L_{t+1} &= \text{floor } \left[bA_t \exp\left(-\frac{c_{ea}}{V}A_t - \frac{c_{el}}{V}L_t\right)\right] \\ P_{t+1} &= \text{floor } \left[(1-\mu_l)(L_t+1)\right] \qquad (3.14) \\ A_{t+1} &= \text{floor } \left[(P_t+1)\exp\left(-\frac{c_{pa}}{V}A_t\right)\right] + \text{floor } \left[(1-\mu_a)(A_t+1)\right], \end{aligned}$$

which we call the mode-LPA model. It turns out that this integerized version of the LPA model also predicts a 6-cycle for the chaos treatment $c_{pa} = 0.35$, and simulations of the Poisson-binomial LPA model (3.13) typically exhibit randomly interspersed patterns of the chaotic attractor predicted by the continuous state space LPA model and this discrete state space 6-cycle.

A final point concerning the analysis of the chaotic treatment $c_{pa} = 0.35$ has to do with the dependency of the model predictions on the habitat size V. Notice first that in the continuous state space LPA model (3.11), V is simply a scaling factor. That is to say, if $(L(t), P(t), A(t))$ is an orbit for $V = 1$, then $(L(t)/V, P(t)/V, A(t)/V)$ is an orbit for any $V > 0$. As a result, orbits and attractors remain unchanged when plotted in density phase space $(L/V, P/V, A/V)$. This fact is not true of the mode-LPA model (3.14) or the Poisson/binomial LPA model (3.13).

If L, P, and A are integers, then the points $(L/V, P/V, A/V)$ in density phase space constitute a finite lattice on which densities orbits of the mode and Poisson/binomial LPA models reside. As V increases, the lattice becomes finer. For each value of V orbits of the mode-LPA model (3.14) attain a periodic cycle in finite time. As $V \to \infty$ we might expect that these lattice cycles converge to an attractor of the continuous state space model (3.11). This can be proved true when the attractor is a hyperbolic cycle. For a chaotic attractor, such as in the treatment $c_{pa} = 0.35$ of the route-to-chaos experiment, this fact has not been proved rigorously, although the numerical simulations in Figure 27 suggest it is true for the LPA model. These simulations suggest that as the habitat size increases lattice effects become less pronounced and the continuous state space attractor should becomes more visible. This suggests that a clearer view of the chaotic attractor in the $c_{pa} = 0.35$ treatment would be obtained if the experiment were conducted in larger volumes of media. ($V = 1$ corresponds to the experimental conditions in which 20g of flour media were used.)

There is another phenomena that also implies a clearer view of the chaotic attractor occurs in larger habitats. Consider the Poisson/binomial model (3.13). Each stage variable L, P and A are random variables. The coefficient of variation κ of a random variable measures the amount of variation relative to the mean; namely, $\kappa = s/m$ is the ratio of the standard deviation $s = \sqrt{v}$ to the mean m. For a Poisson

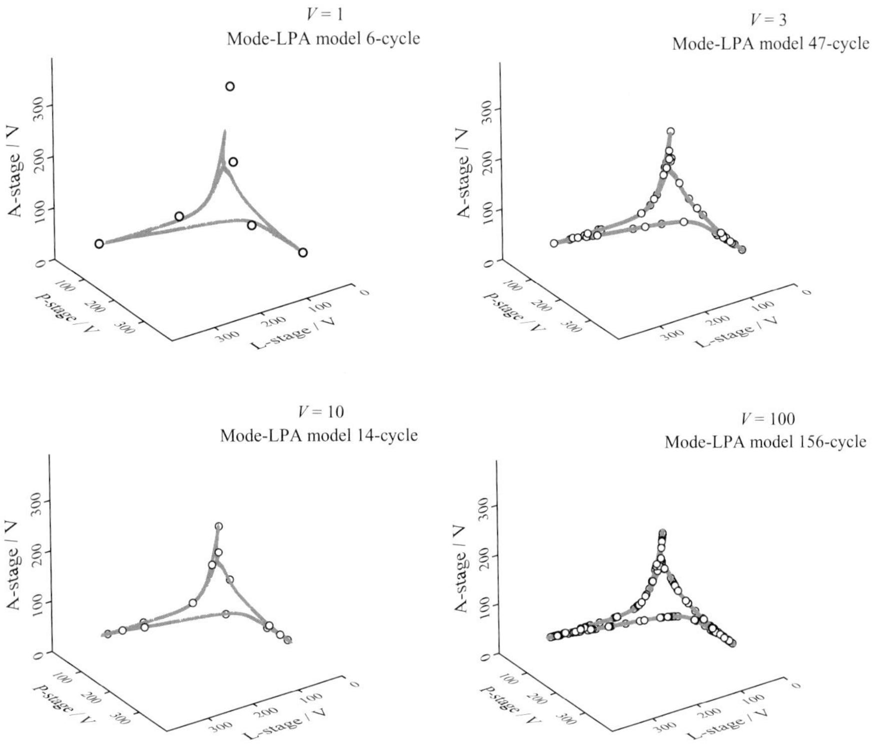

FIGURE 27. From the experimental initial conditions the orbit of the mode-LPA model (3.14) tends to different cycles for different habitat volumes V. Namely, for $V = 1$ the lattice attractor is a 6-cycle, whose points are shown in phase space in relation to the continuous state space chaotic attractor in the upper left plot. For $V = 3$ the lattice attractor is a 47-cycle; for $V = 10$ it is a 14-cycle; and for $V = 100$ it is a 156-cycle. Notice how the lattice cycle attractor seems to converge to the chaotic attractor as V increases.

random variable, the mean and variance are equal and hence $\kappa = 1/\sqrt{m}$. Consider the larval stage variable L in the Poisson-binomial LPA model. A comparison of the coefficient of variation κ_1 of the L component at a point (L, P, A) in volume $V = 1$ with that κ_V of the L component of the corresponding point (LV, PV, AV) in volume V follows from the relationship

$$\begin{aligned} Vm_1 &= VbA\exp\left[-c_{el}L - c_{ea}A\right] \\ &= b\,(AV)\exp\left[-\frac{c_{el}}{V}\,(VL) - \frac{c_{ea}}{V}\,(VA)\right] \\ &= m_V \end{aligned}$$

between the means m_1 and m_V and the calculation

$$\kappa_V = \frac{1}{\sqrt{m_V}} = \frac{1}{\sqrt{V}}\kappa_1.$$

It follows that the coefficient of variation decreases inversely with the square root of habitat volume. Similar calculations show that the same is true for the binomial random variables P and A.

Thus, as habitat volume V increases we expect the demographic noise in model (3.13) to decrease in the sense that the coefficient of variation decreases. This rule for demographic stochasticity is a familiar one to ecologists [**54**]. (It does not hold for environmental stochasticity.) The rule is usually expressed in terms of increased population size (carrying capacity) rather than habitat size, but, as we have observed, for our system population size scales with habitat size V.

For two reasons, namely decreased lattice effects and decreased demographic noise, the Poisson-binomial LPA model (3.13) predicts clearer deterministic signals in larger habitat sizes V. In particular, experimental data obtained from larger habitat volumes should more closely resemble the chaotic attractor in the $c_{pa} = 0.35$ treatment of the route-to-chaos experiment. See Figure 28. A recently conducted experiment [**21**] tested this prediction by repeating the treatment $c_{pa} = 0.35$ of

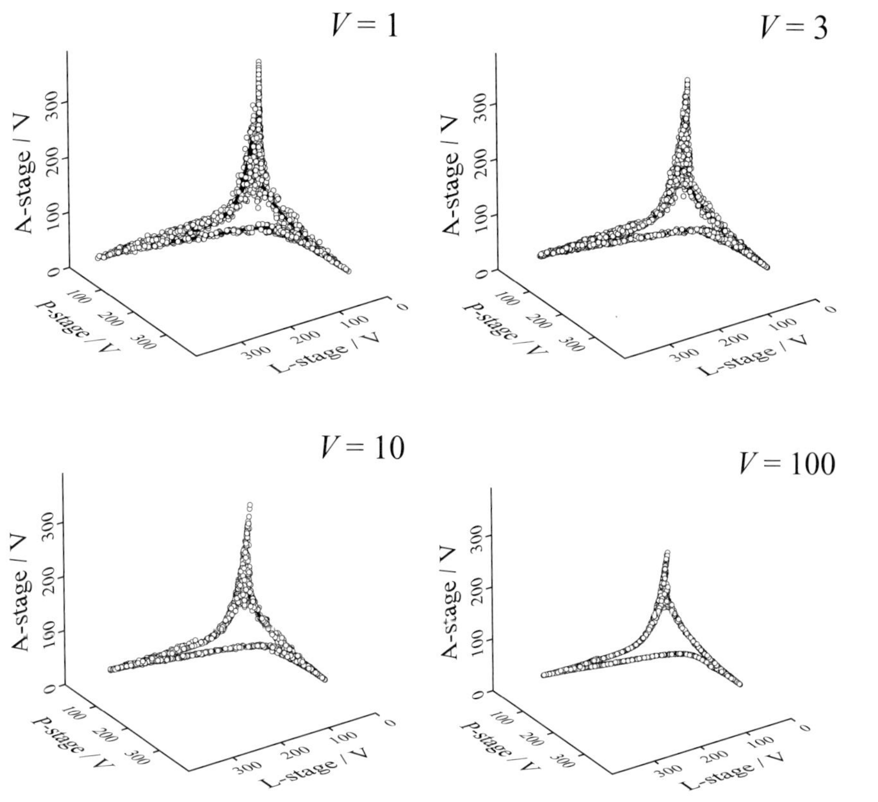

FIGURE 28. Sample simulations of the Poisson-binomial LPA model of length 2,000 show a convergence around the deterministic, continuous state space chaotic attractor predicted by the LPA model for the $c_{pa} =$ 0.35 treatment as habitat volume V Increases.

the hunt-for-chaos experiment in a habitat of $60g$, i.e., $V = 3$. Figure 29 shows the data obtained from the experiment in habitat size $V = 3$ together with that from habitat size $V = 1$ and compares both with the LPA model predicted chaotic attractor. The data from the larger habitat is clearly more tightly clustered around the chaotic attractor than the data from the smaller habitat. This observation can be quantified by calculating the mean distance δ_c of the data points from the

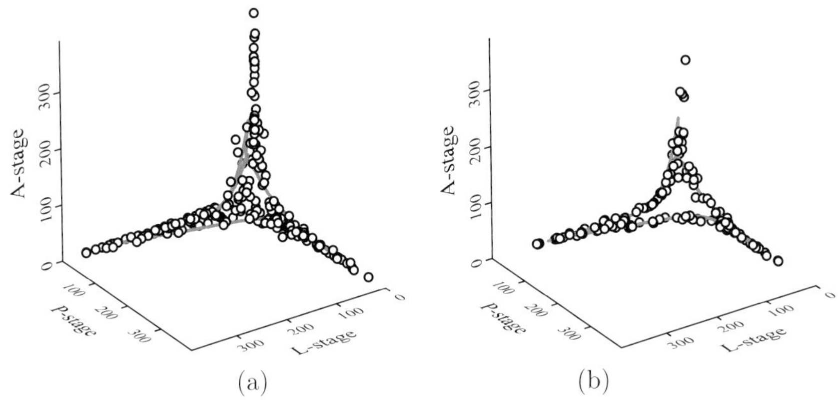

FIGURE 29. (a) The data from one replicate (open circles) of the chaos treatment $c_{pa} = 0.35$ is plotted together with the continuous state space chaotic attractor predicted by the LPA model (in gray). This experiment was conducted in 20g of medium, i.e., $V = 1$ in the LPA model. A follow-up experiment in 60g ($V = 3$) produced the data shown in (b).

chaotic attractor. This distance decreased from $\delta_c = 18.39$ in $V = 1$ to $\delta_c = 5.504$ in $V = 3$. We can also quantify the strengthening of the deterministic signal of the LPA model in the larger habitat by comparing the variances of the one-step residuals from both habitats. This variance decreased from 36.64 in $V = 1$ to 16.45 in $V = 3$, illustrating that the deterministic LPA model had considerably less error in its one-step predictions in the larger habitat.

In summary, in this lecture we considered methods that connect population data with matrix models, and with the theory presented in Lectures 1 and 2, by making use of stochastic versions of deterministic models. Case studies involving experimental cultures of flour beetles illustrate how one can use a parameterized model to design experimental protocols for the study dynamic bifurcations and in particular a route-to-chaos. That study led to new insights into the analysis of complex dynamics in a biological population: how deterministic attractors alone are not sufficient to explain observed dynamic patterns, even in highly controlled circumstances; how a stochastic mix of transients, attractors and unstable invariant sets can provide an adequate description of real data; how discrete state space lattice effects might play a role in observed patterns; and how habitat size plays a significant role in the expression of deterministic dynamic patterns.

EXERCISES

Exercise 14. Consider the environmental stochastic Ricker model

$$x(t+1) = bx(t)e^{-cx(t)}e^{E(t)}, \ b > 0, \ c > 0$$

where $E(t)$ is a normal random variable with mean 0 and variance $v > 0$. Given a set $y(0), y(1), \cdots, y(q)$ of $q+1$ observations, find algebraic formulas for the maximum likelihood estimates of b, c and v.

Exercise 15. In a simple population survivorship model we saw that the variance in the random number of survivors in one unit of time under demographic stochasticity was proportional to the population size n, while under environmental stochasticity it is proportional to n^2. For demographic stochasticity the variance was approximately stabilized by means of the transformation $g(n) = \sqrt{n}$, while under environmental stochasticity it was approximately stabilized by the transformation $g(n) = \ln n$. Suppose we characterize the variability in the presence of both types of stochasticity as a linear combination of n and n^2, i. e., $v(n) = k_1 n + k_2 n^2$. Calculate the variance stabilizing transformation $g(n)$. Choose the arbitrary constants involved in the calculation so that in the formula gives $\ln n$ when $k_1 \to 0$ and $\sqrt{n}$ when $k_2 \to 0$.

LECTURE 4

Periodically fluctuating environments

An interesting experiment involving *T. castaneum* was reported by D. Jillson in [**38**]. In this experiment the habitat volume V was not held fixed in time, but was periodically fluctuated. Jillson used various periodic schedules for V, but one particular schedule caused an unexpected result. All periodic schedules for V were designed to have an average equal to the habitat size in standard, non-fluctuating cultures, namely, the volume occupied by 20g of medium ($V = 1$ in the LPA model (3.11)). In those cultures in which Jillson oscillated the medium with period 2 (between 8g and 32g), the beetle populations markedly increased in numbers. This was not true for schedules involving other periods.

One reason Jillson's result was surprising is that it contradicted a tenet held at the time that fluctuations in habitat size should be deleterious to a population. By this is meant that the population's average abundance would be less in a fluctuating habitat than in it would be in a habitat held constant at the average of the fluctuations [**54**]. This tenet was based on properties of the logistic differential equation with a non-constant carrying capacity. It is also implied by the discrete logistic equation, as we can see by the following. Consider the discrete logistic equation (1.2)

$$x(t+1) = b\frac{1}{1+(b-1)\frac{1}{K}x(t)}x(t), \quad b > 1.$$

If the inherent growth rate b is greater than 1, then all solutions with $x(0) > 0$ tend to the carrying capacity $K > 0$. If the carrying capacity fluctuates periodically, i.e., if $K = K(t)$ is a p-periodic sequence, then

$$x(t+1) = b\frac{1}{1+(b-1)\frac{1}{K(t)}x(t)}x(t), \quad b > 1, \quad K(t+p) = K(t) > 0. \tag{4.1}$$

This p-periodically forced, discrete logistic equation still has the trivial (extinction) equilibrium $x = 0$, but no longer has a positive equilibrium (carrying capacity). Instead all solutions with $x(0) > 0$ tend to a positive p-cycle with an average less than that of $K(t)$.

To see this for $p = 2$, notice that the population density $y(t) = x(2t)$ monitored at even time steps satisfies the equation (obtained from the composition of the right hand side with itself)

$$y(t+1) = b^2\frac{1}{1+(b-1)\left(\frac{K(1)+bK(0)}{K(0)K(1)}\right)y(t)}y(t).$$

All solutions of this autonomous discrete logistic equation with $y(0) = x(0) > 0$ tend to the positive equilibrium, i.e.

$$\lim_{t\to+\infty} y(t) = (b+1)\frac{K(0)K(1)}{K(1)+bK(0)}.$$

A similar argument shows that that the population density $w(t) = x(2t+1)$ sampled at odd times tends to the positive equilibrium of

$$w(t+1) = b^2 \frac{1}{1+(b-1)\left(\frac{K(0)+bK(1)}{K(0)K(1)}\right)y(t)} w(t),$$

that is

$$\lim{}_{t\to+\infty} w(t) = (b+1)\frac{K(0)K(1)}{K(0)+bK(1)}.$$

It follows that all solutions of the 2-periodic logistic (4.1) with $x(0) > 0$ tend to the 2-cycle solution that oscillates between the two values

$$x(0) = (b+1)\frac{K(0)K(1)}{K(1)+bK(0)}, \qquad x(1) = (b+1)\frac{K(0)K(1)}{K(0)+bK(1)}.$$

The average of this 2-cycle

$$\frac{1}{2}(x(0)+x(1)) = (b+1)^2 \frac{K(0)K(1)}{(K(1)+bK(0))(K(0)+bK(1))}\frac{K(0)+K(1)}{2}$$

is strictly less than the average of the carrying capacity $(K(0)+K(1))/2$ if $K(t)$ is not constant, i.e., if $K(0) \neq K(1)$, since the inequality

$$(b+1)^2 \frac{K(0)K(1)}{(K(1)+bK(0))(K(0)+bK(1))} < 1$$

is equivalent to $0 < (K(1)-K(0))^2$. It follows, for a discrete logistically growing population in a habitat with a period $p = 2$ fluctuating carrying capacity, that the average asymptotic population size is less than the average of the carrying capacity.

In fact, it has been proved for all periods p that if $b > 1$ then all solutions of the p-periodic logistic (4.1) with $x(0) > 0$ tend to a positive p-cycle whose average is less than that of $K(t)$ (if $K(t)$ is not constant) [**25, 42**]. Such cycles are called *attenuant* [**30**]. Attenuant cycles have been studied in other models as well [**6, 30, 43, 44**].

It follows that Jillson's experimental observation cannot be explained by the periodically forced logistic equation (4.1). However, attenuance is not a universal property of periodically forced population models. Under some circumstances the opposite can occur, i.e., the average of an attracting p-cycle can exceed that of the carrying capacity, a property called *resonance* [**2**]. Since Jillson's experiment was carried out with populations of *T. castaneum* in periodically fluctuating volumes of culture medium, perhaps we can find an explanation using the LPA model (3.11). To do so, however, would involve letting the model parameter V fluctuate. For example, for Jillson's period 2 protocol we have

$$V = V_0\left(1+\alpha(-1)^t\right)$$

where the average and the relative amplitude are $V_0 = 1$ (the volume occupied by the standard 20g of medium) and $\alpha = 3/5$. Such a change in the LPA model makes the model non-autonomous, specifically, periodically forced with period 2.

The modeling methodology and results in Lectures 1 and 2 involve time autonomous matrix models. The entries in the fertility and transition matrices can change in time only through a dependence on the state variables and not an explicit dependence on t. Such autonomous models are not appropriate if, for example, a vital rate or a environmental parameter does not remain constant in time. This is the case for the LPA model with $V = V_0\left(1 + \alpha(-1)^t\right)$.

Before addressing the "Jillson effect", I give a brief overview of how to generalize the equilibrium theory for autonomous matrix models in Lectures 1 and 2 to a theory of periodic cycles for periodically forced matrix models.

It is necessary to begin with some preliminary facts about periodically forced linear matrix models. Let $Z \triangleq \{0, 1, 2, \cdots\}$ denote the non-negative integers. The set $S_{m,p}$ of p-periodic sequences in R^m $(1 \leq p \in Z)$ is a Hilbert space under the inner product $\langle x, y\rangle \triangleq \Sigma_{t=0}^{p-1} x^*(t)y(t)$ for x and $y \in S_{m,p}$. Let $S_{m,p}^+$ denote the set of positive p-periodic sequences in $S_{m,p}$, i.e., sequences for which $x(t) > 0$ for all $t \in Z$.

Consider the linear matrix equation

$$x(t+1) = P(t)x(t) + h(t), \quad t \in Z \tag{4.2}$$

where $h \in S_{m,p}$ and $P(t)$ is a p-periodic matrix: $P(t+p) = P(t)$, $t \in Z$. Of interest is the existence of p-periodic solutions of this equation. Solutions of (4.2) are provided by the (variation of parameters) formula

$$x(t) = \begin{cases} X(t,0)x(0) + \sum_{i=0}^{t-1} X(t,i+1)h(i) & \text{for } t = 1, 2, \cdots \\ x(0) & \text{for } t = 0 \end{cases} \tag{4.3}$$

where for $s \in Z$

$$X(t,s) \triangleq \begin{cases} P(t-1)P(t-2)\cdots P(s+1)P(s) & \text{for } t = s+1, s+2, \cdots \\ I & \text{for } t = s \end{cases}$$

is the *fundamental solution matrix* [**24**]. It is straightforward to see that a solution is p-periodic if and only if $x(0) = x(p)$, i.e., if and only if $x(0)$ solves the equation

$$\left(I - X(p,0)\right)x(0) = \sum_{i=0}^{p-1} X(t,i+1)h(i). \tag{4.4}$$

If the associated homogeneous system

$$\left(I - X(p,0)\right)x(0) = 0$$

has no nontrivial solution, that is to say if the homogeneous matrix equation

$$x(t+1) = P(t)x(t) \tag{4.5}$$

has no nontrivial p-periodic solution, then there is a unique solution

$$x(0) = \left(I - X(p,0)\right)^{-1} \sum_{i=0}^{p-1} X(t,i+1)h(i)$$

of the periodicity condition (4.4) and therefore a unique p-periodic solution of (4.2). This solution is given by the formula

$$x(t) = X(t,0)\left(I - X(p,0)\right)^{-1} \sum_{i=0}^{p-1} X(t,i+1)h(i) + \sum_{i=0}^{t-1} X(t,i+1)h(i). \tag{4.6}$$

The assumption that the homogeneous equation (4.5) has no nontrivial p-periodic solution is equivalent to the nonsingularity of the matrix $I - X(p,0)$, i.e., that $\lambda = 1$ is not an eigenvalue of $X(p,0)$.

If, on the other hand, the homogeneous equation (4.5) has nontrivial p-periodic solutions, i.e., if 1 is an eigenvalue of $X(p,0)$, then equation (4.4) has a solution if and only if the right hand side is orthogonal to the kernel of the transpose, that is,

$$l\textstyle\sum_{i=0}^{p-1} X(t,i+1)h(i) = 0 \tag{4.7}$$

for all left eigenvectors l of $X(p,0)$ associated with eigenvalue 1.

If 1 is not an eigenvalue of $X(p,0)$, then we can rewrite the formula for the unique p-periodic solution (4.6) as

$$x(t) = \textstyle\sum_{i=0}^{p-1} G(t,i)h(i)$$

where

$$G(t,i) = \begin{cases} (I - X(p,0))^{-1} X(p,i+1) + X(t,i+1) & \text{for } 0 \le i < t \\ (I - X(p,0))^{-1} X(p,i+1) & \text{for } 0 \le t \le i \le p-i \end{cases}$$

is a *Green's function.* This formula defines a solution operator

$$G : S_{m,p} \to S_{m,p}$$

which is linear and bounded (and therefore compact, since $S_{m,p}$ is finite dimensional). For $h \in S_{m,p}$, $x = Gh$ is the unique p-periodic solution of the nonhomogeneous, linear matrix equation (4.2).

The first goal is to establish generalizations of the bifurcation Theorems 2.1 and 2.3 that apply to periodically forced matrix equations. Consider a periodically forced version of the nonlinear matrix equation (2.7)

$$x(t+1) = (A(t) + \mu B(t) + H(t,\mu,x(t)))\, x(t) \tag{4.8}$$

where μ is a real parameter. Here $A(t)$ and $B(t)$ are p-periodic matrices and the components h_{ij} of the matrix $H(t,\mu,x) = (h_{ij}(t,\mu,x))$ are p-periodic in t, are k times differentiable, and are higher order in x near $x = 0$, that is[1]

$$\begin{gathered} h_{ij} \in C^k \left(Z \times R^1 \times R^m \to R^1\right) \text{ for some } k \in Z \\ |h_{ij}(t,\mu,x)| = O\left(|x|\right) \text{ near } x = 0 \\ h_{ij}(t+p,\cdot,\cdot) = h_{ij}(t,\cdot,\cdot). \end{gathered} \tag{4.9}$$

Consider the linear equation (the linearization of (4.8) at $x = 0$):

$$x(t+1) = (A(t) + \mu B(t))\, x(t). \tag{4.10}$$

A real number μ_c is a *characteristic value* if (4.10), with $\mu = \mu_c$, has a nontrivial p-periodic solution $x \in S_{m,p}$. Let $X_\mu(t,x)$ denote the fundamental matrix of (4.10). A characteristic value is *simple* if $\dim\ker\left(I - X_{\mu_c}(p,0)\right) = 1$, i.e., 1 is an eigenvalue of $X_{\mu_c}(p,0)$ with (geometric) multiplicity 1. This means there is only one independent nontrivial p-periodic solution.

Suppose $\mu = 0$ is not a characteristic value of (4.10) and let G be the Green's function of $x(t+1) = A(t)x(t)$. Solving the periodically forced, nonlinear matrix equation (4.8) for a p-periodic solution $x \in S_{m,p}$ is equivalent to solving the equation

$$x = \mu Lx + g(\mu,x) \tag{4.11}$$

where $L \triangleq GB$ is a linear operator and $g : R^1 \times S_{m,p} \to S_{m,p}$, defined by $g(\mu,x) \triangleq GH(t,\mu,x)x$, is $O\left(|x|^2\right)$ near $x = 0$ in $S_{m,p}$.

[1] $|x| \triangleq \langle x,x\rangle^{1/2}$.

Define $\mathcal{S}$ to be the set of all nontrivial p-periodic solution pairs $(\mu, x) \in R \times S_{m,p}$, $x \neq 0$, of (4.11) (equivalently (4.8)). If $(\mu_c, 0) \in \bar{\mathcal{S}}$ (the closure of $\mathcal{S}$), then $(\mu_c, 0)$ is a bifurcation point. A solution pair (μ, x) is *positive* if $x(t)$ for all t (i.e., if $x \in S^+_{m,p}$).

A necessary condition that $(\mu_c, 0)$ is a bifurcation point is that μ_c be a characteristic value of L (a reciprocal of a nonzero eigenvalue), that is to say, $x = \mu_c Lx$ for some $0 \neq x \in S_{m,p}$. By definition, a characteristic value of the linear operator L is a characteristic value of (4.10). If μ_c is a characteristic value of L, a *characteristic solution* x is a nontrivial p-periodic solution of the linear equation (4.10), namely, $x = X_{\mu_c}(t, 0)v_0$ where v_0 is a right eigenvector of $X_{\mu_c}(p, 0)$ associated with eigenvalue 1.

The following theorem results from the well-known global bifurcation theorems of Rabinowitz applied to the operator equation (4.11) [**39, 61**].

Theorem 4.1. *In the matrix equation (4.8) assume A and B are p-periodic matrices and $H = (h_{ij})$ satisfies (4.9). Suppose*

(1) $\mu = 0$ is not a characteristic value of the linearization (4.10);

(2) μ_c is a simple characteristic value of (4.10) for which there is an associated positive characteristic p-periodic solution[2].

Then there exists a continuum $\mathcal{C}^+$ in $\bar{\mathcal{S}}$ that contains the bifurcation point $(\mu_c, 0)$ and satisfies one of the following alternatives:

(a) $\mathcal{C}^+ / \{(\mu_c, 0)\}$ contains only positive p-periodic solution pairs (μ, x) for which $\mu > 0$ and $\mathcal{C}^+$ is unbounded (in $R \times S^+_{m,p}$);

(b) $\mathcal{C}^+ / \{(\mu_c, 0)\}$ contains a nontrivial, non-negative p-periodic solution pair (μ^*, x^*) for which $x^*(t) \geq 0$ has a 0 component at some time t;[3]

(c) $\mathcal{C}^+ / \{(\mu_c, 0)\}$ contains a point $(\mu_c^*, 0)$ where $\mu_c^* \neq \mu_c$ is a characteristic value of (4.10) associated with a non-negative characteristic p-periodic solution.[4]

Alternative (b) says that the continuum of nontrivial equilibria bifurcating from $(\mu_c, 0)$ leaves the positive cone. Often in applications one can rule this alternative out by showing that no nontrivial p-periodic solution can lie on the boundary of the positive cone; that is to say, in many applications one can show that

$$\text{if } x(t) \geq 0 \text{ solves (4.8) then either } x \equiv 0 \text{ or } x(t) > 0 \text{ for all } t. \tag{4.12}$$

Alternative (c) says that the continuum of nontrivial equilibria bifurcating at the characteristic value μ_c also bifurcates from another characteristic value $\mu_c^* \neq \mu_c$ of (4.10) that is associated with a non-negative characteristic vector. This alternative is ruled out, of course, if no other characteristic value of L is associated with a non-negative characteristic vector (as is the case, for example, when the Perron-Frobenius Theorem applies).

Theorem 4.2 ([**7**]). *In the matrix equation (4.8) assume A and B are p-periodic matrices and $H = (h_{ij})$ satisfies (4.9). Assume*

(1) $\mu = 0$ is not a characteristic value of the linearization (4.10);

(2) μ_c is a simple characteristic value of (4.10) for which there is an associated positive characteristic p-periodic solution;

---

[2]That is, $X_{\mu_c}(p, 0)$ has eigenvalue 1 and an associated positive eigenvector $v_0 > 0$.

[3]That is to say, x^* lies on the boundary of the positive cone $S^+_{m,p}$.

[4]That is to say, $X_{\mu_c^*}(p, 0)$ has eigenvalue 1 and an associated non-negative eigenvector $v_0^* \geq 0$.

(3) (4.10) has no other characteristic value with a non-negative p-periodic solution;

(4) condition (4.12) holds.

Then there exists an unbounded continuum $\mathcal{C}^+$ in $\bar{S}$ such that $\mathcal{C}^+/\{(\mu_c, 0)\}$ contains only positive p-periodic solution pairs (μ, x) with $\mu > 0$. Moreover, $\mathcal{C}^+$ is unbounded (in $R \times S^+_{m,p}$).

That $\mathcal{C}^+$ is unbounded means that either the set of positive p-cycles associated with $\mathcal{C}^+$ is unbounded or the spectrum $\sigma(\mathcal{C}^+) \triangleq \{\mu : (\mu, x) \in \mathcal{C}^+\}$ is unbounded (or both).

Stability properties of cycles of p-periodic matrix models are defined in terms of the stability properties of fixed points of the ($p-1$ fold) composite of the equation. A solution $x(t)$ of a p-periodic equation $x(t+1) = f(t, x(t))$ (i.e., $f(t, x) = f(t+p, x)$ for all t and x) is a p-cycle if and only if $x(0) = x(p)$, that is to say, if and only if $x(0)$ is an equilibrium (fixed) point of the composite map $f^{(p-1)}(x) \triangleq f(p-1, f(p-2, \dots, f(1, f(0, x))))$. A p-cycle $x_p(t)$ is *stable* if $x_p(0)$ is a stable equilibrium of the autonomous (composite) equation $y(t+1) = f^{(p-1)}(y(t))$. Similarly, a p-cycle is an attractor if $x_p(0)$ is an attractor of the composite equation, and a p-cycle is asymptotically stable if $x_p(0)$ is an asymptotically stable equilibrium of the composite equation.

The fundamental solution matrix $X(t, s)$ of the linearization (4.10) at $x = 0$ of the nonlinear, periodically forced matrix model (4.8) depends on μ. A subscript indicates this dependence:

$$X_\mu(p, 0) = \prod_{i=1}^{p} \left(A(p-i) + \mu B(p-i) \right). \tag{4.13}$$

A stability analysis of the equilibrium $x = 0$ of (4.8) is possible by means of the linearization principle [**24**]. Define

$$\delta \triangleq l_0 \frac{d}{d\mu} X_\mu \left. (p, 0) \right|_{\mu=\mu_c} v_0 \tag{4.14}$$

where l_0 and v_0 are the left and right eigenvalues of $X_{\mu_c}(p, 0)$ normalized so that $l_0 v_0 = 1$. The following theorem gives conditions under which the trivial (extinction) equilibrium $x = 0$ of the nonlinear, periodically forced equation (4.8) loses stability at the bifurcation point in Theorem 4.1.

Theorem 4.3 ([**29**]). *In the matrix equation (4.8) assume A and B are p-periodic matrices and $H = (h_{ij})$ satisfies (4.9) with $k \geq 1$. Suppose μ_c is a simple characteristic value of (4.10) with the property that 1 is a strictly dominant eigenvalue of $X_{\mu_c}(p, 0)$ and let l_0 and v_0 denote left and right eigenvectors of $X_{\mu_c}(p, 0)$ normalized so that $l_0 v_0 = 1$. Further, suppose 0 is not an eigenvalue of $X_{\mu_c}(p, 0)$.*

If $\delta > 0$ then the trivial solution $x = 0$ of the nonlinear, periodically forced equation (4.8) loses stability as μ increases through μ_{cr}.

If $\delta < 0$ then $x = 0$ gains stability as μ increases through μ_{cr}.

An analysis of the properties of the positive p-cycles near the bifurcation point $(\mu_c, 0)$ guaranteed by Theorem 4.1 is possible by means of a parametrization of the bifurcating branch. To do this we represent the bifurcating positive, p-cycle pairs $(\mu, x(t))$ as functions of a branch parameter ε and calculate the lowest order terms in the expansions

$$x(t) = x_1(t)\varepsilon + O(\varepsilon^2) \tag{4.15}$$
$$\mu = \mu_c + \mu_1\varepsilon + O(\varepsilon^2)$$

(under the assumption that $k \geq 2$ in (4.9)). These expansions provide approximations to the positive p-cycles near the bifurcation point and allow us to make approximations to the Jacobian of the composite and its dominant eigenvalue

$$\lambda = 1 + \varepsilon\lambda_1 + O(\varepsilon^2),$$

which determines the stability of the cycles. Here $x_1(t)$ is the positive p-cycle solution of the linearization (4.10) with $\mu = \mu_{cr}$, namely

$$x_1(t) = X_{\mu_c}(t, 0)v_0$$

and the positive p-cycles on the branch correspond to $\varepsilon > 0$.

To calculate the coefficients in (4.15) we substitute these expansions into the nonlinear equation (4.8) and equate coefficients of like powers of ε from both sides of the resulting expressions. This leads to linear matrix equations from which, with the aid of the orthogonality condition (4.7), we can calculate formulas for μ_1 and λ_1.The results are as follows [**29**].

Define $\gamma_{ij}(t)$ to be the gradient of h_{ij} with respect to x evaluated at $(\mu, x) = (\mu_c, 0)$:

$$\gamma_{ij}(t) \triangleq \nabla_x h_{ij}(t, \mu, x)|_{(\mu,x)=(\mu_c,0)}$$

and define the scalars

$$d_{ij}(t) \triangleq \gamma_{ij}^*(t)x_1(t).$$

Form the matrix $D(t) = (d_{ij}(t))$ and define

$$\kappa \triangleq -l_0\textstyle\sum_{t=0}^{p-1} X_{\mu_c}(p, t+1)D(t)X_{\mu_c}(t, 0)v_0.$$

Then it turns out

$$\mu_1 = \frac{1}{2}\frac{\kappa}{\delta}, \qquad \lambda_1 = -\frac{1}{2}\kappa \tag{4.16}$$

where δ is given by (4.14). Note that the bifurcation is supercritical if $\mu_1 > 0$ and subcritical if $\mu_1 < 0$. The dominant eigenvalue λ is less than 1 (for $\varepsilon > 0$ small) if $\mu_1 > 0$ and greater than 1 if $\mu_1 < 0$.

The bifurcation at $\mu = \mu_c$ described in Theorem 4.1 is called stable if the p-periodic solutions from the positive solutions pairs near the bifurcation point $(\mu, x) = (\mu_c, 0)$ are (locally asymptotically) stable. If these positive periodic solutions are unstable, then the bifurcation is unstable.

Theorem 4.4 ([**29**]). *In addition to the assumptions in Theorem 4.3 assume* $H = (h_{ij})$ *satisfies (4.9) with* $k \geq 2$ *and that* $\kappa \neq 0$.

Suppose $\delta > 0$, *i.e., the trivial (extinction) equilibrium* $x = 0$ *loses stability as* μ *increases through* μ_c. *Then for* μ *sufficiently close to* μ_c *the bifurcation of positive p-cycles is supercritical and stable if* $\kappa > 0$ *and is subcritical and unstable if* $\kappa < 0$.

Suppose $\delta < 0$, *i.e., the trivial (extinction) equilibrium* $x = 0$ *loses stability as* μ *decreases through* μ_c. *Then for* μ *sufficiently close to* μ_c *the bifurcation of positive p-cycles is supercritical and stable if* $\kappa < 0$ *and is subcritical and unstable if* $\kappa > 0$.

Theorems 4.1, 4.3 and 4.4 are generalizations to periodically forced equations of Theorems 2.1, 1.7, and 2.3 for autonomous equations. They provide some general conditions under which a periodic matrix equation has a positive p-cycle and conditions under which it is stable or unstable. The expansions (4.15) also provide a way to study the properties of the p-cycles (averages, amplitudes, phases, etc.), at least near the bifurcation point. If need be we can also calculate (by the same procedure described above) the expansion of the p-cycle to higher order in ε :

$$x(t) = x_1(t)\varepsilon + x_2(t)\varepsilon^2 + O\left(\varepsilon^3\right).$$

Example 4.5. *The matrix model*

$$\begin{pmatrix} J(t+1) \\ A(t+1) \end{pmatrix} = \begin{pmatrix} 0 & b\frac{1}{1+A(t)} \\ \tau_{21} & \tau_{22} \end{pmatrix} \begin{pmatrix} J(t) \\ A(t) \end{pmatrix}$$

is a variant of the discrete logistic model in which a juvenile stage J is included. With constant coefficients b (inherent juvenile recruitment rate per adult), τ_{21} (juvenile survivorship) and τ_{22} (adult survivorship), the model is autonomous and a (global) continuum of positive equilibria bifurcates from the critical value $b_c = (1-\tau_{22})/\tau_{21}$ (or equivalently at the critical value 1 of the inherent net reproductive number $n = b\tau_{21}/(1-\tau_{22})$).

Suppose instead that the recruitment rate b oscillates periodically with period 2. Suppose this oscillation has mean μ and amplitude α so that

$$b = \mu\left(1 + \alpha\left(-1\right)^t\right), \qquad 0 \leq \alpha < 1.$$

The model takes the form (4.8) with

$$A(t) = \begin{pmatrix} 0 & 0 \\ \tau_{21} & \tau_{22} \end{pmatrix}, \qquad B(t) = \begin{pmatrix} 0 & 1+\alpha(-1)^t) \\ 0 & 0 \end{pmatrix}$$

$$H = \begin{pmatrix} 0 & -\mu(1+\alpha(-1)^t)\frac{A}{1+A} \\ 0 & 0 \end{pmatrix}.$$

From (4.13) we find

$$X_\mu(2,0) = \begin{pmatrix} 0 & \mu(1+\alpha) \\ \tau_{21} & \tau_{22} \end{pmatrix} \begin{pmatrix} 0 & \mu(1-\alpha) \\ \tau_{21} & \tau_{22} \end{pmatrix}$$
$$= \begin{pmatrix} \tau_{21}\mu\left(1+\alpha\right) & \tau_{22}\mu\left(1+\alpha\right) \\ \tau_{22}\tau_{21} & \tau_{22}^2 + \tau_{21}\mu\left(1-\alpha\right) \end{pmatrix}$$

is a positive matrix. Perron's Theorem implies that $X_\mu(2,0)$ has a positive, strictly dominant simple eigenvalue λ with positive right and left eigenvectors $v_0 > 0$ and $l_0 > 0$ and that no other eigenvalue has a non-negative eigenvector [**26, 37**]. *Specifically, the positive eigenvalue is*

$$\lambda = \tau_{21}\mu + \frac{1}{2}\tau_{22}^2 + \frac{1}{2}\sqrt{\tau_{22}^4 + 4\tau_{22}^2\tau_{21}\mu + 4\tau_{21}^2\mu^2\alpha^2}.$$

This eigenvalue equals 1 when the mean birth rate μ equals the critical value

$$\mu_c \triangleq \frac{1-\sqrt{\left(1-\alpha^2\right)\tau_{22}^2+\alpha^2}}{\tau_{21}\left(1-\alpha^2\right)}.$$

From (4.14) we calculate

$$\delta = l_0\frac{d}{d\mu}\left. X_\mu(2,0)\right|_{\mu=\mu_c} v_o = l_0 \begin{pmatrix} \tau_{21}\left(1+\alpha\right) & \tau_{22}\left(1+\alpha\right) \\ 0 & \tau_{21}\left(1-\alpha\right) \end{pmatrix} v_0 > 0.$$

Theorem 4.3 implies that the extinction equilibrium $J = A = 0$ *loses stability as the average recruitment rate increases through the critical value* μ_c. *Theorem 4.2 implies the existence of an unbounded continuum of positive 2-cycles with a spectrum (interval) of positive average recruitment rates* $\mu > 0$ *that bifurcates from the extinction equilibrium at the critical value* μ_c. *Further calculations show*

$$D(0) = \begin{pmatrix} 0 & -\mu_c(1+\alpha) \\ 0 & 0 \end{pmatrix}, \qquad D(1) = \begin{pmatrix} 0 & -\mu_c(1-\alpha)(\tau_{21}+\tau_{22}) \\ 0 & 0 \end{pmatrix}$$

and

$$\begin{aligned}\kappa &= -l_0\left[(A(1)+\mu_c B(1))\, D(0) + D(1)\,(A(0)+\mu_c B(0))\right] v_0 \\ &= \frac{2}{3}\tau_{21}\mu_c(1+\alpha) + \frac{1}{3}\mu_c(\tau_{22}+\tau_{21})^2(1-\alpha).\end{aligned}$$

Since $\kappa > 0$, *Theorem 4.4 implies the bifurcation at* μ_c *is supercritical and stable.*

As a numerical example, take $\tau_{21} = \tau_{22} = 1/2$ *and* $\alpha = 1/10$. *Then*

$$\mu_c = \frac{10}{99}\left(20 - \sqrt{103}\right) \approx 0.9951$$

and

$$X_{\mu_c}(2,0) = \begin{pmatrix} \frac{1}{2} & \frac{1}{2} \\ \frac{1}{4} & \frac{3}{4} \end{pmatrix}$$

has a dominant eigenvalue 1 *with positive eigenvectors*

$$v_0 = \begin{pmatrix} 1 \\ 1 \end{pmatrix}, \qquad l_0 = \begin{pmatrix} \frac{1}{3} & \frac{2}{3} \end{pmatrix}$$

that satisfy $l_0 v_0 = 1$. *Then* $\delta = 2/3$ *and* $\kappa = 20\left(20 - \sqrt{103}\right)/297 \approx 0.6634$.

It is interesting to note that the bifurcation point μ_c *is a decreasing function of the amplitude* α. *This means that a lower average recruitment rate* μ *is needed for population survival in the periodically fluctuating habitat than in the constant habitat. In this sense, an oscillatory habitat is advantageous to the population. Although the decrease in* μ_c *in the numerical example above is not large (from* 1 *to* 0.9951*) if the amplitude is changed to* $\alpha = 9/10$ *the decrease is over 20%, namely from* 1 *to* 0.7788. *See Figure 30.*

In [**31**] Theorems 4.1, 4.3 and 4.4 are applied to the LPA model with a 2-periodic habitat volume V with average $V_0 = 1$ and amplitude α $(0 < \alpha < 1)$:

$$\begin{aligned} L(t+1) &= bA(t)\exp\left(-\frac{c_{el}}{1+\alpha(-1)^t}L(t) - \frac{c_{ea}}{1+\alpha(-1)^t}A(t)\right) \\ P(t+1) &= (1-\mu_l)\, L(t) \qquad (4.17) \\ A(t+1) &= P(t)\exp\left(-\frac{c_{pa}}{1+\alpha(-1)^t}A(t)\right) + (1-\mu_a)\, A(t). \end{aligned}$$

With $\mu = b$, this model has the form (4.8) and satisfies the hypotheses of Theorem 4.3 (for any integer k) with

$$X_b(p,0) = \begin{pmatrix} 0 & b & b(1-\mu_a) \\ 0 & 0 & b(1-\mu_l) \\ 1-\mu_l & 1-\mu_a & (1-\mu_a)^2 \end{pmatrix}$$

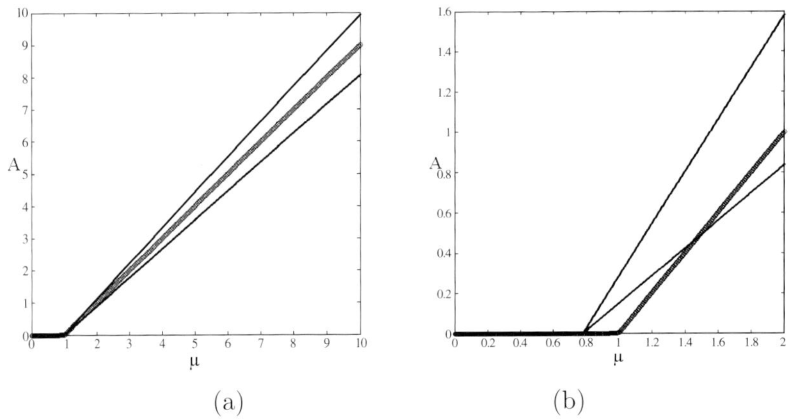

FIGURE 30. The plots show bifurcation diagrams for the 2-periodic, juvenile-adult logistic model in Example 4.5 with $\tau_{21} = \tau_{22} = 1/2$. The adult component A of the stable, positive 2-cycle is plotted against the average recruitment rate μ. (a) The thin lines are the maximum and minimum of the bifurcating 2-cycles when $\alpha = 1/10$. The thick line is the branch of bifurcating equilibria when $\alpha = 0$. (b) When $\alpha = 9/10$ the bifurcation point μ_c for the 2-cycles is less than it is for the equilibrium case $\alpha = 0$.

and

$$b_c = \frac{\mu_a}{1 - \mu_l}.$$

(Equivalently we could use $n = b(1 - \mu_l)/\mu_a$ as the bifurcation parameter μ, in keeping with the general equilibrium theory in Lectures 1 and 2, in which case $\mu_c = 1$, but we follow the analysis in [**31**].) The matrix

$$X_{b_c}(2,0) = \begin{pmatrix} 0 & \frac{\mu_a}{1-\mu_l} & \frac{\mu_a}{1-\mu_l}(1-\mu_a) \\ 0 & 0 & \frac{\mu_a}{1-\mu_l}(1-\mu_l) \\ 1-\mu_l & 1-\mu_a & (1-\mu_a)^2 \end{pmatrix}$$

has two complex eigenvalues $\mu_a(\mu_a - 2)/2 \pm i\mu_a\sqrt{\mu_a(4-\mu_a)}/2$ of magnitude $\mu_a < 1$ and a dominant eigenvalue 1 with eigenvectors

$$l_0 = \begin{pmatrix} \frac{1-\mu_l}{1+2\mu_a} & \frac{1}{1+2\mu_a} & \frac{1}{1+2\mu_a} \end{pmatrix}, \qquad v_0 = \begin{pmatrix} b_c \\ \mu_a \\ 1 \end{pmatrix}.$$

A calculation shows

$$\delta = l_0 \frac{d}{db} X_b(2,0)|_{b=b_c} v_0 = \frac{2(1-\mu_l)}{1+2\mu_a} > 0$$

and hence Theorem 4.3 implies the extinction equilibrium $x = \mathrm{col}(L, P, A) = \mathrm{col}(0, 0, 0)$ loses stability as b increases through the critical value b_c. In fact it is shown in [**31**] that the extinction equilibrium $x = 0$ is a global attractor for $b < b_c$ and the periodic LPA model (4.17) is uniformly persistent for $b > b_c$.

It is straightforward to show that if a component of a 2-cycle solution of (4.17) equals 0 at some time t, then all three components equal 0 for all time t, i.e., such a 2-cycle must in fact be the extinction equilibrium $x = 0$. As a result condition (4.12) holds and Theorem 4.2 implies the existence of an unbounded continuum of positive 2-cycles that bifurcates from $x = 0$ at $b = b_c$. Since $x = 0$ is a global attractor for $b < b_c$ it follows that the bifurcation is supercritical. This, and the stability of the bifurcating branch of positive 2-cycles near b_c, also follows from Theorem 4.4 provided $\kappa \neq 0$. In fact a calculation shows

$$\kappa = \frac{4\mu_a}{(1-\alpha^2)(1+2\mu_a)} > 0$$

which proves the bifurcation is supercritical and stable.

The expansions (4.15) permit some analysis of the bifurcating 2-cycles and, in particular, imply they are attenuant. A calculation of the expansions for both the bifurcating 2-cycles $(L_\alpha(t), P_\alpha(t), A_\alpha(t))$ and the bifurcating equilibria (L_0, P_0, A_0) (Corollary 2.1 and Theorem 2.3) shows this. Let ε_α and ε_0 denote the expansion parameters for the branches respectively. Then [**31**]

$$\begin{aligned} L_\alpha(t) &= b_c\varepsilon_\alpha + O\left(\varepsilon_\alpha^2\right)(t) \\ P_\alpha(t) &= \mu_a\varepsilon_\alpha + O\left(\varepsilon_\alpha^2\right)(t) \\ A_\alpha(t) &= \varepsilon_\alpha + O\left(\varepsilon_\alpha^2\right)(t) \\ b &= b_c + \left(\frac{b_c\left(c_{ea} + b_c c_{el} + c_{pa}\right)}{1-\alpha^2}\right)\varepsilon_\alpha + O\left(\varepsilon_\alpha^2\right) \end{aligned}$$

for $\varepsilon_\alpha > 0$ small and

$$\begin{aligned} L_0 &= b_c\varepsilon_0 + O\left(\varepsilon_0^2\right) \\ P_0 &= \mu_a\varepsilon_0 + O\left(\varepsilon_0^2\right) \\ A_0 &= \varepsilon_0 + O\left(\varepsilon_0^2\right) \\ b &= b_c + b_c\left(c_{ea} + b_c c_{el} + c_{pa}\right)\varepsilon_0 + O\left(\varepsilon_0^2\right) \end{aligned}$$

for $\varepsilon_0 > 0$ small. A comparison of the averages of the 2-cycle with the equilibrium at the same value of b, to first order, requires $\varepsilon_\alpha = (1-\alpha^2)\varepsilon_0$. Then

$$\begin{aligned} L_0 - \langle L_\alpha(t)\rangle &= \alpha^2 b_c\varepsilon_0 + O\left(\varepsilon_0^2\right) > 0 \\ P_0 - \langle P_\alpha(t)\rangle &= \alpha^2\mu_a\varepsilon_0 + O\left(\varepsilon_0^2\right) > 0 \\ A_0 - \langle A_\alpha(t)\rangle &= \alpha^2\varepsilon_0 + O\left(\varepsilon_0^2\right) > 0 \end{aligned}$$

for $\varepsilon_0 > 0$ small. Here $\langle L_\alpha(t)\rangle \triangleq \left(L_\alpha(0) + L_\alpha(1)\right)/2$ denotes the average of the L-stage component of the 2-cycles, etc.

Note that all three life cycle stages are attenuant and, as a consequence, the total population size $L(t) + P(t) + A(t)$ is also attenuant. It follows that near the bifurcation point b_c the bifurcating 2-cycles *cannot* explain the Jillson effect observed in the *T. castaneum* experiment. If resonance is to occur in the periodic LPA model, it must occur for larger values of b.

One approach that has been used to study periodic solutions of periodically forced difference equations utilizes perturbation theory. This approach views a periodic solution as a function of a small parameter appearing in the model equations, which are known to have a solution of a particular type (e.g., an equilibrium or a cycle) when the parameter equals 0. The periodic solutions when the parameter is small (but not equal to 0) are considered perturbations of this known solution. Taylor expansions with respect to the small parameter lead to useful approximations of the perturbed solutions.

For example, consider the relative amplitude α as a small parameter in the periodic LPA model (4.17). When $\alpha = 0$ the model is autonomous and has (for $b > b_c$) a positive equilibrium. We expect this equilibrium to perturb to a 2-cycle for $\alpha > 0$. Relevant mathematical questions are: does such a perturbed 2-cycle exist? when is it stable? is it resonant or attenuant?

A study of these questions for a general class of difference equations with periodic coefficients appears in [**30**]. Consider the equation $x(t+1) = F(c, x(t))$ in which a coefficient c has been selected to oscillate periodically with an amplitude α around an average which can be taken, without loss in generality, to be 1. Consider the periodically forced equation

$$x(t+1) = F(1 + \alpha\beta(t), x(t)) \tag{4.18}$$

where β is a p-periodic sequence with mean $\langle\beta\rangle = 0$. Assume

$$F : R \times R^m \to R^m \text{ is } k \geq 2 \text{ times continuously differentiable.} \tag{4.19}$$

p-periodic solutions of (4.18) satisfy the operator equation

$$K(\alpha, x) = 0 \tag{4.20}$$

where the operator $K : R \times S_{m,p} \to S_{m,p}$ is defined by

$$K(\alpha, x) \triangleq \{x(t+1) - F(1 + \alpha\beta(t), x(t))\}_{t \in Z} \in S_{m,p}.$$

The Implicit Function Theorem applies to equation (4.20) and yields a solution $(\alpha, x(\alpha))$, $x(\alpha) \in S_{m,p}$, in the neighborhood of a known solution $(0, x(0))$ provided the derivative of K with respect to x at this point is nonsingular. Specifically, suppose the autonomous equation (4.18) when $\alpha = 0$ has an equilibrium $x_0 = F(1, x_0)$. Let $F_x(\alpha, x)$ denote the Jacobian of F with respect to x evaluated at (α, x). Then the Implicit Function Theorem applies to (4.18) at $(\alpha, x) = (0, x_0)$ provided $I - F_x(1, x_0)$ is invertible.

Theorem 4.6. *Consider the p-periodically forced difference equation (4.18) in which β is a p-periodic sequence with mean $\langle\beta\rangle = 0$ and for which (4.19) holds. Assume the equation has an equilibrium when $\alpha = 0$ and that 1 is not an eigenvalue of $F_x(1, x_0)$. Then for each α of sufficiently small magnitude there exists a p-periodic solution $x_\alpha(t)$ of (4.18). The p-cycle $x_\alpha(t)$ is k times continuously differentiable in α and satisfies $\lim_{\alpha \to 0} x_\alpha(t) = x_0$.*

Assume $F_x(1, x_0)$ has no eigenvalue of magnitude equal to 1 (x_0 is hyperbolic). If the equilibrium x_0 is locally asymptotically stable (or unstable), then the p-cycle $x_\alpha(t)$ is locally asymptotically stable (unstable) for $|\alpha|$ sufficiently small.

The assertion about stability in Theorem 4.6 follows from the continuity of the Jacobian of the composites of F and its eigenvalues as functions of α.

If $k \geq 3$ we can write

$$x_\alpha(t) = x_0 + y(t)\alpha + z(t)\alpha^2 + O\left(\alpha^3\right)(t)$$

and calculate the p-periodic coefficients y, z by a substitution of this expansion into equation (4.18). From coefficients of like powers of α we obtain linear difference equations whose solution is in principle straightforward, but in practice is often tedious (if not intractable), particularly for the coefficients of terms of order two or higher.

Theorem 4.6 applies to the 2-periodic LPA model (4.17), which has a positive equilibrium (L_0, P_0, A_0) when $\alpha = 0$. The α-expansions

$$\begin{aligned} L_\alpha(t) &= L_0 + L_1(t)\alpha + L_2(t)\alpha^2 + O\left(\alpha^3\right)(t) \\ P_\alpha(t) &= P_0 + P_1(t)\alpha + P_2(t)\alpha^2 + O\left(\alpha^3\right)(t) \\ A_\alpha(t) &= A_0 + A_1(t)\alpha + A_2(t)\alpha^2 + O\left(\alpha^3\right)(t) \end{aligned}$$

of the perturbing 2-cycle $(L_\alpha(t), P_\alpha(t), A_\alpha(t))$ are calculated and studied in [**31**]. The results show that the averages of the first order coefficients equal zero and hence

$$\begin{aligned} \langle L_\alpha(t)\rangle - L_0 &= \langle L_2(t)\rangle\, \alpha^2 + O\left(\alpha^3\right) \\ \langle P_\alpha(t)\rangle - P_0 &= \langle P_2(t)\rangle\, \alpha^2 + O\left(\alpha^3\right) \\ \langle A_\alpha(t)\rangle - A_0 &= \langle A_2(t)\rangle\, \alpha^2 + O\left(\alpha^3\right). \end{aligned}$$

The second order coefficients in these α-expansions determine the resonance or attenuance of each stage of the 2-cycle. The calculations necessary to obtain closed form formulas for these coefficients and their averages are intractable. However, conditions are derived in [**31**] sufficient to show that resonance can occur in the periodic LPA model (4.17) in individual life cycle stages and in total population size $T_\alpha(t) = L_\alpha(t) + P_\alpha(t) + A_\alpha(t)$, for sufficiently large $b > 0$ and sufficiently small amplitudes $\alpha > 0$. See Table 1.

For $b > 0$ large and $\alpha > 0$ small
$A_\alpha(t)$ is resonant
Both $L_\alpha(t)$ and $P_\alpha(t)$ are resonant if $\dfrac{\mu_a^2}{1-\mu_a} > 2\dfrac{c_{ea}}{c_{pa}}$
Both $L_\alpha(t)$ and $P_\alpha(t)$ are attenuant if $\dfrac{\mu_a^2}{1-\mu_a} < 2\dfrac{c_{ea}}{c_{pa}}$
Total population size $T_\alpha(t)$ is resonant if $\dfrac{\mu_a^2}{1-\mu_a} > 2\dfrac{c_{ea}}{c_{pa}} - \left(\dfrac{2}{2-\mu_l}\right)\dfrac{c_{el}}{c_{pa}}$
Total population size $T_\alpha(t)$ is attenuant if $\dfrac{\mu_a^2}{1-\mu_a} < 2\dfrac{c_{ea}}{c_{pa}} - \left(\dfrac{2}{2-\mu_l}\right)\dfrac{c_{el}}{c_{pa}}$

TABLE 1.

Maximum likelihood parameter estimates for the 2-periodic LPA model (4.17) obtained from Jillson's data are [**2**]:

$$\begin{aligned} &b = 4.445, \quad \mu_l = 4.794 \times 10^{-1}, \quad \mu_a = 1.524 \times 10^{-1} \\ &c_{ea} = 5.785 \times 10^{-3}, \quad c_{el} = 5.841 \times 10^{-2}, \quad c_{pa} = 1.053 \times 10^{-2}. \end{aligned} \tag{4.21}$$

and the positive equilibrium of the resulting model is

$$\begin{pmatrix} L_0 \\ P_0 \\ A_0 \end{pmatrix} = \begin{pmatrix} 30.5914 \\ 15.9259 \\ 57.2116 \end{pmatrix}. \tag{4.22}$$

The inequality for a resonant total population size in Table 1 is satisfied (as is the inequality for L and P stage attenuance). However, several points keep this result from providing a satisfactory explanation of Jillson's resonance effect: neither the amplitude $\alpha = 3/5$ used by Jillson is particularly small nor is $b = 4.445$ particularly large. Moreover, the equilibrium is unstable and, as a result, in the application to Jillson's data the 2-cycle perturbing from the equilibrium is unstable.

The equilibrium (4.22) is unstable because one of the eigenvalues (-1.7639, 0.8047, and -0.0718) of the Jacobian at the equilibrium is greater than 1 in magnitude. The stable equilibria bifurcating at $b = b_c \approx 0.29$ destabilize and give rise to a 2-cycle bifurcation at approximately $b = 1.6$. At the estimated parameter value $b = 4.445$ for Jillson's experiment the autonomous ($\alpha = 0$) LPA model (4.17) has a stable 2-cycle that oscillates between the stage vectors

$$\begin{pmatrix} L(0) \\ P(0) \\ A(0) \end{pmatrix} = \begin{pmatrix} 5.037 \times 10^{-5} \\ 138.3 \\ 119.0 \end{pmatrix}, \quad \begin{pmatrix} L(1) \\ P(1) \\ A(1) \end{pmatrix} = \begin{pmatrix} 265.7 \\ 2.622 \times 10^{-5} \\ 140.4 \end{pmatrix}. \tag{4.23}$$

What becomes of this 2-cycle when $\alpha > 0$?

Theorem 4.6 is a special case of a more general theorem proved by Henson [**30**] in which it assumed that the equation (4.18) has, when $\alpha = 0$, a periodic solution $x_0(t)$ of period p. Each phase shift of $x_0(t)$ is also a p-cycle. Denote these phases by

$$x_0^i(t) = x_0(t+i), \quad i = 0, 1, \cdots, p-1.$$

We anticipate that each phase perturbs to a p-cycle solution of (4.18) when $\alpha \neq 0$. Implicit function theorem methods yield the following theorem.

Theorem 4.7 ([**30**]). *Consider the p-periodically forced difference equation (4.18) in which β is a p-periodic sequence with mean $\langle\beta\rangle = 0$ and for which (4.19) holds. Assume the equation has a p-periodic solution $x_0(t)$ when $\alpha = 0$ and that 1 is not an eigenvalue of $\Pi_{t=p-1}^{0} F_x(1, x_0^0(t))$. Then for each phase $x_0^i(t)$ and each α of sufficiently small magnitude there exists a p-periodic solution $x_\alpha^i(t)$ of (4.18). Each p-cycle $x_\alpha^i(t)$ is k times continuously differentiable in α and satisfies $\lim_{\alpha\to 0} x_\alpha^i(t) = x_0^i(t)$.*

Assume $\Pi_{t=p-1}^{0} F_x(1, x_0^0(t))$ has no eigenvalue of magnitude equal to 1 ($x_0(t)$ is hyperbolic). If the p-periodic solution $x_0(t)$ is locally asymptotically stable (or unstable), then all the p-periodic solutions $x_\alpha^i(t)$ are locally asymptotically stable (unstable) for $|\alpha|$ sufficiently small.

The p-cycles x_α^i are not necessarily phase shifts of one another when $\alpha \neq 0$ (although in some cases this is a possibility). Theorem 4.7 does not require that p be the minimal period of either $x_0(t)$ or $\beta(t)$. If the minimal period of $x_0(t)$ is q and the minimal period of β is r, then Theorem 4.7 can be applied with the common period $p = \operatorname{lcm}(q, r)$. The following theorem addresses the relationships among the phases of the perturbed p-periodic solutions.

Theorem 4.8 ([**30**]). *In addition to the assumptions of Theorem 4.7 assume*

$$F\left(\eta_1, x_\alpha^i(t)\right) = F\left(\eta_2, x_\alpha^i(t)\right) \Longrightarrow \eta_1 = \eta_2 \tag{4.24}$$

for all t, all $i = 0, 1, \cdots, p-1$, and all sufficiently small $|\alpha|$. Then for $|\alpha|$ small the perturbed p-cycles $x_\alpha^i(t)$ have minimal period $p = \operatorname{lcm}(q, r)$. Moreover, modulo phase shifts, $\gcd(q, r)$ of these perturbed cycles are distinct.

An example given in [**30**] shows (4.24) cannot be dropped from this theorem.

An interesting consequence of Theorem 4.8 is that periodically forcing an equation that has a (non-equilibrium) cycle in general leads to multiple (cyclic) attractors.

With regard to the p-periodic LPA model and Jillson's experiment, Theorems 4.7 and 4.8 with $p = 2$ imply the existence and stability of two distinct 2-cycle solutions that perturb from the two phase shifts of the 2-cycle (4.23). Are either of the perturbed 2-cycles resonant?

In general, the periodic coefficients $y^i(t)$ in the expansions

$$x_\alpha^i(t) = x_0^i(t) + y^i(t)\alpha + O\left(\alpha^2\right) \tag{4.25}$$

of the perturbed p-cycles from Theorem 4.7 determine, to lowest order in the amplitude α, the relationship among the properties of the unperturbed cycle x_α^i (such as cycle average) and those of the perturbed cycle x_0^i. Let $\lceil x \rceil$ denote the sum of the m components of $x \in R^m$. In the population model context in which x is a distribution vector of stage classes $\lceil x \rceil$ is the total population size. The perturbed cycle x_α^i is *resonant (or attenuant) at* $\alpha = 0$ if there exists $\delta > 0$ such that $\langle \lceil x_\alpha^i \rceil \rangle > \langle \lceil x_0^i \rceil \rangle$ (or $\langle \lceil x_\alpha^i \rceil \rangle < \langle \lceil x_0^i \rceil \rangle$) for all $0 < \alpha < \delta$. In [**30**] Henson shows that $\Sigma_{i=0}^{p-1} \lceil y^i \rceil = 0$. Therefore, if $\lceil y^i \rceil \neq 0$ for at least $i = 0, 1, \cdots p-1$, it follows that there must be at least one $\lceil y^i \rceil > 0$ and one $\lceil y^i \rceil < 0$.

Theorem 4.9 ([**30**]). *Under the assumptions of Theorem 4.7, if there exists at least one* $i = 0, 1, \cdots p-1$ *such that*

$$\lceil y^i \rceil = \left\lceil \left. \frac{d}{d\alpha} x_\alpha^i \right|_{\alpha=0} \right\rceil \neq 0$$

then at least one of the perturbed cycles x_α^i *is resonant at* $\alpha = 0$ *and at least one is attenuant at* $\alpha = 0$.

It is tedious to calculate the expansion (4.25) for the 2-periodic LPA model with parameter values (4.21). However, with the help of an algebraic computer program to carry out the calculations, we find that one of the 2-cycles that perturbs from the 2-cycle (4.23) is resonant while the 2-cycle that perturbs from its phase shift is attenuant. Since both 2-cycles are stable, a population tends to one or the other depending on its initial condition.

This analysis is valid, however, only for small amplitudes α. Are there two stable 2-cycles in Jillson's experiment when $\alpha = 3/5$? Figure 31 shows a numerically calculated bifurcation diagram for the 2-periodic LPA model with parameters (4.21). This diagram shows that the attenuant 2-cycle disappears for $\alpha > 0.45$. This is because the attenuant 2-cycle collides with the unstable 2-cycle that perturbs from the unstable equilibrium in a saddle-node bifurcation at $\alpha = 0.45$. For larger values of α, including $\alpha = 3/5$ that Jillson used, this bifurcation leaves only the stable resonant 2-cycle.

Theorem 4.9 indicates that, in general, the result of periodically forcing an oscillating population is a multiple attractor situation (as in Figure 31(b)). Multiple attractors for the dynamics of a biological population are intriguing. For *Tribolium* the possibility is especially intriguing because there is no reported evidence of multiple attractors in single species cultures, despite over a half century of research utilizing these insects. Moreover, one of the 2-cycles predicted by the 2-periodic

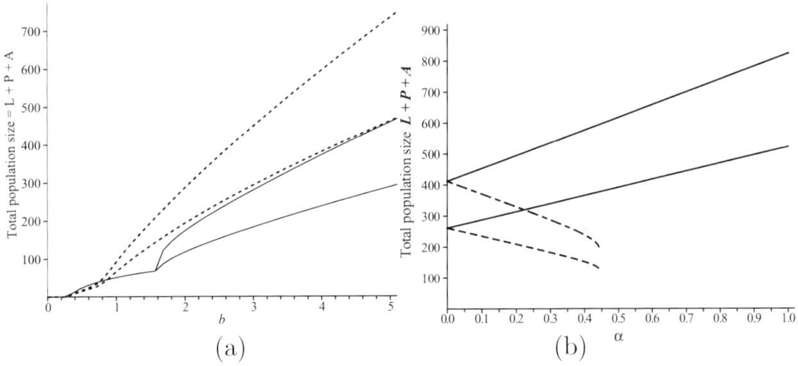

FIGURE 31. (a) A bifurcation diagram for the 2-periodic LPA model with ML parameters (4.21) is shown using b as the bifurcation parameter. The solid line is for the autonomous case when the amplitude $\alpha = 0$. Notice the period doubling bifurcation at $b \approx 1.6$. The dashed lines are the maxima and minima of 2-cycles for the case $\alpha = 3/5$ used in Jillson's experiment. A close look near the bifurcation point $b_c \approx 0.29$ shows that the bifurcating 2-cycles are attenuant for b near b_c. As b increases they become resonant. The ML estimated value of b for Jillson's experiment is 4.445 where resonance clearly occurs. (b) The solid lines are the maxima and minima of the resonant, perturbed 2-cycles for the 2-periodic LPA model with ML parameters (4.21) as a function of the relative amplitude α of the habitat oscillation. The dashed lines are the maxima and minima of the attenuant, perturbed 2-cycles. These dashed lines disappear at approximately $\alpha = 0.45$ due to a (reverse) saddle-node bifurcation with the 2-cycles that perturb from the unstable equilibrium (not shown). [Reprinted from [**33**], with permission from Springer.]

LPA model (namely the attenuant cycle in Figure 31(b)) has dynamic characteristics that do not seem biologically feasible with regard to the synchronization of the life cycle oscillations with those of the habitat. Will beetle cultures really exhibit multiple attractor dynamics with small amplitude oscillations in habitat volume? Or is this prediction just a spurious prediction of an overly simplified mathematical model?

In [**33**] there appears a report and analysis of an experiment carried out to test the multiple attractor prediction implied by Figure 31(b). The highlights of that analysis are as follows. Replicated cultures of *T. castaneum* were grown in 2-periodic fluctuating habitats with $\alpha = 0.4$ and 0.6 as well as the constant habitat case $\alpha = 0$ (as a control). For the (model predicted) multiple attractor case at $\alpha = 0.4$, one set of cultures (with replicates) initiated in the predicted basin of attraction of the resonant 2-cycle and another set (with replicates) initiated in the predicted basin of the attenuant 2-cycle[5].

Beside resonance and attenuance, notable characteristics of the predicted 2-cycles include larva oscillations that, in the resonance case, are large and out-of-phase with the oscillations in the habitat volume V. In the attenuant case, the oscillations in larvae numbers are suppressed and in-phase with the habitat volume.

[5]The proper phase space in which to view the dynamics is that of the composite map. Fixed points of the composite correspond to 2-cycles and their basins of attraction are those of the 2-cycles.

See Figure 32(a) for the model predicted dynamics of the larval stages. Figure 32(b) shows plots of the larval data from selected experimental replicates for each basin at attraction. These plots exhibit these predicted characteristics of the two different 2-cycle attractors.

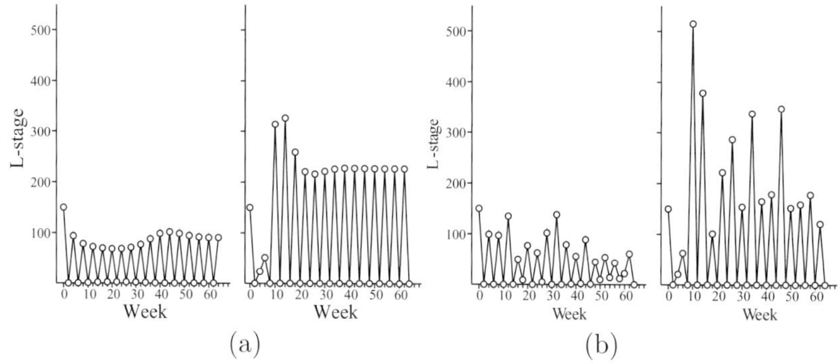

FIGURE 32. (a) The plots show the L-stage of two model orbits that approach the two stable 2-cycles predicted by the 2-periodic LPA model (4.17) in the experiment reported in [**33**]. In that experiment the model parameter values were $b = 6.598$, $\mu_l = 0.2055$, $\mu_a = 0.1000$, $c_{ea} = 0.01000$, $c_{el} = 0.1000$, $c_{pa} = 0.004700$ and $\alpha = 0.4000$. The graph on the left is for the initial condition $(L(0), P(0), A(0)) = (150, 0, 150)$, which lies in the basin of attraction of the attenuate 2-cycle. For this 2-cycle the L-stage oscillates in-phase with the oscillations in habitat volume V. The graph on the right is for the initial condition $(L(0), P(0), A(0)) = (150, 200, 150)$, which lies in the basin of attraction of the resonant 2-cycle. For this 2-cycle the L-stage oscillates out-of-phase with the oscillations in habitat volume V. (b) The plots of the L-stages of one experimental replicate for each initial conditions show the predicted oscillatory characteristics of the two different attractors in (a).

The multi-attractor experiment lasted longer than the 64 weeks shown in Figure 32. Figure 33(a) shows the L-stage of the attenuate 2-cycle data for the 140 weeks of the experiment. There is an startling occurrence: at week 66 the culture "jumps" to the resonant -cycle! A stochastic event apparently caused the population to hop into the basin of attraction of the resonant 2-cycle. This in fact occurred in all replicates started in the basin of attraction of the attenuant 2-cycle. Moreover, the reverse switch from the resonant cycle basin to the attenuant cycle basin never occurred.

It is interesting that a (environmental) stochastic version of the 2-periodic LPA model predicts this "basin jumping" phenomenon; see [**33**]. Thus, although the deterministic model predicts a multiple attractor dynamic, the stochastic version of the model predicts, in effect, a single "attractor". In 50,000 simulations of the stochastic version of the 2-periodic LPA model, with ML parameter estimates obtained from the experiment, all orbits starting from the experimental initial condition $(L(0), P(0), A(0)) = (150, 0, 150)$ lying in the attenuant 2-cycle basin of attraction had moved to the resonant 2-cycle's basin of attraction within 100 times steps. A histogram of the basin jump times appears in Figure 33(b). The mean basin jump

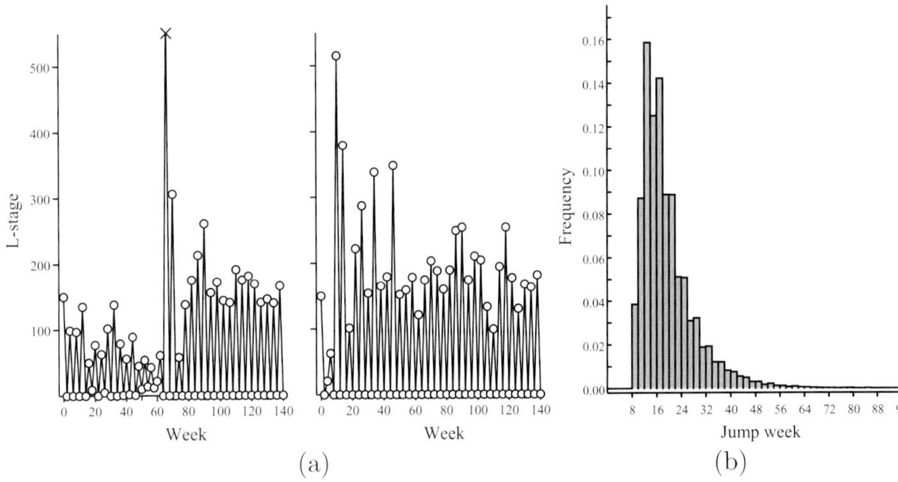

FIGURE 33. (a) The time series data shown in Figure 32(b) for 64 weeks is shown here for 140 weeks. The data appearing in the graph on the left abandons the dynamic pattern of the attenuant 2-cycle at week 66 and takes on that of the resonant 2-cycle. (b) The histogram shows the results of calculating the time at which each of 50,000 simulated orbits of the stochastic 2-periodic LPA model (with ML parameter estimates obtained from the experiment), starting from the initial condition $(L, P, A) = (150, 0, 150)$ lying in the attenuant 2-cycle basin of attraction, jumped into the basin of attraction of the resonant 2-cycles.

time in these simulations was 18.55 time steps. In the experiment, the replicate shown in Figures 31 and 32 is an unusual one in this regard. The other two replicates both jumped basins at week 22, which is quite close to the model predicted mean jump time.

A more detailed experimental study of attractor basin jumping, as well as the role of the unstable saddle 2-cycle that lies on the basin boundary, appears in [**33**]. In that study characteristics of the saddle 2-cycle, and its stable manifold, are also observed in the experimental data.

Thus, we have another example of the importance of stochastic events in data (even those collected from highly controlled experiments). To explain the observed patterns, it is not sufficient to consider deterministic model attractors alone, but transients and unstable invariant sets mixed together by stochasticity. For more on these issues see [**36**].

In summary, this lecture generalized the equilibrium theory presented in Lecture 2 to periodically forced matrix models. The results of an experiment conducted by Jillson [**38**], which exhibited a resonance effect, motivated this generalization. The theory and methods, when applied to a periodically forced version of the LPA model, provided an explanation for Jillson's observation. Furthermore, the study of this experimental result led to further theoretical insights into the effects that periodically fluctuating habitats can have on the dynamics of a population. These insights stimulated new experiments that corroborated some non-intuitive predictions concerning multiple attractors predicted by theoretical models.

EXERCISES

Exercise 16. Using the average recruitment rate μ as the bifurcation parameter, analyze the juvenile-adult model in Example 2.2

$$\begin{aligned} J(t+1) &= cbA(t)e^{-A(t)}J(t) + be^{-A(t)}A(t) \\ A(t+1) &= \tau_{21}J(t) + \tau_{22}A(t) \end{aligned}$$

when the recruitment rate $b = \mu\left(1 + \alpha\left(-1\right)^t\right)$ is 2-periodic. How does the bifurcation point μ_c depend on the amplitude α?

Exercise 17. Derive the variation of constants formula 4.3.

Exercise 18. (a) Find a formula for the initial condition $x(0)$ of the 2-cycle solutions of the discrete logistic

$$x(t+1) = b(t)\frac{1}{1+x(t)}x(t)$$

when $b(t)$ is a positive 2-periodic sequence. (b) Repeat (a) when $b(t)$ is 3-periodic. (c) By induction, repeat (a) when $b(t)$ is p-periodic.

Exercise 19. (a) By definition, a stable p-cycle $x_p(t)$ of a p-periodic equation $x(t+1) = f(t, x(t))$, $x_p(0)$ is a stable fixed point of the autonomous composite equation $y(t+1) = f^{(p-1)}(y(t)) \triangleq f(p-1, f(p-2, \ldots, f(1, f(0, x))))$. Show that this implies for each $\varepsilon > 0$ there exists a $\delta > 0$ such that $|x(0) - x_c(0)| < \delta$ implies $\Sigma_{i=t}^{t+p-1} |x(i) - x_c(i)| < \varepsilon$ for all $t \geq 0$. (b) By definition, a p-cycle $x_p(t)$ is an attractor if $x_p(0)$ is an attractor of the composite equation. Show that this implies there exists a $\delta > 0$ such that $|x(0) - x_c(0)| < \delta$ implies $\lim_{t\to+\infty}\Sigma_{i=t}^{t+p-1} |x(i) - x_c(i)| = 0$.

Exercise 20. Consider the scalar ($m = 1$) linear equation $x(t+1) = bx(t) + \tau x(t)$ where the per capita birth rate $b > 0$ and the survivorship rate satisfies $0 \leq \tau < 1$. The extinction equation $x = 0$ is stable if $b > 1 - \tau$ and unstable if $b < 1 - \tau$. The critical bifurcation value of the birth rate is $b_{cr} = 1 - \tau$. Suppose the birth rate oscillates with period 2, relative amplitude α ($0 \leq \alpha < 1$) and mean $b > 0$ and replace b in the equation by $b\left(1 + \alpha\left(-1\right)^t\right)$. Find the critical bifurcation value b_c of the mean birth rate b. Treating $b_c = b_c(\alpha)$ as a function of the amplitude α show $b_c(\alpha) > b_c(0) = 1 - \tau$. Interpret this result biologically.

Exercise 21. The extinction equation of the linear, juvenile-adult, semelparous model

$$\begin{pmatrix} J(t+1) \\ A(t+1) \end{pmatrix} = \begin{pmatrix} 0 & b \\ \tau & 0 \end{pmatrix} \begin{pmatrix} J(t) \\ A(t) \end{pmatrix}$$

is stable if $b < 1/\tau$ and is unstable if $b > 1/\tau$. Thus, the critical bifurcation value is $b_c = 1/\tau$. Suppose the birth rate oscillates with period 2 and replace b by $b\left(1 + \alpha\left(-1\right)^t\right)$ where b is now the mean birth rate and α is the amplitude ($0 < \alpha < 1$). Find the critical bifurcation value b_c of the mean birth rate b. Treating $b_c = b_c(\alpha)$ as a function of the amplitude α show $b_c(\alpha) < b_c(0) = 1/\tau$. Interpret this result biologically.

LECTURE 5

Competitive interactions

The general theory of structured population dynamics presented in Lectures 1, 2, and 4, and the case studies in Lecture 3 deal with populations of a single species. Matrix models of the form (1.7) can also describe interactions of two or more structured species when the projection matrix $P = T + F$ of each species depends on the state variables of the other species [**4**]. Classical classifications of ecological interactions, which historically are based on models without structuring, are limited to a small number of basic types: mainly, competition, predation, and mutualism (symbiosis). Structured models, on the other hand, allow for more complicated interactions that take into account changes in the relationships and interactions among species as individuals pass through life cycle stages. For example, two species might compete as juveniles, but as adults one species might become a prey of the other. See [**66**] for biological examples of mixed interactions of these kinds. Matrix models for mixed ecological interactions remain largely unexplored.

Laboratory experiments by G. F. Gause (using paramecia) and T. Park (using flour beetles) helped establish the competitive exclusion principle, which during the first half of the twentieth century was hotly debated. The fundamental concept underlying this principle is that in order to coexist two species must find a way to decrease their competition for resources. Competition can take many forms, ranging from direct confrontations and conflicts among individuals to more indirect struggles for access to resources of limited availability. This principle finds expression in the notions of ecological niche and limiting similarity among species. The principles of competitive exclusion and ecological niche are so dogmatic today that they often are not explicitly stated in ecological studies and are instead implicitly assumed in force. It is well-known that the famous Lotka-Volterra differential equations played a significant role in the historical development of competition theory and in the formulation of the competitive exclusion principle. It is less well known that discrete time models also played an important role.

In his experimental studies of the competitive exclusion principle, T. Park collaborated with P. H. Leslie and J. C. Gower in formulating a mathematical model for two competing species. To interpret and explain Park's experimental data they used the system of difference equations

$$
\begin{aligned}
x(t+1) &= b_1 \frac{1}{1 + c_{11}x(t) + c_{12}y(t)} x(t) \\
y(t+1) &= b_2 \frac{1}{1 + c_{21}x(t) + c_{22}y(t)} y(t)
\end{aligned}
\tag{5.1}
$$

to account for the dynamics of two competing species. In their application of this *Leslie-Gower competition model* to Park's experimental data, x and y are densities of adults from two different species of *Tribolium*. This model is an extension of the discrete logistic model and is naturally regarded as the discrete analog of the Lotka-Volterra differential equation model.

The competition model (5.1) has been thoroughly analyzed mathematically (it defines what is called a monotone flow [**64**]) and it turns out that its dynamic possibilities are exactly the same as those of the Lotka-Volterra differential model [**64, 12**]. Only four phase portraits are possible, as shown in Figure 34. The asymptotic dynamics are only equilibrium dynamics. Both species survive (i.e., the positive equilibrium is attracting) if and only if $c_{12}c_{21} < c_{11}c_{22}$, that is to say, if competition between the species, as measured by the product $c_{12}c_{21}$ of the interspecific competition coefficients, is small (relative to intraspecific competition as measured by the product $c_{11}c_{22}$).

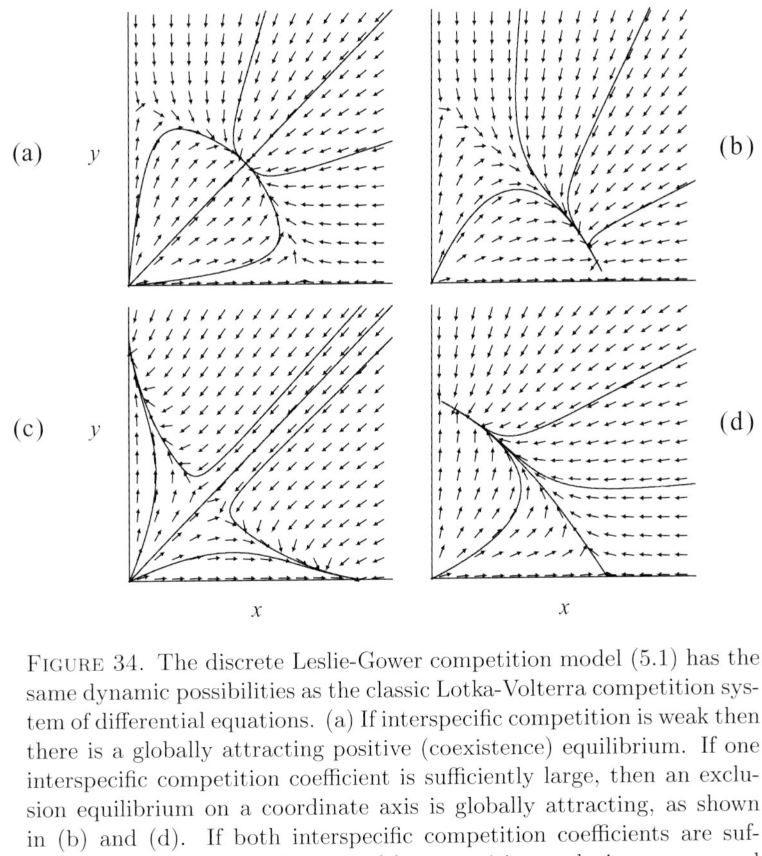

FIGURE 34. The discrete Leslie-Gower competition model (5.1) has the same dynamic possibilities as the classic Lotka-Volterra competition system of differential equations. (a) If interspecific competition is weak then there is a globally attracting positive (coexistence) equilibrium. If one interspecific competition coefficient is sufficiently large, then an exclusion equilibrium on a coordinate axis is globally attracting, as shown in (b) and (d). If both interspecific competition coefficients are sufficiently large, then as shown in (c) competitive exclusion occurs and which species wins depends on initial conditions (except for the stable manifold of the saddle equilibrium). It is this saddle case that played a prominent role in Park's *Tribolium* experiments.

The Leslie-Gower model does not model structured species, despite the fact that it was derived and utilized in the analysis of experiments with insects that have significant life cycle stages (flour beetles). What happens to the competition theory based on the Lotka-Volterra phase portraits in Figure 34 when one or both species are given, say, juvenile stages in the model equations? It is still true that strong competitive interactions as measured by large competition coefficients will lead to the exclusion of one of the species?

For a study of the Leslie-Gower model when one species of the two species has a juvenile stage in the model see [**12, 13**]. Since (as pointed out by Darwin) competition is most likely strongest among similar species, of more interest would be a Leslie-Gower model extended so as to include juvenile stages for both species:

$$\begin{aligned} J(t+1) &= b_A \frac{1}{1+A(t)+c_j j(t)} A(t) \\ A(t+1) &= (1-\mu_J)J(t) + (1-\mu_A)A(t) \\ j(t+1) &= b_a \frac{1}{1+a(t)+c_J J(t)} a(t) \\ a(t+1) &= (1-\mu_j)j(t) + (1-\mu_a)a(t). \end{aligned} \tag{5.2}$$

In this model the competition between the (upper case and the lower case distinguished) species occurs between the juvenile classes J and j and results in a reduction in juvenile recruitment. The intensity of the interspecific competition is measured by the magnitude of the competition coefficients c_j, $c_J \geq 0$.

In the absences of one species the dynamics of the other species are governed by the juvenile-adult version

$$\begin{aligned} x_1(t+1) &= b \frac{1}{1+x_2(t)} x_2(t) \\ x_2(t+1) &= (1-\mu_1)x_1(t) + (1-\mu_2)x_2(t) \end{aligned} \tag{5.3}$$

of the discrete logistic equation. This equation has a locally asymptotically stable positive equilibrium

$$\begin{pmatrix} x_1 \\ x_2 \end{pmatrix} \triangleq \begin{pmatrix} \frac{(1-\mu_1)b-\mu_2}{1-\mu_1} \\ \frac{(1-\mu_1)b-\mu_2}{\mu_2} \end{pmatrix}$$

provided the inherent net reproductive number exceeds 1: that is,

$$\frac{b(1-\mu_1)}{\mu_2} > 1.$$

Therefore, both species of the competition model (5.2) have, in the absence of the other species, stable positive equilibria

$$\begin{pmatrix} J_e^0 \\ A_e^0 \end{pmatrix} \triangleq \begin{pmatrix} \frac{\mu_A}{1-\mu_J}\left(n_J^0 - 1\right) \\ n_J^0 - 1 \end{pmatrix}, \quad \begin{pmatrix} j_e^0 \\ a_e^0 \end{pmatrix} \triangleq \begin{pmatrix} \frac{\mu_a}{1-\mu_j}\left(n_j^0 - 1\right) \\ n_j^0 - 1 \end{pmatrix}$$

provided

$$n_J^0 \triangleq b_A \frac{1-\mu_J}{\mu_A} > 1, \qquad n_j^0 \triangleq b_a \frac{1-\mu_j}{\mu_a} > 1. \tag{5.4}$$

We assume these two inequalities hold. The quantities n_J^0 and n_j^0 are the inherent net reproductive numbers of the two species in the absence of competition. These equilibria give rise to the *exclusion equilibria*

$$(5.5) \qquad \begin{pmatrix} J_e^0 \\ A_e^0 \\ 0 \\ 0 \end{pmatrix} \triangleq \begin{pmatrix} \frac{\mu_A}{1-\mu_J}\left(n_J^0 - 1\right) \\ n_J^0 - 1 \\ 0 \\ 0 \end{pmatrix}$$

$$(5.6) \qquad \begin{pmatrix} 0 \\ 0 \\ j_e^0 \\ a_e^0 \end{pmatrix} \triangleq \begin{pmatrix} 0 \\ 0 \\ \frac{\mu_a}{1-\mu_j}\left(n_j^0 - 1\right) \\ n_j^0 - 1 \end{pmatrix}$$

of the competition model (5.2). If the first of these equilibria is locally asymptotically stable, then the (j, a) species cannot survive (invade the (J, A) species) when starting from small initial numbers. Similarly, if the second of these equilibria is locally asymptotically stable, then the (J, A) species cannot survive (invade the (j, a) species) when starting from small initial numbers.

Conditions for the local stability of the exclusion equilibria (5.5) and (5.6) derive from an investigation of the eigenvalues of the Jacobian of the system (5.2) evaluated at these equilibria. These Jacobians are 4×4 matrices and the analysis seems daunting until we notice that these matrices are block diagonal. For example, the Jacobian at the exclusion equilibrium (5.5) has the form

$$\begin{pmatrix} B_1 & B_2 \\ 0 & B_3 \end{pmatrix}$$

and therefore its eigenvalues are those of the two 2×2 matrices B_1 and B_3. It turns out that B_1 is the Jacobian of the juvenile-adult, logistic (5.3) that governs the dynamics of the (J, A) species in the absence of the (j, a) species. By assumption (5.4) this matrix has eigenvalues of magnitude less than 1. Therefore, the stability of the exclusion equilibrium of the competitive system is determined by the eigenvalues of the matrix

$$B_3 = \begin{pmatrix} 0 & b_a \frac{1}{1+c_J J_e^0} \\ 1 - \mu_j & 1 - \mu_a \end{pmatrix}.$$

The eigenvalues of this matrix have magnitude less than 1 if (and only if)

$$n_j \overset{\circ}{=} b_a \frac{1 - \mu_j}{\mu_a} \frac{1}{1 + c_J J_e^0} < 1.$$

The number n_j is the inherent net reproductive number of species (j, a) when species (J, A) is at its inherent equilibrium (J_e^0, A_e^0). If $n_j > 1$ the exclusion equilibrium is unstable, which suggests that the species (j, a) will not go extinct and can successfully invade the species (J, A).

The loss of stability of the exclusion equilibrium is consistent with a transcritical bifurcation at $n_j = 1$ from the exclusion equilibrium (5.5). In this example, we can verify this bifurcation by algebraically solving the equilibrium equations

$$\begin{aligned} J &= b_A \tfrac{1}{1+A+c_j j} A \\ A &= (1 - \mu_J) J + (1 - \mu_A) A \\ j &= b_a \tfrac{1}{1+a+c_J J} a \\ a &= (1 - \mu_j) j + (1 - \mu_a) a \end{aligned}$$

for

$$\begin{pmatrix} J_e \\ A_e \\ j_e \\ a_e \end{pmatrix} = \begin{pmatrix} \left(\left(n_J^0 - 1\right) \frac{1-\mu_j}{\mu_a} - \left(n_j^0 - 1\right) c_j\right) \Delta^{-1} \\ \frac{1-\mu_J}{\mu_A} \left(\left(n_J^0 - 1\right) \frac{1-\mu_j}{\mu_a} - \left(n_j^0 - 1\right) c_j\right) \Delta^{-1} \\ \left(n_j - 1\right) \frac{n_j^0}{n_j} \frac{1-\mu_J}{\mu_A} \Delta^{-1} \\ \left(n_j - 1\right) \frac{n_j^0}{n_j} \frac{1-\mu_j}{\mu_a} \frac{1-\mu_J}{\mu_A} \Delta^{-1} \end{pmatrix} \tag{5.7}$$

where

$$\Delta \triangleq \frac{1-\mu_J}{\mu_A} \frac{1-\mu_j}{\mu_a} - c_j c_J.$$

The j_e and a_e components of this equilibrium vanish when $n_j = 1$, at which point a calculation shows $J_e = J_e^0$ and $A_e = A_e^0$. In other words, this equilibrium bifurcates from the exclusion equilibrium (5.5) at $n_j = 1$. For n_j near 1, the components J_e and A_e remain positive while the sign of the j_e and a_e components is the same as the sign of $(n_j - 1)/\Delta$. Thus, a supercritical bifurcation of positive equilibria (coexistence equilibria) occurs if $\Delta > 0$ and a subcritical bifurcation occurs if $\Delta < 0$. The inequality $\Delta > 0$ means that competition is weak in the sense that the competition coefficients are small enough; $\Delta < 0$ means competition is strong.

The exclusion equilibrium (5.5) loses stability as n_j increases through 1. It follows by the exchange of stability property of transcritical bifurcations (see [**4**]) that the coexistence equilibria are stable when they bifurcate supercritically ($\Delta > 0$) and unstable when they bifurcate subcritically ($\Delta < 0$). Thus, at low population levels the species (j, a) can successfully invade species (J, A) if the competition between the species is not strong and if the inherent net reproductive number of (j, a) at the equilibrium level of (J, A) exceeds 1.

The model scenario is symmetric and analogous conclusions hold for the invasion of (j, a) by (J, A).

The analysis of the juvenile-adult, Leslie-Gower model (5.2) illustrates general theorems that guarantee the bifurcation of a coexistence equilibria from the equilibrium of a resident as an appropriate parameter varies. For example, consider two matrix models of the form

$$\begin{aligned} x(t+1) &= \left(F_1(x(t), y(t)) + T_1(x(t), y(t))\right) y(t) \\ y(t+1) &= \left(F_2(x(t), y(t)) + T_2(x(t), y(t))\right) y(t) \end{aligned} \tag{5.8}$$

in which F_i and T_i are the fertility and transition matrices of two species x and y. The two species do not necessarily have the same number of classes (i.e., x is an m_1-dimensional vector, y is an m_2-dimensional vector and it is allowed that $m_1 \neq m_2$). Consider y as the resident species in a habitat and assume it has a stable equilibrium $y_e > 0$ in the absence of species x. For notational convenience define

$$P(x, y) \triangleq F_2(x, y) + T_2(x, y)$$

Then the equilibrium y_e satisfies the equation

$$y = P(0, y)y.$$

The equilibrium equations for the interacting system (5.8) are

$$x = (F_1(x,y) + T_1(x,y))\,y$$
$$y = P(x,y)y.$$

Under sufficient smoothness assumptions (say, P is $k \geq 1$ times continuously differentiable in x and y) the Implicit Function Theorem applies to the second equilibrium equation and results in a solution $y = \xi(x)$, $\xi(0) = y_e$, provided $I - P_y(0, y_e)$ is nonsingular (here P_y is the Jacobian of P with respect to y). This solution $\xi(x)$ is defined for x in a neighborhood of $0 \in R^{m_1}$ and is k times continuously differentiable in x. A substitution of $\xi(x)$ into the first equilibrium equation yields the algebraic matrix equation

$$x = (F(x) + T(x))\,x$$

for x where $F(x) \triangleq F_1(x, \xi(x))$ and $T(x) \triangleq T_1(x, \xi(x))$ are smooth functions in a neighborhood of $x = 0$. We can apply the bifurcation methods of Lectures 1 and 2 to this equation. To do this write $F(x) = n\Phi(x)$ where n is the inherent net reproductive of species x when species y is at equilibrium y_e. Provided the necessary assumptions on F and T hold, the methods and results of Lectures 1 and 2 imply the bifurcation of a branch of positive equilibria $(x, y(x))$ near $(0, y_e)$ at $n = 1$. With this approach there is a proviso, however. The implicit function theorem provides only a local solution $y(x)$, i.e., a solution for x near 0, and as a result we obtain only a local bifurcation at $n = 1$ (a bifurcating branch whose existence is guaranteed only near the bifurcation points $(0, y_e)$). For details relating the direction of bifurcation and stability properties of the bifurcating branch for multi-species matrix models see [**4**].

This equilibrium bifurcation result applies to the general two species matrix model (5.8). In specific applications, one often has in mind a model of a specific type of interaction based on classical classifications: for example, a competition or predator-prey model. However, the general bifurcation result applies to models that describe any kinds of interaction, including mixed types such as those mentioned above.

On the other hand, for one important class of matrix models it is possible to devise a classification scheme, based on the classical types of interactions, in terms of each species' net reproductive number. Suppose that the dependencies of the fertility and transition matrices for both models depend on two weighted population sizes

$$p_1(t) = \sum_{i=1}^{m_1} \omega_i x_i(t), \quad p_2(t) = \sum_{i=1}^{m_2} \chi_i y_i(t)$$

($\omega_i, \chi_i \geq 0$, $\Sigma_i \omega_i \neq 0$), so that (5.8) becomes

$$(5.9)\qquad \begin{aligned} x(t+1) &= (F_1(p_1(t), p_2(t)) + T_1(p_1(t), p_2(t)))\,x(t) \\ y(t+1) &= (F_2(p_1(t), p_2(t)) + T_2(p_1(t), p_2(t)))\,y(t). \end{aligned}$$

Under assumptions (2.2) the species have net reproductive numbers $n_1 = n_1(p_1, p_2)$ and $n_2 = n_2(p_1, p_2)$ that depend on p_1 and p_2. Both reproductive numbers equal 1 at an equilibrium

$$n_1(p_1, p_2) = 1$$
$$n_2(p_1, p_2) = 1$$

where p_1 and p_2 are the weighted population sizes at equilibrium.

Let n be the *inherent* net reproductive number of species y at the *extinction* equilibrium $p_1 = p_1^e > 0$, $p_2 = 0$ and write $n_2(p_1, p_2) = n\nu_2(p_1, p_2)$ where $\nu(p_1^e, 0) = 1$. Then along the branch of equilibria bifurcating from this extinction equilibrium we have the identities

$$n_1(p_1, p_2) = 1$$
$$n\nu_2(p_1, p_2) = 1.$$

From these invariants along the bifurcating continuum of positive equilibria, we can calculate conditions that determine the direction of bifurcation (i.e., whether $n > 1$ or $n < 1$ near the bifurcation point). When the Implicit Function Theorem is applied to the first equation, we obtain $p_1 = p_1(p_2)$, $p_1(0) = p_1^e$, provided the partial derivative $\partial n_1/\partial p_2$ evaluated at $(p_1, p_2) = (p_1^e, 0)$ is nonzero. Denote this derivative by $\partial_2 n_1^0 \neq 0$. Then the second invariant near bifurcation becomes

$$n\nu_2(p_1(p_2), p_2) = 1. \tag{5.10}$$

The bifurcation will be supercritical if $\nu_2(p_1(p_2), p_2)$ is decreasing in p_2 at $p_2 = 0$ and will be subcritical if it is increasing. For example, the bifurcation is supercritical (stable) if

$$\partial_1 \nu_1^0 p_1'(0) + \partial_2 \nu_2^0 < 0$$

which, since $p_1'(0) = -\partial_2 n_1^0/\partial_1 n_1^0$ (by implicit differentiation of (5.10)), is the same as

$$\det M > 0, \quad M \doteq \begin{bmatrix} \partial_1 n_1^0 & \partial_2 n_1^0 \\ \partial_1 n_2^0 & \partial_2 n_2^0 \end{bmatrix}.$$

If $\det M < 0$, then the bifurcation is subcritical (unstable).

Using the net reproductive numbers for the model (5.9) we can define competition and predator-prey interactions according to the inequalities:

$$\begin{array}{lll} \text{predator-prey:} & \partial_2 n_1(p_1, p_2) > 0, & \partial_1 n_2(p_1, p_2) < 0 \\ \text{competition:} & \partial_2 n_1(p_1, p_2) < 0, & \partial_1 n_2(p_1, p_2) < 0 \end{array}$$

If each population has self density regulation, then

$$\partial_1 n_1(p_1, p_2) < 0, \quad \partial_2 n_2(p_1, p_2) < 0.$$

Under this assumption, $\det M > 0$ for a predator-prey interaction and the bifurcation is supercritical (stable). For a competitive interaction the bifurcation is supercritical (species y can invade the resident x) if

$$\det M = \partial_1 n_1^0 \partial_2 n_2^0 - \partial_2 n_1^0 \partial_1 n_2^0 > 0$$

i.e., interspecific competition is weaker that intraspecific competition as measured by the products $\partial_2 n_1^0 \partial_1 n_2^0$ and $\partial_1 n_1^0 \partial_2 n_2^0$ respectively. The bifurcation is subcritical (species y cannot invade the resident x) if the reverse inequality holds, i.e., if interspecific competition is strong.

While this analysis provides a nice classification scheme for models of the type (5.9), not all two species matrix models are of this type. The competition model (5.2), for example, is not.

Historically competition theory is almost exclusively an equilibrium theory. For structured populations, however, the possibility of non-equilibrium dynamics is high and we are led to wonder in what way and to what extent the competitive exclusion principle (founded on equilibrium dynamics) holds true in non-equilibrium scenarios.

As a tractable example, consider the juvenile-adult competition model (5.2) when adult survivorships are equal to 0 :

$$\begin{aligned}
J(t+1) &= b_A \frac{1}{1 + A(t) + c_j j(t)} A(t) \\
A(t+1) &= (1 - \mu_J) J(t) \\
&\qquad\qquad\qquad\qquad\qquad\qquad (5.11) \\
j(t+1) &= b_a \frac{1}{1 + a(t) + c_J J(t)} a(t) \\
a(t+1) &= (1 - \mu_j) j(t).
\end{aligned}$$

Biologically this model concerns two competing semelparous populations (both populations in model (5.2) are iteroparous when μ_A and $\mu_a < 1$). It was shown above that the exclusion equilibrium (5.5) is locally stable if $n_j < 1$. This inequality is equivalent to

$$c_J > (1 - \mu_J) \frac{n_j^0 - 1}{n_J^0 - 1}.$$

Thus, if the interspecific competition coefficient c_J is sufficiently large, then species (j, a) cannot invade species (J, A) when starting at low population density. By a symmetric analysis the exclusion equilibrium (5.6) is locally stable if

$$c_j > (1 - \mu_j) \frac{n_J^0 - 1}{n_j^0 - 1}.$$

If the interspecific competition coefficient c_j is sufficiently large, then species (J, A) cannot invade species (j, a) when starting at low population density.

It follows that if both competition coefficients are large, then both exclusion equilibrium are locally stable. Moreover, in this case the coexistence equilibrium (5.7) is positive. These facts are commensurate with the saddle case in the Leslie-Gower (Lotka-Volterra) competition scheme shown in Figure 34(c). It is proved in [**14**] that the coexistence equilibrium (5.7) is a saddle in this case.

Unlike the Leslie-Gower model (and the Lotka-Volterra differential model), however, equilibrium dynamics are not the whole story for the juvenile-adult competition model (5.11).

The single species, semelparous juvenile-adult model

$$\begin{aligned}
x_1(t+1) &= b \frac{1}{1 + x_2(t)} x_2(t) \qquad (5.12) \\
x_2(t+1) &= (1 - \mu_1) x_1(t)
\end{aligned}$$

has a non-negative 2-cycle

$$\begin{pmatrix} x_1(0) \\ x_2(0) \end{pmatrix} = \begin{pmatrix} 0 \\ b(1-\mu_1)-1 \end{pmatrix}, \quad \begin{pmatrix} x_1(0) \\ x_2(0) \end{pmatrix} = \begin{pmatrix} \frac{b(1-\mu_1)-1}{1-\mu_1} \\ 0 \end{pmatrix}$$

when the inherent net reproductive number $b(1-\mu_1)$ exceeds 1. This cycle is called *synchronous* because the juvenile and adult stages are synchronized so as to never appear together (the generations are non-overlapping) [**8, 9**]. This cycle gives rise to *two exclusion 2-cycles* of the competition model (5.11)

$$\begin{pmatrix} J(0) \\ A(0) \\ j(0) \\ a(0) \end{pmatrix} = \begin{pmatrix} 0 \\ n_J^0 - 1 \\ 0 \\ 0 \end{pmatrix}, \quad \begin{pmatrix} J(1) \\ A(1) \\ j(1) \\ a(1) \end{pmatrix} = \begin{pmatrix} \frac{n_J^0-1}{1-\mu_J} \\ 0 \\ 0 \\ 0 \end{pmatrix}$$

$$\begin{pmatrix} J(0) \\ A(0) \\ j(0) \\ a(0) \end{pmatrix} = \begin{pmatrix} 0 \\ 0 \\ 0 \\ n_j^0 - 1 \end{pmatrix}, \quad \begin{pmatrix} J(1) \\ A(1) \\ j(1) \\ a(1) \end{pmatrix} = \begin{pmatrix} 0 \\ 0 \\ \frac{n_j^0-1}{1-\mu_j} \\ 0 \end{pmatrix} \tag{5.13}$$

when

$$n_J^0 = b_A(1-\mu_J) > 1, \qquad n_j^0 = b_a(1-\mu_j) > 1.$$

There are also two *coexistence 2-cycles*

$$\begin{pmatrix} J(0) \\ A(0) \\ j(0) \\ a(0) \end{pmatrix} = \begin{pmatrix} 0 \\ n_J^0 - 1 \\ 0 \\ n_j^0 - 1 \end{pmatrix}, \quad \begin{pmatrix} J(1) \\ A(1) \\ j(1) \\ a(1) \end{pmatrix} = \begin{pmatrix} \frac{n_J^0-1}{1-\mu_J} \\ 0 \\ \frac{n_j^0-1}{1-\mu_j} \\ 0 \end{pmatrix} \tag{5.14}$$

$$\begin{pmatrix} J(0) \\ A(0) \\ j(0) \\ a(0) \end{pmatrix} = \begin{pmatrix} 0 \\ \frac{(n_J^0-1)c_J - c_j(n_j^0-1)(1-\mu_J)}{c_J - c_j(1-\mu_j)(1-\mu_J)} \\ (1-\mu_j)\frac{(n_J^0-1)c_J - c_j(n_j^0-1)(1-\mu_J)}{c_J - c_j(1-\mu_j)(1-\mu_J)} \\ 0 \end{pmatrix}$$

$$\begin{pmatrix} J(1) \\ A(1) \\ j(1) \\ a(1) \end{pmatrix} = \begin{pmatrix} \frac{1}{1-\mu_J}\frac{(n_J^0-1)c_J - c_j(n_j^0-1)(1-\mu_J)}{c_J - c_j(1-\mu_j)(1-\mu_J)} \\ 0 \\ 0 \\ (1-\mu_J)(1-\mu_j)\frac{(n_J^0-1)c_J - c_j(n_j^0-1)(1-\mu_J)}{c_J - c_j(1-\mu_j)(1-\mu_J)} \end{pmatrix}. \tag{5.15}$$

By the linearization principle we can study the stability of a 2-cycle by examining the eigenvalues of the Jacobian of the composite map (which turns out to equal the product of the Jacobians of the map evaluated at $t = 0$ and $t = 1$). A computer algebra program makes tractable such a stability analysis of the 2-cycles (5.13), (5.14) and (5.15). The results imply that all 2-cycles are unstable, except the coexistence 2-cycle (5.14) which is stable if the competition coefficients c_J and c_j are sufficiently *large*, specifically if the competition coefficients satisfy the inequalities

$$c_J > (1-\mu_J)\frac{n_j^0-1}{n_J^0-1}, \qquad c_j > (1-\mu_j)\frac{n_J^0-1}{n_j^0-1}. \tag{5.16}$$

Note that these are precisely the inequalities that imply both exclusion equilibria are stable. If one of the inequalities is reversed the 2-cycle is unstable.

Thus, in model (5.11) coexistence is possible in a non-equilibrium (2-cycle) way when interspecific competition is strong enough. When inequalities (5.16) hold there are three locally stable attractors, the two exclusion equilibria and a coexistence 2-cycle. See Figure 35.

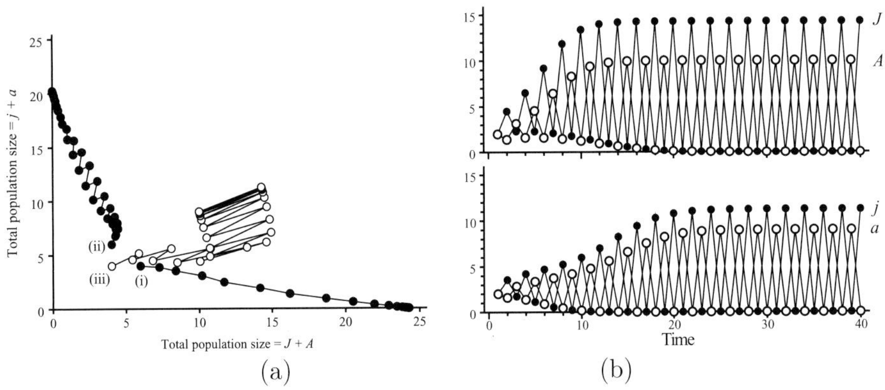

FIGURE 35. (a) The total population sizes of three orbits of the semelparous juvenile-adult, Leslie-Gower model (5.11) with different initial conditions that lead to three different asymptotic outcomes. The model parameters are $c_j = c_J = 2$, $\mu_J = 0.3$, $\mu_j = 0.2$, $n_J = 11$, $n_j = 10$ ($b_A = 110/7 \approx 15.71, b = 100/8 \approx 12.50$). Initial conditions (i) $(J, A, j, a) = (4, 2, 2, 2)$ and (ii) $(J, A, j, a) = (2, 2, 4, 2)$ lead to competitive exclusion, with species (J, A) winning in the first case and (j, a) winning in the second case. Initial condition (iii) $(J, A, j, a) = (2, 2, 2, 2)$ leads to coexistence in a 2-cycle. (b) The coexistence 2-cycle is synchronous in that the juvenile and adult stages do not overlap.

In this model the fact that increased competition intensity can lead to coexistence seemingly goes against the competitive exclusion principle, which asserts interspecific competition must be avoided if two species are to coexist. In the case of the competition model (5.11) there is a reconciliation, however. Note that when oscillating according to the coexistence 2-cycle (5.14) the two species actually avoid competition altogether because of the synchrony of the oscillations in the life cycle stages. In this way the two species coexist even when the intensity of (potential) competition is high.

Another interesting observation concerning the competition model (5.11) is that the avoidance 2-cycle is available for all parameter values, including those for which there is a stable coexistence equilibrium. This fact does not support a competition principle that asserts species strive (or are selected) to avoid competition. In this case, the species "choose" to compete at equilibrium rather than avoid competition altogether in a 2-cycle.

The competition model (5.11) is only a "toy" model. However, one bit of evidence for non-equilibrium coexistence under increased competitive intensity occurs in *Tribolium* species. In T. Park's famous competition experiments, one example of two species coexistence unexpectedly occurred and a speculative explanation based on a competition version of the LPA model (3.11) is given by Edmunds et al. [**23**]. Their explanation is based on the existence of robust coexistence 2-cycles when realistic parameter values are used and interspecific cannibalistic coefficients sufficiently large. It is suggestive that Park in fact reported increased cannibalistic voracity in his experiments. The competition LPA model is a $m = 6$ dimensional system for which extensive analysis is intractable (although a basic theory, particularly of the exclusion equilibria, is worked out in [**22**]). Numerical studies show, however, that the model can exhibit very exotic dynamics and coexistence non-equilibrium attractors (even multiple coexistence attractors).

Although derived in application to a specific experimental system, the LPA model is a rather generic, three life cycle stage model, not uncommon in biological species (especially insects). A competition theory based on that model is far richer than that derived from the limited equilibrium dynamic possibilities of Lotka-Volterra type models. To what extent these complicated dynamics necessitates a modification of classical competition tenets, or to what extent they can be reconciled with classical theory, remains an open question.

EXERCISES

Exercise 22. The system

$$\begin{aligned} J(t+1) &= b_1 \frac{1}{1 + A(t) + c_1 y(t)} A(t) \\ A(t+1) &= \tau J(t) \\ y(t+1) &= b_2 \frac{1}{1 + y(t) + c_2 J(t)} y(t) \\ b_1, b_2, c_1, c_2 > 0, &\qquad 0 < \tau < 1 \end{aligned}$$

is a version of the Leslie-Gower competition model in which one of the species has a juvenile stage. (a) Show the (J, A) species goes extinct if $n \triangleq \tau b_1 < 1$ and that the y species goes extinct if $b_2 < 1$. (b) Assume $b_1, b_2 > 1$. Find formulas for the two exclusion equilibria and determine conditions on the competition coefficients c_1 and c_2 under which each is locally asymptotically stable and conditions when each is unstable. Interpret your answers biologically. (c) Find a formula for a synchronous exclusion 2-cycle and determine a conditions on c_2 under which it is locally asymptotically stable and conditions when it is unstable.

Exercise 23. Find all equilibrium of the Leslie-Gower competition model (5.1)

$$\begin{aligned} x(t+1) &= b_1 \frac{1}{1 + c_{11} x(t) + c_{12} y(t)} x(t) \\ y(t+1) &= b_2 \frac{1}{1 + c_{21} x(t) + c_{22} y(t)} y(t). \end{aligned}$$

Show the extinction equilibrium $(x, y) = (0, , 0)$ is globally asymptotically stable if $b_1 < 1$ and $b_2 < 1$. Do a linearization stability analysis of the competitive exclusion equilibria $(x, 0)$ and $(0, y)$ and determine conditions on the competition coefficients c_{ij} under which they are stable and conditions under which they are unstable. Under what conditions is there a positive (competitive coexistence) equilibrium $(x, y) > 0$? Under those conditions do a linearization stability analysis on the equilibrium and determine conditions on the competition coefficients c_{ij} under which they are stable and conditions under which they are unstable (difficult).

Bibliography

[1] H. Caswell, *Matrix Population Models: Construction, Analysis and Interpretation*, Second Edition, Sinauer Associates, Inc. Publishers, Sunderland, Massachusetts, 2001

[2] R. F. Costantino, J. M. Cushing, B. Dennis, R. A. Desharnais, S. M. Henson, Resonant population cycles in temporally fluctuating habitats, *Bulletin of Mathematical Biology* 60, No. 2 (1998), 247-275

[3] R. F. Costantino, Robert A. Desharnais, J. M. Cushing, Brian Dennis, S. M. Henson, and A. A. King, The flour beetle Tribolium as an effective tool of discovery, *Advances in Ecological Research*, Volume 37, 2005, 101-141

[4] J. M. Cushing, *An Introduction to Structured Population Dynamics*, CBMS-NSF Regional Conference Series in Applied Mathematics, Vol. 71, SIAM, Philadelphia, 1998

[5] J. M. Cushing, The LPA model, *Fields Institute Communications* 43 (2004), 29-552

[6] J. M. Cushing, Oscillatory population growth in periodic environments, *Theoretical Population Biology* 30 (1987), 289-308

[7] J. M. Cushing, Periodically forced nonlinear systems of difference equations, *Journal of Difference Equations and Applications* 3 (1998), 547-561

[8] J. M. Cushing, Cycle chains and the LPA model, *Journal of Difference Equations and Applications* 9, No. 7 (2003), 655-670

[9] J. M. Cushing, Nonlinear Semelparous Leslie Models, *Mathematical Biosciences and Engineering*, 3, No. 1 (2006), 17-36

[10] J. M. Cushing, R. F. Costantino, B. Dennis, R. A. Desharnais, and S. M. Henson, Nonlinear population dynamics: models, experiments, and data, *Journal of Theoretical Biology* 194, No. 1 (1998), 1-9

[11] J. M. Cushing, R. F. Costantino, Brian Dennis, R. A. Desharnais, S. M. Henson, *Chaos in Ecology: Experimental Nonlinear Dynamics*, Theoretical Ecology Series, Academic Press/Elsevier, San Diego, 2003

[12] J. M. Cushing, S. LeVarge, N. Chitnis and S. M. Henson, Some discrete competition models and the competitive exclusion principle, *Journal of Difference Equations and Applications* 10, No. 13-15 (2004), 1139-1151

[13] J. M. Cushing and S. LeVarge, Some Discrete Competition Models and the Principle of Competitive Exclusion, *Proceedings of the* 9th *International Conference on Difference Equations and Applications*, World Scientific

[14] J. M. Cushing, S. M. Henson and L.-I. Roeger, A competition model for populations with juvenile-adult structure, Journal of Biological Dynamics 1, No. 2 (2007), 201-231

[15] J. M. Cushing and Z. Yicang, The net reproductive value and stability in matrix population models, *Natural Resource Modelling* 8 (1994), 297-333

[16] B. Dennis, Allee effects: population growth, critical density, and the chance of extinction. *Natural Resource Modeling* 3 (1989), 481-538

[17] B. Dennis and M. L. Taper, Density dependence in time series observations of natural populations: estimation and testing, *Ecological Monographs* 64 (1994), 205-224

[18] B. Dennis, R. A. Desharnais, J. M. Cushing and R. F. Costantino, Nonlinear demographic dynamics: mathematical models, statistical methods and biological experiments, *Ecological Monographs* 65, No. 3 (1995), 261-281

[19] B. Dennis, R. A. Desharnais, J. M. Cushing, and R. F. Costantino, Transitions in population dynamics: equilibria to periodic cycles to aperiodic cycles, *Journal of Animal Ecology* 66 (1997), 704-729

[20] B. Dennis, R. A. Desharnais, J. M. Cushing, S. M. Henson, and R. F. Costantino, Estimating chaos and complex dynamics in an insect population, *Ecological Monographs* 71, No. 2 (2001), 277-303

[21] R. A. Desharnais, R. F. Costantino, J. M. Cushing, S. M. Henson, B. Dennis, and A. A. King, Experimental confirmation of the scaling rule for demographic stochasticity, Ecological Letters 9 (2006), 537-547

[22] J. Edmunds, A Study of a Stage-Structured Model of Two Competing Species. Ph.D. dissertation, University of Arizona, Tucson, 2001

[23] J. Edmunds, J. M. Cushing, R. F. Costantino, S. M. Henson, B. Dennis and R. A. Desharnais, Park's *Tribolium* competition experiments: a non-equilibrium species coexistence hypothesis, *Journal of Animal Ecology* 72 (2003), 703-712

[24] S. N. Elaydi, *An Introduction to Difference Equations*, Springer-Verlag, New York, 1996

[25] S. Elaydi and R. J. Sacker, Global stability of periodic orbits of non-autonomous difference equations in population biology and the Cushing/Henson conjectures, *Proceedings of the 8th International Conference of Difference Equations and Applications*, pp. 113-126, Chapman & Hall/CRC, Boca Raton, FL, 2005

[26] F. R. Gantmacher, *The Theory of Matrices*, Volume 2, Chelsea Publishing Company, New York, 1960

[27] J. Guckenheimer and P. Holmes, *Nonlinear Oscillations, Dynamical Systems and Bifurcations of Vector Fields*, Springer-Verlag, Berlin, 1983

[28] M. P. Hassell, J. H. Lawton and R. M. May, Patterns of dynamical behavior in single species populations, *Journal of Animal Ecology* 45 (1976), 471-486

[29] S. M. Henson, Existence and stability of nontrivial periodic solutions of periodically forced discrete dynamical systems, *Journal of Difference Equations and Applications*, 2(1996), 315-331

[30] Multiple attractors and resonance in periodically forced populations models, *Physica D* 140 (2000), 33-49

[31] S. M. Henson and J. M. Cushing, The effect of periodic habitat fluctuations on a nonlinear insect population model, *Journal of Mathematical Biology* 36 (1997), 201-226

[32] S. M. Henson, J. M. Cushing, R. F. Costantino, B. Dennis, and R. A. Desharnais, Phase switching in biological population, *Proceedings of the Royal Society* 265 (22 November 1998), 2229-2234

[33] S. M. Henson, R. F. Costantino, J. M. Cushing, B. Dennis and R. A. Desharnais, Multiple attractors, saddles, and population dynamics in periodic habitats, *Bulletin of Mathematical Biology* 61 (1999), 1121-1149

[34] S. M. Henson, R. F. Costantino, J. M. Cushing, R. A. Desharnais, B. Dennis, and A. A. King, Lattice effects observed in chaotic dynamics of experimental populations, *Science* 294 (19 Oct 2001), 602-605

[35] S. M. Henson, A. A. King, R. F. Costantino, J. M. Cushing, B. Dennis, and R. A. Desharnais, Explaining and predicting patterns in stochastic population systems, *Proceedings of the Royal Society*, London B 270 (2003), 1549-1553

[36] S. M. Henson, J. R. Reilly, S. L. Robertson, M. C. Schu, E. W. Davis and J. M. Cushing, Predicting irregularities in population cycles, *SIAM Journal on Applied Dynamical Systems* 2, No. 2 (2003), 238-253

[37] R. A. Horn and C. R. Johnson, *Matrix Analysis*, Cambridge University Press, Cambridge, 1985

[38] D. Jillson, Insect populations respond to fluctuating environments, *Nature* 288 (1980), 699-700

[39] H. Keilhöfer, *Bifurcation Theory: An Introduction with Applications to PDEs*, Applied Mathematical Sciences 156, Springer, New York, 20041

[40] A. A. King, R. A. Desharnais, S. M. Henson, R. F. Costantino, J. M. Cushing, and B. Dennis, Random perturbations and lattice effects in chaotic population dynamics, *Science* 297, (27 September, 2002), p. 2163

[41] A. A. King, R. F. Costantino, J. M. Cushing, S. M. Henson, R. A. Desharnais, and B. Dennis, Anatomy of a chaotic attractor: subtle model-predicted patterns revealed in population data, *Proceedings of the National Academy of Sciences* 101, No. 1 (2003), 408-413

[42] V. L. Kocic, A note on the nonautonomous Bervton-Holt model, Journal of Difference Equations and Applications 11 (2005), 415-422

[43] Ryusuke Kon, A note on attenuant cycles of population models with periodic carrying capacity, *Journal of Difference Equations and Applications* 10, No. 8 (2004), 791-793

[44] Ryusuke Kon, Attenuant cycles of population models with periodic carrying capacity, *Journal of Difference Equations and Applications*, 11 (2005), 423-430

[45] R. Kon, Y. Saito and Y. Takeuchi, Permanence of single-species stage-structured models, *Journal of Mathematical Biology* 48 (2004), 515-528

[46] L. P. Lefkovitch, The study of population growth in organisms grouped by stage, *Biometrics* 21 (1965), 1-18

[47] P. H. Leslie, On the use of matrices in certain population mathematics, *Biometrika* 33 (1945), 183-212

[48] P. H. Leslie, Some further notes on the use of matrices in population mathematics, *Biometrika* 35 (1948), 213-245

[49] E. G. Lewis, On the generation and growth of a population, *Sankhya* 6 (1942), 93-96

[50] C.-K. Li and H. Schneider, Applications of Perron-Frobenius theory to population dynamics, *Journal of Mathematical Biology* 44 (2002), 450-462

[51] T. Y. Li and J. A. Yorke, Period three implies chaos, *American Mathematical Monthly* 82 (1975), 985-992

[52] R. M. May, Biological populations with nonoverlapping generations: stable points, stable cycles and chaos, *Science* 186 (1974), 645-647

[53] R. M. May, Simple mathematical models with very complicated dynamics, *Nature* 261 (1976), 459-467

[54] R. M. May, *Stability and Complexity in Model Ecosystems*, Princeton Landmarks in Biology, Princeton University Press, Princeton, New Jersey, 2001

[55] J. A. J. Metz and O. Diekmann, *The Dynamics of Physiologically Structured Populations*, Lecture Notes in Biomathematics, Volume 68, Springer-Verlag, Berlin, 1986

[56] L. D. Mueller and A. Joshi, *Stability in Model Populations*, Monographs in Population Biology 31, Princeton University Press, Princeton, New Jersey, 2000

[57] R. Pearl, *The Biology of Population Growth*, Alfred A. Knopf, New York, 1925

[58] J. N. Perry, R. H. Smith, I. P. Woiwod and D. R. Morse, *Chaos in Real Data: the Analysis of Nonlinear Dynamics from Short Ecological Time Series*, Kluwer Academic Publishers, Dordrecht, The Netherlands, 2000

[59] E. C. Pielou, *Mathematical Ecology*, Wiley-Interscience, John Wiley & Sons, New York, 1974

[60] E. C. Pielou, *Population and Community Ecology: Principles and Methods*, Gordon & Breach Science Publishers, New York, 1977

[61] P. H. Rabinowitz, Some global results for nonlinear eigenvalue problems, *Journal of Functional Analysis* 7, No 3 (1971), 487-513

[62] W. E. Ricker, Stock and recruitment, *Journal of the Fisheries Research Board of Canada* 11 (1954), 559-623

[63] R. C. Robinson, *An Introduction to Dynamical Systems: Continuous and Discrete*, Pearson/Prentice Hall, Upper Saddle River, New Jersey, 2004

[64] H. L. Smith, Planar competitive and cooperative difference equations, *Journal of Difference Equations and Applications* 3 (1998), 335-357

[65] H. Tong, *Nonlinear Time Series: A Dynamical System Approach*, Oxford University Press, Oxford, England, 1990

[66] E. E. Werner and J. F. Gilliam, The ontogenetic niche and species interactions in size-structured populations, *Ann. Rev. Ecol. Syst.* 15 (1984), 393-425

[67] S. Wiggins, *Introduction to Applied Nonlinear dynamical Systems and Chaos*, Texts in Applied Mathematics 2, Springer, New York 1990

[68] E. O. Wilson, *The Future of Life*, Alfred A. Knopf, New York, 2002

Mathematical Epidemiology of Infectious Diseases

David J. D. Earn

IAS/Park City Mathematics Series
Volume 14, 2005

Mathematical Epidemiology of Infectious Diseases

David J. D. Earn

1. Introduction

Mathematical modelling has contributed significantly to our understanding of the epidemiology of infectious diseases [**AM91**], and is beginning to have a substantial impact on research at the immunological and molecular levels [**NM00**]. The earliest mathematical study of an infectious disease was undertaken more than 200 years ago by Daniel Bernoulli, who attempted to influence public health policy concerning smallpox [**Ber1760, BB04**]. Nevertheless, a firm theoretical basis for infectious disease modelling was not established until the 20th century. Some of the first analyses provided fundamental insights about thresholds for epidemics to occur and the incidence patterns that result [**Ros11, KM27**], while more recent work has elucidated effects of demographic stochasticity [**Bar60, Bai75, Kur80, AB00**], heterogeneity in age, genetics, social and spatial structure [**AM91**]. The ultimate goal of most of this work is to contribute to efforts to develop improved strategies for eradication and control of many infectious diseases.

If we hope to build models that can help predict the effects of changes in control strategies, we must first demonstrate that models can successfully predict patterns of epidemics that have been observed in the past. Fortunately, medical and public health administrators in many countries have kept records of reported deaths from (and cases of) infectious diseases for decades and, in some places, centuries. These records, together with immunological and genetic data that have become available more recently, provide an empirical backbone for the mathematical study of infectious disease transmission dynamics and control.

These lectures will introduce you to the subject of mathematical epidemiology from an applied perspective. By "applied" I mean that we will try to make sense of specific examples of data, and we will try to draw inferences that could influence decisions of public health policymakers. We will begin with a few examples of epidemiological data that present intriguing puzzles, and some elementary statistical tools for quantifying the patterns in these data. We will then examine two very different approaches for investigating these patterns — statistical time series modelling and mechanistic mathematical modelling — before considering how we might

Department of Mathematics and Statistics, McMaster University, Hamilton, Ontario, Canada L8S 4K1
E-mail address: earn@math.mcmaster.ca
I was supported by the Natural Sciences and Engineering Research Council of Canada (NSERC) and the Canadian Institutes of Health Research (CIHR). It is a pleasure to thank the organizers of the PCMI summer school for inviting me to contribute, and the course participants for their enthusiasm and insightful questions

predict changes in patterns of epidemics and how we might manipulate infectious disease dynamics so as to reduce expected morbidity and mortality.

2. Describing epidemics

2.1. Plague

One of the most famous examples of an epidemic of an infectious disease in a human population is the Great Plague of London, which took place in 1665 [**MM04**]. We know quite a lot about the progression of the Great Plague because weekly bills of mortality from that time have been retained. A photograph of such a bill is shown in Figure 1. Note that the report indicates that the number of deaths from plague (5533) was more than 37 times the number of births (146) in the week in question, and that wasn't the worst week! An even worse plague — known as the *Black*

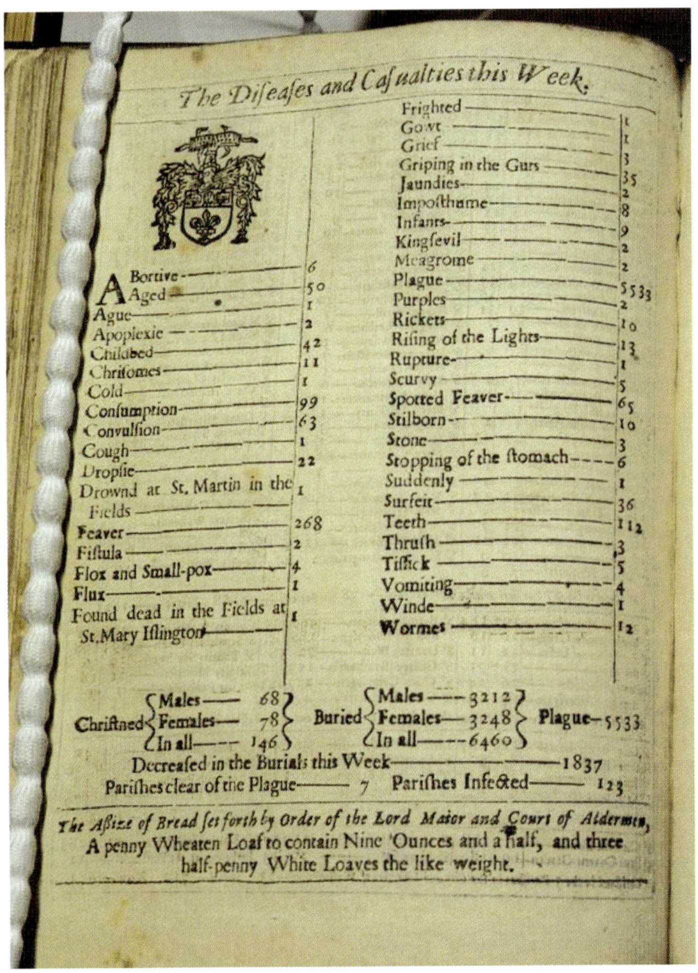

The Diseaſes and Caſualties this Week.

Abortive	6	Frighted	1
Aged	50	Gowt	1
Ague	1	Grief	3
Apoplexie	2	Griping in the Guts	35
Childbed	42	Jaundies	2
Chriſomes	11	Impoſthume	8
Cold	1	Infants	9
Conſumption	99	Kingſevil	2
Convulſion	63	Meagrome	2
Cough	1	Plague	5533
Dropſie	22	Purples	2
Drownd at St. Martin in the Fields	1	Rickets	10
Feaver	268	Riſing of the Lights	13
Fiſtula	2	Rupture	1
Flox and Small-pox	4	Scurvy	5
Flux	1	Spotted Feaver	65
Found dead in the Fields at St. Mary Iſlington	1	Stilborn	10
		Stone	3
		Stopping of the ſtomach	6
		Suddenly	1
		Surfeit	36
		Teeth	112
		Thruſh	3
		Tiſſick	5
		Vomiting	4
		Winde	1
		Wormes	12

Chriſtned: Males 68, Females 78, In all 146
Buried: Males 3212, Females 3248, In all 6460
Plague 5533

Decreaſed in the Burials this Week 1837

Pariſhes clear of the Plague 7 Pariſhes Infected 123

The Aſſize of Bread ſet forth by Order of the Lord Maior and Court of Aldermen, A penny Wheaten Loaf to contain Nine Ounces and a half, and three half-penny White Loaves the like weight.

FIGURE 1. A bill of mortality for the city of London, England, for the week of 26 September to 3 October 1665. This photograph was taken by Claire Lees at the Guildhall in London, England, with the permission of the librarian.

Death — occurred in the 14th century, but the remaining records of that epidemic are much less detailed.

Putting together the weekly counts of plague deaths from all the relevant mortality bills, we can obtain the *epidemic curve* for the Great Plague, which I've plotted in the top left panel of Figure 2. The bottom left panel of Figure 2 shows weekly mortality from plague in London over a period of 70 years. The Great Plague is the rightmost (and highest) peak in the plot. You can see that on a longer timescale, there was a complex pattern of plague epidemics including extinctions and re-emergences. At present, however, we have no direct evidence that all those plagues were really caused by the same pathogenic organism.

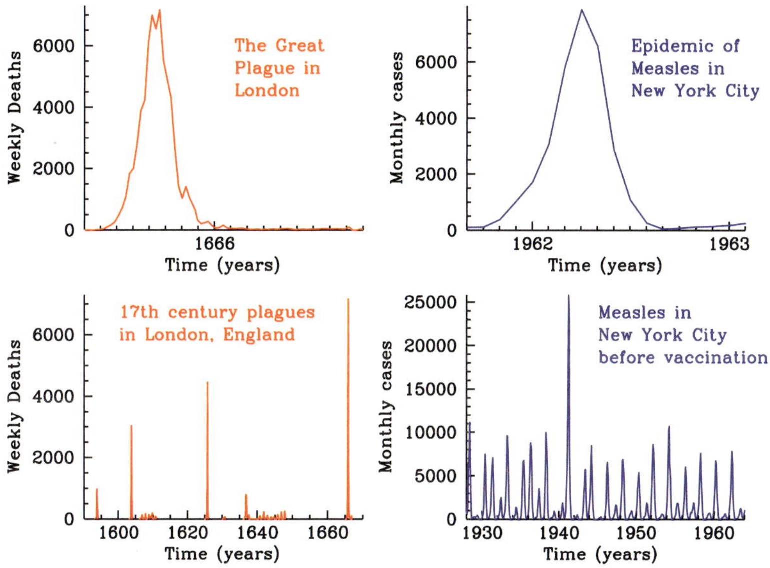

FIGURE 2. Epidemic curves for plague in London (left panels) and measles in New York City (right panels). For plague, the curves show the number of deaths reported each week. For measles, the curves show the number of cases reported each month. In the top panels, the small ticks on the time axis occur at monthly intervals.

2.2. Measles and other childhood diseases

A less contentious example is given by epidemics of measles, which are definitely caused by a well-known virus that infects the respiratory tract in humans and is transmitted by airborne particles [**AM91**]. Measles gives rise to characteristic red spots that are easily identifiable by physicians who have seen many cases, and parents are very likely to take their children to a doctor when such spots are noticed. Consequently, the majority of measles cases in developed countries end up in the office of a doctor (who, in many countries, is required to report observed measles cases to a central body). The result is that the quality of reported measles case data is unusually good.

An epidemic curve for measles in New York City in 1962 is shown in the top right panel of Figure 2. The period shown is 17 months, exactly the same length of time shown for the Great Plague of London in the top left panel. The 1962 measles epidemic in New York took off more slowly and lasted longer than the Great Plague of 1665. The bottom right panel of Figure 2 shows reported measles cases in New York City for a 36 year period, the end of which includes the 1962 epidemic. The cyclicity in the epidemic pattern is striking and much more regular than the series of plague epidemics in the bottom left panel.

There are also striking differences between the epidemic patterns of measles and those of other common childhood diseases. Figure 3 shows examples of sequences of recorded epidemics of measles, whooping cough, rubella and chicken pox. While all of these diseases display recurrent epidemics, the detailed structure of the temporal patterns could hardly be more varied. This is especially puzzling given that all these diseases are caused by respiratory infections with many clinical similarities, and all are observed primarily in children.

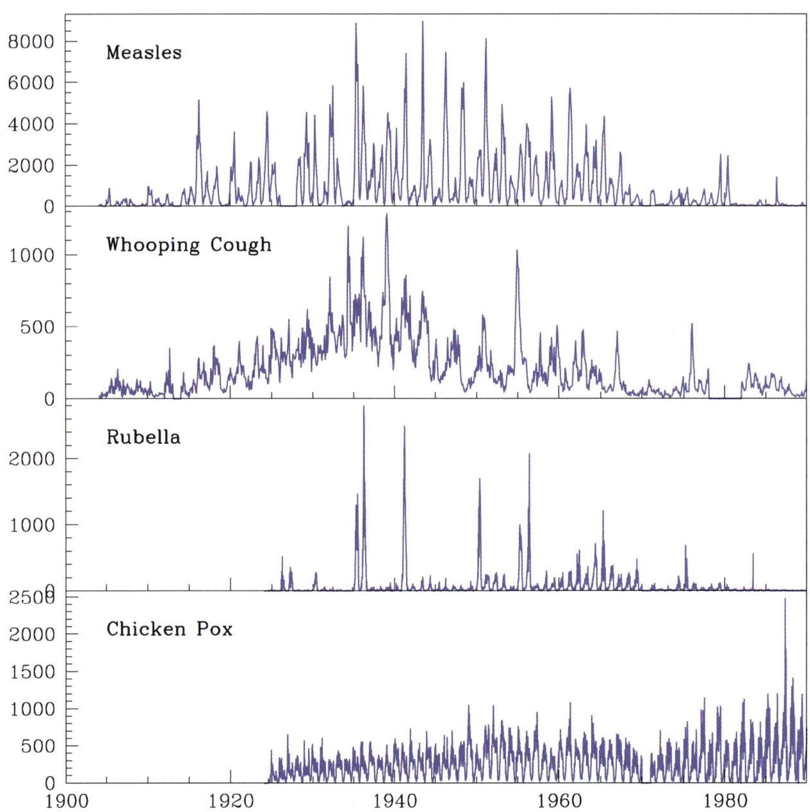

FIGURE 3. Reported cases of childhood infectious diseases in Ontario, Canada. For measles and whooping cough the data are monthly and begin in 1904. For rubella and chicken pox the data are weekly and begin in 1924.

2.3. Influenza

One of the diseases that we encounter most often is influenza, which is a major cause of morbidity in people of all ages and mortality in the elderly [**EDL02, DPV$^+$06**]. The strong seasonality of influenza epidemics is clearly evident in the main panel, (a), of Figure 4, but this plot hides some critical features of influenza incidence patterns. The plot inset on the top right of panel (a) shows the first 20 years of data on a much larger vertical scale and emphasizes the true magnitude of the 1918 influenza pandemic. The other panels of this figure show that different indicators of the "influenza season" are consistent (panel b), that influenza epidemics occur in the winter in both the northern and southern hemispheres (panel c), that mortality data mask more complex patterns of incidence of different influenza types and subtypes (panel d), and that influenza evolves rapidly by a process known as *antigenic drift* (panel e) allowing the virus to evade our immune system again within a few years of infection.

None of the graphs in the figure reveal the most dangerous property of influenza, that every few decades (three times in the last century) it has undergone major changes which have left most of the world's human population suddenly susceptible to the newly invading strain. All three *antigenic shifts* in the last century (in 1918, 1957 and 1968) have originated from influenza viruses circulating in aquatic birds, which is why reports of "bird flu" infecting humans cause serious concern.

2.4. Statistical description of epidemic time series

All of the examples above have in common that epidemics of the disease in question are *recurrent*. If we wish to understand these epidemic patterns, then we must establish a precise way of quantifying the notion of "recurrence". Methods of *time series analysis* [**Cha04**] provide a standard set of tools to describe the temporal structure of recurrent epidemics.

The most natural way to begin analysis is with a *time plot*, like the graphs shown for each of the diseases discussed above. For a single epidemic, such as the Great Plague of London shown in top left panel of Figure 2, the time plot is an excellent descriptor. For recurrent epidemics, it is still the best starting point but is often difficult to interpret, especially if processes are occurring on a number of different scales. Figure 5 shows the original time series of cases of measles reported in England and Wales from 1944 to 1994 (top panel), together with two time plots of *transformed data*, which reveal much more structure (and noise) in the later parts of the time series when the overall number of cases had been greatly reduced as a result of mass vaccination.

Awareness of noise is important, but noise can mask important features that take place on longer timescales. For this reason, it is useful to make time plots of *smoothed data*, examples of which are shown in Figure 6. The plots shown here are of the England and Wales measles time series smoothed with a *moving average*, i.e., by replacing the original data points x_t with averages of nearby points,

$$x_t \to \frac{1}{2a+1}\sum_{i=-a}^{a} x_{t+i}\,. \tag{2.1}$$

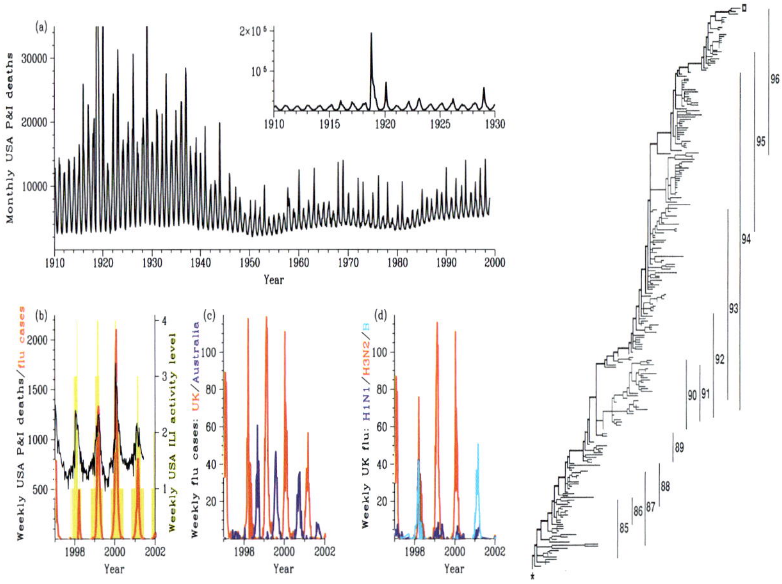

FIGURE 4. Influenza ecology and evolution. (a) Monthly mortality attributed to pneumonia and influenza (P&I deaths) in the USA in the 20th century. The inset plot shows the period 1910–1930 on a much larger scale, revealing the magnitude of the three peaks that extend beyond the top of the main panel: 1918–1919, 1919–1920 and 1928–1929. (b) Weekly measures of influenza incidence in the USA: P&I mortality in 122 cities (black curve), laboratory-confirmed cases (red curve), and influenza-like illness (ILI activity) on a five-point scale from 0 to 4 (yellow shading). (c) Weekly laboratory-confirmed cases in the UK (red curve) and Australia (blue curve). (d) Weekly patterns of relative prevalence of the three major groups of strains (H1N1, blue curve; H3N2, red curve; B, cyan curve) in the UK. (e) Maximum parsimony phylogenetic reconstruction of the evolution of the HA1 domain of the H3N2 subtype of influenza A from 1985 to 1997. The thick line running from the lower left (*=root) to the upper right (open square) is called the trunk and represents the successful H3N2 lineage. The vertical lines indicate the range of isolates from each flu year (1 October to 30 September). [Panels (a)–(d) reprinted from Earn et al. [**EDL02**]; copyright © (2002), with permission of Elsevier. Panel e reprinted from Fitch et al [**FBBC97**]; copyright © (1997) National Academy of Sciences, USA.] Panel (d) gives a limited impression of the level of aggregation of data in the other panels. The coarsest data are those in panel (a), where the monthly records span periods during which different subtypes of influenza A have been present in human populations: 1918–1957 (H1N1), 1957–1968 (H2N2), 1968–1977 (H3N2), 1977–present (both H1N1 and H3N2). Data sources: US Vital Statistics [**USVS**], US CDC 122 cities data base (`http://www.cdc.gov/epo/dphsi/121hist.htm`), WHO FluNet data base (`http://qamapserver.who.int/GlobalAtlas/`)

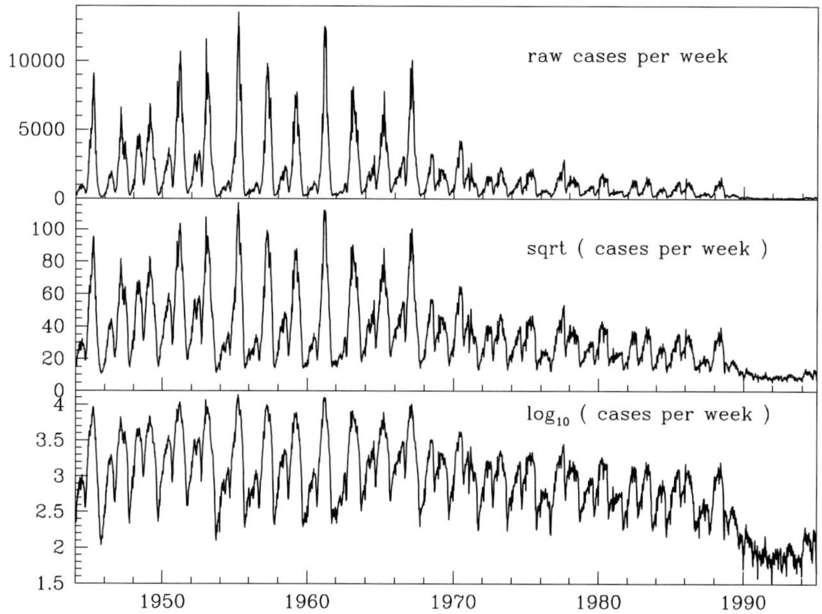

FIGURE 5. Transformed time plots of measles incidence in England and Wales from 1944 to 1994.

A moving average is the simplest example of a *linear filter*,

$$x_t \to \sum_{i=-\infty}^{\infty} \lambda_i x_{t+i}\,, \tag{2.2}$$

where the *weights* λ_i can be nonlinear functions of i [**Cha04**]. Linear filters are especially useful for extracting long-term trends that are clouded by noise or seasonality.

A useful way to think about recurrent patterns of epidemics is that the number of cases or deaths at a given time is correlated with the number at other times in the past or the future. As is probably familiar to you, given N pairs of observations of different quantities, $\{(x_i, y_i) : i = 1, \ldots, N\}$, the *correlation coefficient* is defined to be

$$r = \frac{\sum_{i=1}^{N}(x_i - \bar{x})(y_i - \bar{y})}{\sqrt{\sum_{i=1}^{N}(x_i - \bar{x})^2 \sum_{i=1}^{N}(y_i - \bar{y})^2}}\,, \tag{2.3}$$

where $\bar{x}$ and $\bar{y}$ are the means of $\{x_i\}$ and $\{y_i\}$, respectively. Given a single sequence of observations $\{x_t : t = 1, \ldots, N\}$, we can compute the correlation of each observation with the observation k time steps in the future. Thus, we consider the pairs of observations $\{(x_t, x_{k+t}) : t = 1, \ldots, N-k\}$ and define the *autocorrelation coefficient at lag k* to be

$$r_k = \frac{\sum_{t=1}^{N-k}(x_t - \bar{x}_{1,N-k})(x_{k+t} - \bar{x}_{k+1,N})}{\sqrt{\sum_{t=1}^{N-k}(x_t - \bar{x}_{1,N-k})^2 \sum_{t=1}^{N-k}(x_{k+t} - \bar{x}_{k+1,N})^2}}\,, \tag{2.4}$$

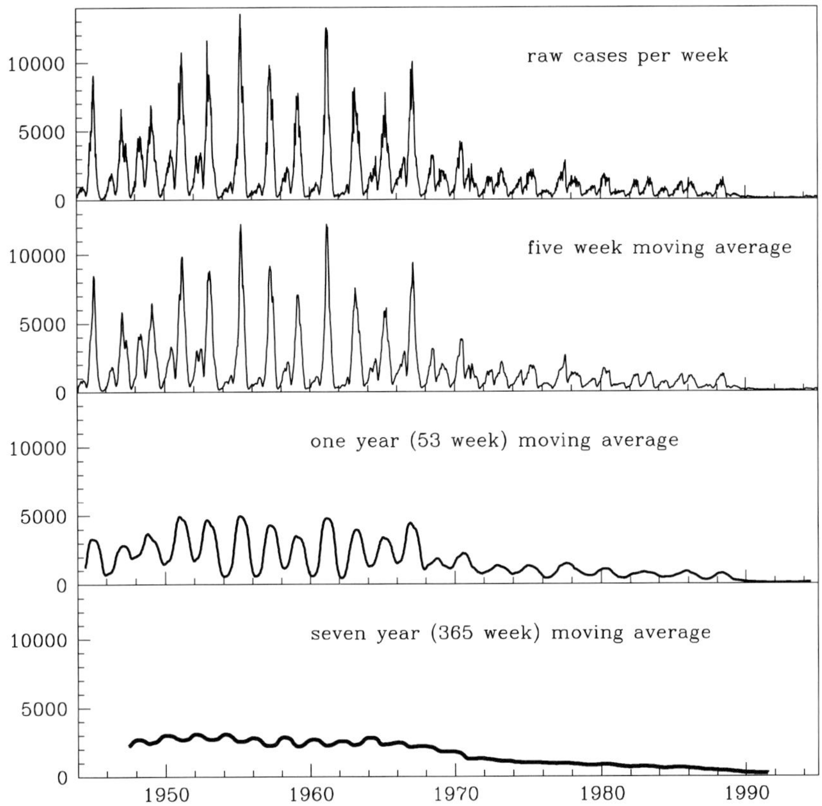

FIGURE 6. Smoothed time plots of measles incidence in England and Wales from 1944 to 1994.

where $\bar{x}_{1,N-k}$ and $\bar{x}_{k+1,N}$ are the means of first and last $N-k$ observations, respectively. If N is large and $k \ll N$ then $\bar{x}_{1,N-k} \simeq \bar{x}_{k+1,N} \simeq \bar{x}$, so

$$r_k \simeq \frac{\sum_{t=1}^{N-k}(x_t - \bar{x})(x_{k+t} - \bar{x})}{\sqrt{\sum_{t=1}^{N-k}(x_t - \bar{x})^2 \sum_{t=1}^{N-k}(x_{k+t} - \bar{x})^2}} \tag{2.5a}$$

$$\simeq \frac{\sum_{t=1}^{N-k}(x_t - \bar{x})(x_{k+t} - \bar{x})}{\frac{N-k}{N}\sum_{t=1}^{N}(x_t - \bar{x})^2}\,. \tag{2.5b}$$

Because this approximation depends on the assumption that $k \ll N$, the factor $N/(N-k) \sim 1$ and is usually ignored, yielding the standard formula

$$r_k \simeq \frac{\sum_{t=1}^{N-k}(x_t - \bar{x})(x_{k+t} - \bar{x})}{\sum_{t=1}^{N}(x_t - \bar{x})^2}\,. \tag{2.5c}$$

The factor $N/(N-k)$ increases with k, so ignoring it leads to an artificial reduction in r_k as k increases, but if k is large enough that this could affect any inferences of interest then a greater concern might be the approximation of the means of the first and last $N-k$ observations by the mean of all observations. In general, it is worth

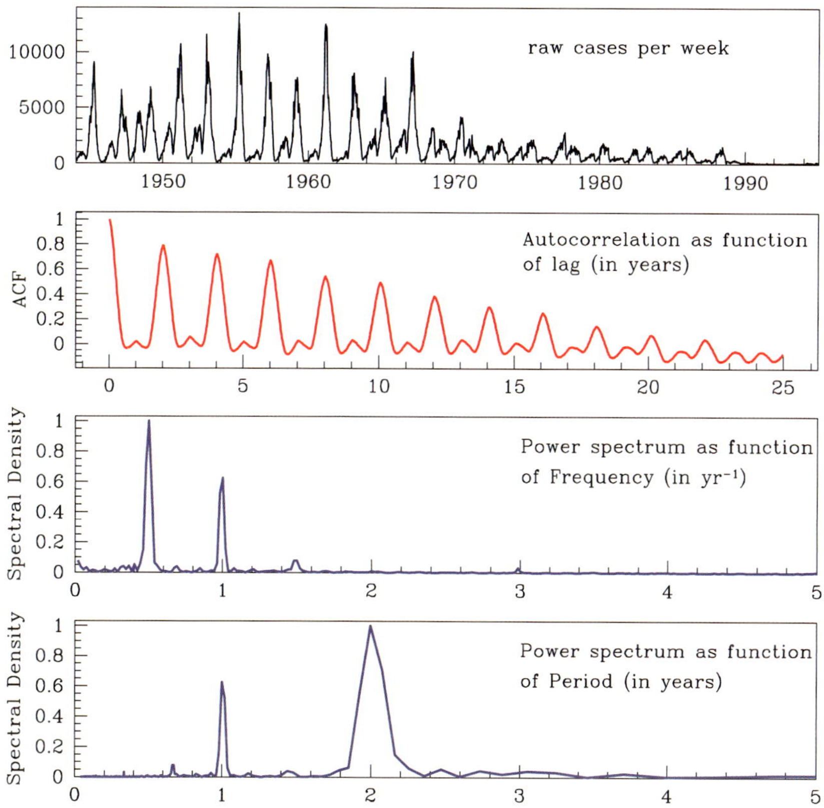

FIGURE 7. Time series analysis tools applied to measles incidence in England and Wales from 1944 to 1994. From top to bottom: the time plot, correlogram, periodogram, and period periodogram.

being aware that if the standard approximation (2.5c) is used then the precision of the autocorrelation coefficients decreases with lag.

A plot of the autocorrelation as a function of lag is known as the *correlogram* of the time series (see the second panel of Figure 7). The correlogram is a valuable indicator of the nature of the serial dependence in a time series. Nevertheless, since we are often more interested in the periodicities in the time series than in the degree to which correlation decays with lag, a frequency representation of the correlogram is usually preferred. The discrete Fourier transform of the correlogram is called the *periodogram*,

$$I(\omega_p) = \frac{1}{\pi}\left(r_0 + 2\sum_{k=1}^{N-1} r_k \cos\omega_p k\right), \tag{2.6}$$

where $\omega_p = 2\pi p/N$, $p = 1, \dots, N/2$. The periodogram can also be written directly in terms of the original time series. For example,

$$I(\omega_p) = \frac{1}{\pi N \mathrm{var}[x_t]}\left|\sum_{t=1}^{N} x_t e^{i\omega_p t}\right|^2 . \tag{2.7}$$

You may wonder why this quantity is of greater interest than the Fourier transform of $\{x_t\}$. The answer is simply that it has a much more natural interpretation. Unlike the Fourier transform of $\{x_t\}$, the periodogram partitions the variance in $\{x_t\}$, i.e., $I(\omega_p)$ is the proportion of the variance associated with the harmonic $\cos\omega_p t$ (hence $\sum_{p=1}^{N/2} I(\omega_p) = 1$). Thus, while the positions of the peaks in the periodogram will coincide with those in the Fourier transform of $\{x_t\}$, the magnitude of these peaks is much easier to interpret in the periodogram.

In spite of its name, the periodogram is a function of frequency. It is often more convenient to plot it as a function of period, in which case it is called the *period periodogram*. Figure 7 shows the time plot, correlogram, periodogram and period periodogram for the measles case series in England and Wales. Note that while it is easiest to identify the *periods* of the peaks from the period periodogram, the area under the plotted curve is not straightforward to interpret. Area under the normal (frequency) periodogram for a range of frequencies is equal to the proportion of the variance of the time series associated with those frequencies.

Although the time series that we analyze contain discrete data points that are separated in time, the underlying processes occur in continuous time. We therefore think of the correlogram as an estimate of the underlying continuous *autocorrelation function* and of the periodogram as an estimate of the underlying (continuous) *power spectral density* or *power spectrum*, which is defined for all frequencies (all $\omega > 0$ rather than only $\omega_p = 2\pi p/N$ for $p = 1, \ldots, N/2$).

Because any observed time series that we analyze must be finite, the autocorrelation function is always less well sampled at longer lags (at lag k, the number of pairs of points being compared is $N - k$). Consequently, one might wish to put greater weight on the coefficients at shorter lags, which can be achieved by linearly filtering the autocorrelation coefficients in the calculation of the periodogram, i.e., by replacing Eq. (2.6) with

$$I(\omega_p) = \frac{1}{\pi}\left(\lambda_0 r_0 + 2\sum_{k=1}^{M} \lambda_k r_k \cos\omega_p k\right). \tag{2.8}$$

Here, the set of weights $\{\lambda_k\}$ is called the *lag window* and M is known as the *truncation point*. Weighting coefficients "appropriately" is a bit of a black art, but a ubiquitous filter is the *Tukey window*,

$$\lambda_k = \frac{1}{2}\left(1 + \cos\frac{\pi k}{M}\right), \qquad k = 0, 1, \ldots, M. \tag{2.9}$$

You can gain valuable intuition by applying this procedure to a time series of interest and looking at the effects of adjusting the truncation point (M). A variety of procedures for estimating the power spectrum of a time series are discussed and compared by Chatfield [**Cha04**].

A serious limitation of the periodogram is that it does not account for changes in the structure of a time series over time. If different segments of a time series correspond to different external conditions (such as different birth rates) then we must break up the time series ourselves before computing the periodogram. More sophisticated spectral methods that overcome this limitation do exist. *Wavelet analysis* [**TC98, Nie01, GBK01**] provides a method for frequency decomposition that is local in time, so you can see changes in the spectrum over time without having to identify distinct temporal segments yourself.

2.5. Exercises

The best way to learn the techniques described above is to try to apply them to an interesting time series.

(1) Download a time series of recurrent epidemics from the International Infectious Disease Data Archive [**IIDDA**]. Nice examples to try first are the monthly incidence of measles in New York City from 1928 to 1964 or the weekly incidence of measles in England and Wales from 1944 to 1994.
(2) Analyze the time series as a whole by constructing (i) time plots of raw and transformed data, (ii) correlograms, (iii) periodograms, (iv) period periodograms. Calculate the periodograms with and without a Tukey window, and examine the effects of different truncation points.
(3) Repeat these analyses for segments of the time series that look different by eye.
(4) Is there evidence for: Frequency components not evident by eye? Changes in the frequency structure over long time scales? If so, why might this have occurred? What if you remove trend and/or seasonality?
(5) If you're ambitious, explore the time series using wavelet analysis, e.g., via the MATLAB *Wavelet Toolbox* or Torrence and Compo's software [**TC98**] available at `http://atoc.colorado.edu/research/wavelets/`.

If you are so inclined, there are enough time series on the IIDDA web site to keep you busy making interesting discoveries for a long time. You can also find scripts on the IIDDA web site that will enable you to do all the analyses suggested in the exercises without programming anything yourself, but I encourage you to write your own code — you will learn more that way.

3. Modelling epidemics

As we have seen, the character of epidemic patterns differs for different diseases and even for the same disease in different places or at different times. Can models help us predict these differences, forecast future epidemics, or develop an understanding of the mechanisms responsible for the various patterns? Specifically, can models help us understand the structure of an epidemic time series as displayed by the time plot, correlogram or periodogram?

It is not clear at the outset how we should even begin trying to develop potentially useful models. Where you start will depend on what background and experience you have with other problems. I will discuss two standard approaches, which have their own strengths and weaknesses. The first approach is useful if you have little to go on beyond a time series of cases or deaths that you would like to understand, while the second approach is generally more powerful if you have more knowledge of how the fundamental biological processes might be related to observed events and some information in addition to a case or death time series.

3.1. Statistical modelling

Epidemic processes are ultimately stochastic. They depend on contacts between individuals (which occur randomly at least to some extent) and infections (which occur upon contact with some probability less than 1). We can build a *statistical model* using nothing but the fact there there is a stochastic process involved, and that the observed data we have are in the form of a time series $\{x_t\}$.

As a first attempt, we could imagine that the number of cases at time t, X_t, is simply a random variable with a known distribution. Our model for the epidemic process would then be simply

$$X_t = \mu + Z_t\,, \tag{3.1}$$

where μ is the time average number of cases and Z_t is a sequence of independent and identically distributed (IID) random variables with zero mean. The parameters of this model are μ and the moments (of order 2 and higher) of the distribution of Z_t. We could then fit the model parameters to the mean, variance, etc. of our observed data. This might be a reasonable model for the importation of new, infectious individuals into a focal community, but it clearly fails to take into account that infection is transmitted from one individual to another, so there must be some correlation between the number of individuals in the focal community who are infected now and the number who will be infected in the near future.

So, imagine that the time series is generated by random processes, but that successive data points are correlated. For example, perhaps the data are generated by an *autoregressive* or "AR(p)" process,

$$X_t - \mu = \alpha_1(X_{t-1} - \mu) + \alpha_2(X_{t-2} - \mu) + \cdots + \alpha_p(X_{t-p} - \mu) + Z_t\,, \tag{3.2}$$

where the α_i are constants that determine the degree of correlation along the time series and Z_t is as in Eq. (3.1). Alternatively, the data might be generated by a *moving average* or "MA(q)" process,

$$X_t - \mu = \beta_0 Z_t + \beta_1 Z_{t-1} + \cdots + \beta_q Z_{t-q}\,, \tag{3.3}$$

where the β_i are constants that define a weighted average. More generally, the data might be generated by an *autoregressive moving average* or "ARMA(p, q)" process,

$$X_t - \mu = \alpha_1(X_{t-1} - \mu) + \alpha_2(X_{t-2} - \mu) + \cdots + \alpha_p(X_{t-p} - \mu) \tag{3.4a}$$

$$+ \beta_0 Z_t + \beta_1 Z_{t-1} + \cdots + \beta_q Z_{t-q}\,. \tag{3.4b}$$

Of course, that the data *might* be generated by such a process is not especially enlightening. Can we identify a "good" model of our data from the infinite class of ARMA(p, q) processes?

You can develop some intuition about time series models by plotting the correlogram and periodogram for a variety of ARMA(p, q) processes with specific parameter values. The autocorrelation function of a first-order AR process declines exponentially, while the autocorrelation of an MA(q) process drops to zero at lag q. Unfortunately, for a more general ARMA(p, q) the theoretical autocorrelation function is complicated so, unless the correlogram of our observed time series is as simple as that of a pure AR(1) or MA(q) process, we need a quantitative method to identify the best ARMA(p, q) model.

But what do we mean by a "best" model? The usual answer is the most *parsimonious* model, i.e., the model that gets as close as possible to the data using the fewest parameters. If we think an observed time series $\{x_t\}$ was generated by an AR(p) process then we can attempt to find the best model by minimizing the residual sum of squares,

$$S = \sum_{t=p+1}^{N} \left[(x_t - \mu) - \alpha_1(x_{t-1} - \mu) - \cdots - \alpha_p(x_{t-p} - \mu)\right]^2. \tag{3.5}$$

If we compute the least-squares fit for a range of orders (p) and plot S against p, we will likely find that there are diminishing returns for increasing the order beyond a certain point, which might suggest a "most parsimonious" AR(p) model. Of course, this model can be considered "best" only within the constraint that the distribution of noise (Z_t) is taken as given. Experience shows, moreover, that fewer parameters are typically required to model a process as ARMA(p, q) rather than pure AR(p) or MA(q) processes. So we really need to compute the residual sum of squares for ranges of p and q and for a variety of possible noise distributions, and consider the number of parameters involved in each model. This is not straightforward. We really need a quantitative criterion that we can apply blindly to a wide variety of models. Unsurprisingly, the development of such a criterion is something to which statisticians have devoted much time and effort.

We know that statistical models involving more parameters will generally yield fits that are closer — in the least squares sense — to observed data. We need a formal away to penalize models for using more parameters so we can arrive at an objective measure of the statistical significance of the fit. There are many ways to do this, the most common being the *Akaike Information Criterion* (AIC). The AIC depends on the notion of the *likelihood* of the model given the data, which is defined to be the probability of observing the data given the model. If $L(\theta|x_t)$ denotes the likelihood of the model with parameter values θ, given the observed time series $\{x_t\}$, then the AIC is

$$\text{AIC} = -2\ln\left[\max_{\theta} L(\theta|x_t)\right] + 2k\,, \tag{3.6}$$

where k is the number of independent parameters that are fitted for the model being assessed (i.e., the dimension of the vector θ). By choosing the model with the lowest value of the AIC we balance the closeness of the fit (maximum likelihood) with the complexity of the model (minimum number of parameters). It is often assumed that measurement errors in the data are normally distributed about the "true" model, in which case maximizing the likelihood is equivalent to minimizing the residual sum of squares [**PTVF92**] (e.g., Eq. (3.5) for an AR(p) model). With this assumption, Eq. (3.6) becomes

$$\text{AIC} = N\ln(S/N) + 2k\,, \tag{3.7}$$

where N is the number of observations in the time series.

Once we have arrived at a "best fit" model, we can use it to *forecast* future observations. The natural way to test this procedure is to fit a model to part of the observed time series and then use the resulting fit to predict the remainder of the data. How successful is this likely to be for an infectious disease time series like one of those discussed above?

The answer depends on which time series we are interested in and which segment of the data we start with. Consider, for example, weekly chicken pox in Ontario (Figure 3). If we fit an ARMA(p, q) model using the first 20 years of the 50 year time series and then use the model to forecast the remaining 30 years of data, we can expect to do reasonably well. On the other hand, if we fit an ARMA(p, q) model to the New York City monthly measles reports from 1928 to 1945 (top right panel of Figure 2), we will fail miserably at predicting the monthly case reports from 1946 to 1964. The reason is that the first and second halves of the New York City measles time series have completely different structure (apparently irregular until 1945 and strictly biennial after 1945). We cannot expect to predict the second half given only the first half. We need further information — such as the fact that the birth rate rose dramatically after 1945 — and we need to know how to alter our fitted ARMA(p, q) model in light of this information. Unfortunately, we will never have the information required to alter the fitted parameters of our statistical time series model, because the parameters of the model have no biological meaning. This is a serious problem that is not easy to overcome.

You might hope that you could do better with a more sophisticated time series model. ARMA(p, q) models are relatively simple and you can often obtain better fits to data (with lower AIC) using more elaborate statistical models [**Scu03**]. In particular, a time series analyst would probably advise you to first fit trends and cycles to the observed data and remove these features before attempting to fit an ARMA(p, q) or more sophisticated model. But this emphasizes the problem, because we are usually more interested in understanding trends and changes in cyclicity than in modelling small stochastic fluctuations in detail.

Ultimately, statistical time series models are useful as forecasting tools only if the underlying stochastic processes are *stationary*, which means roughly speaking that the processes are identical for the period during which the parameters are fit and the period over which predictions are sought. If external conditions change in a way that affects the processes that generate the data we observe, then we need a different modelling tool.

Much more detail on statistical modelling of time series can be found in the excellent text by Chatfield [**Cha04**].

3.2. Mechanistic modelling

A fundamental problem with statistical time series models is that they use the observed time series itself to estimate parameters. It would be much better to have a model that we can parameterize from independently collected data and then explore whether that model can explain the observed time series. This more desirable approach can be pursued only if we have a lot more information than the time series we are trying to explain. Fortunately, when studying time series of infectious disease epidemics this is precisely the situation in which we usually find ourselves.

Rather than beginning with a time series, let's consider what we know about the biological processes that must ultimately give rise to the observed cases or deaths, i.e., the mechanisms of disease transmission and spread. To be specific, think about a childhood disease such as measles or chicken pox. Initially, perhaps after a period of maternally-acquired immunity has waned, each individual is *susceptible* to the disease in question. If some individuals in the population have been exposed to the disease-causing agent and are currently at the *infectious* stage then they can transmit the pathogen to susceptible individuals. For typical childhood diseases, recovery entails essentially lifelong immunity, so regardless of whether infected individuals eventually recover or die, we can consider them to be *removed* from the transmission process.

We are thus naturally led to a simple model in which the host population (humans) is divided into three compartments, each containing individuals who are identical in terms of their status with respect to the disease in question. In this "SIR model", which is depicted as a flow chart in Figure 8, the three compartments are:

- *Susceptible*: individuals who have no immunity to the infectious agent, so might become infected if exposed.
- *Infectious*: individuals who are currently infected and can transmit the infection to susceptible individuals who they contact.
- *Removed*: individuals who are immune to the infection, and consequently do not affect the transmission dynamics in any way when they contact other individuals.

It is traditional to denote the number of individuals in each of these compartments as S, I and R, respectively. The total host population size is $N = S + I + R$.

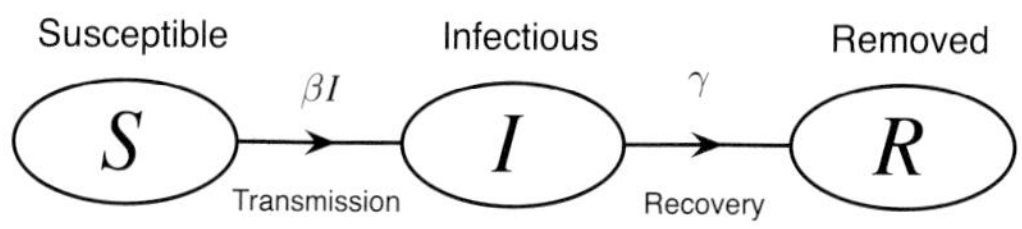

FIGURE 8. The basic SIR model. Individuals in the host population are in one of three possible states: *Susceptible*, *Infectious*, or *Removed*. The rates of flow between compartments are specified relative to the number of individuals in the state in question. The differential equation version of this model is given in Eq. (3.8).

Having compartmentalized the host population, we now need a set of equations that specify how the sizes of the compartments change over time. Solutions of these equations will yield, in particular, $I(t)$, the size of the infectious compartment at time t. A plot of $I(t)$ should bear a strong resemblance to observed epidemic curves if this is a reasonable model.

The numbers of individuals in each compartment must be integers, of course, but if the host population size N is sufficiently large we can treat S, I and R as continuous variables and express our model for how they change in terms of a system of differential equations,

$$\frac{dS}{dt} = -\beta SI\,, \tag{3.8a}$$

$$\frac{dI}{dt} = \beta SI - \gamma I\,. \tag{3.8b}$$

Here, the *transmission rate* (per capita) is β and the *recovery rate* is γ (so the *mean infectious period* is $1/\gamma$). In writing Eqs. (3.8) we are assuming implicitly that contacts among individuals are random and, moreover, that the population is homogeneously mixed (like an ideal gas). Note that we do not need a differential equation for the number of removed individuals. The appropriate equation is $dR/dt = \gamma I$ (outflow from the I compartment goes into the R compartment) but since R does not appear in equations (3.8) the equation for dR/dt has no effect on the dynamics of S and I (formalizing the fact that removed individuals cannot affect transmission). This basic SIR model has a long history [**KM27**] and is now so standard that you can even find it discussed in some introductory calculus textbooks [**HHGL$^+$06**].

If everyone is initially susceptible ($S(0) = N$), then a newly introduced infected individual can be expected to infect other people at the rate βN during the expected infectious period $1/\gamma$. Thus, this first infective individual can be expected to infect $\mathcal{R}_0 = \beta N/\gamma$ individuals. The number $\mathcal{R}_0$ is called the *basic reproduction number* and is unquestionably the most important quantity to consider when analyzing any epidemic model for an infectious disease. In particular, $\mathcal{R}_0$ determines whether an epidemic can occur at all; to see this for the basic SIR model, note in equations (3.8) that I can never increase unless $\mathcal{R}_0 > 1$. This makes intuitive sense, since if each individual transmits the infection to an average of less than one individual then the number of cases must decrease with time.

Intuitively, if you introduce a pathogen with $\mathcal{R}_0 > 1$ into a fully susceptible population then you expect the number of cases to grow exponentially until a large fraction of the susceptible have been infected. The epidemic curve must eventually turn over. Simple analyses of the basic SIR equations (3.8) with $\mathcal{R}_0 > 1$ show that the SIR model does exactly this (see exercises below). But it is also easy to show that Eqs. (3.8) always lead to exactly one maximum in $I(t)$ so we can never predict more than one epidemic. This means that the basic SIR model cannot account for the recurrent epidemics described above. In fact, it can't even explain persistence of a disease in a population. We must include more structure in our model, which we can do by considering more carefully the biological processes involved.

One aspect of disease transmission that we have ignored is that there is a delay between the moment at which an individual is infected and the beginning the period of infectiousness (the *latent period*). To account for this delay, an *exposed* compartment is usually added ("exposed" individuals are infected but not yet infectious), yielding an "SEIR model". In this way, we introduce one more equation and one more parameter, the rate (σ) of flow from the exposed to the infectious compartment. The mean latent period ($1/\sigma$) can often be estimated, so adding this parameter doesn't get us into trouble, but the dynamics of the basic SEIR model are almost identical to the SIR model. We still cannot explain persistence, let alone cyclical recurrence. We are still missing something fundamental.

The key element that is absent from our discussion so far is a source of susceptible individuals, the most obvious being births. If we are considering timescales over which births are important then we also need to account for deaths (from causes other than the disease in question). The streams of births and deaths together are known as the *vital dynamics* of the system. Of course, only births of individuals who are not immune affect the transmission dynamics of the disease in question. For the childhood diseases I have mentioned, many young children are vaccinated before ever being exposed to the associated pathogens, so we can consider the proportion of the population who has been vaccinated to have been born into the removed compartment. If the birth rate is ν ("nu for natality"), the *per capita* death rate is μ ("mu for mortality") and the proportion vaccinated is p, then we arrive at the "SEIR model with vital dynamics and vaccination" depicted as a flow chart in Figure 9. Expressing this model using differential equations we obtain

$$\frac{dS}{dt} = \nu(1-p) - \beta SI - \mu S\,, \tag{3.9a}$$

$$\frac{dE}{dt} = \beta SI - \sigma E - \mu E\,, \tag{3.9b}$$

$$\frac{dI}{dt} = \sigma E - \gamma I - \mu I\,, \tag{3.9c}$$

$$\frac{dR}{dt} = \nu p + \gamma I - \mu R\,. \tag{3.9d}$$

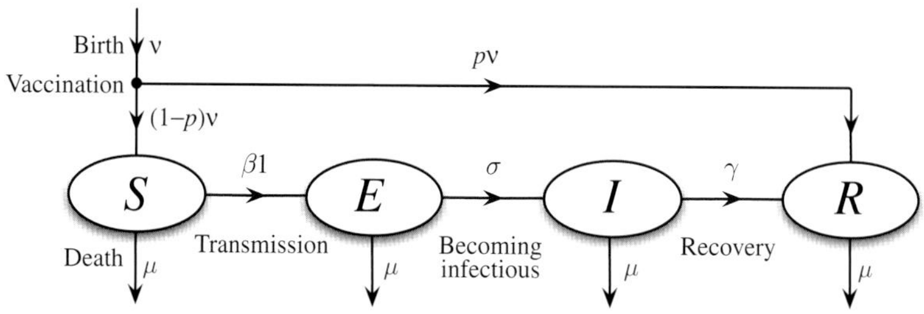

FIGURE 9. The SEIR model with vital dynamics and vaccination. This expands the model depicted in Figure 8 to include a class of individuals who have been *Exposed* to the pathogen and are currently infected, but are not yet infectious. In addition, births occur at rate ν, deaths occur at *per capita* rate μ, and the rate at which exposed individuals become infectious is σ (so $1/\sigma$ is the *mean latent period*, the mean time from the moment of initial infection to the beginning of the infectious period). This model is represented as a system of differential equations in Eq. (3.9).

If births exactly balance deaths ($\nu = \mu N$) then the population size ($N = S+E+I+R$) is constant and one of these equations is superfluous. In this expanded model, the basic reproduction number $\mathcal{R}_0$ is still approximately, but no longer exactly, $\beta N/\gamma$ (see exercises below).

Equilibrium and stability analyses of Eqs. (3.9) lead to important conclusions. For any parameter set, there is always at least one equilibrium, the *disease free equilibrium* or DFE corresponding to no infected individuals at all ($E = I = 0$). If $\mathcal{R}_0(1-p) < 1$ then the DFE is globally asymptotically stable. If $\mathcal{R}_0(1-p) > 1$ then the DFE is unstable and there is a second equilibrium, the *endemic equilibrium* corresponding to a constant number of exposed and infected individuals ($E > 0$ and $I > 0$), which is globally asymptotically stable [**LM95, KM04**].

The existence of a globally stable endemic equilibrium immediately provides an explanation for persistence of infectious diseases. However, it is the condition for global stability of the DFE that is in fact most important. Since the proportion vaccinated is under our control, we can re-arrange the condition $\mathcal{R}_0(1-p) < 1$ to find a critical vaccination proportion,

$$p_{\text{crit}} = 1 - \frac{1}{\mathcal{R}_0}, \tag{3.10}$$

above which the disease cannot persist. Consequently, if we can estimate $\mathcal{R}_0$ for a given disease then we can also estimate the threshold vaccination level that we must reach in order to eradicate the infection. This is probably the single most important conclusion that has been drawn from mathematical analysis of epidemic models, and it helps us understand why smallpox (with $\mathcal{R}_0 \sim 4$) was successfully eradicated in the 1970s, while measles (with $\mathcal{R}_0 \sim 20$) is still with us. But from the point of view of our present discussion, the critical vaccination level is just a bonus. We were actually aiming to understand something else, the recurrent pattern of epidemics seen in most of the data we discussed.

A standard local stability analysis (see exercises below) reveals that the endemic equilibrium is normally approached by damped oscillations with period

$$T = 2\pi\sqrt{\frac{1/\sigma + 1/\gamma}{\nu(\mathcal{R}_0 - 1)}} \tag{3.11}$$

This suggests an explanation of cyclical epidemic patterns and a method to predict the inter-epidemic period. However, as the examples discussed above show, epidemic cycles seen in practice are often undamped, and the inter-epidemic period can change over time and differ from place to place. So, what are we missing?

3.3. Demographic stochasticity

One thing we have glossed over is the presence of noise. While it is true that for sufficiently large population size N it is reasonable to treat S, E, I and R as continuous variables, it is not true that the genuine discreteness of the number of individuals in each compartment has no observable effect.

If we re-cast the SEIR model as a stochastic process [**Bai75**] then our ability to derive conclusions analytically rapidly evaporates. Fortunately, we can simulate stochastic models on a computer and numerically explore the significance of demographic noise. Gillespie [**Gil76**] showed that a simple algorithm provides rigorously correct realizations of the sorts of stochastic processes that arise from systems like the SIR and SEIR models that concern us here. So, we can run many realizations and infer the expected behaviour of genuinely stochastic epidemics.

Gillespie's algorithm works as follows. Let a_1, a_2, ... , be the rates at which the various processes occur. For example, a_1 might be the birth rate, a_2 might be the rate of going from susceptible to exposed, a_3 might be the rate of going from infectious to removed (recovering) and so on. Let a_0 be the overall event rate, i.e., $a_0 = \sum_i a_i$, so the average time between events is $1/a_0$. If the transitions between the various states are Poisson processes then the time spent in any given state is exponentially distributed and the probability that the next event occurs between times t and $t + dt$ is $a_0 e^{-a_0 t} dt$. If we let $u = e^{-a_0 t}$ then $u \in [0, 1]$ and $du = a_0 e^{-a_0 t} dt$, so u is uniformly distributed in $[0, 1]$. Thus, the time t to the next event can be obtained by sampling u from a uniform distribution in $[0, 1]$ and setting $t = \frac{1}{a_0} \ln(1/u)$. The nature of the event is then determined by sampling a point from a uniform distribution on $[0, a_0]$ and setting the event to type i if the uniform deviate lies in the ith interval in the following list.

$$[0, a_1),\ [a_1, a_1 + a_2),\ \ldots,\ [a_1 + \cdots + a_{i-1}, a_1 + \cdots + a_i),\ \ldots \tag{3.12}$$

Using Gillespie's algorithm, we can easily confirm the discovery made 50 years ago by Bartlett [**Bar57**] that a relatively small amount of noise is sufficient to prevent the oscillations of the SIR model from damping out. In the stochastic SIR model [**Bai75, Gil76**] demographic noise sustains the oscillations that damp out in the deterministic model (Figure 10).

Again, this is progress that has arisen from an important mechanistic insight. But we are left with another puzzle. If you look carefully at the New York City measles reports in the bottom right panel of Figure 2 you'll see that before about 1945 the epidemics were fairly irregular, whereas after 1945 they followed an almost perfect two-year cycle. Even with oscillations sustained by noise, the SIR model cannot explain why the measles epidemic pattern in New York City changed in this way. Have we missed another important mechanism?

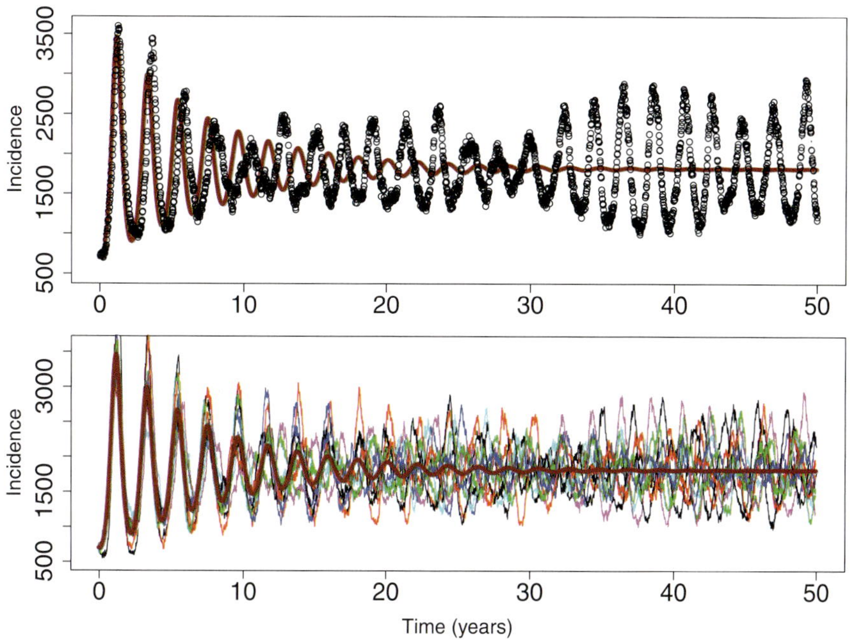

FIGURE 10. Simulations of the SEIR model model with vital dynamics, for a population of 5 million individuals. The birth and death rates are $\nu = \mu = 0.02$, the basic reproduction number is $\mathcal{R}_0 = 17$, and the mean latent and infectious periods are 8 and 5 days, respectively. In both panels, the heavy solid red line is the solution of the deterministic model (3.9). The top panel also shows a single realization of the associated stochastic process computed with the Gillespie algorithm; it is clear that a small amount of noise prevents the deterministically damped oscillations from decaying. The bottom panel shows 10 Gillespie realizations starting from the same initial conditions.

3.4. Seasonal forcing

So far, we have been assuming implicitly that the transmission rate β (or, equivalently, the basic reproduction number $\mathcal{R}_0$) is simply a constant and, in particular, that it does not change in time. Let's think about that assumption. The transmission rate is really the product of the rate of contact among individuals and the probability that a susceptible individual who is contacted by an infectious individual will become infected. But the contact rate is *not* constant throughout the year. To see that, consider the fact that in the absence of vaccination the average age at which a person is infected with measles is about five years [**AM91**], hence most susceptibles are children. Children are in closer contact when school is in session, so the transmission rate must vary seasonally.

A crude approximation of this seasonality is to assume that β varies sinusoidally,

$$\beta(t) = \langle\beta\rangle(1 + \alpha \cos 2\pi t)\,. \tag{3.13}$$

Here, $\langle\beta\rangle$ is the mean transmission rate, α is the *amplitude* of seasonal variation ($0 \le \alpha \le 1$) and time t is assumed to be measured in years. If, as above, β is assumed to be a periodic function (with period one year) then the SIR model is said to be *seasonally forced*.

You might think that seasonal forcing is just a minor tweak of the model. In fact, this forcing has an enormous impact on the epidemic dynamics that the model predicts. If you've ever studied the forced pendulum then you might already have some intuition for this. A pendulum with some friction will exhibit damped oscillations and settle down to an equilibrium. But if you tap the pendulum with a hammer periodically then it will never settle down and it can exhibit quite an exotic range of behaviours including chaotic dynamics [**Gle87, Str94**] (oscillations that look random). Similarly complex dynamics can occur in the seasonally forced SIR model.

Importantly, with seasonal forcing the SIR model displays *un*damped oscillations, even in the absense of stochasticity. More than that, for different parameter values, the seasonally forced SIR model can produce all the different types of oscillatory measles patterns that have ever been seen in real data. The range of possible dynamical behaviour is summarized by the *bifurcation diagram* shown in Figure 11.

We are definitely making good progress. With a fixed set of parameters we can explain epidemic cycles of many different periods (whereas, for given set of parameters, the unforced model always predicted the same oscillation period given by Eq. (3.11)). However, as noted in the previous section, the observed measles epidemics in New York City show very clearly that the dynamical pattern changed over time (bottom right panel of Figure 2) and other significant qualitative changes have been observed in measles case series in other places [**ERBG00**]. How can we explain *changes over time* in the *pattern* of epidemics?

If you are familiar with chaotic dynamics, then one possible solution might occur to you, namely *intermittency* [**PST93**]. Some chaotic dynamical systems appear to switch randomly from one type of oscillation to another. Perhaps transitions from cycles of various kinds to irregular epidemics are manifestations of intermittency in a system that is fundamentally chaotic? It has been proposed that measles dynamics are chaotic [**SK85, OS90, SKTO93**], but there is a serious problem with the chaotic solutions of the seasonally forced SIR model: as is evident from Figure 11, they often reach absurdly small infected populations (e.g., proportion infected $< 10^{-15}$). In practice, this means that if the epidemic pattern is chaotic then the pathogen can be expected to go extinct in a relative short time, which is not observed.

What biological mechanism that we have not yet included might repair the incorrect prediction of rapid extinction? One possibility might be that the population is not really homogeneously mixed. In particular, young children mix a lot more with others of their own age than they do with teenagers or adults. The influence of age-dependent contact structures on measles dynamics was investigated by Schenzle [**Sch84**], who developed a *realistically age-structured* (RAS) model involving 84 differential equations (SEIR in each of 21 age classes). Analysis of this model showed that it did seem to avoid predicting rapid extinctions and that it seemed to explain the detailed temporal dynamics of individual measles time series [**BG93**].

However, later analysis indicated that the critical ingredient in the RAS model was not really that contact rates are age structured, but that they are higher during school terms [**ERBG00**]. This may seem like a step backwards — after all, the point of introducing seasonal forcing was precisely to model the seasonal variation

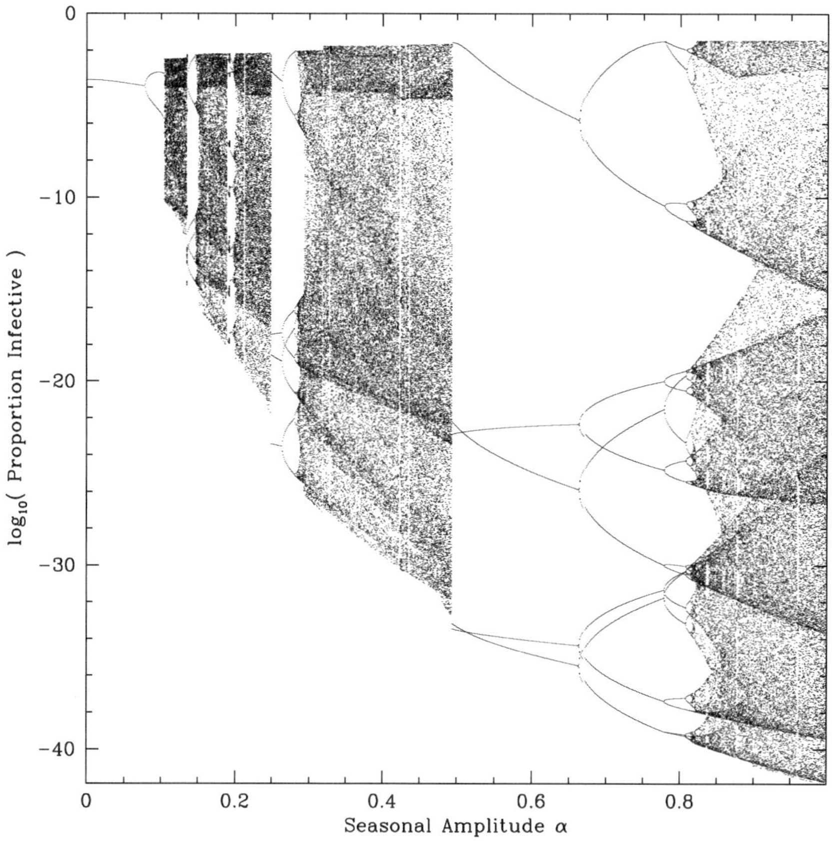

FIGURE 11. Bifurcation diagram for the sinusoidally forced SIR model with $\mathcal{R}_0 = 18$, $1/\gamma = 5$ days, $\mu = 0.02\,\text{yr}^{-1}$. The seasonal amplitude α is defined by Eq. (3.13). For each value of α, the sinusoidally forced SIR equations were integrated using the standard fourth order Runge-Kutta scheme [**PTVF92**] with a time-step of 0.5 day. After ignoring a 1000 year transient, the proportion of the population that was infectious (I) was plotted once a year for 200 years. Thus, a single point above a given value of α indicates an annual cycle, two points correspond to a biennial cycle, etc., and a broad band of points indicates aperiodic behaviour (chaos). Because only one solution is plotted for each α, this bifurcation diagram is incomplete. For some ranges of α there are, in fact, multiple co-existing stable cycles [**SS83**].

in contact rates due to school terms. But the true variation in contact rates is not sinusoidal. Schenzle's RAS model includes not only 21 age classes, but genuine *term-time forcing*, i.e., rather than Eq. (3.13), the RAS model uses

$$\beta(t) = \begin{cases} \beta_{\text{H}} & \text{school days,} \\ \beta_{\text{L}} & \text{non-school days.} \end{cases} \tag{3.14}$$

i.e., the transmission rate is high on days when school is in session (β_{H}) and low otherwise (β_{L}). For the purposes of comparing results with those obtained with the sinusoidally forced model, it is more convenient to express the term-time forcing function in terms of the mean transmission rate $\langle\beta\rangle$ and the amplitude of seasonal forcing α. If p_{s} is the proportion of days in school then the mean transmission rate is

$$\langle\beta\rangle = p_{\mathrm{s}}\beta_{\mathrm{H}} + (1 - p_{\mathrm{s}})\beta_{\mathrm{L}} \tag{3.15}$$

and we define the amplitude of seasonality to be

$$\alpha = \frac{1}{2}\Big(\frac{\beta_{\mathrm{H}} - \beta_{\mathrm{L}}}{\langle\beta\rangle}\Big). \tag{3.16}$$

In terms of $\langle\beta\rangle$ and α we can write the transmission rate

$$\beta(t) = \begin{cases} \big[(1 + 2(1 - p_{\mathrm{s}})\alpha\big]\langle\beta\rangle & \text{school days,} \\ \big[1 - 2p_{\mathrm{s}}\alpha\big]\langle\beta\rangle & \text{non-school days.} \end{cases} \tag{3.17}$$

This form for $\beta(t)$ is more convenient than (3.14) because it is $\langle\beta\rangle$ that corresponds to β in the unforced model (for which $\alpha = 0$). Note that since $\beta(t) \geq 0$, it is clear from equation (3.17) that the range over which the seasonal amplitude α can be varied is

$$0 \leq \alpha \leq \frac{1}{2p_{\mathrm{s}}}. \tag{3.18}$$

At the extreme of $\alpha = 1/2p_{\mathrm{s}}$, $\beta_{\mathrm{L}} = 0$, i.e., $\beta(t) = 0$ on non-school days. As an example, note that if we apply Eq. (3.18) to England and Wales, where a typical year contains 198 school days, we must have $0 \leq \alpha \leq 365/396 \simeq 0.9217$.

In terms of its ability to account for observed measles dynamics, the term-time forced SEIR model is just as successful as the full RAS model, and avoids the need to estimate—or guess the values of—many new parameters [**ERBG00**]. The model predicts that the disease will persist and it reproduces many detailed features of the biennial pattern of measles epidemics observed in New York City and London. But it does perhaps *too good* a job of reproducing this behaviour! No other epidemic patterns are predicted, so we are back to a situation in which we cannot explain changes in the cyclical structure of the time series. We will need to think again if we want to predict the observed dynamical transitions.

3.5. Exercises

If you are not familiar with the notions of local and global asymptotic stability, or you need to refresh your memory, you can consult any standard text on dynamical systems (e.g., [**GH83, Str94, Wig03**]) or mathematical ecology (e.g., [**BCC01**]).

(1) Find an exact analytical formula for the phase portrait of solutions to the basic SIR equations (3.8). *Hint:* Consider dI/dS.
(2) In the basic SIR model (3.8), show that the number of cases must eventually decrease, regardless of initial conditions or parameter values.
(3) Consider the SIR and SEIR models with vital dynamics, but without vaccination.
    (a) Use biological reasoning to derive an exact expression for $\mathcal{R}_0$ and show that $\mathcal{R}_0 \simeq \beta N/\gamma$. *Hint:* Consider the average length of time spent in each compartment.

(b) Prove that there is always a disease free equilibrium (DFE) and there is an endemic equilibrium if and only if $\mathcal{R}_0 > 1$.

(c) Show that the DFE is globally asymptotically stable if $\mathcal{R}_0 < 1$ and unstable if $\mathcal{R}_0 > 1$. Show that the endemic equilibrium is locally asymptotically stable if $\mathcal{R}_0 > 1$, and that for parameter values that could correspond to a childhood disease the equilibrium is approached by damped oscillations (the relevant eigenvalue has a non-zero imaginary part) with period given by Eq. (3.11).

(4) Use a numerical ODE solver to explore solutions of the sinusoidally forced SIR model for measles ($\mathcal{R}_0 = 18$, mean infectious period $1/\gamma = 5$ days) for a range of seasonal forcing amplitudes. Compare the character of the solution with what you would have guessed from the bifurcation diagram in Figure 11.

4. Predicting epidemics

We have converged on a model—the term-time forced SEIR model—that does an excellent job of reproducing measles dynamics in a few places at particular times (e.g., New York City from 1945 to 1964, bottom right panel of Figure 2). To explain the occurrence of different measles patterns in other places and at other times, we do not need to include more structure in our model. We simply need to recognize that we have assumed implicitly that all the parameters of the model are fixed constants, whereas some of these parameters might change over time for reasons that are external to the process of disease transmission. In particular, the birth rate (ν) is not constant, and neither is the proportion of the population that is vaccinated (p).

We might, therefore, be able to predict different dynamics based on different birth rates and vaccination levels. Before attempting to do so, however, it is extremely useful to note [**ERBG00**] that changes in ν or p are equivalent—in terms of their dynamical influence—to changes in the mean transmission rate $\langle\beta\rangle$ (or, alternatively, the basic reproduction number $\mathcal{R}_0$). To see this, suppose that a proportion of newborns is vaccinated (after maternally acquired immunity wanes) so $p > 0$ in Eqs. (3.9a) and (3.9d). Now consider a simple change of variables,

(4.1)
$$S = S'(1-p)\,, \qquad E = E'(1-p)\,, \qquad I = I'(1-p)\,, \qquad R = R'(1-p) + \frac{\nu}{\mu}p\,.$$

The dynamical equations for the primed variables are exactly the SEIR equations (3.9) *without vaccination*, except that the transmission rate β is everywhere replaced by $\beta(1-p)$. Thus, except for an overall reduction in the number of cases, when a proportion of a population is vaccinated the dynamics are identical to those of an unvaccinated SEIR system with a smaller mean transmission rate, $\langle\beta\rangle \to \langle\beta\rangle(1-p)$.

Since the birth rate ν enters Eq. (3.9a) in exactly the same place as vaccination, a similar argument applies to changes in birth rate. A change in birth rate from ν to ν' is dynamically equivalent to a change in mean transmission rate from $\langle\beta\rangle$ to $\langle\beta\rangle(\nu'/\nu)$. More generally, if the birth rate changes from ν to ν' and the vaccination proportion changes from p to p' then the resulting dynamics are identical to those based on no change in ν or p but a change in the mean transmission rate,

$$\langle\beta\rangle \to \langle\beta\rangle \frac{(1-p')}{(1-p)} \frac{\nu'}{\nu} . \tag{4.2}$$

This strongly suggests that we will be able to predict changes in dynamics based on demographic data (which tell us how ν and p change with time) and a single bifurcation diagram, where the control parameter is the mean transmission rate $\langle\beta\rangle$. The predicted dynamics will depend on the *susceptible recruitment rate*, $\nu(1-p)$.

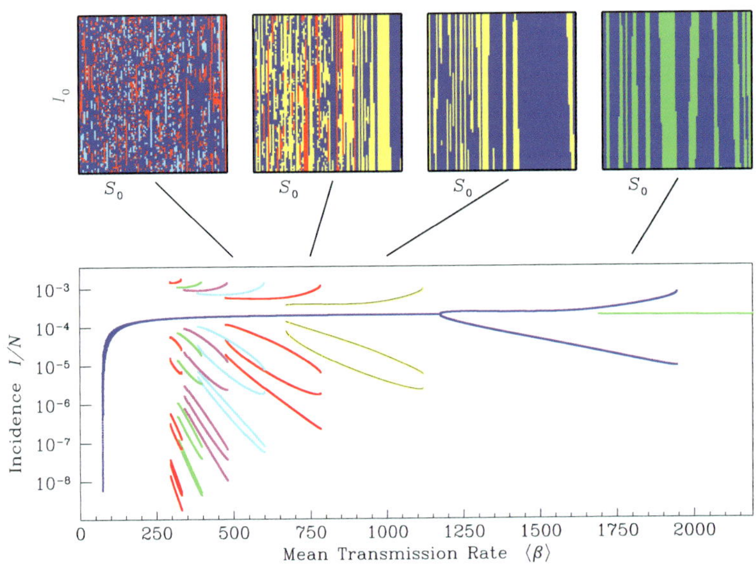

FIGURE 12. The main panel is the bifurcation diagram for the term-time forced SEIR model, showing incidence on 1 January, normalized by (constant) population size; the control parameter is the mean transmission rate $\langle\beta\rangle$. The fixed parameter values are $1/\gamma = 5$ days, $1/\sigma = 8$ days, $\mu = \nu/N = 0.02\,\text{yr}^{-1}$ and $\alpha = 0.25$ (Eq. (3.17)). Each attractor is identified with a different color. For sufficiently high $\langle\beta\rangle$, there is a unique (annual) attractor. As $\langle\beta\rangle$ is reduced, biennial, 3-, 4-, 5-, 6-, 7-, and 8-year cycles all occur, before all but the annual attractor are extinguished. The term-time forcing function used to produce this figure corresponds to school terms in England and Wales [**BE03**]. In the United States, the summer holiday is longer; this does not affect the structure of the bifurcation diagram, but with a longer summer holiday each of the various bifurcations occurs at lower $\langle\beta\rangle$. Above the bifurcation diagram, basins of attraction (initial susceptibles, $0 < S_0/N < 0.1$, versus initial infectives, $0 < I_0/N < 0.0001$, with $E_0/N = 0.0001$) are shown for the various attractors at four particular values of $\langle\beta\rangle$. Figure 13 identifies regions of this diagram that correspond to the dynamics observed at various times and places. [Reproduced from Earn et al. [**ERBG00**]. Copyright © (2000) American Association for the Advancement of Science.]

Figure 12 presents the type of bifurcation diagram that we need, for the term-time forced SEIR model (Figure 9) with fixed parameters chosen to correspond to measles. The control parameter (on the abscissa) is the mean transmission rate $\langle\beta\rangle$. The ordinate shows measles incidence on 1 January of each year, so annual cycles are represented by a single curve, biennial cycles by double curves, and so on. Different colors correspond to different stable solutions of the model, which attract different sets of initial conditions (basins of attraction). For several values of the mean transmission rate $\langle\beta\rangle$, basins of attraction of the various attractors are shown above the bifurcation diagram in Figure 12. Where multiple stable solutions coexist, stochasticity can induce complicated dynamics due to shifts among attractors [**SS83, Sch85, Aro90**]. The upper panels of Figure 12 show that the basins of coexisting attractors are more intermixed if $\langle\beta\rangle$ is smaller, so we might expect the effects of stochasticity to be greater for smaller $\langle\beta\rangle$ (or, equivalently, when the effective $\langle\beta\rangle$ is reduced by vaccination or a decrease in birth rate).

Figure 13 shows measles incidence in four representative large cities; major dynamical transitions are evident in each case. Superimposed on the time series, each panel shows births (in red). In the vaccine era, susceptible recruitment (light blue) is lower than births. Above the incidence data, horizontal lines indicate periods of annual, biennial, and more complex dynamics that are predicted (Figure 12) on the basis of the observed exogenous variables (birth and vaccination rates). The horizontal lines in Figure 13 are color-coded according to the corresponding attractors in Figure 12.

London (Figure 13A) experienced biennial cycles of measles epidemics from 1950 to 1968; the estimated mean transmission rate for this period [**AM91**] is $\langle\beta\rangle \simeq 1240$, corresponding to a biennial attractor (dark blue, see Figure 12). Before 1950, epidemics were roughly annual; over the same brief period the birth rate was much higher, which greatly increased the effective mean transmission rate, allowing attraction to an annual cycle (green, far right of Figure 12). After 1968, recruitment rates steadily decreased because of mass vaccination (for example, when vaccine uptake reached 60%, the effective mean transmission rate was reduced to $\langle\beta\rangle \simeq 500$); this brought the system into the parameter region where there are multiple coexisting attractors with extremely intermixed basins, suggesting an explanation for the irregular epidemics in the vaccine era. Alternatively, or in addition, irregular dynamics in this region may arise from stochastic interactions with a chaotic repellor [**RW91**].

In Liverpool (Figure 13B) the birth rate was much higher than the mean in England and Wales throughout the post-war period until 1968 [**FG98**] (the birth rate in Liverpool is drawn as a dotted red line in Figure 13A as well, for comparison with the birth rate in London). This explains the roughly annual cycle of measles epidemics over the same period. After 1968, the combination of vaccination and a lower birth rate brought Liverpool, like London, into the regime where multiple co-existing cyclical attractors exist and might lead to irregular dynamics in the presence of noise.

Birth rates in the United States were relatively low during the Great Depression. Throughout this period, measles epidemics were irregular in New York and Baltimore (Figure 13C and D), consistent with stochastic switching between densely intermixed attractors or a chaotic repellor. After World War II, birth rates rose

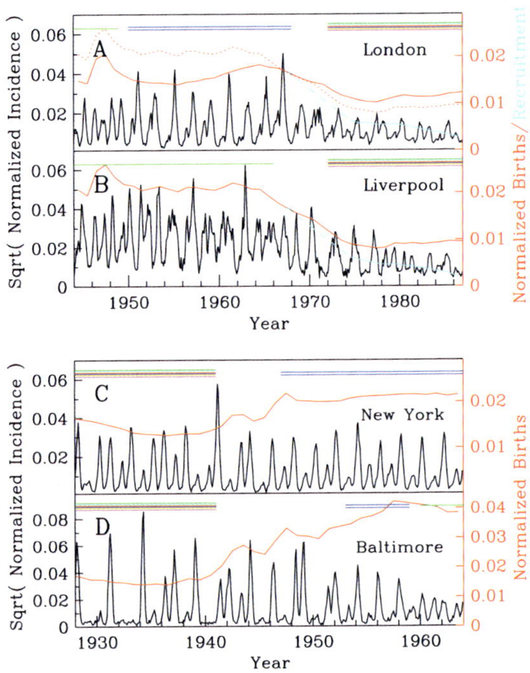

FIGURE 13. Measles dynamics in four large cities: London, Liverpool, New York, and Baltimore. In each panel, the black time series shows the square root of measles incidence normalized by the population size in 1960; the UK data are 4-weekly (13 points per year) whereas the US data are monthly (12 points per year). The red curve shows annual births normalized by the population size in 1960. In London and Liverpool, the light blue curve shows (normalized) annual susceptible recruitment (which is lower than the birth rate in the vaccine era). Together with the mean transmission rate $\langle\beta\rangle$, the recruitment rate determines the nature of the dynamics; see Eq. (4.2) (effects of changes in birth rate will be delayed by initial maternally acquired immunity and the age of school entry). The vertical scale for Baltimore is higher because reporting rates were twice as high [see London and Yorke [**LY73**]]. Note that the American data cover an earlier period than the English data; perhaps because of increases in mobility with time (as well as differences in population density), estimated values of $\langle\beta\rangle$ are significantly lower for the American cities [**And82, BE03**]. This (and the length of the summer holiday; see caption to Figure 12) must be borne in mind when comparing measles incidence data with various regions of Figure 12. The colored lines at the top of each panel correspond to the region of the bifurcation diagram in Figure 12 that applies to each section of the measles time series. Many colored lines together indicate the region of Figure 12 containing many coexisting attractors with intermixed basins. Two horizontal blue lines indicate the biennial attractor in Figure 12, while the single green line indicates the annual attractor that exists for high transmission rates. Gaps, where no region of Figure 12 is indicated, correspond to transient periods when the recruitment rate changed relatively rapidly. [Reproduced from Earn et al. [**ERBG00**]. Copyright © (2000) American Association for the Advancement of Science.]

dramatically, pulling the system out of the regime with multiple co-existing attractors. In New York, the birth rate quickly reached a plateau, apparently fixing the system on the biennial attractor. In Baltimore, the birth rate continued to rise, eventually enough to bring the effective $\langle\beta\rangle$ into the region of Figure 12 where either biennial or annual cycles are possible (far right of Figure 12).

We now appear to be doing extremely well. We can successfully predict all of the dynamical transitions evident in each the measles time series in Figure 13 (and every other measles time series that I am familiar with). Note, incidentally, that this rules out chaotic intermittency as a mechanism for the dynamical transitions: if chaotic intermittency were the correct explanation for changes in periodicity of childhood disease epidemics then it would not be possible to predict the times at which changes in cycle structure occur.

Our confidence in the term-time forced SEIR model is now considerable, based on analysis of measles dynamics. Naturally, the next question we would like to address is whether the model is equally good at predicting the dynamics of other childhood diseases. We have every reason to anticipate success with other similar diseases, since the only relevant difference should be the values of the disease-specific parameters ($\mathcal{R}_0$, σ, γ).

Unfortunately, the analysis that has worked so well at explaining measles dynamics fails miserably for many other childhood infections!! In particular, with parameters estimated for rubella or whooping cough, the term-time forced SEIR model yields a unique attractor—an annual cycle—throughout the relevant range of $\langle\beta\rangle$. Yet the observed dynamics of these diseases is much more complicated than an annual cycle. Figure 14 shows examples of time series of case notifications for rubella and whooping cough, together with periodograms computed separately for two segments of each time series. It is clear from the time plots that the dynamics are not strictly annual for either disease. Moreover, there are strong spectral peaks at frequencies not predicted by the *asymptotic analysis*, i.e., not displayed by the (strictly annual) attractor of the term-time forced SEIR model.

Ugh! Could our success with measles have been extraordinarily lucky? Why doesn't the same approach lead to correct predictions of the dynamics of other childhood diseases?

Before giving up hope, it is quite telling that the transitions in the dynamics of rubella and whooping cough observed in Figure 14 correspond to substantial changes in the susceptible recruitment rate (plotted in green), which is exactly the parameter that we predicted would control dynamical transitions.

Once again, the truly discrete nature of disease transmission comes to our rescue. As Bartlett [**Bar57**] discovered for the unforced SIR model, demographic stochasticity can sustain oscillations that would be damped out in the absence of noise. In the seasonally forced model, noise can prevent the decay of transient oscillations at a period different from that of the attractor. We call the period associated with the attractor the *resonant period*, because it is arises from resonance with the seasonal forcing period of one year. In contrast, we say the period of transient oscillations is *non-resonant* because it is stimulated by demographic stochasticity and need not be an integer multiple of the period of seasonal forcing [**BE03**].

The resonant period is predicted by the asymptotic analysis that identifies the attractors of the model and worked so well for measles. The non-resonant period

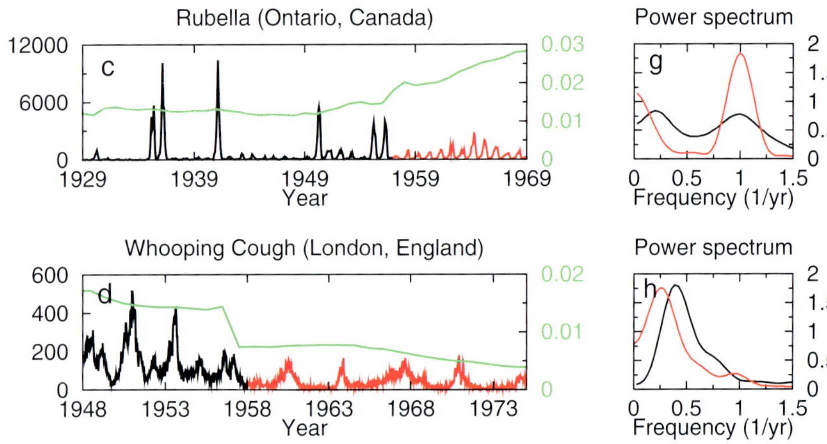

FIGURE 14. Case notification time series for rubella in Ontario, Canada, and whooping cough in London, UK. In each panel, the green line shows annual susceptible recruitment, $\nu(1-p)$, where ν denotes births normalized by 1955 population size and p denotes the proportion vaccinated; recruitment is shown displaced forward in time by the mean age at infection to account for the typical delay between birth and infection (11 years for rubella and 4 years for whooping cough). Time-series are divided into sections based on substantial differences in recruitment rates; the corresponding periodograms are not sensitive to the precise point at which the time-series is divided. [Reproduced from Bauch and Earn [**BE03**]. Copyright © (2003) The Royal Society.]

can be predicted by a *perturbation analysis*. Rather than a simple linear stability analysis of an equilibrium solution, we must consider perturbations of non-trivial attractors. Conducting such a perturbation analysis is not straightforward, but standard (numerical) methods exist for doing this [**Kuz98, BE03**].

Both the resonant and non-resonant periods can be expected to correspond to peaks in the periodogram of the disease time series. In every case I have ever considered, a resonant peak in the periodogram is correctly predicted by asymptotic analysis (and the prediction is always exact, which is perhaps not too surprising given that seasonal forcing will tend to lock the system onto an exact integer multiple of the forcing period). Moreover, a non-resonant peak in the periodogram is also predicted with great accuracy by perturbation analysis, for all of the childhood diseases we have investigated; see Figure 15 and Bauch and Earn [**BE03**].

We are now back in business. We have developed a quantitative method for predicting the structure of the periodogram of disease time series, and it works!

You may wonder why we didn't need to worry about sustaining transient dynamics in the case of measles. The answer is that the non-resonant period associated with the biennial attractor is too long to be observable in the relatively short time series we have. In this sense, we were indeed lucky in our analysis of measles dynamics using only the asymptotic analysis. On the other hand, restricting attention to bifurcation theory did lead us to what now appears to be an incorrect inference. At times when measles is not strictly biennial, we find that the periodogram has two peaks, a resonant peak at one year and non-resonant peak at the frequency

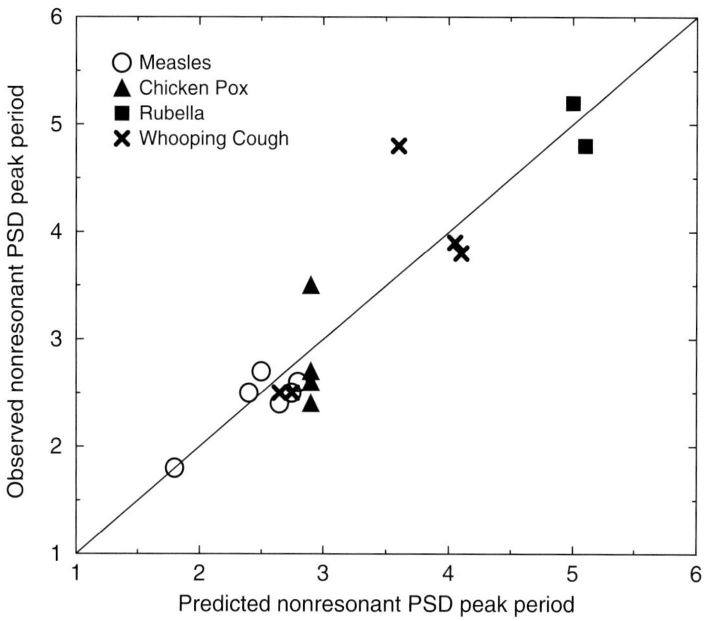

FIGURE 15. The correlation between predicted and observed periods of non-resonant periodogram peaks is positive and highly significant ($r^2 = 0.83$, $t = 8.69$, d.f. $= 16$, $p < 10^{-7}$). If the observations agreed perfectly with our predictions (as they do for resonant periodogram peaks) then the corresponding points would lie precisely on the line of slope 1 shown on the graph (the least-squares linear fit of the data has a slope of 1.02 and an intercept of 20.11). See electronic Appendix A of Bauch and Earn [**BE03**] for details. [Reproduced from Bauch and Earn [**BE03**]. Copyright © (2003) The Royal Society.]

associated with transient oscillations around the annual attractor. Apparent irregularity of measles epidemics does not, in fact, seem to arise from stochastic switching among co-existing attractors. We appear to be seeing simply the effects of noise preventing decay of transients.

While the approach we have adopted successfully predicts the positions of both peaks in the periodogram, it cannot explain the heights of the peaks. For that, we return again to demographic stochasticity. If the non-resonant peak occurs because noise sustains transient dynamics then that peak should be relatively larger if there is more noise. In other words, we expect that time series of infectious disease dynamics in smaller populations will show relatively larger non-resonant peaks, and this is what we observe, both in observed data and stochastic simulations [**BE03**].

4.1. Exercises

(1) Show, by scaling of the usual variables in the SEIR model as in Eq. (4.1), that a change in vaccination proportion from p to p' together with a change in birth rate from ν to ν' yields exactly the same dynamics as a change in basic reproduction number from $\mathcal{R}_0$ to $\mathcal{R}_0(1-p')\nu'/[(1-p)\nu]$, as indicated in Eq. (4.2).

(2) In the SIR model, show that if the birth rate ν is different from the death rate μ, and the contact term is $\beta SI/N$, rather than βSI, then the equations for the *proportions* of the population in each state are the same as the usual SIR model with both the birth rate and the death rate equal to ν. What if ν is a function of time? Is this consistent with the equivalence you derived in the previous exercise?

5. Manipulating epidemics

Vaccination can be viewed as a method of manipulating the structure of epidemic patterns. As a result of the equivalence (4.2) discussed in the previous section, vaccination at a given level will induce a predictable transition in the character of recurrent epidemics. Of course, if we vaccinate a sufficiently large proportion of the population (3.10) then we will eradicate the infection. But can we somehow build from our success in predicting dynamical transitions to develop a vaccination strategy that beats the classical eradication threshold (3.10)?

All of our analysis has been of temporal patterns of epidemics. What about spatial patterns? Can a spatially explicit version of the SEIR model lead us to suggest improved strategies for control or eradication of measles of other infectious diseases?

Before getting involved with a spatial model, it would make sense to look at some spatial epidemic data and see whether the data inspire any particular ideas. One potentially promising idea is suggested by a comparison of measles and whooping cough dynamics in England and Wales during the second half of the 20th century. Before mass vaccination for measles was initiated in 1968, measles epidemics in UK cities were strongly synchronized, but the degree of spatial correlation was greatly reduced in the vaccine era [**BG96**]. In sharp contrast, whooping cough epidemics in the identical population were not synchronized in the pre-vaccine era but became strongly synchronized upon implementation of mass vaccination [**REG99**]. These examples indicate that vaccination can induce spatial dynamical transitions (in the degree of synchony), in addition to the temporal transitions discussed in the previous section.

Attempting to manipulate the degree of synchrony of epidemics might be worthwhile. Increasing synchrony has the potential to reduce the threshold vaccination level required for eradication [**ERG98**], because synchrony reduces the "rescue effect" [**BKB77**] whereby regions in which populations are small are saved from extinction by dispersal of individuals from other areas. Can we devise a vaccination strategy that will synchronize measles epidemics worldwide?

One way to synchronize epidemics might be through "global pulse vaccination", i.e., intense vaccination efforts that are simultaneous around the world and repeated periodically [**ERG98**], or by supplementing continuous vaccination with periodic pulses [**GH97**]. Intuitively, the introduction of globally synchronized periodic forcing might synchronize the troughs between epidemics, stimulating the desired "anti-rescue effect". But this tantalizing prospect requires careful analysis. As we saw for the simple SIR model, periodic forcing can have complex dynamical effects, and we can't be sure *a priori* that the probability of synchronous oscillations will definitely be increased by global pulse vaccination.

The theory of synchronization is important in many areas of science [**PRK01, Str03**]. Most mathematical research on the subject has focused on the local stability of synchronous solutions of dynamical systems, which determines whether particular synchronous solutions attract nearly synchronous initial states with some non-zero probability [**Bue97**]. For epidemiological purposes, we really need a global theory, so we can determine conditions under which all initial states will lead to synchronous solutions [**ELR00**]. While we have made some valuable progress on a fully global theory of synchrony [**EL06**], it remains to be seen whether it can be successfully exploited to discover synchronizing vaccination strategies. Of course, a great deal might be learned from intermediate approaches that attempt to find conditions that yield synchrony with high probability but not certainty [**MEms**]. Simulations will also play an important role, though the parameter space that must be explored is high-dimensional and involves substantial computational challenges.

6. Conclusions and Take-home messages

The detailed temporal patterns of recurrent epidemics can be usefully described and analyzed using tools of time series analysis, especially the time plot and the periodogram. In terms of their ability to forecast patterns of recurrence of future epidemics, time series models such as ARMA(p, q) models are of limited value. Mechanistic mathematical models, on the other hand, are extremely powerful in this respect. (Note that there is scope for combining time series models with mechanistic modelling, which leads to models that can fit observed epidemic time series with remarkable precision [**FG00**].)

Mathematical and computational analysis of a simple mechanistic model (the term-time forced SEIR model; Figure 9, Eqs. (3.9) and (3.17)) explains the observed complex dynamics of epidemics of a variety of childhood diseases. We arrived at this model after ruling out a number of simpler models, and I hope that I have convinced you that successful mathematical modelling of biological systems is likely to proceed in this way. By adding model structure in small steps, we are most likely to discover which biological features have the greatest impact on observed behaviour. It is also worth highlighting that apparent limitations of a model can sometimes reflect more about our failure to think of the right analysis tool than about any inherent inadequacy of the model itself.

We are lucky that medical and public health personnel have painstakingly conducted surveillance of infectious diseases for centuries. This has created an enormous wealth of valuable data [**IIDDA**] with which to test hypotheses about disease spread using mathematical models. Building from models that successfully explain past patterns of epidemics, we can hope to design improved strategies for control or eradication of many infectious diseases.

Bibliography

[AB00] Andersson, H. and Britton, T., *Stochastic epidemic models and their statistical analysis*, Lecture notes in Statistics, vol. 151, Springer-Verlag, New York, 2000.

[AM91] Anderson, R. M. and May, R. M., *Infectious diseases of humans: Dynamics and control*, Oxford University Press, Oxford, 1991.

[And82] Anderson, R. M. (ed.), *The population dynamics of infectious diseases: Theory and applications*, Population and Community Ecology, Chapman and Hall, London, 1982.

[Aro90] Aron, J. L., *Multiple attractors in the response to a vaccination program*, Theoretical Population Biology **38** (1990), 58–67.

[Bai75] Bailey, N. T. J., *The mathematical theory of infectious diseases and its applications*, Second ed., Hafner Press, New York, 1975.

[Bar57] Bartlett, M. S., *Measles periodicity and community size*, Journal of the Royal Statistical Society, Series A **120** (1957), 48–70.

[Bar60] ———, *Stochastic population models in ecology and epidemiology*, Methuen's Monographs on Applied Probability and Statistics, vol. 4, Spottiswoode, Ballantyne & Co. Ltd., London, 1960.

[BCC01] Brauer, F. and Castillo-Chavez, C., *Mathematical models in population biology and epidemiology*, Texts in Applied Mathematics, vol. 40, Springer-Verlag, New York, 2001.

[BE03] Bauch, C. T. and Earn, D. J. D., *Transients and attractors in epidemics*, Proceedings of the Royal Society of London, Series B **270** (2003), no. 1524, 1573–1578.

[Ber1760] Bernoulli, D., *Essai d'une nouvelle analyse de la mortalité causée par la petite vérole et des advantages de l'inoculation pour la prévenir*, Mém Mathematical Physics Academy Royal Science Paris (1760), 1–45.

[BB04] Bernoulli, D., *An attempt at a new analysis of the mortality caused by smallpox and of the advantages of inoculation to prevent it*, Reviewed by Blower, S., Rev. Med. Virol. **14** (2004), 275-288, http://www.semel.ucla.edu/biomedicalmodeling/pdf/Bernoulli&Blower.pdf

[BG93] Bolker, B. M. and Grenfell, B. T., *Chaos and biological complexity in measles dynamics*, Proceedings of the Royal Society of London, Series B **251** (1993), 75–81.

[BG96] ———, *Impact of vaccination on the spatial correlation and persistence of measles dynamics*, Proceedings of the National Academy of Sciences, USA **93** (1996), 12648–12653.

[BKB77] Brown, J. H. and Kodric-Brown, A., *Turnover rates in insular biogeography: effect of immigration on extinction*, Ecology **58** (1977), 445–449.

[Bue97] Buescu, J., *Exotic attractors: From Liapunov stability to riddled basins*, Progress in Mathematics, vol. 153, Birkhäuser Verlag, Basel, 1997.

[Cha04] Chatfield C., *The analysis of time series: An introduction*, sixth ed., Texts in Statisitical Science Series, Chapman & Hall/CRC, Boca Raton, 2004.

[DPV+06] Dushoff, J., Plotkin, J. B., Viboud, C., Earn, D. J. D., and Simonsen, L., *Mortality due to influenza in the United States - an annualized regression approach using multiple-cause mortality data*, American Journal of Epidemiology **163** (2006), no. 2, 181–187.

[EDL02] Earn, D. J. D., Dushoff, J., and Levin, S. A., *Ecology and evolution of the flu*, Trends in Ecology and Evolution **17** (2002), no. 7, 334–340.

[EL06] Earn, D. J. D. and Levin, S. A., *Global asymptotic coherence in discrete dynamical systems*, Proc. Natl. Acad. Sci. USA **103** (2006), no. 11, 3968–3971, [*Correction:* PNAS **103** (2006), no. 51, 19605].

[ELR00] Earn, D. J. D., Levin, S. A., and Rohani, P., *Coherence and conservation*, Science **290** (2000), no. 5495, 1360–1364.

[ERBG00] Earn, D. J. D., Rohani, P., Bolker, B. M., and Grenfell, B. T., *A simple model for complex dynamical transitions in epidemics*, Science **287** (2000), no. 5453, 667–670.

[ERG98] Earn, D. J. D., Rohani, P., and Grenfell, B. T., *Persistence, chaos and synchrony in ecology and epidemiology*, Proceedings of the Royal Society of London, Series B **265** (1998), no. 1390, 7–10.

[FBBC97] Fitch, W. M., Bush, R. M., Bender, C. A., and Cox, N. J., *Long term trends in the evolution of H(3) HA1 human influenza type A*, Proceedings of the National Academy of Sciences, USA **94** (1997), no. 15, 7712–7718.

[FG98] Finkenstädt, B. F. and Grenfell, B. T., *Empirical determinants of measles metapopulation dynamics in england and wales*, Proceedings of the Royal Society of London, Series B **265** (1998), 211–220.

[FG00] Finkenstädt, B. F. and Grenfell, B. T., *Time series modelling of childhood diseases: A dynamical systems approach*, Applied Statistics **49** (2000), no. 2, 187–205.

[GBK01] Grenfell, B. T., Bjornstad, O. N., and Kappey, J., *Travelling waves and spatial hierarchies in measles epidemics*, Nature **414** (2001), no. 6865, 716–723.

[GH83] Guckenheimer, J. and Holmes, P., *Nonlinear oscillations, dynamical systems, and bifurcations of vector fields*, Applied Mathematical Sciences, vol. 42, Springer-Verlag, New York, 1983.

[GH97] Grenfell, B. and Harwood, J., *(meta)-population dynamics of infectious diseases*, Trends in Ecology and Evolution **12** (1997), 395–399.

[Gil76] Gillespie, D. T., *A general method for numerically simulating the stochastic time evolution of coupled chemical reactions*, Journal of Computational Physics **22** (1976), 403–434.

[Gle87] Gleick, J., *Chaos*, Abacus, London, 1987.

[HHGL$^+$06] Hallett, D. H., Gleason, A. M., Lock, P. F., Flath, D. E., Lomen, D. O., Lovelock, D., McCallum, W. G., Osgood, B. G., Quinney, D., Rhea, K., Tecosky-Feldman, J., and Tucker, T. W., *Applied calculus*, third ed., John Wiley & Sons, New Jersey, 2006.

[IIDDA] IIDDA, *The International Infectious Disease Data Archive*, http://iidda.mcmaster.ca.

[KM27] Kermack, W. O. and McKendrick, A. G., *A contribution to the mathematical theory of epidemics*, Proceedings of the Royal Society of London, Series A **115** (1927), 700–721.

[KM04] Korobeinikov, A. and Maini, P. K., *A Lyapunov function and global properties for SIR and SEIR epidemiological models with nonlinear incidence*, Mathematical Biosciences and Engineering **1** (2004), no. 1, 57–60.

[Kur80] Kurtz, T. G., *Relationships between stochastic and deterministic population models*, Lecture Notes in Biomathematics **38** (1980), 449–467.

[Kuz98] Kuznetsov, Y. A., *Elements of applied bifurcation theory*, Second ed., Applied Mathematical Sciences, vol. 112, Springer-Verlag, New York, 1998.

[LM95] Li, M. Y. and Muldowney, J. S., *Global stability for the SEIR model in epidemiology*, Mathematical Biosciences **125** (1995), no. 2, 155–164.

[LY73] London, W. and Yorke, J. A., *Recurrent outbreaks of measles, chickenpox and mumps. I. Seasonal variation in contact rates*, American Journal of Epidemiology **98** (1973), no. 6, 453–468.

[MEms] McCluskey, C. and Earn, D. J. D., *Attractivity of coherent manifolds in metapopulation models*, (in prep).

[MM04] Moote, A. L. and Moote, D. C., *The Great Plague: The story of London's most deadly year*, John Hopkins University Press, Baltimore, 2004.

[Nie01] Nievergelt, Y., *Wavelets made easy*, Birkhäuser, Boston, 2001.

[NM00] Nowak, M. A. and May, R. M., *Virus dynamics: Mathematical principles of immunology and virology*, Oxford University Press, Oxford, 2000.

[OS90] Olsen, L. F. and Schaffer, W. M., *Chaos versus noisy periodicity: alternative hypotheses for childhood epidemics*, Science **249** (1990), 499–504.

[PRK01] Pikovsky, A., Rosenblum, M., and Kurths, J. (eds.), *Synchronization: A universal concept in nonlinear sciences*, Cambridge Nonlinear Science Series, vol. 12, Cambridge University Press, Cambridge, 2001.

[PST93] Platt, N., Spiegel, E. A., and Tresser, C., *On-off intermittency: a mechanism for bursting*, Physical Review Letters **70** (1993), no. 3, 279–282.

[PTVF92] Press, W. H., Teukolsky, S. A., Vetterling, W. T., and Flannery, B. P., *Numerical recipes in C: The art of scientific computing*, 2nd ed., Cambridge University Press, New York, 1992.

[REG99] Rohani, P., Earn, D. J. D., and Grenfell, B. T., *Opposite patterns of synchrony in sympatric disease metapopulations*, Science **286** (1999), no. 5441, 968–971.

[Ros11] Ross, B., *The prevention of malaria*, 2nd ed., Murray, London, 1911.

[RW91] Rand, D. A. and Wilson, H. B., *Chaotic stochasticity: a ubiquitous source of unpredictability in epidemics*, Proceedings of the Royal Society of London, Series B **246** (1991), 179–184.

[Sch84] Schenzle, D., *An age-structured model of pre- and post-vaccination measles transmission*, IMA Journal of Mathematics Applied in Medicine and Biology **1** (1984), 169–191.

[Sch85] Schwartz, I. B., *Multiple stable recurrent outbreaks and predictability in seasonally forced nonlinear epidemics*, Journal of Mathematical Biology **21** (1985), 347–361.

[Scu03] Scuffham, P. A., *Estimating influenza-related hospital admissions in children and adults – a time series analysis*, Disease Management and Health Outcomes **11** (2003), no. 4, 259–269.

[SK85] Schaffer, W. M. and Kot, M., *Nearly one dimensional dynamics in an epidemic*, Journal of Theoretical Biology **112** (1985), 403–427.

[SKTO93] Schaffer, W. M., Kendall, B. E., Tidd, C. W., and Olsen, L. F., *Transient periodicity and episodic predictability in biological dynamics*, IMA Journal of Mathematics Applied in Medicine and Biology **10** (1993), 227–247.

[SS83] Schwartz, I. B. and Smith, H. L., *Infinite subharmonic bifurcation in an SEIR model*, Journal of Mathematical Biology **18** (1983), 233–253.

[Str94] Strogatz, S. H., *Nonlinear dynamics and chaos*, Addison Wesley, New York, 1994.

[Str03] ______, *Sync: The emerging science of spontaneous order*, Hyperion, New York, 2003.

[TC98] Torrence, C. and Compo, G. P., *A practical guide to wavelet analysis*, Bulletin of American Meteorological Society **79** (1998), no. 1, 61–78. http://atoc.colorado.edu/research/wavelets/.

[USVS] United States Bureau of the Census, *Vital statistics of the united states (before 1937: Mortality statistics)*, 1900-1992.

[Wig03] Wiggins, S., *Introduction to applied nonlinear dynamical systems and chaos*, 2nd ed., Texts in applied mathematics, vol. 2, Springer-Verlag, New York, 2003.

Topological Approaches to Biological Dynamics

Leon Glass

IAS/Park City Mathematics Series
Volume 14, 2005

Topological Approaches to Biological Dynamics

Leon Glass

The growth of computer power and biological data bases is revolutionizing the application of mathematics to biology. Mathematical models of biological systems are being developed at all levels of organization from the folding of protein molecules, to the development of cells and tissues, to the beating of a heart, to the evolution of ecosystems. In some cases, the mathematical model simulates the biological behavior at an extraordinary level of detail and accuracy. Yet, such models are invariably complicated and though the behavior might be similar to the biological system, the underlying principles behind the dynamics may be opaque. Of course, for some purposes it would make no difference whether or not we understand the observed behavior in the model. Thus, a sufficiently accurate model of the heart might enable us to test the effects of drugs in silico, rather than in vivo. Such a model would be useful in screening for drugs that could lead to fatal arrhythmias. There are many groups that are working towards realistic models and the potential benefit of such models will be huge.

In this series of lectures, I will focus on mathematical approaches that are useful for thinking about the underlying principles for the organization of biological dynamics. The approach is topological and qualitative, but in some cases, can make contact with and in fact require high precision data, as in the consideration of the experimental resetting of biological oscillators. The topological approach was first championed in the 1970s and 1980s by giants like Art Winfree and René Thom, whose influence and contributions are gratefully acknowledged at the outset. Winfree's masterpiece, *The Geometry of Biological Time* defines the approach and covers a huge range of topics [**W02**].

I do not try to give a comprehensive review of all uses of topology in biology, but mostly focus on a series of results found by myself and colleagues related to fixed points and continuity in biological dynamics. I will not attempt to be rigorous, but will try to present the material in a fashion that I hope can be understood and applied without advanced mathematics. I will not give elaborate proofs or definitions. I will assume the reader has a basic background in elementary concepts in nonlinear dynamics at the level in [**KG95**]. Whenever possible, I will give experimental or clinical biological examples to illustrate the ideas.

3655 Promenade Sir William Osler, Department of Physiology, McGill University, Montreal, Quebec, Canada H3G 1Y6
E-mail address: glass@cnd.mcgill.ca
I thank NSERC, CIHR, MITACS, CHSF and the NIH Research Resource for Complex Physiologic Signals for support of my research.

All of the material is based on previous publications which I cite and which should be consulted for full details and extensive references to earlier work. Laboratory exercises illustrating some of the concepts can be found in [**BGMT03**] (downloadable MATLAB code for these exercises is at `http://www.cnd.mcgill.ca/ebook/`). In the last section, I give the projects that were suggested to the students during the summer school.

LECTURE 1

Linear stability and the Poincaré–Hopf theorem applied to biology

One of the fundamental techniques in the analysis of nonlinear dynamical systems is to determine the stability of a nonlinear system in the neighborhood of a fixed point. The basic technique is to find the fixed point, linearize the system in the neighborhood of the fixed point, and determine the eigenvalues of the resulting linear equation.

Consider the nonlinear ordinary differential equation,

$$\frac{dx_i}{dt} = f_i(x), \quad \text{for } i = 1, \ldots, n \tag{1.1}$$

where f_i is a nonlinear function of x. A fixed point, x^* is a point for which

$$f_i(x^*) = 0$$

for all i. The elements of the Jacobian matrix, A are determined by computing

$$a_{ij} = \left.\frac{\partial f_i}{\partial x_j}\right|_{x^*}. \tag{1.2}$$

The eigenvalues of of A are found by solving the equation

$$\det \mid A - \lambda I \mid = 0. \tag{1.3}$$

If the real parts of all the eigenvalues are different from zero the fixed point is called hyperbolic. I will assume that all fixed points are hyperbolic. If the real parts of all the eigenvalues of a fixed point are negative, then the fixed point is locally stable. This means that all initial conditions in the neighborhood of the fixed point will approach it in the limit $t \to \infty$. If one or more eigenvalues of a fixed point are positive, the fixed point is not locally stable. Some or all of the initial conditions starting in the neighborhood of the fixed point, will diverge from it as $t \to \infty$.

If the nonlinear equation is in 2 dimensions, then the eigenvalues are just the solution of a quadratic equation. If both eigenvalues are real and the same sign, then the fixed point is called a node. If the eigenvalues are real and different signs, then the fixed point is called a saddle point. Finally if the eigenvalues are complex, then the fixed point is called a focus. In the neighborhood of the fixed points, the geometry of the flow is different for each type of fixed point.

The Poincaré–Hopf Index Theorem places restrictions on the numbers and types of fixed points in ordinary differential equations embedded on manifolds. It is based on the Euler theorem which states that the number of faces plus the number of vertices minus the number of edges for a convex polytope is equal to 2. [**KG95**] gives Poincaré's original proof and some simple examples. A wonderful formal mathematical treatment is in [**GP74**].

Assume that there is a differential equation with hyperbolic fixed points. Call π_i the number of eigenvalues of the ith fixed point whose real part is positive and μ_i the number of eigenvalues of the ith fixed point whose real part is negative. For a differential equation defined on a manifold M with k fixed points, the Poincaré–Hopf Index Theorem states

$$\Sigma_{i=1}^{k}(-1)^{\mu_i} = \chi(M), \tag{1.4}$$

where $\chi(M)$ is the Euler Characteristic of the manifold M. For example, the surface of sphere has an Euler characteristic of 2 and the Euler characteristic of a torus is 0. In general, for an oriented surface of genus k (i.e. a "torus" with k holes), $\chi(M) = 2 - 2k$.

Exercise 1.1. To help you understand this result try to draw the following vector fields:

- A vector field on the surface of a sphere S^2 with two stable fixed points.
- A vector on the surface of a sphere S^2 with two unstable fixed points. What else can you say about the asymptotic dynamics in this case?
- Three different vector fields with no fixed points on the torus.
- A vector field on a torus with 4 fixed points?

Exercise 1.2. Assume the Poincaré–Hopf Index Theorem is correct, and use it to derive the Euler characteristic for a surface of genus k by drawing a legal vector field on the surface as in [**GP74**].

What does any of this have to do with biology? Many biological models are formulated as ordinary differential equations. Since in a real system, trajectories cannot escape to infinity, it should be possible to embed an equation with n variables in an n-dimensional ball, B^n (i.e. the set of points defined by $|x|^2 < a$, where $|x|^2 = \Sigma x_i^2$), where all trajectories on the boundary of the ball are pointing into the ball. The result below applies to all equations of this sort. It is then possible to identify the boundary of B^n with an unstable source for which $\pi_i = n$ of the South Pole of the n-sphere, S^n (S^n is the set of points in $n+1$ dimensions defined by $|x| = a$). The Euler–Poincaré characteristic for S^n is:

$$\chi(S^n) = 1 + (-1)^n. \tag{1.5}$$

By combining this result with Eq. (1.4) we are led to the following result.

Theorem 1.3. *For an ordinary differential equation defined on an attracting ball in Euclidean space, with k fixed points,*

$$\Sigma_{i=1}^{k}(-1)^{\pi_i} = 1. \tag{1.6}$$

where π_i is the number of eigenvalues of the ith fixed point whose real part is positive [**G75, G75a**].

The result provides a classification of dynamical systems based on their fixed point structure independent of the number of interacting variables. For example, the classic boring situation is a single stable, globally attracting fixed point for which $\pi_1 = 0$, Figure 1A. More interesting is the situation with a single unstable fixed point for which $\pi_1 = 2$, and an attractor which is a stable limit cycle oscillation. Figure 1B depicts a three-dimensional system with a single hyperbolic fixed point with two eigenvalues with positive real parts and one eigenvalue with negative real parts, and a stable limit cycle oscillation.

Another example is a dynamical system with three 3 fixed points, two of which are stable nodes and the third is a saddle. Figure 1C depicts this flow in a three-dimensional system. This type of geometry is found often in ecological models with competitive exclusion, or in genetic models with mutual inhibition or mutual activation.

Finally, we can also have two stable attractors: a stable fixed point and a stable limit cycle and no other fixed points.

Exercise 1.4. This exercise deals with constructing flows in phase space for dynamical systems with a single stable fixed point and a single stable limit cycle. In two dimensions, there must be an unstable limit cycle oscillation separating the basins of attraction of the limit cycle and the fixed point. What are possible invariant sets for dynamical systems in three and higher dimensions in which for there are two stable attractors: a fixed point and a limit cycle? Can you develop a topological method to analyze the invariant sets. I do not know how to do this, so if you do know, please let me know or write a paper about it.

Of course, Eq. (1.6) is incredibly limited because it does not report on the dynamics outside the neighborhood of the fixed points of the system. But I do like the notion that we can come up with a qualitative picture of the structure of the allowed flows in a host of biological systems without knowing anything about the system other than it does not escape to infinity.

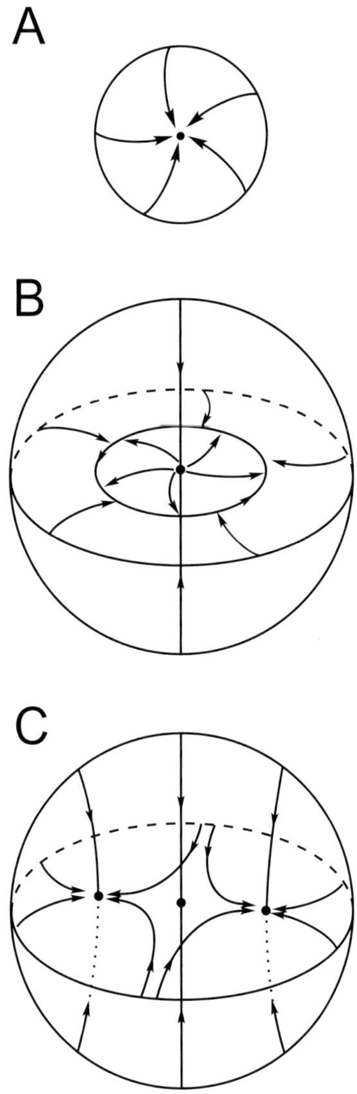

FIGURE 1. Some flows commonly found in ordinary differential equation models of biological systems. We assume that all flows on the boundaries transversely cross the boundaries. A. A stable attracting fixed point in a ball, B^2. All trajectories asymptotically approach the fixed point. B. A stable limit cycle oscillation, and a single fixed point with one eigenvalue with a negative real part, and two eigenvalues with positive real parts in B^3. All trajectories with the exception of the stable manifold of the fixed point (which is shown), approach the limit cycle in the limit $t \to \infty$. C. Two stable nodes and a fixed point with two eigenvalues with negative real parts and one eigenvalue with a positive real part in B^3.

LECTURE 2

Discontinuous phase resetting

I expect that in all fields of biology in which there are oscillations, experiments have been carried out to test the effect of a stimulus delivered to the oscillator. There are several possibilities: (i) the stimulus can have no effect; (ii) the oscillation can be established following some transient but the phase of oscillation may be shifted; (iii) the original oscillation may be destroyed and replaced by a quiescent dynamics or some other type of regular or irregular dynamics. These different behaviors may be of great practical importance. For example, when we travel from one time zone to another, we hope that our internal circadian rhythm is reset. If we have an implanted anti–tachycardia pacemaker, it will deliver one or more stimuli directly to the heart when it detects an rapid abnormal ventricular rhythm with the intention of annihilating that rhythm so that the normal sinus node pacemaker can resume its pacemaking function.

We imagine that the biological system is described by some appropriate set of dynamical equations such as nonlinear ordinary or partial differential equations. It is usual to associate the oscillation with a stable limit cycle in some appropriate nonlinear theoretical model of the oscillation [**W02**]. Recall that a *stable limit cycle* is a periodic solution of a differential equation that is attracting in the limit of $t \to \infty$ for all points in the neighborhood of the limit cycle. Say that the period of the oscillation is T_0. We will designate a particular event to be the fiducial event, designated as phase, $\phi = 0$. The *phase* at any subsequent time $t > 0$ is defined to be $\phi = t/T_0 \pmod 1$. The phase here is defined to lie between 0 and 1; to convert it to radians multiply it by 2π.

The set of all initial conditions that attract to the limit cycle in the limit $t \to \infty$ is called the *basin of attraction* of the limit cycle. Let $x(t)$ be on a limit cycle at time t and $y(t)$ be in the basin of attraction of the limit cycle. Denote the distance between a and b by $d[a, b]$. Let the phase of x at $t = 0$ be ϕ. Then if in the limit $t \to \infty$,

$$d[x(t), y(t)] = 0,$$

the *latent* or *asymptotic* phase of $y(t)$ is also ϕ. We say that $y(t)$ is on the same *W-isochron* as $x(t)$.

The development of the concept of W-isochrons and the recognition of their significance is due to Winfree [**W02**]. Many important mathematical results concerning W-isochrons were established by Guckenheimer [**Gu75**], who considered dynamical systems in n-dimensional Euclidean space. He proved the existence of isochrons and showed that every neighborhood of every point on the frontier of the basin of attraction of a limit cycle, intersects every W-isochron. Moreover, the dimension of the frontier of the basin of attraction $\geq n - 2$.

We now consider the effects of perturbations delivered to the biological oscillation. Assume that a perturbation delivered to an oscillator at phase ϕ shifts the oscillation to the latent phase $g(\phi)$. The function $g(\phi)$ is called the *phase resetting curve*.

The *Continuity Theorem* summarizes important aspects of the effects of perturbations on limit cycle oscillations in ordinary differential equations and partial differential equations.

Theorem 2.1. *If a perturbation delivered at any phase of a limit cycle oscillation leaves the state point in the basin of attraction of the asymptotically stable limit cycle, then the resetting curves characterizing the effects of the stimuli, will be continuous*[**GG98**].

In general, the phase resetting curve $g(\phi)$ is a circle map $g : S^1 \to S^1$.

Circle maps can be continuous or discontinuous. Continuous circle maps can be characterized by their *(topological) degree* or *winding number*. The degree of a continuous circle map measures the number of times, $g(\phi)$ wraps around the unit circle as ϕ goes around the circle once. For example, for oscillations associated with stable limit cycle oscillations in differential equations, for very weak perturbations in general $g(\phi) \approx \phi$, and the degree is 1. In many instances, as Winfree discusses [**W02**], the degree of the resetting curve is 0 when the stimulation is strong. If the degree of the resetting curve is 1 for weak stimuli and 0 for strong stimuli, there must be an intermediate stimulus (or stimuli) that will perturb the system outside of the basin of attraction of the limit cycle – though whether the limit cycle is eventually reestablished depends on whether the stimulus perturbs the system to the basin of attraction of another stable attractor. Similarly, if the resetting curve is discontinuous there must be a stimulus phase or range of stimulus phases that will perturb the system outside of the basin of attraction of the limit cycle[**GG98**].

The prototypical oscillator for studies of resetting is the Poincaré oscillator (which is also called the radial isochron clock) [**KH82, GG82, W02**]. The Poincaré oscillator is most conveniently written in a polar coordinate system where r is the distance from the origin and ϕ is the angular coordinate. The equations are written

$$\begin{aligned} \frac{dr}{dt} &= kr(1-r), \\ \frac{d\phi}{dt} &= 2\pi, \end{aligned} \tag{2.1}$$

where k is a positive parameter. Starting at any value of r, except $r = 0$, there is an evolution until $r = 1$. The parameter k controls the relaxation rate. The phase, $\phi = 0$, corresponds to the fiducial phase of an oscillation such as the upstroke of an action potential or the onset of contraction.

We assume that perturbations are modeled by a horizontal translation to the right by a distance b, Figure 2. This theoretical model facilitates analytical work because of its comparatively simple analytical form. The phase resetting curve, $g(\phi)$ is given by

$$g(\phi) = \frac{1}{2\pi} \arccos \frac{\cos 2\pi\phi + b}{(1 + b^2 + 2b\cos 2\pi\phi)^{1/2}} (\text{mod } 1). \tag{2.2}$$

Exercise 2.2. Derive Eq. (2.2).

In computations using equation (2.2), in evaluating the arccosine function, take $0 < g(\phi_i) < 0.5$ for $0 < \phi < 0.5$, and $0.5 < g(\phi_i) < 1$ for $0.5 < \phi_i < 1$.

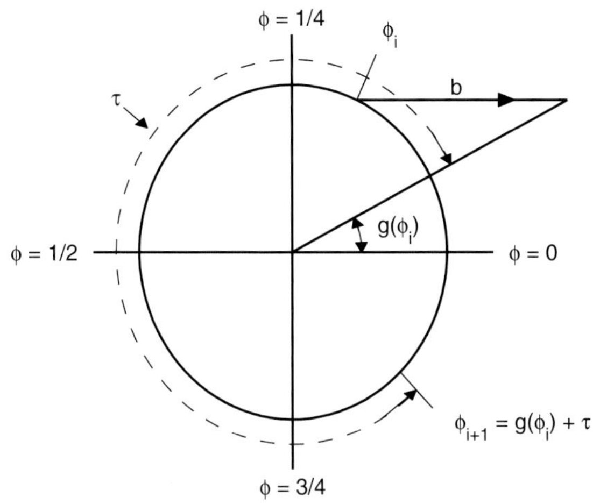

FIGURE 2. Resetting and entrainment of the Poincaré oscillator. Following a stimulus delivered when the oscillator is at phase ϕ_i, represented by a horizontal translation of magnitude b, the phase is reset to $g(\phi_i)$. In the case of rapid relaxation to the oscillation, see Lecture 3, the effects of the periodic stimulation with a period τ, can be computed from the relationship $\phi_{i+1} = g(\phi_i) + \tau$, where ϕ_i is the phase before the ith stimulus. [Adapted from [**GM88**]. Reprinted by permission of Princeton University Press.]

The main theoretical insight into this formula is that for low stimulus amplitude $|b| < 1$, there is resetting where g has topological degree (winding number 1), and for $|b| > 1$, there is resetting where g has topological degree 0 [**W02**].

These abstract notions are directly related to experiments. The phase resetting curve can be measured experimentally. Assume once again that the marker event of an oscillation is defined as $t = 0$, $\phi = 0$. Assume that in response to a perturbation delivered at phase ϕ marker events recur at successive time $T_1(\phi), T_2(\phi), \ldots, T_n(\phi)$. Let us assume that for all j sufficiently large, the limit cycle is asymptotically approached so that $T_j(\phi) - T_{j-1}(\phi) = T_0$, where T_0 is the control cycle length.

The phase resetting curve can be determined from the values of $T_j(\phi)$ determined experimentally under the assumption that the original oscillation is reestablished eventually and that noise is negligible (of course these are mathematical assumptions, not always true in practice). It is given by

$$g(\phi) = \phi - \frac{T_j(\phi)}{T_0} \pmod 1 \tag{2.3}$$

where we take j sufficiently large that $g(\phi)$ converges.

There were many early reports of resetting curves that seemed to show discontinuities in phase resetting curves, see the listing in [**GW84**]. However, carrying out careful experiments requires very careful measurement of the effects of a stimulus as a phase is varied. For oscillations with long periods such as the 24 hour circadian rhythm, such experiments are virtually impossible to carry out. Further, since in all real physical and biological systems, noise is inevitable, it can be difficult to measure small discontinuities if they are buried in noise. Finally, the issue of whether or not a resetting curve is continuous or discontinuous seems mostly of mathematical interest and the issue has not been carefully probed experimentally.

However, in the early 1980s, Michael Guevara carried out an experimental study of resetting in spontaneously beating cells from embryonic chick heart in which he measured resetting curves that were not continuous [**GSG86**]. Furthermore, no stimuli led to the annihilation of the oscillation. These experiments were carried out as carefully as possible. They are shown in Figure 3. A stimulus delivered at

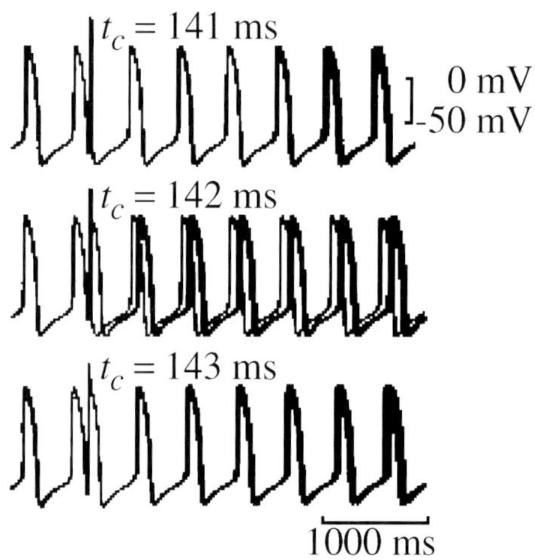

FIGURE 3. Resetting a spontaneously beating aggregate of chick embryonic heart cells. A stimulus delivered at 141 ms after an action potential leads to a delay in the timing of the next action potential, and a stimulus delivered at 143 ms after an action potential leads to and immediate action potential. A stimulus delivered at 142 ms after an action potential will either lead to a delayed or advanced action potential. The observation that the envelopes of the two responses remain distinct after several beats indicates that the experimentally resetting curve is discontinuous. [Reprinted from [**GSG86**]. Used with permission from The American Physiological Society.]

141 ms after the start of an action potential always leads to a delayed next action whereas a stimulus delivered at 143 ms after the start of an action potential always leads to an advanced action potential. However, a stimulus delivered at 142 ms after the start of an action potential sometimes leads to an advanced action potential, and sometimes a delayed action potential, and the resetting would be discontinuous even after several cycles. These were disturbing since they seemed to indicate that the robust qualitative predictions of the topological approach were wrong.

A resolution for this problem has been proposed by Krogh–Madsden et al. [**KGDG04**]. According to our current understanding of cardiac electrophysiology, in the real biological system, there are ion channels that might open and close stochastically. It is usually assumed that there are so many of these channels that it is possible to average and that deterministic equations can be used. However, when carrying through stochastic, as opposed to deterministic, models of resetting of a model oscillator, the tiny differences in current that arise from the opening and closing of a single channel combined with the flows in the differential equation are adequate to lead to discontinuities in simulated resetting curves, even though no stimulus leads to an annihilation of the rhythm. From a theoretical perspective, I believe this result is important since it shows that the topological predictions for resetting do not robustly hold for resetting of all biological oscillations, contrary to what many had assumed. Further, the necessary conclusion is that the random opening or closing of a single channel can lead to grossly different dynamics (action potential or no action potential in a group of cells). I believe that this insight can help explain the apparently stochastic dynamics that are sometimes observed in recording of cardiac activity in people [**SAIGGS01, LKGG07**].

Another example where continuity properties of resetting is practically important involves the effects of stimuli delivered directly to the heart during certain types of cardiac arrhythmias in which the heart beats abnormally rapidly (tachycardias). It is often possible to terminate a tachycardia by delivering a single pulse directly to the heart, or a sequence of several pulses directly to the heart. To consider what happens in this case, consider the simplest model of tachycardia, a pulse of activity circulating on a one dimensional ring. I assume that the ring represents an excitable medium. Although real hearts are not homogeneous, also assume that the ring is homogeneous, so that if a pulse of activity was initiated at a point on the ring, it would propagate in two directions until the waves collided and annihilated at the antipodal point on the ring.

One of the basic models for excitable media is the FitzHugh–Nagumo equations,

$$\begin{aligned}\frac{\partial v}{\partial t} &= -w - v(v-0.139)(v-1) + D\frac{\partial^2 v}{\partial R^2} + I(R),\\ \frac{\partial w}{\partial t} &= 0.008(v-2.54w)\end{aligned} \qquad (2.4)$$

where D represents the diffusion coefficient, $I(R)$ represents injected current (the perturbation) at position R [**GJ95**]. In the homogeneous equations ($D = 0$) with no injected current, there is a globally stable steady state at $v = 0, w = 0$. In response to a sufficiently large perturbation a large transient of v and w away from 0 can be elicited. This equation supports a circulating pulse on the ring, It may be surprising, but this very simple model is very close to the working model that many cardiologists have for some instances of such dangerous cardiac arrhythmias rhythms as monomorphic ventricular tachycardia and atrial flutter. The mathematical significance is that the cardiologists try to ablate a portion of the ring by using radio frequency radiation to burn the heart tissue. If this procedure is successful, then the arrhythmia is cured. In carrying out this procedure, cardiologists think and act like topologists (they do real surgery on imagined manifolds in the heart).

I now consider the resetting of a circulating pulse on the ring [**GJ95, NG96, GG98**]. By considering the effect of a pulse delivered to an idealized excitable media at different phases of the cycle, it appears that the resetting curves should be discontinuous. However, according to the continuity theorem, this could only occur if there are stimuli that lead to annihilation of the oscillation, and in the FitzHugh-Nagumo equation above, annihilation of the rhythm was found for a critical stimulus

[**GJ95**]. It would be great to have a general theorem for all excitable media but this is not yet possible [**GG98**]. Of course in real hearts, there can be heterogeneities which will play a role in the annihilation of reentrant rhythms [**SC02**].

LECTURE 3

Fixed points and the entrainment of biological oscillations

The last chapter discussed the effect of a single stimulus delivered to an oscillator. Now I consider the effects of repeated stimuli. This topic is also of practical biological significance. For example, in normal hearts a specialized region of the heart (the sinus node) sets the pace. The heart is entrained or phase locked to the sinus pacemaker. However, in some circumstances, there can be rhythms in which different regions of the heart beat with different frequencies, as in 4:3 heart block where the atria have 4 contractions, for every 3 contractions of the ventricles. A discussion of some of the mathematics with lots of biological applications is in [**G01**].

The results on resetting oscillations discussed in Lecture 2 can be used immediately to study the effects of periodic stimulation provided *the oscillation is immediately reestablished following the stimulus, or in Eq. (2.1) in the limit* $k \to \infty$, (i.e. the transients induced by the stimulus are short compared to the time between stimuli). In this case, we can derive an appropriate one-dimensional map to predict the effects of periodic stimulation with period t_s on a limit cycle with intrinsic period T_0. Call ϕ_i the phase of stimulus i. Then, if the phase resetting curve is $g(\phi_i)$, the effects of periodic stimulation are given by

$$\phi_{i+1} = g(\phi_i) + \tau \pmod 1 \equiv f(\phi_i, \tau), \tag{3.1}$$

where, $\tau = t_s/T_0$. Starting from an initial condition ϕ_0 we generate the sequence of points $\phi_1, \phi_2, \dots, \phi_n$.

The sequence $\{\phi_i\}$ is well–defined, provided no stimulus results in a resetting to a point outside the basin of attraction of the limit cycle. If $\phi_n = \phi_0$ and $\phi_i \neq \phi_0$ for $1 \leq i < n$ where i and n are positive integers, there is a periodic cycle of period n. A cycle of period 1 is by definition a fixed point.

A periodic cycle of period n is stable if

$$\left| \frac{\partial f^n(\phi_0)}{\partial \phi} \right| = \left| \prod_{i=0}^{n-1} \left. \frac{\partial f}{\partial \phi} \right|_{\phi_i} \right| < 1. \tag{3.2}$$

If this product is equal to 0, the cycle is called superstable.

The *rotation number*, ρ, gives the average increase in ϕ per iteration. Calling

$$\Delta_{i+1} = g(\phi_i) + \tau - \phi_i \tag{3.3}$$

we have

$$\rho = \limsup_{N \to \infty} \frac{1}{N} \sum_{i=1}^{N} \Delta_i. \tag{3.4}$$

Stable periodic orbits are associated with *phase locking*. In $n : m$ phase locking, there is a periodic orbit of consisting of n stimuli and m cycles of the oscillator. This orbit has a rotation number m/n. For periodically forced oscillators neither the periodicity nor the rotation number alone is adequate to characterize the dynamics. Periodic stimulation of the Poincaré oscillator leads to an unbelievably complicated organization of locking zones, as well as zones in which there are cascades of period-doubling bifurcations, and chaotic dynamics. Extensive studies carried out in the 1980s of the effects of periodic stimulation of embryonic chick heart cells have demonstrated that many of these exotic types of dynamical behavior can be observed in experimental systems [**GGS81, GGBS84**].

Consistent with the theme of these lectures, I give two fixed point theorems related to the phase locking of nonlinear oscillations. As discussed above, the effects of periodic stimulation of nonlinear oscillations can be described by maps. First assume that the oscillation is rapidly reestablished so that the map is a one dimensional circle map. The typical situation for sufficiently low amplitude stimuli is that the map is an invertible differentiable map of the circle [**Ar83**]. An *Arnold tongue of rotation number m/n* is defined as the union of values in parameter space for which there is unique attracting $n : m$ phase locking for all initial conditions. For invertible diffeomorphisms of the circle of the form in Eq. (3.1), if there is $n : m$ phase locking for τ and $n' : m'$ phase locking for τ', then there exists a value $\tau < \tau^* < \tau'$, leading to $n + n' : m + m'$ phase locking. Usually, the range of values of τ associated with a given Arnold tongue covers an open interval in parameter space. For a given set of parameters the rotation number is unique. If it is rational, there is phase locking, and if it is irrational there is *quasiperiodicity*. Perhaps the classic example is the sine circle map

$$x_{n+1} = x_n + \tau + b \sin 2\pi x_n, \tag{3.5}$$

where b and τ are constants (think of τ as the time between periodic stimuli). The organization of phase locking zones in the (τ, b) plane $0 \le b < 1/2\pi$ shown in Figure 4 is typical and is called the *classic Arnold tongue structure*. The periodic

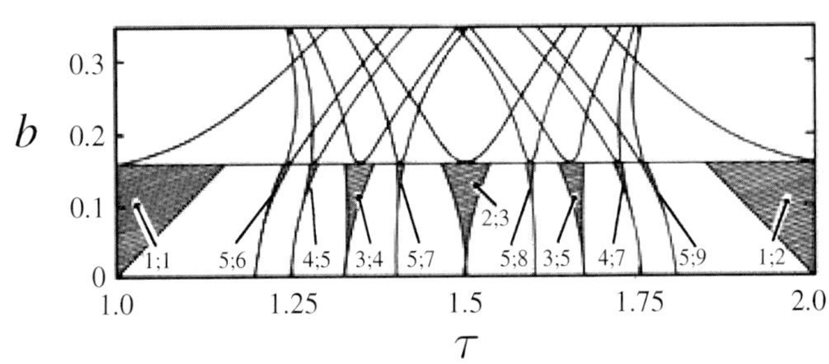

FIGURE 4. Arnold tongue structure for the degree 1 sine circle map, Eq. (3.5) for $0 < b < 1/2\pi$, and the locus of the superstable cycles for $1/2\pi < b$. [Reprinted Fig. 2 with permission from [**GP82**]. Copyright 1988 by the American Physical Society.]

orbits lose stability via a tangent bifurcation. Arnold's classic analysis stops however at $b = 1/2\pi$. For greater values of b there is a remarkable structure of complex bifurcations that are still incompletely understood [**GP82**].

However, based on the continuity properties of the circle map, we can assert the following theorem:

Theorem 3.1. *For $x_{n+1} = g(x_n) + \tau$ where g is a circle map of degree 1, with a single maximum and minimum, there are at least two values of τ at which there exist superstable cycles for each rational rotation number* [**GP82**].

For $b > 1/2\pi$, Figure 4 shows the locus of the superstable cycles in the (τ, b) plane. This result thus allows the Arnold tongues to be continued even when circle maps become non-monotonic. Interestingly, before the 1980s most mathematicians who studied circle maps did not venture into the territory where the degree one circle map becomes non–monotonic (e.g. see Fig. 78 in [**Ar83**]).

A second fixed point result can be found for some classes of periodically stimulated two–dimensional (or higher) ordinary differential equations. An example is given by the Poincaré oscillator for the case in which there is not an instantaneous relaxation back to the limit cycle. For Eq. (2.1), we assume once again the stimulus is a δ function perturbation to the state point leading to an instantaneous increase of the x-coordinate by a magnitude b. Following the stimulus, the equations of motion take over and the analytic formula for the resetting can be computed. Consequently, if a stimulus is given when the system is at state point (r_i, ϕ_i) we have

$$\begin{aligned} r'_i &= (r_i^2 + b^2 + 2br_i \cos 2\pi\phi_i)^{1/2}, \\ \phi'_i &= \frac{1}{2\pi} \arccos \frac{r_i \cos 2\pi\phi_i + b}{r'_i}. \end{aligned} \tag{3.6}$$

where (r'_i, ϕ'_i) are the coordinates immediately after the stimulus.

Following a time interval τ, we are at a new state point (r_{i+1}, ϕ_{i+1}) given by

$$\begin{aligned} r_{i+1} &= \frac{r'_i}{(1 - r'_i)\exp(-k\tau) + r'_i}, \\ \phi_{i+1} &= \phi'_i + \tau (\text{mod } 1). \end{aligned} \tag{3.7}$$

Based on these formulae, and choosing an initial circular domain sufficiently large, we show that the initial circular domain maps into itself, so that by the Brouwer fixed point theorem and the formulae in Eq. (3.7) we have:

Theorem 3.2. *For the 2-dimensional Poincaré oscillator, subjected to a periodic pulsatile (δ function) stimulus, there will be a period-1 cycle for any amplitude or frequency of the periodic forcing* [**GS94**].

This result indicates that there are major qualitative differences in the Arnold tongue structure between systems described by one dimensional maps, and systems described by two dimensional maps, since in the 2 dimensional system, there may appear to be the same type of Arnold tongue organization, but the locking is not longer the global attractor for all initial conditions.

Exercise 3.3. Compute the boundaries of the 1:1 entrainment for the sine circle map Eq. (3.5) [**PG82**]. Where is stability lost by a tangent bifurcation? By a period–doubling bifurcation?

Exercise 3.4. Compute the boundaries of the 1:1 entrainment for the Poincaré oscillator Eq. (2.1) [**GG82, KH82**] using the phase transition curve in Eq. (2.2). Where is stability lost by a tangent bifurcation? By a period–doubling bifurcation?

LECTURE 4

Fixed points in phase maps with applications to development and spiral waves on spheres

In many situations it is convenient to label each point in some topological space with a point on the unit circle. Such a representation is called a phase map. An example is the longitudinal coordinates on the globe. For phase maps, there may be points where the phase is not defined. These are called phase singularities. An early biological paper that used phase maps was the "clockface" model for developmental biology [**FBB76**], Figure 5. French and colleagues assumed the

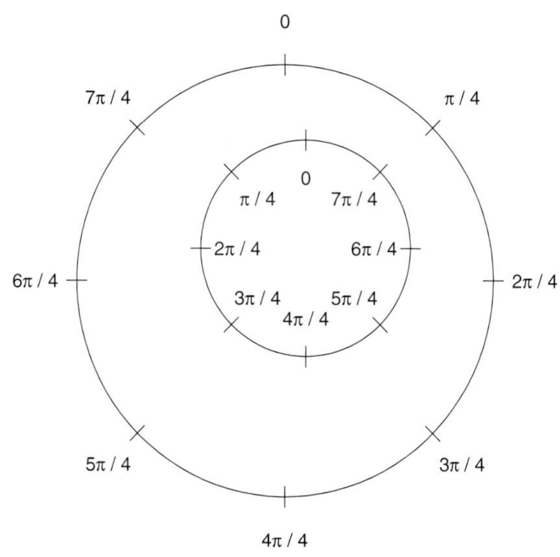

FIGURE 5. Schematic diagram of the clockface model [**FBB76**]. Points on the circumference of a circle are associated with a angular coordinate, where a left limb is associated with coordinates increasing in a clockwise sense and a right limb is associated with coordinates increasing in a counterclockwise sense. The diagram represents transplantation of a right limb on a left stump. The "clockface" polar coordinate system of positional information cannot fill in smoothly without generating singularities in the space between the two circles. In the biological system the singularities would be manifest as supernumerary limbs.

positions around the periphery of a limb was coded by a phase, where the phases for right limbs decreased in a clockwise direction and those for the left limbs increased in a clockwise direction. They supposed that after transplantation of a right limb on a left stump, it was necessary for the phase values to fill in continuously. However, they observed that this could not happen, and that it would be necessary to generate two supernumerary left limbs. I presented an analysis of this work early on [**G77**], but the topological aspects have not been well developed and are still of some interest [**DGK04**], see also Cruz-White's recent Ph.D. thesis [**CW03**].

In this Lecture, I will only talk about phase maps on surfaces, but phase maps in 3 dimensions are also of great importance, for example, see the work by Strogatz and Winfree e.g. see [**WS83**] and subsequent papers as discussed in [**W02**].

The mathematical description of spiral waves is based on the notion of phase which in turn allows one to characterize spiral waves by an index. From this description, a number of topological results placing restrictions on spiral wave dynamics can be derived [**G77, WS83, S87, W02, CW03**]. The current discussion is based on work originally presented in [**G77**], but follows the presentation in [**DGK04**] which should be consulted for more details and references.

With the exception of a finite number of singular points, with each point in an orientable and compact two-dimensional differentiable manifold M we identify a unique phase lying on the unit circle, $\Phi \in S^1$. The resulting phase map or phase field is assumed to be continuously differentiable, except at the singular points. The manifold can be triangulated (subdividing it into a set of polygons), where none of the edges or vertices of the polygons pass through a singularity. The index, I (sometimes also called the topological charge or winding number) of a curve C bounding a polygon is found by computing the line integral

$$2\pi I = \oint_C \nabla\Phi \cdot d\mathbf{l}, \tag{4.1}$$

where the polygon is always traversed in a clockwise orientation. By continuity of $\nabla\Phi$, I must be an integer. The index of a singular point is uniquely defined as the index of any curve C provided that C encircles the point but no other singular points. The index of a curve that does not enclose any singular points is obviously zero.

If the manifold M has no boundaries, each edge of the triangulation is an edge of two polygons. Since the line integral adds up the change in phase along the various edges of the polygon, the sum of the indices of the singular points for a phase field in M can be computed.

Theorem 4.1. *For a phase map on an oriented two dimensional manifold with with a finite number of isolated singularities*

$$\Sigma_{i=1}^k I_i = 0, \tag{4.2}$$

where the sum is over all the singular points.

This follows since the contribution of the change in phase of each edge to the total integral is counted twice, but since the edge is traversed in opposite directions each time, the net contribution of each edge is zero. However, unlike the more familiar Poincaré–Hopf index theorem that I discussed in Lecture 1, the sum of the indices of the singular points does not depend on the genus of the surface.

This index theorem for manifolds without boundaries can be extended to manifolds with boundaries. In the following, we will consider the case of structures that arise from puncturing orientable and compact two-dimensional differentiable manifolds. The index of a hole can be uniquely defined as the index of a curve C provided that C encircles the hole but no other holes or singular points and C is positively oriented with respect to the domain which contains the hole and is bounded by C. Applying this definition and taking the summation in Eq. (4.2) over the singularities and the hole, or the holes if there is more than one hole, the index theorem can be proved by the same line of arguments as for the case of manifolds without boundaries.

A sphere with a hole is topologically equivalent to a disk, and, indeed, the results for disks and for spheres with holes are consistent: For the disk D^2, bounded by a curve C,

$$\Sigma_{i=1}^{k} I = \oint_C \nabla\Phi \cdot d\mathbf{l}, \tag{4.3}$$

so that the sum of the indices of the singular points in the disk is equal to the index of the curve C bounding the disk. If there is a single singular point on the disk, with an index of $+1$, the index of the curve bounding the disk will also be $+1$. Imagine now the boundary of the disk to be brought together (like a draw-string bag) so that the boundary of the disk now defines a hole in the sphere. In this geometry the curve C will be traversed in an opposite orientation (the hole is now inside C) from the direction it was traversed when it was the boundary of the disk. Now if there is a singular point with an index of $+1$ on the sphere, the index of the hole is -1, so that the sum of the indices is again zero.

Since it is necessary to conserve the sum of the indices, singularities of index ± 1 usually arise and are destroyed in pairs of opposite sign [**WS83**]. An exception occurs when singularities are destroyed by collision into a boundary, so that the index of the singular point and the index of the boundary both change simultaneously. Another exception occurs if there are singularities with index different from one. In such cases interactions between different singularities can lead to destruction or creation of odd numbers of singularities [**ZP02**].

There are many different applications of these results in biology and chemistry. My original interest in this problem related to clockface model for regeneration [**FBB76**], Figure 5. Other applications involve chemical and biological waves on spherical surfaces [**DGK04**]. Perhaps the most exciting application is the analysis of cardiac arrhythmias. In the normal heart the rhythm is set by a specialized region, the sino-atrial node that sets the rhythm under neural control. However, in some circumstances, there is an abnormal pattern of excitation so that the period of the oscillation is set by the length of time it takes for the excitation to travel in a reentrant pathway. Although sometimes the reentrant pathway might encircle an anatomical feature, in other cases there may be a rotating spiral wave, for extensive discussion and references, see [**W02**]. In the current context, rotating spiral waves would be associated with singularities in phase maps. In particular, fibrillation is a fatal cardiac arrhythmia that many believe is associated with rotating reentrant spiral waves on the heart. If this is the case, then stimuli that terminate fibrillation must be capable of eliminating all phase singularities from the heart [**K04, GC05**]. The strong shocks needed to do this are quite different from the small shocks that can be capable of terminating some types of tachycardia which were discussed in Lecture 3.

LECTURE 5

Unique stable limit cycles in model genetic networks

The material in this section has been presented in many places earlier and is based on a result by Joel Pasternack and myself [**GP78**]. The current presentation is taken mostly from introductory sections of a recent paper [**GPMSE05**].

A gene is a sequence of DNA that codes for a sequence of amino acids that constitute a protein. Even though all cells have the DNA code to make a vast array of different proteins, a variety of regulatory mechanisms, still not completely understood, determine which proteins will be synthesized in each cell. One mode of gene regulation is through a class of proteins called transcription factors. Transcription factors bind to the DNA, turning "on" or "off" the synthesis of specific mRNA molecules. These are in turn translated into proteins, some of which may be transcription factors and thus may influence mRNA synthesis. There are many different mathematical models of genetic networks. I discuss a highly idealized model that reduces to a piecewise linear equation. Consequently, analytic computation is possible.

Let x_i represent the concentration of chemical species i in a cell. The time rate of change of x_i is

$$\frac{dx_i}{dt} = h_i(\mathbf{x}) - \frac{x_i}{\tau_i}, \qquad i = 1, \dots, N, \tag{5.1}$$

where there are N chemical species, $\mathbf{x}$ is a vector giving their concentrations, h_i is a function giving the control of the synthesis of the ith chemical species by the others, and τ_i is a decay constant.

In what follows, I adopt a piecewise linear form for the functions h_i. To each continuous variable $x_i(t)$, we associate a discrete variable $X_i(t)$,

$$X_i(t) = 0 \text{ if } x_i(t) < \theta_i; \text{ otherwise } X_i(t) = 1 , \tag{5.2}$$

where θ_i is an arbitrary real-valued threshold. The key restriction we place on Eq. (5.1) is that the synthesis rates depend only on discrete states of chemical species.

$$\frac{dx_i}{dt} = f_i(\mathbf{X}(t)) - \frac{x_i}{\tau_i}, \qquad i = 1, \dots, N, \tag{5.3}$$

where $\mathbf{X}(t)$ is the Boolean vector of discrete states of the chemical species at time t. In the biochemical context, in which the x_i are meant to represent concentrations, we assume all x_i are nonnegative and all τ_i and θ_i are positive. In general, however, we need not restrict the ranges of the x_i or θ_i, and we refer to the x_i as elements in an interacting network.

Equation (5.3) is piecewise linear in $\mathbf{x}$, with the pieces defined by the threshold hyperplanes $x_i = \theta_i$, and hence is easily integrated. As long as variables cross threshold hyperplanes transversely, there will be a finite or countably-infinite sequence of switching times, $\{t_1, t_2, t_3, \dots\}$, at which some element of the network crosses its threshold. We can obtain the solution of Eq. (5.3) for each variable x_i for $t_j < t < t_{j+1}$:

$$x_i(t) = x_i(t_j)\, e^{-(t-t_j)/\tau_i} + \tau_i f_i(\mathbf{X}(t)) \left(1 - e^{-(t-t_j)/\tau_i}\right). \tag{5.4}$$

Thus, for any point in the orthant defined by the Boolean vector $\mathbf{X}$, element i asymptotically approaches $\tau_i f_i(\mathbf{X})$. We call the vector of these values the *focal point* for the orthant. If the focal point is within the orthant, then it is an asymptotically stable fixed point. If it is within another orthant, then some variable will cross its threshold, and the trajectory continues in the adjacent orthant. We assume that for all $\mathbf{X}$, $\tau_i f_i(\mathbf{X}) \neq \theta_i$, so that no focal points lie on threshold hyperplanes.

The flow is therefore piecewise focused, with the focal points depending in general on the current orthant. For example, Figure 6 shows a flow in a two-

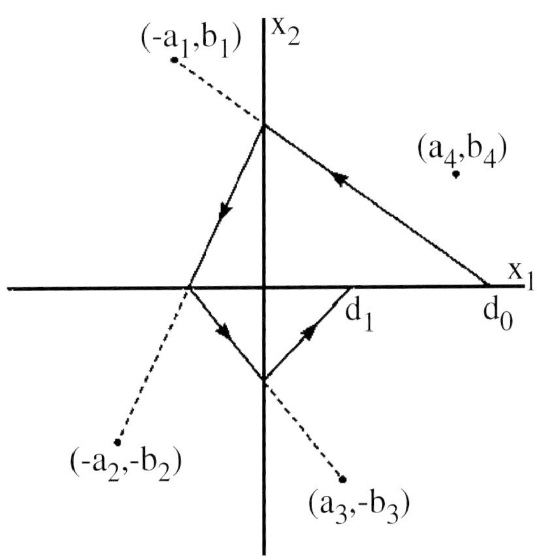

FIGURE 6. The phase plane portrait for a piecewise linear, piecewise focused flow in two dimensions. Starting at distance d_0 on the positive x_1 axis, there is a return to the x_1 axis at a distance d_1 from the origin. The map $d_1 = f(d_0)$, is called the Poincaré map. It is shown in [**GP78**] that this map is a linear fractional map, of the form in Eq. (5.5). The limiting properties of linear fractional maps upon iteration can be readily determined, even in higher dimensions.

dimensional system. Starting at a distance d_0 from the origin on the positive x_1-axis, there is a return to the positive x_1-axis at a distance d_1 from the origin. The map that relates d_1 to d_0 can be readily calculated using the methods described below [**GP78**].

As a consequence of variables crossing thresholds, the logical state changes and so might the focal point that the solution approaches. It is possible to piece together the trajectories and determine the dynamics for future times once the logical structure of the network and the parameters are known.

The logical structure of the differential equation can be represented as a directed graph on an N-dimensional hypercube (N-cube). The N-cube has 2^N vertices, each corresponding to a logical state, $\mathbf{X}$, of the network and hence to a region of state space. A directed edge between two adjacent vertices indicates a change in the logical state of the network that can be observed from some initial condition. Thus, edges in this representation represent flows across the boundaries of adjacent orthants of state space. The orientations of the directed edges in the graph can be determined by checking, for each $\mathbf{X}$ and i, whether $\tau_i f_i(\mathbf{X})$ is greater than or less than θ_i.

In certain special cases, knowledge about the dynamics in the differential equation follows immediately from the N-cube representation. A *stable vertex* on the N-cube is a vertex with no outgoing edges. A cycle is a directed path on the N-cube that starts and ends on the same vertex. A *cyclic attractor* is a cycle on the N-cube in which each vertex on the cycle has $N-1$ edges directed towards it, and one edge directed away from it. If an N-cube has no cycles, then in the limit $t \to \infty$, solutions of Eq. (5.3) approach a stable fixed point corresponding to one of the stable vertices in the N-cube. Which vertex is reached depends on the initial condition when there is more than one stable fixed point.

A cycle on the N-cube is a necessary condition for a cycle in Eq. (5.3). For situations in which there are cycles, it is useful to compute the Poincaré map describing the return to a hyperplane between two adjacent orthants.

If all the decay constants τ_i are equal, the maps that take the flows from one orthant boundary to the next have a simple form called a linear fractional map

$$M(\mathbf{x}) = \frac{A\mathbf{x}}{1 + \langle \phi, \mathbf{x} \rangle}, \tag{5.5}$$

where the thresholds are translated to zero, $\mathbf{x} \in \mathbf{R}^N$ is a point on the initial orthant boundary (in an N-dimensional network), A is an $N \times N$ matrix, $\phi \in \mathbf{R}^N$, and $\langle \phi, \mathbf{x} \rangle$ represents a vector dot product between ϕ and $\mathbf{x}$. This form for Eq. (5.5) follows directly from the solution of Eq. (5.4), where A and ϕ depend on the focal point coordinates f_i of the flow for the orthants being traversed. The composition of two linear fractional maps of the same dimension is once again a linear fractional map. As a consequence of this property, if there is a cycle we can analytically (usually with the assistance of a computer) compute the return map for a given cycle of orthants starting on a particular orthant boundary crossing. The return map is often called the Poincaré map, which can be represented as in Eq. (5.5), but where A is an $(N-1) \times (N-1)$ matrix, and ϕ and $\mathbf{x}$ are $(N-1)$ vectors [**GP78**].

If the directed graph on the N-cube displays a cyclic attractor, by a change of coordinates we can represent A as a positive matrix and ϕ as a positive vector. The limiting dynamics of Eq. (5.5) under iteration can be analyzed by application of the Perron–Frobenius theorem [**GP78**].

Theorem 5.1. *For a differential equation of the form Eq. (5.3) for which there is a cyclic attractor in N-cube representation, then one of the following two situations holds:*

(1) *There is a unique stable limit cycle in phase space which passes through the orthants in the same sequence and order as the cyclic attractor in the*

state transition diagram. This is attracting for all regions of phase space associated by the cyclic attractor.

(2) *The trajectories in the regions of phase space associated with the cyclic attractor asymptotically spiral towards the point defined by the intersection of the threshold hyperplanes.*

The first case applies if the leading eigenvalue of the matrix A is greater than 1. Otherwise the second case applies. This result enables us to identify large numbers of stable limit cycle oscillations in high dimensional dynamical systems.

An interesting biological example is provided by the design of a genetic circuit in *E. coli* in which there are several genes in a feedback loop where each gene controls the synthesis of the next in the loop by an inhibitory interaction.

As a simplified model, we assume that there are N genes governed by the differential equations [**G75a**]

$$\frac{dx_i}{dt} = \frac{{\theta_i}^n}{{\theta_i}^n + x_{i-1}^n} - x_i \quad i = 1, \ldots, N. \tag{5.6}$$

where θ_i is a threshold constant, which we will assume is 0.5, and we identify $x_0 = x_N$.

When $N = 2$, the equations represent mutual inhibition, and provided $n > 2$, there are two stable fixed points and an unstable saddle point illustrating one of the generic geometries for phase planes (see Lecture 1). Gardner and colleagues constructed a bacterium with this circuit and it behaved like a toggle switch [**GCC00**].

When $N = 3$, this equation displays a stable limit cycle oscillation provided $n > 4$. Elowitz and Leibler constructed a bacterium with this circuit and it displayed periodic oscillations [**EL00**]. They called this network the "repressilator". Interestingly, it has been known for a long time that an inhibitory loop with three elements is capable of giving rise to stable oscillations. An analysis of the Hopf bifurcation in this model is in [**G75**]. For the case where $n \to \infty$, $\theta = 0.5$, it is possible to analytically prove stability of the resulting limit cycle and to compute its period [**GP78**]. Before looking up the results, try these exercises. Also, as a way to track if anyone actually reads this, please send me an email if you read this sentence.

Exercise 5.2. For $N = 2$, analytically compute the value for one fixed point and determine the stability of this as a function of n. Sketch the trajectories in the two dimensional phase plane. Compare the phase plane structures that you find with those allowed by the results in Lecture 1.

Exercise 5.3. For $N = 3$, analytically compute the fixed point, and determine the value of n where a Hopf bifurcation occurs leading to instability. Now numerically compute the period of the oscillation as a function of n.

Exercise 5.4. For $N = 3$, the theoretical value for the period in the limit $n \to \infty$ when $\theta_i = 0.5$ is $-6\ln(\frac{-1+\sqrt{5}}{2}) = 2.88727\ldots$ Derive this result. You need to use the analytic solution of the equations, Eq. (5.4). Shift the fixed point to 0, and use symmetry to make the problem simpler. If you do this correctly, you will find that calculating the period involves the solution of a quadratic equation.

Exercise 5.5. If you enjoy doing these problems, you could write a computer program to integrate the piecewise linear equation, and study randomly generated networks searching for ones that give chaotic dynamics, e.g. see [**EG00**].

LECTURE 6

Projects: Resetting and entraining biological oscillations

Both of these projects relate to resetting and entraining biological oscillators. Good results would probably merit publication. The projects involve computer simulations, but there is also lots of room for analysis and topological thinking. The work is also relevant to experiments and clinical work.

Exercise 6.1. Entraining the 2D Poincaré oscillator. Perhaps the simplest ordinary differential equation that has been used to model biological oscillations is the Poincaré oscillator Eq. (2.1) which has already been presented and discussed in Lecture 2 in the infinite time relaxation limit,

This problem deals with finite relaxation times. Using Eq. (3.7) compute the different locking regions as a function of (k, b, τ). What is the maximum number of fixed points and periodic points possible for any given choice of parameters? For what parameters and initial conditions is there chaos? Can you prove chaos? Considering the locking zones in (b, τ) for k fixed. What is the organization of the zones for some fixed value of k and how does picture evolve as k increases. The limit for k large but finite should be fascinating in itself.

The locking zones in the 1D map that arises in the $k \to \infty$ limit are discussed in detail in [**KG84**]. Partial results for the case where k is finite are in [**GS94**].

Exercise 6.2. Resetting reentrant excitation oscillations in different geometries. This project is related to material in Lecture 2 on the continuity of phase resetting curves in the FitzHugh–Nagumo Eq. (2.4). Assume the circumference is $2 \times \sqrt{5}$ cm, $D = 1$ cm^2/sec, and cyclic boundary conditions. If you have a pulse circulating on a 1D ring, there is a stable oscillation with a period $T_0 = 356.1$ msec. If you prefer you could assume any other model for an excitable medium.

Your project is to study the continuity of the resetting curves assuming a delta function stimulus for a pulse circulating on a ring, or better in some other geometry such as an annulus or better yet, a "ring with a tail" when the stimulus is given in the ring, or someplace on the tail.

The issue involves the continuity of the resetting curves. In [**GJ95**] we argued that there should be critically timed stimuli that lead to the annihilation of the circulating pulse, and indeed a range of phases that lead to annihilation of the circulating pulse were found. Theorem 2.1 is relevant [**GG98**]. According to this theorem, if a stimulus delivered to a limit cycle oscillation at any phase of the cycle always leaves you in the basin of attraction of the limit cycle, then the phase resetting curve is continuous. However, results in [**SC02**] indicate that it might be necessary to revisit issues about the continuity of the resetting curves, and that it might be possible to get real discontinuities in resetting curves, without annihilating the reentry rhythm. The mathematical issues may be applicable to real life since anti-tachycardia pacemakers are used to terminate reentrant arrhythmias by giving one or more pulses to heart. Though such devices are often used clinically it is not known how to find an optimal location to place the wires in the heart that will deliver stimuli to the heart.

Acknowledgments. Over the years, I have benefited greatly from interactions with my colleagues, and especially in the work mentioned here from Joel Pasternack, Michael Guevara, Alvin Shrier, James Keener, Tomáš Gedeon, and Trine Krogh-Madsden. I thank the PCMI administration for their care in the arrangements, James Keener and Mark Lewis for their role in organizing the graduate summer school in mathematical biology, and the students for their interest and enthusiasm. I have benefited from research support from NSERC, CIHR, MITACS, the Canadian Heart and Stroke Foundation, and the NIH Research Resource for Complex Physiologic Signals.

Bibliography

[Ar83] V. I. Arnold. *Geometric Methods in the Theory of Ordinary Differential Equations* (Springer-Verlag, New York, 1983).

[BGMT03] A. Beuter, L. Glass, M. C. Mackey, M. S. Titcombe, Eds. *Nonlinear Dynamics in Physiology and Medicine* (Springer-Verlag, New York, 2003).

[CW03] I. Cruz-White. *Topology of Spiral Waves in Excitable Media* Ph.D. Thesis, Florida State University (2003).

[DGK04] J. Davidsen, L. Glass, R. Kapral. Topological constraints on spiral wave dynamics in spherical shells with inhomogeneous excitability. Physical Review E **70**, 056203 (2004).

[EG00] R. Edwards and L. Glass, Combinatorial explosion in model gene networks. *Chaos* **10**, 691–704 (2000).

[EL00] M. B. Elowitz and S. Leibler, A synthetic oscillatory network of transcriptional regulators. *Nature* **403**, 335–338 (2000).

[FBB76] V. French, P. J. Bryant, and S. V. Bryant. Pattern regulation in epimorphic fields. *Science* **193**, 969-981 (1976).

[GCC00] T. S. Gardner, C. R. Cantor, and J. J. Collins, Construction of a genetic toggle switch in *Escherichia coli. Nature* **403**, 339–342 (2000).

[GG98] T. Gedeon, L. Glass. Continuity of resetting curves for FitzHugh-Nagumo equations on the circle. In: *Fields Institute Communications Volume 21: Differential Equations with Applications to Biology*, 225-236 (1998).

[G75] L. Glass. A topological theorem for nonlinear dynamics in chemical and ecological networks. *Proceedings of the National Academy of Sciences (U.S.)* **72**, 2856-2857 (1975).

[G75a] L. Glass. Combinatorial and topological methods in nonlinear chemical kinetics. *Journal of Chemical Physics* **63**, 1325-1335 (1975).

[G77] L. Glass. Patterns of supernumerary limb regeneration. *Science* **198**, 321-322 (1977)

[GP78] L. Glass, J.S. Pasternack. Stable oscillations in mathematical models of biological control systems. *Journal of Mathematical Biology* **6**, 207-223 (1978).

[GP82] L. Glass, R. Perez. Fine structure of phase-locking. *Physical Review Letters* **48**, 1772-1775 (1982).

[GW84] L. Glass, A.T. Winfree. Discontinuities in phase resetting experiments. *American Journal of Physiology* **246** *(Regulatory, Integrative Comprehensive Physiology* **15**), R251-R258 (1984).

[GGBS84] L. Glass, M.R. Guevara, J. Bélair, A. Shrier. Global bifurcations of a periodically forced biological oscillator. *Physical Review A* **29**, 1348-1357 (1984).

[GM88] L. Glass, M.C. Mackey. *From Clocks to Chaos: The Rhythms of Life.* (Princeton University Press, Princeton, 1988).

[GS94] L. Glass, J. Sun. Periodic forcing of a limit cycle oscillator: Fixed points, Arnold tongues, and the global organization of bifurcations. *Physical Review E* **50**, 5077-5084 (1994).

[GJ95] L. Glass, M.E. Josephson. Resetting and annihilation of reentrant abnormally rapid heartbeat. *Physical Review Letters* **75**, 2059-2063 (1995).

[G01] L. Glass. Synchronization and rhythmic processes in physiology. *Nature* **410**, 277-284 (2001).

[GPMSE05] L. Glass, T. J. Perkins, J. Mason, H. T. Siegelmann, R. Edwards. Chaotic dynamics in an electronic model of a genetic network. Journal of Statistical Physics **121**, 969–994 (2005).

[GC05] R. A. Gray, N. Chattipakorn. Termination of spiral waves during cardiac fibrillation via shock-induced phase resetting. *Proc. Natl. Acad. Sci. USA* **102** 4672-4677 (2005).

[Gu75] J. Guckenheimer. Isochrons and phaseless sets. *J. Math. Biol.* **1**, 259-273 (1975).

[GGS81] M.R. Guevara, L. Glass, A. Shrier. Phase-locking, period-doubling bifurcations and irregular dynamics in periodically stimulated cardiac cells. *Science* **214**, 1350-1353 (1981).

[GG82] M.R. Guevara, L. Glass. Phase-locking, period-doubling bifurcations and chaos in a mathematical model of a periodically driven biological oscillator: A theory for the entrainment of biological oscillators and the generation of cardiac dysrhythmias. *Journal of Mathematical Biology* **14**, l-23 (1982).

[GSG86] M.R. Guevara, A. Shrier, L. Glass. Phase resetting of spontaneously beating embryonic ventricular heart cell aggregates. *American Journal of Physiology* **251** (*Heart Circ. Physiol.* **20**), H1298-H1305 (1986).

[GP74] V. Guillemin, A. Pollock. *Differential Topology* (Prentice-Hall, Englewood Cliffs, NJ, 1974).

[KG95] D.T. Kaplan, L. Glass. *Understanding Nonlinear Dynamics* (Springer-Verlag, New York, 1995).

[KH82] J. Keener, F. Hoppensteadt. Phase locking of biological clocks. *Journal of Mathematical Biology* **15**, 339-349 (1982).

[KG84] J. Keener, L. Glass. Global bifurcations of a periodically forced nonlinear oscillator. *Journal of Mathematical Biology* **21**, 175-190 (1984).

[K04] J. P. Keener. The topology of defibrillation. *J. Theoretical Biology* **230** 459-473 (2004).

[KGDG04] T. Krogh-Madsen, L. Glass, E. J. Doedel, M. R. Guevara. Apparent discontinuities in the phase-setting response of cardiac pacemakers. Journal of Theoretical Biology **230**, 499-519 (2004).

[LKGG07] C. Lerma, T. Krogh-Madsen, M. R. Guevara, L. Glass. Stochastic aspects of cardiac arrhythmias. Journal of Statistical Physics, **128**, 347–374, 2007.

[NG96] T. Nomura, L. Glass. Entrainment and termination of reentrant wave propagation in a periodically stimulated ring of excitable media. *Physical Review E* **53**, 6353-6360 (1996).

[PG82] R. Perez, L. Glass. Bistability, period doubling bifurcations and chaos in a periodically forced oscillator. *Physics Letters* **90A**, 441-443 (1982).

[SAIGGS01] V. Schulte-Frohlinde, Y. Ashkenazy, P. Ch. Ivanov, L. Glass, A. L. Goldberger, H. E. Stanley. Noise effects on the complex patterns of abnormal heartbeats. *Physical Review Letters*, **87**, 068104 (2001).

[SC02] S. Sinha and D. J. Christini. Termination of reentry in an inhomogeneous ring of model cardiac cells. *Phys. Rev. E* **66** 061903 (2002).

[S87] D. Sumners. In: *Graph Theory and Topology in Biology and Chemistry* ed. by R. B. King and D. Rouvray. (Elsevier, Amsterdam, 1987), p. 3.

[WS83] A. T. Winfree and S. H. Strogatz. Singular filaments organize chemical waves in three dimensions. Geometrically simple waves. *Physica D* **8** 35-49, (1983).

[W02] A. T. Winfree. *The Geometry of Biological Time. Second Edition* (Springer-Verlag, New York, 2002).

[ZP02] R. M. Zaritski and A. M. Pertsov. Stable spiral structures and their interaction in two-dimensional excitable media. *Phys. Rev. E* **66**, 066120 (2002).

Mathematical Modelling of Solid Tumour Growth: from Avascular to Vascular, via Angiogenesis

Helen Byrne

IAS/Park City Mathematics Series
Volume 14, 2005

Mathematical Modelling of Solid Tumour Growth: from Avascular to Vascular, via Angiogenesis

Helen Byrne

1. Introduction

Over the past ten years, the number of theoreticians studying solid tumour growth has increased so rapidly that it now represents a central component of research in mathematical biology, *c.f.* the appearance of books and review articles devoted to the subject [**3, 15, 104, 110, 123, 127, 157**]. It is difficult to know whether this activity has been stimulated by the fact that most people will have experience of cancer (either directly or indirectly) at some stage during their lives or whether it is due to a growing awareness of the challenging mathematical problems that it has generated and its potential as a rich source of future research projects.

This lecture series is intended to provide not only an introduction to the field and a working knowledge of the main areas of mathematical research but also an appreciation of the current and emerging directions for theoretical research. Rather than reviewing the entire field in little detail, we have decided to be selective and concentrate on several pieces of work that we believe are representative and illustrate how the underlying models have been improved and extended to accommodate new biological knowledge. A common feature of these models is that they can, in general, be formulated as systems of ordinary and partial differential equations. Thus we must postpone a discussion of the excellent work on stochastic and discrete models to a future date (for reviews, see [**157, 110**]).

The remainder of the lecture series is structured as follows. In Section 2 we provide a brief introduction to solid tumour growth, introducing the relevant biological terminology and describing how tumours progress from being localised and noninvasive to malignant masses that can invade the tissue that surrounds them and colonise (or metastasise to) other organs. In Section 3 we describe some simple ODE models that can be used to describe the growth of solid tumours and their response to chemo- and radiotherapy. In Sections 4, 5 and 6 we discuss spatially-structured models of increasing complexity, starting in Section 4 with one-dimensional models of avascular tumour growth. The stability of these models to symmetry-breaking perturbations is considered in Section 5 and then used to determine conditions under which a radially-symmetric tumour remains compact and localised (*i.e.* stability) and conditions under which it becomes irregularly shaped and invasive (*i.e.* instability). In Section 6 we use a multiphase modelling framework to extend the

---

Centre for Mathematical Medicine, School of Mathematical Sciences, University of Nottingham
E-mail address: helen.byrne@nottingham.ac.uk
HMB gratefully acknowledges financial support from the EPSRC through an Advanced Research Fellowship and from Queensland University of Technology, Australia.

models from Sections 4 and 5 to allow for tumour heterogeneity. In so doing, we not only provide justification for the use of Darcy's law to describe cell motion (this law states that cells move down pressure gradients) but also enable other descriptions of the tumour's material properties to be incorporated. Section 7 contains a review of models of angiogenesis and vascular tumour growth and the lecture series concludes in Section 8 with a discussion of the models presented and the theoretical challenges that lie ahead. Where possible, the theoretical material is supported by exercises designed to reinforce the accompanying analysis.

2. Background biology

The progression of a particular cancer from a harmless mass into a life-threatening disease is a multistage process which starts when one or more mutations occur within an individual cell [**65**]. Tumourigenesis, the process by which normal cells transform into cancer cells, is associated with the progressive loss of function of a range of regulatory genes, including repair genes that correct mutations and DNA damage before cell division and tumour suppressor genes that signal for cell-cycle arrest or induce programmed cell death (apoptosis) if substantial genetic damage is detected. The uncontrolled division of a mutated cell leads to the growth of a (small) avascular tumour, which may then remain dormant unless, and until, it acquires its own blood system by the process of angiogenesis. The resulting vascular tumour is relatively well-supplied by nutrients and may grow to a much larger size. Finally, malignant tumours are able to invade the surrounding tissue, leading to metastatic spread, with secondary tumours arising elsewhere in the host. Further genetic mutations within the tumour mass may endow it with drug resistance and facilitate the later stages of development.

Once a solid tumour has developed and been detected, it may be categorised according to the tissue and cell-type of origin, the fractal dimension of its periphery [**95**], whether it is benign or malignant and whether it possesses a blood supply (vascular tumours have a blood supply whereas avascular ones do not). During avascular growth, nutrients that enter the tumour are consumed by live, proliferating cells as they diffuse towards the tumour centre. As the tumour grows the amount of nutrient reaching the centre declines until there is insufficient to sustain viable cells. There ensues the formation of a central core of dead (necrotic) cellular material whose size increases as the tumour continues to grow. Thus a well-developed avascular tumour comprises an outer rim of nutrient-rich, proliferating cells and a central core of nutrient-starved, necrotic debris. These regions may be separated by a layer of oxygen-poor (hypoxic) cells which are quiescent (viable but nonproliferating). Since diffusion controls the delivery of nutrients (e.g. oxygen and glucose) to, and the removal of waste products from, avascular tumours [**63, 141**], the diameter to which they may grow is limited to several millimetres and they are relatively harmless. Vascular tumours, however, are life-threatening for two main reasons. Firstly, being connected to the host's blood supply, they have access to an almost limitless supply of nutrients. The consequent rapid growth of such tumours may impair the function of neighbouring vital organs. Second, tumour fragments that enter the vasculature may be transported to other parts of the body where they may establish secondary tumours (metastases) that further jeopardise the host. The switch from avascular to vascular growth is effected by angiogenesis [**60, 61, 111**]. During this process the tumour cells secrete a range of diffusible chemicals (e.g. vascular endothelial growth factor and tumour necrosis factor-β) which are known collectively as angiogenic factors. The angiogenic factors stimulate endothelial cells from neighbouring blood vessels to proliferate and migrate towards the tumour, eventually furnishing it with a circulating blood supply so that vascular growth may commence. Invasion of the surrounding tissue is another key feature of solid tumours: contact with the tissue stimulates the production of enzymes such as matrix metallo-proteases which digest the tissue. This creates spaces into which the tumour cells may then migrate [**137**]. At a cellular level, the sequence of events that are needed to establish a well-developed, vascularised tumour may be associated with genetic mutations. For example, mutations in the tumour suppressor gene p53 have been linked with tumour angiogenesis and cell immortality, the latter being a hallmark of many tumours [**128, 132**].

From the discussion thus far, it is clear that solid tumour growth is a complex, multiscale process, involving phenomena that operate across a spectrum of spatial and temporal scales. Additionally, there is increasing evidence that these phenomena are coupled. Consider, for example, a vascularised tumour growing *in vivo* and the mechanisms that operate on the subcellular, cellular and macroscales. These include the expression of a range of cell-cycle proteins and angiogenic factors at the subcellular level, competition for space between normal and cancer cells at the cellular level and vascular adaptation, blood flow and nutrient transport at the macroscale. Now, in response to environmental stresses, such as hypoxia, tumour cells may alter their subcellular behaviour by slowing (or halting) the rate at which they progress through the cell cycle and/or producing angiogenic factors such as VEGF. When released by the cell, the angiogenic proteins diffuse through the region until they reach neighbouring vessels where they stimulate vascular adaptation and, thereby, increase the blood flow and supply of nutrients to the tumour. In this way, phenomena at the subcellular and macroscale may be interlinked. We discuss these ideas further in Section 7 where we present a multiscale model of vascular tumour growth.

Having explained how solid tumours grow, we now describe the main modes of treatment currently available, together with alternative approaches that are currently being developed. Typically, a tumour that is detected using routine screening or because the patient has developed symptoms (*e.g.* a lump in the breast or impaired function in another organ) will be vascularised and may have spread to other organs. Surgery, to excise as much of the abnormal mass as possible, is the first method of choice (although, in some cases, a tumour's location may preclude surgery). Because of difficulties associated with identifying the extent of a tumour's spread and in an attempt to prevent the tumour regrowing, surgery is often followed by a course of chemo- or radiotherapy. Radiotherapy can be focussed on the affected area and is typically administered on a daily basis over a period of several weeks, sometimes in combination with hyperoxygenation to enhance the response. As with most cancer treatments, unpleasant side-effects including hair loss and nausea often accompany radiotherapy. In Section 3 we present a simple model which describes a tumour's response to radiotherapy and the damage that it causes to healthy tissue. We then use the model to show that a daily schedule may not be optimal for all cancers and suggest instead that the interval between doses should be reduced for tumours with shorter doubling times (and conversely for tumours with longer doubling times). Chemotherapy (*i.e.* the administration of cytotoxic drugs) is also frequently used following, or in place of, surgery. Most drugs are designed to target rapidly dividing cells and, as such, act specifically on those tumour cells that have a higher division rate than their normal counterparts. There are a large number of chemotherapeutic agents in clinical use and they differ in their precise mode of action: some may act throughout a cell's division cycle while others act only during a particular phase. In addition to suffering similar side-effects as those experienced during radiotherapy, the main problems associated with chemotherapy are due to poor drug delivery and the emergence over time of tumour cells which are resistant to a particular drug. The rapid rate of cell division in a tumour creates adverse pressure gradients which hinder the diffusion of drugs from blood vessels to tumour regions. While cells in close proximity to a vessel may respond well to therapy, cells located further from the vessel will be exposed to lower drug concentrations because of the adverse pressure gradient and progressive uptake of the drug as it diffuses through the tumour. Further, cells at distance from blood vessels are often hypoxic and respond to this stress by reducing their rates of division so that any drugs penetrating hypoxic tumour regions may be ineffective.

As stated above, a tumour will usually have undergone angiogenesis by the time it is detected clinically. Many new therapies are being designed to target and destroy the tumour's vasculature and, in so doing, to prevent it from metastasising. These anti-angiogenic and vascular-targeting agents include angiostatin, combretastatin and endostatin and, while their precise mode of action remains to be determined, they are believed to act by killing proliferating endothelial cells and/or by reducing blood flow to the tumour region.

Other therapeutic approaches that show promise involve the use of viral and non-viral vectors to deliver treatment to solid tumours. For example, experimentalists are currently exploiting the tendency of macrophages to migrate to hypoxic tumour regions by genetically engineering them to release anti-angiogenic or cytotoxic drugs under low oxygen [**112**]. Alternatively, viruses that can infect and replicate insider tumour cells, eventually causing cell death, have shown promising results: virus particles released during cell lysis are able to infect other cells and, thereby, to establish alternating periods of infection, proliferation and cell death as the virus spreads through the tumour [**54, 155**].

As our understanding of the biology underpinning solid tumour growth increases it is likely that new, more sophisticated treatments will be developed to exploit differences between normal cells and their pathological counterparts. Ideally they will be used in combination with existing approaches to maximise the damage to the tumour while minimising harmful side-effects, a nontrivial optimisation problem! Further, and perhaps in the longer term, techniques for analysing the genetic composition of tumour biopsies should enable clinicians to characterise in more detail a particular cancer and, as a result, to devise individual-based treatment protocols. Armed with these new treatments, we face the future optimistic that the prognosis for cancer patients will only improve.

3. ODE models

3.1. Introduction

We start by discussing some of the simplest and earliest mathematical models of solid tumour growth. Spatial effects are neglected and the models formulated as systems of ordinary differential equations (ODEs). As we show, models of this type may be used to describe how the numbers of proliferating, quiescent and dead cells contained within a tumour change over time. They may also be used to compare the response of a tumour to chemotherapy or radiotherapy.

The section is structured as follows. In Subsection 3.2 we discuss several models of homogeneous tumour growth which we use in Subsections 3.3 and 3.4 to predict the tumour's response to chemo- and radiotherapy. In Subsection 3.5 we consider more realistic models in which the tumour is assumed to contain different cell types (*e.g.* proliferating, quiescent and necrotic cells). We then conclude with a critique of the spatially-uniform models.

3.2. Growth of homogeneous solid tumours

Arguably the simplest model that describes how the number of cells $N(t)$ within a solid tumour evolves over time is the exponential growth law which states

$$\frac{dN}{dt} = kN, \quad \text{with } N(t=0) = N_0, \tag{1}$$

with solution

$$N(t) = N_0 e^{kt}.$$

In Equation (1), $k > 0$ represents the net rate of cell proliferation and N_0 denotes the initial number of tumour cells. In this model, there are no constraints on cell growth: all nutrients and growth factors are assumed to be in abundance. In consequence, the model predicts that the population undergoes unbounded growth.

Whilst providing an accurate description of the early stages of a tumour's development, the exponential growth law is clearly unable to capture the periods of reduced growth and eventual saturation that are observed when avascular tumours are cultured *in vitro* or when vascular tumours develop *in vivo*. In practice the tumour's growth rate slows because, as it increases in size, competition for nutrients and other vital resources, such as space, can no longer be neglected. A simple modification of (1) which accounts for competition for resources (without specifying what those resources are) is the logistic growth law:

$$\frac{dN}{dt} = kN\left(1 - \frac{N}{\theta}\right), \quad \text{with } N(t=0) = N_0 > 0, \tag{2}$$

with solution

$$N(t) = \frac{\theta N_0}{N_0 + (\theta - N_0)e^{-kt}} \to \theta \text{ as } t \to \infty.$$

In (2), $\theta > 0$ represents the carrying capacity of the population

Whilst equation (2) predicts exponential growth of small tumours and growth saturation when the tumour reaches its carrying capacity ($N = \theta$), it is not particularly flexible when used to fit experimental data. A more realistic model which approaches the carrying capacity more (or less) rapidly, depending on the choice of the parameter $\alpha > 0$, is given by

$$\frac{dN}{dt} = \frac{k}{\alpha} N\left[1 - \left(\frac{N}{\theta}\right)^{\alpha}\right], \quad \text{with } N(t=0) = N_0, \tag{3}$$

with solution

$$N(t) = \theta\left(\frac{N_0^{\alpha}}{N_0^{\alpha} + (\theta^{\alpha} - N_0^{\alpha})e^{-kt}}\right)^{1/\alpha}.$$

We remark that equation (3) reduces to (2) when $\alpha = 1$ and to the Gompertzian growth law in the limit as α approaches zero from above.

In Figure 1 we compare the three models described above, presenting plots of the growth laws specified by equations (1), (2) and (3) for fixed values of k and θ (and two choices of α).

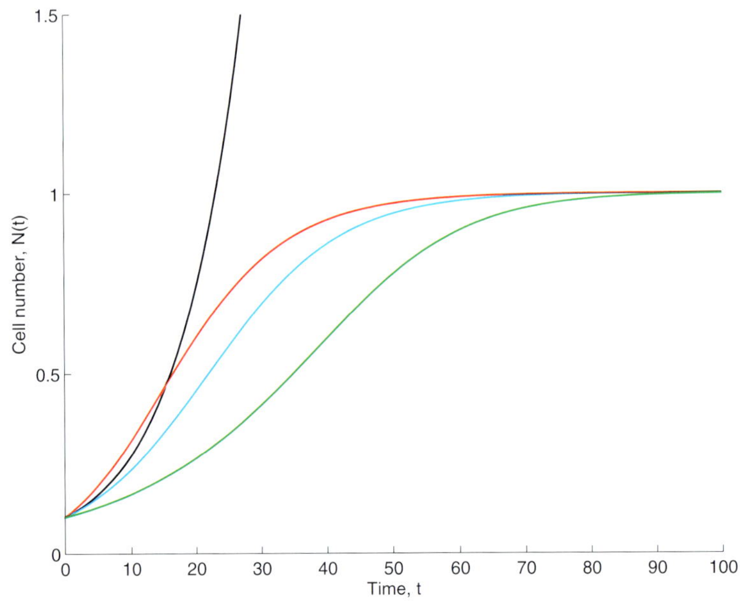

FIGURE 1. Series of curves illustrating how the tumour's dynamics depend upon the choice of growth law. Key: exponential growth law (black); logistic growth law (cyan); Verhulst growth law ($\alpha = 0.5$, red and $\alpha = 2.0$, green). Parameter values: $k = 0.1, \theta = 1.0, N_0 = 0.1$, with either $\alpha = 0.5$ or $\alpha = 2.0$.

Many other growth laws have been used to describe solid tumour growth (see, for example [**108, 109**]). A common weakness of models of this type is the difficulty of relating the model parameters (e.g., k, θ and α in equation (3)) to the behaviour of individual cells. We return to this matter in Section 4 where we discuss spatially-structured models of avascular tumour growth. First, however, we consider three simple extensions to the models reviewed thus far and show how they can provide insight into the impact that cellular heterogeneity, chemotherapy and radiotherapy have on solid tumour growth.

3.3. Chemotherapy

Consider a tumour that undergoes logistic growth *in vivo* when no therapy is applied. We focus on the tumour's dynamics when a chemotherapeutic drug is injected into the patient's bloodstream. Our model describes how the number of tumour cells $N(t)$ and the drug concentration $A(t)$ within the tumour evolve over time. We assume that the drug kills tumour cells and is itself degraded at rates which are proportional to the product AN, with constants of proportionality μ and γ respectively. We assume further that the drug undergoes natural decay with half-life λ and that it is delivered to the tumour at some rate $a(t)$. Combining these features, we obtain the following ODEs for N and A:

$$\frac{dN}{dt} = kN\left(1 - \frac{N}{\theta}\right) - \mu AN \equiv f(N, A), \tag{4}$$

$$\frac{dA}{dt} = a(t) - \lambda A - \gamma AN \equiv g(N, A), \tag{5}$$

with

$$N(t = 0) = N_0 \quad \text{and} \quad A(t = 0) = A_0. \tag{6}$$

The functional form of $a(t)$ depends on the treatment protocol. We will consider two alternatives:

$$\textbf{Continuous Infusion:} \quad a(t) = a_\infty \ \forall \, t \geq 0;$$

$$\textbf{Periodic Infusion:} \quad a(t) = \begin{cases} a_\infty & n < t < n + \tau \\ 0 & n + \tau < t < n + 1, \end{cases}$$

for positive constants a_∞ and τ.

3.3.1. *Continuous infusion.* Given that when no drug is applied (*i.e.* $a_\infty = 0$) equation (4) reduces to (2) and the tumour evolves to its carrying capacity, it is natural to ask whether the tumour evolves to an equilibrium when $a_\infty > 0$ and, if so, how that equilibrium depends on the drug's activity. To do this, we use linear stability analysis to identify and classify the critical points (or equilibria) of equations (4) and (5).

When $\frac{d}{dt} = 0$, equations (4) and (5) reduce to give

$$0 = kN\left(1 - \frac{N}{\theta} - \frac{\mu}{k}A\right) \quad \text{and} \quad 0 = a_\infty - \lambda A - \gamma NA,$$

so that either $N = 0$ and $A = a_\infty/\lambda$ or $A = k/\mu(1 - N/\theta)$ and

$$0 = N^2 + \frac{\lambda}{\gamma}\left(1 - \frac{\gamma\theta}{\lambda}\right)N + \frac{\lambda\theta}{\gamma}\left(\frac{a_\infty \mu}{\lambda k} - 1\right) \quad \text{and} \quad A = \frac{k}{\mu}\left(1 - \frac{N}{\theta}\right). \tag{7}$$

For a particular tumour and a specific drug, the parameters k, θ, μ, λ and γ will be fixed: the only parameter which can be manipulated is a_∞ and, hence, we use it as our key bifurcation parameter.

It is straightforward to show that (7) possesses no real roots if

$$a_\infty > a_\infty^{max} \equiv \frac{\lambda k}{\mu}\left[1 + \frac{\lambda}{4\gamma\theta}\left(1 - \frac{\gamma\theta}{\lambda}\right)^2\right].$$

Thus, if the administered dose is such that $a_\infty > a_\infty^{max}$ then only the trivial, tumour-free solution ($N = 0$) exists.

Further analysis of (7) indicates that for intermediate values of a_∞ ($0 < a_\infty \leq a_\infty^{max}$) the non-trivial steady state solutions depend on the parameter grouping $\gamma\theta/\lambda$. Referring to Figure 2, we note that if $\gamma\theta/\lambda > 1$ then there is a finite range of a_∞ ($\lambda k/\mu < a_\infty < a_\infty^{max}$) for which (7) possesses two, physically realistic solutions: for smaller values of a_∞ ($0 \leq a_\infty < \lambda k/\mu$) there is a single, nontrivial solution. We note also that if $\gamma\theta/\lambda < 1$ then for moderate values of a_∞ ($0 \leq a_\infty < \lambda k/\mu$) there is a single, nontrivial solution: for larger values of a_∞ ($\lambda k/\mu < a_\infty < a_\infty^{max}$) there are no physically realistic, nontrivial equilibrium solutions.

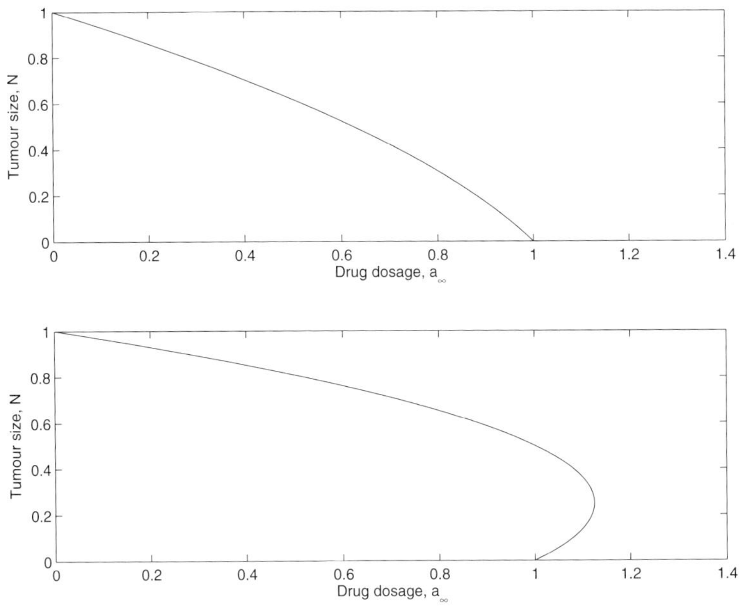

FIGURE 2. Pair of bifurcation diagrams revealing how varying a_∞, the drug dosage, and the parameter grouping $\gamma\theta/\lambda$ influence the tumour's equilibrium size. In case (a) we have $\gamma\theta/\lambda < 1$ while in case (b) $\gamma\theta/\lambda > 1$. Parameter values: (a) $\theta = \lambda = \mu = k = 1, \gamma = 0.5$; (b) $\theta = \lambda = \mu = k = 1, \gamma = 2$.

Having identified the equilibrium solutions, we must now characterise their (local) stability so that we can predict whether the equilibria will be realised in practice, this being particularly important in situations for which multiple equilibria arise. We start by considering the tumour-free solution $(N, A) = (0, a_\infty/\lambda)$.

We linearise about $(N, A) = (0, a_\infty/\lambda)$ by introducing the small parameter $\epsilon \ll 1$ and writing

$$N(t) = \epsilon N_1(t) \quad \text{and} \quad A(t) = \frac{a_\infty}{\lambda} + \epsilon A_1(t). \tag{8}$$

We substitute with (8) in equations (4)-(5). Equating coefficients of $O(\epsilon)$ we obtain the linearised equations

$$\frac{dN_1}{dt} = \left(k - \frac{\mu a_\infty}{\lambda}\right) N_1,$$
$$\frac{dA_1}{dt} = \lambda A_1 - \frac{\gamma a_\infty}{\lambda} N_1.$$

Solving for N_1 and A_1 we deduce

$$N_1(t) = \bar{N}(0)e^{(k-\mu a_\infty/\lambda)t},$$

and

$$A_1(t) = \left(\bar{A}(0) + \frac{\gamma a_\infty \bar{N}(0)}{\lambda^2 + k\lambda - \mu a_\infty} \right) e^{-\lambda t} - \frac{\gamma a_\infty \bar{N}(0)}{\lambda^2 + k\lambda - \mu a_\infty} e^{(k-\mu a_\infty/\lambda)t}.$$

We conclude that if $a_\infty > \lambda k/\mu$ then $N_1(t), A_1(t) \to 0$ and we say that in this case the trivial solution is linearly stable. By contrast, if $a_\infty < \lambda k/\mu$, then $N_1(t), |A_1(t)| \to \infty$ as $t \to \infty$, our asymptotic expansion breaks down and the trivial solution is said to be linearly unstable.

Similar analysis can be used to characterise the nontrivial equilibrium solutions of (4)-(5), although the algebra is more involved. For example, to determine the stability of a nontrivial solution $(N, A) = (N_\infty, A_\infty)$, where N_∞ and A_∞ solve (7), we seek solutions of the form

$$N(t) = N_\infty + \epsilon N_1(t) \quad \text{and} \quad A(t) = A_\infty + \epsilon A_1(t) \quad \text{where } \epsilon \ll 1.$$

Substituting with these trial solutions in Equations (4)-(5) and equating coefficients of $O(\epsilon)$ we find

$$\frac{d}{dt}\begin{pmatrix} N_1 \\ A_1 \end{pmatrix} = \begin{pmatrix} \frac{\partial f}{\partial N} & \frac{\partial f}{\partial A} \\ \frac{\partial g}{\partial N} & \frac{\partial g}{\partial A} \end{pmatrix} \begin{pmatrix} N_1 \\ A_1 \end{pmatrix}, \tag{9}$$

where partial derivatives are evaluated at $(N, A) = (N_\infty, A_\infty)$. Equations (9) admit solutions of the form

$$(N_1, A_1) = (\tilde{N}, \tilde{A})e^{\sigma t}.$$

For nontrivial solutions $(\tilde{N}, \tilde{A}) \neq 0$ and so we deduce from (9) that the growth rate σ satisfies the dispersion relation

$$0 = \sigma^2 - \left(\frac{\partial f}{\partial N} + \frac{\partial g}{\partial A} \right)\sigma + \left(\frac{\partial f}{\partial N}\frac{\partial g}{\partial A} - \frac{\partial f}{\partial A}\frac{\partial g}{\partial N} \right).$$

Further we have linear stability if $\Re(\sigma) < 0$. It is straightforward to show that this will be the case if

$$\frac{\partial f}{\partial N} + \frac{\partial g}{\partial A} < 0 < \frac{\partial f}{\partial N}\frac{\partial g}{\partial A} - \frac{\partial f}{\partial A}\frac{\partial g}{\partial N}. \tag{10}$$

Worked example

Consider our model of continuous drug infusion with

$$\theta = \lambda = \mu = k = 1, \quad \gamma = 1/2 \quad \text{and} \quad a_\infty \in (0, 1).$$

From (7) we know that nontrivial equilibrium solutions have $A_\infty = 1 - N_\infty$ where

$$0 = N_\infty^2 + N_\infty - 2(a_\infty - 1) \quad \Rightarrow \quad 2N_\infty = -1 \pm \sqrt{9 - 8a_\infty}.$$

Thus for $a_\infty \in (0, 1)$ there is a unique, physically-realistic (i.e. positive) solution . Further, the Jacobean is

$$\begin{pmatrix} \dfrac{\partial f}{\partial N} & \dfrac{\partial f}{\partial A} \\ \\ \dfrac{\partial g}{\partial N} & \dfrac{\partial g}{\partial A} \end{pmatrix}_{(N_\infty, A_\infty)} = \begin{pmatrix} 1 - 2N_\infty - A_\infty & -N_\infty \\ \\ -\dfrac{A_\infty}{2} & -1 - \dfrac{N_\infty}{2} \end{pmatrix}$$

so

$$\frac{\partial f}{\partial N} + \frac{\partial g}{\partial A} = -\left(\frac{3N_\infty}{2} + A_\infty \right) < 0 < \frac{\partial f}{\partial N}\frac{\partial g}{\partial A} - \frac{\partial f}{\partial A}\frac{\partial g}{\partial N} = N_\infty^2 + \frac{N_\infty}{2} > 0.$$

Thus equation (10) holds and we deduce that the nontrivial equilibrium solution is locally stable.

In Figure 3 we characterise the stability of the steady state solutions. If $\gamma\theta/\lambda < 1$ then the nontrivial steady state is stable where it exists and the trivial solution is stable only where no nontrivial steady solution exists. If $\gamma\theta/\lambda > 1$ then, where multiple, nontrivial steady state solutions exist, the smaller steady state is unstable while the larger solution and the trivial solution are stable; the smaller steady state delineates the basins of attraction of the two stable solutions.

3.3.2. *Periodic infusion.* The analysis presented above suggests that if the dose is large enough (*i.e.* $a_\infty > k/\mu$) then the tumour will be eliminated. Continuous infusion is often not practical due to side-effects, affecting, in particular, the skin, hair and liver where there is rapid cell turnover. Instead, chemotherapy is given in a series of short infusions or boluses, allowing the patient's organs (and, unfortunately, the tumour) time to recover between consecutive doses.

We now investigate the impact of periodic infusion on our solid tumour growth model. For simplicity we study the following simple model

$$\frac{dN}{dt} = kN\left(1 - \frac{N}{\theta} - \frac{\mu}{k}A\right), \quad \text{with } N(0) = N_0, \tag{11}$$

$$\text{and} \quad A(t) = \begin{cases} a_\infty & n < t < n+\tau \\ 0 & n+\tau < t < n+1, \end{cases} \tag{12}$$

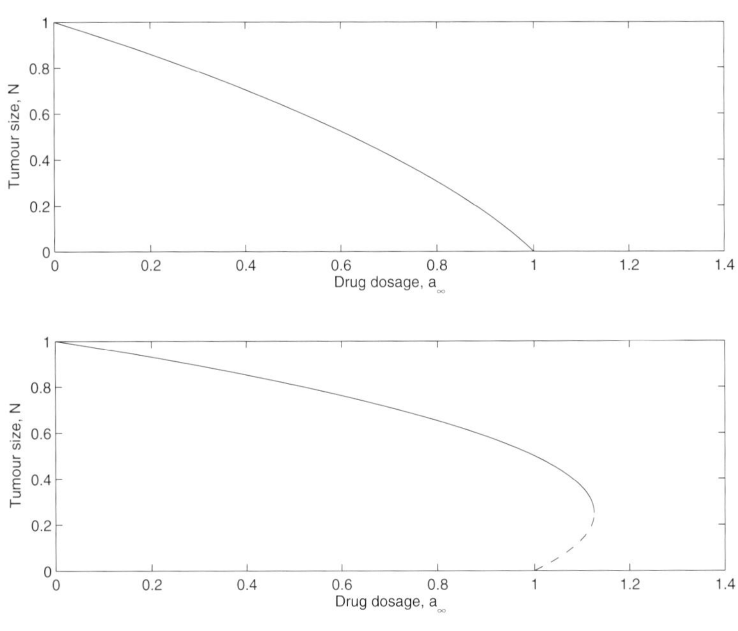

FIGURE 3. Pair of diagrams showing how changing a_∞ influences the stability of the equilibrium solutions presented in Figure 2. Parameter values: as per Figure 2. Key: stable solutions (solid lines); unstable solutions (dashed lines).

for $n = 0, 1, 2, \dots$ [1]

Since $A(t)$ is piecewise constant, equation (11) can be viewed as a logistic equation, with a variable carrying capacity that switches between $\theta(1 - \mu a_\infty/k\theta)$ and θ as $A(t)$ switches between a_∞ and 0. By assuming continuity of $N(t)$ at the switching times $t = n, n+\tau \, (n = 0, 1, 2, \dots)$ we can construct the following solution for $N(t)$:

$$N(t) = \begin{cases} \dfrac{\theta\Lambda N_n}{N_n + [\theta\Lambda - N_n]\exp[-k\Lambda(t-n)]} & n < t < n+\tau, \\ \\ \dfrac{\theta N_{n+\tau}}{N_{n+\tau} + (\theta - N_{n+\tau})\exp[-k(t-n-\tau)]} & n+\tau < t < n+1, \end{cases} \tag{13}$$

where $N_n = N(t = n), N_{n+\tau} = N(t = n+\tau)$, $\Lambda = 1 - \mu a_\infty/k\theta$ and

$$N_{n+\tau} = \frac{\theta\Lambda N_n}{N_n + [\theta\Lambda - N_n]\exp[-k\Lambda\tau]}.$$

Further, continuity of $N(t)$ at $t = n+1$ means that the coefficients N_n satisfy the following recurrence relation:

$$N_{n+1} = \frac{\theta\Lambda N_n}{\Lambda N_n + [(1-\Lambda)N_n + (\theta\Lambda - N_n)\exp[-k\Lambda\tau]]\exp[-k(1-\tau)]}. \tag{14}$$

In Figure 4 we show how the sequence $\{N_n\}_{n=0}^{\infty}$ evolves and how the limiting behaviour depends on a_∞. In all cases, the system evolves to an equilibrium and this decreases in magnitude as a_∞ increases, [2] with elimination of the tumour if a_∞ is large enough. The dose required to achieve eradication is greater with periodic infusion than with continuous infusion because in the former case the tumour cells are not always exposed to the drug.

Recall that when we studied the tumour's response to continuous drug infusion, the tumour eventually settled to a time-independent equilibrium solution. By contrast, as the results of Figure 4 indicate, when there is periodic infusion, the system may also evolve to a (nontrivial) periodic solution for which $N(t) = N(n+t)$ and hence $N_n = N_{n+1} \equiv N_\infty > 0$, say. Using Equation (14) we deduce that nontrivial periodic solutions will arise if

$$N_\infty = \frac{\theta\Lambda(1 - e^{-k(1-\tau)}e^{-k\Lambda\tau})}{\Lambda + (1-\Lambda)e^{-k(1-\tau)} - e^{-k(1-\tau)}e^{-k\Lambda\tau}}.$$

In Figure 5 we show how N_∞ decreases as a_∞ increases.

3.4. Radiotherapy

In this subsection we discuss a simple ODE model of radiotherapy due to Wheldon and coworkers [**156**] and use it to compare the response of tumours with different doubling times to two different protocols which we term conventional and optimal radiotherapy.

---

[1] Equation (11) can be derived from (4)-(5) by introducing a small parameter $0 < \delta \ll 1$, assuming $a(t) = a_\infty/\delta, \lambda = 1/\delta$ and $\gamma = O(1)$ and focussing on a timescale such that $t = O(1)$.

[2] Note that even though the sequence $\{N_n\}_{n=0}^{\infty}$ evolves to an equilibrium, the system settles to a periodic solution, with $N = N(t)$ for $n < t < n+1$.

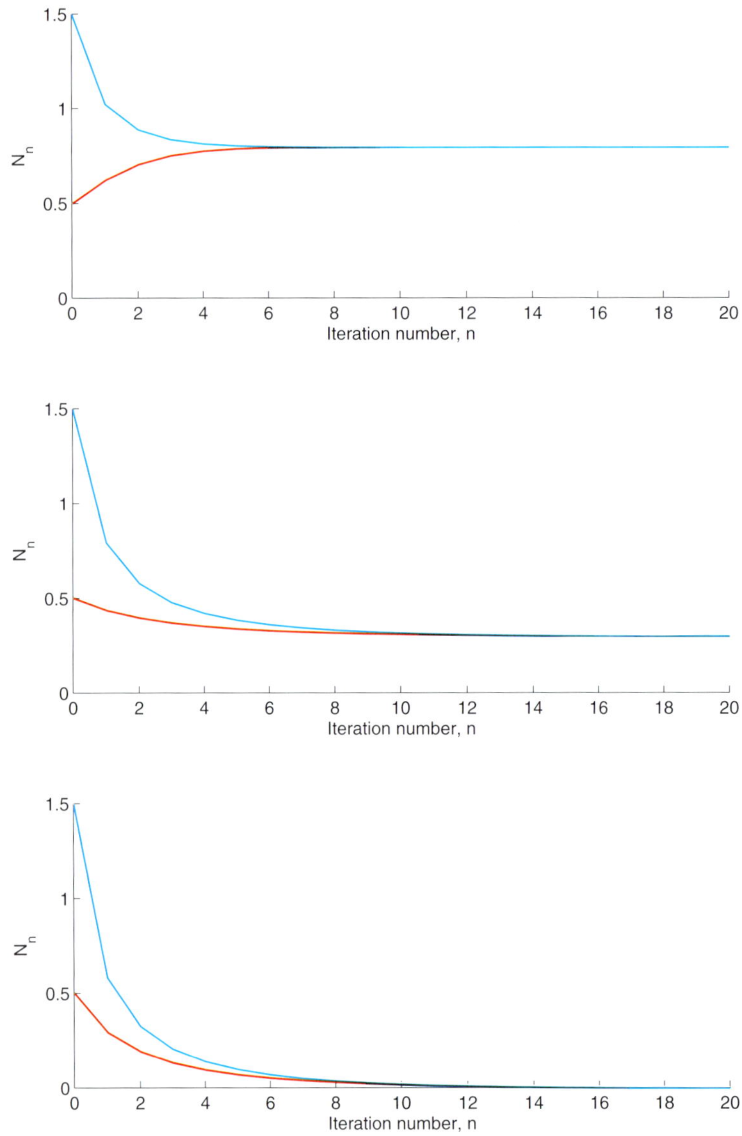

FIGURE 4. Series of diagrams showing the tumour's response to periodic infusion at different drug dosages for different initial conditions. Key: top panel, $a_\infty = 0.5$; middle panel, $a_\infty = 1.5$; bottom panel, $a_\infty = 2.5$. Parameter values: $\theta = 1 = \mu = k, \tau = 0.5$. The cyan curves correspond to $N(0) = N_0 = 1.5$, the red to $N(0) = N_0 = 0.5$.

Conventional radiotherapy

We denote by $N(t)$ the number of tumour cells at time t and assume that $N(t^+)$, the number of cells surviving a dose D of radiotherapy given at time t, is related to $N(t^-)$, the number before treatment, by the Linear-Quadratic model [**156**] so that

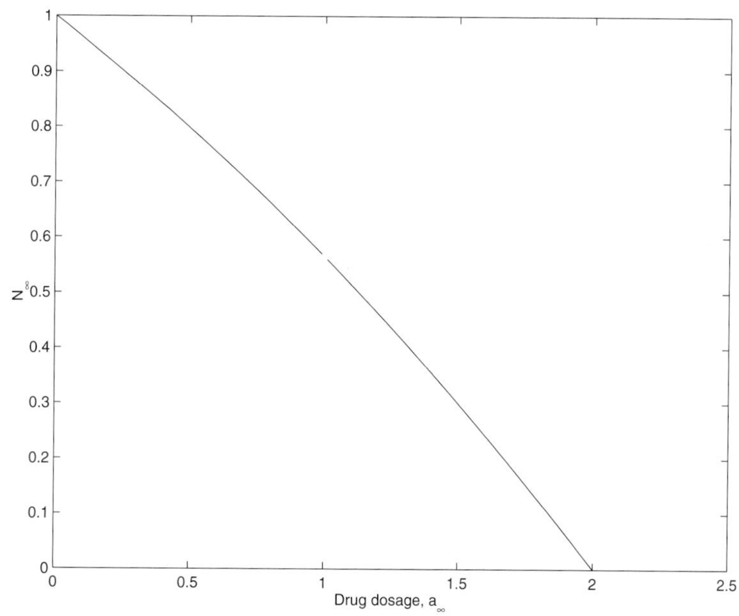

FIGURE 5. Diagram showing how N_∞ varies with a_∞ when periodic solutions emerge. Parameter values: $\theta = 1 = \mu = k, \tau = 0.5$.

$$N(t^+) = N(t^-)\, e^{-(\alpha D+\beta D^2)}, \tag{15}$$

where the parameters α and β characterise the tumour's response to treatment. We assume further that the tumour is irradiated at times $t = \Delta t, 2\Delta t, \ldots$ and that the tumour grows exponentially, with growth rate g (and the associated doubling time $t_2 = \ln 2/g$), between treatments so that

$$N(t + \Delta t^-) = N(t^+)\, e^{g\Delta t}. \tag{16}$$

By combining equations (15) and (16), we obtain an expression relating the tumour's size after the nth and $(n-1)^{th}$ doses of radiotherapy:

$$N(n\Delta t^+) = N(n\Delta t^-)\, e^{-(\alpha D+\beta D^2)} = N((n-1)\Delta t^+)\, e^{g\Delta t-(\alpha D+\beta D^2)},$$

where we have fixed $t = n\Delta t$ in (15) and then substituted for $N(n\Delta t^-)$ using (16). In a similar manner, it is possible to show that

$$N(n\Delta t^+) = N((n-2)\Delta t^+)\, e^{2[g\Delta t-(\alpha D+\beta D^2)]} = \ldots = N(t=0)\, e^{n[g\Delta t-(\alpha D+\beta D^2)]},$$

$$\Rightarrow \frac{N(n\Delta t^+)}{N(t=0)} = \left(\begin{array}{c} \text{fraction of tumour} \\ \text{cells that remain} \\ \text{after } n \text{ doses} \end{array} \right) = e^{n[g\Delta t-(\alpha D+\beta D^2)]}. \tag{17}$$

In Figure 6 we use equation (17) to compare the response of two tumours with different doubling times to one week of radiotherapy. The survival fraction of the faster growing tumour (*i.e.* the shorter doubling time) is greater than that of the slower growing tumour. These results suggest that for more aggressive tumours the radiation dose D should be increased, a longer course of treatment given (larger n) or the interval between treatments (Δt) reduced. In practice, radiologists must also consider the damage to healthy tissue when specifying the control parameters $n, \Delta t$ and D. Tissues that will be most susceptible are those with rapid turnover (*e.g.* normal connective tissue). In the next subsection we explain how our radiotherapy model may be used to predict optimal protocols designed to minimise damage to healthy tissue.

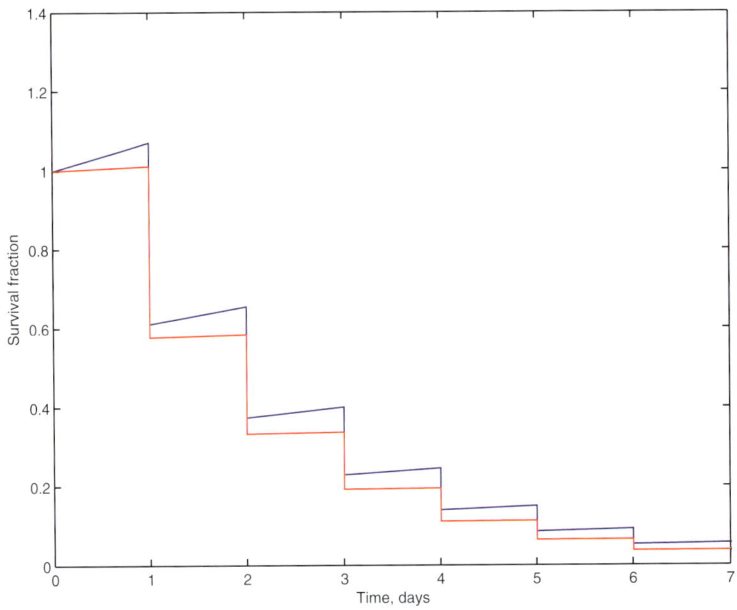

FIGURE 6. Change in survival fraction following 1 week of conventional treatment ($D = 200$ rads) administered to tumours with different doubling times. Key: $t_2 = 10$ days (black line), $t_2 = 60$ days (red line)). In both cases, $\alpha = 2 \times 10^{-3}$ rad^{-1} and $\beta = 4.0 \times 10^{-6}$ rad^{-2}.

Optimal radiotherapy

Following [**156**], we use the following empirical formula to estimate the damage to healthy tissue caused by n rounds of radiotherapy:

$$\text{Damage} = D\, n^a\, (\Delta t)^{-b}. \tag{18}$$

In (18) the parameters a and b have been estimated for tissues worst affected by radiotherapy (*e.g.* for connective tissue, $a = 0.65$ and $b = 0.11$) and the maximum damage that can be tolerated (D_{max}) is specified (typically, $D_{max} = 1800$ rads).

To determine an optimal schedule, we choose $n, \Delta t$ and D to minimise the survival fraction (see equation (17)) while ensuring that the damage to the normal tissue does not exceed D_{max}. Using calculus, it is straightforward to show that the optimal schedule has

(19)

$$D = \frac{\alpha}{\beta}\frac{[1-(a+b)]}{[2(a+b)-1]}, \qquad \Delta t = \frac{b(\alpha D + \beta D^2)}{g(a+b)}, \qquad n = \left(\frac{D_{max}}{D}\right)^{1/a} (\Delta t)^{b/a}.$$

In Table 1 we use expressions (19) to calculate the optimal schedules for tumours with different doubling times (noting that $t_2 = \ln 2/g$). We observe that as the doubling time increases (and the growth rate falls) the interval between doses increases from approximately daily when $t_2 = 10$ days to fortnightly when $t_2 = 90$ days. The number of doses also increase with t_2 but the dose D remains constant, depending only on parameters associated with the responses of the tumour and healthy tissue to radiotherapy (*i.e.* α, β, a and b). We note that the optimal therapy outperforms the conventional therapy in all cases.

Doubling Time t_2 (days)	Dose D (rads)	Interval Δt (days)	Number of doses n	Survival Fraction (Optimal)	Survival Fraction (Conventional)
1	230.8	0.14	17	5.8×10^{-5}	5.4×10
10	230.8	1.4	25	5.5×10^{-7}	4.0×10^{-7}
30	230.8	4.23	30	2.9×10^{-8}	1.0×10^{-7}
60	230.8	8.45	34	3.3×10^{-9}	7.1×10^{-8}
90	230.8	12.70	36	8.3×10^{-10}	6.4×10^{-8}

Table 1: table highlighting the differences between conventional and optimal radiotherapy schedules for tumours with different doubling times. Parameter values: $\alpha = 2 \times 10^{-3}$ rad^{-1}, $\beta = 4 \times 10^{-6}$ rad^{-2}, $a = 0.65$, $b = 0.11, D_{max} = 1800$.

The above analysis reveals how a simple model and elementary mathematics can be used to motivate the need for tailoring treatments to individuals. Of course, the model is overly simplistic. For example, not all tumours undergo exponential growth, nor are they homogeneous entities. Further, since a tissue's response to radiotherapy is strongly influenced by the local oxygen concentration and oxygen is not uniformly distributed across the tumour, the cell kill will be localised in well-oxygenated regions. Possible model modifications that would provide further insight into a tumour's response to radiotherapy include the following: (a) considering different tumour growth laws (*e.g.* Gompertz, logistic), (b) introducing multiple tumour cell populations, each with different radio-sensitivity, or (c) modelling the tumour's spatial structure, together with the distribution of oxygen. It should be possible to explore extensions (a) and (b) using techniques discussed earlier in this lecture series. Possible approaches for tackling (c) will be clearer after we have studied some spatially-structured models of tumour growth (see Section 4).

As a final remark, when extending (or developing) mathematical models, it is worth remembering that it is often difficult to obtain accurate estimates of all the model parameters. In such cases, where it is important to generate precise, physically-meaningful predictions, the theoretician may be forced to compromise, using a simple model based on parameters that can be reliably and accurately estimated.

3.5. Heterogeneous tumour growth

Thus far we have assumed that all cells within a tumour are identical. In practice, this is not the case. For example, vascular tumours growing *in vivo* contain blood vessels that supply nutrients to the tumour, extracellular matrix and immune cells such as leukocytes and macrophages (these white blood cells can play a key role in solid tumour growth [**91**]). Genetic mutations associated with tumour progression mean that the tumour cell population may also contain functionally distinct subpopulations, characterised by different proliferation and death rates, their ability to produce growth factors such as extracellular-matrix degrading proteases to facilitate their invasion and angiogenic factors to stimulate the ingrowth of new blood vessels to the tumour mass. Certain subpopulations may also be resistant to chemotherapy.

At a more fundamental level and in an attempt to account for the spatial structure within solid tumours, we may introduce cellular heterogeneity by characterising tumour cells as either proliferating, quiescent or necrotic. Typically, all three cell types will be present within a well-developed avascular tumour and a vascular tumour [**63, 141**]. The spatially-structured models that we study in Section 4 will enable us to distinguish between these different cell types on the basis of the local nutrient concentration. However, in order to gain preliminary insight into how the proportion of each cell type changes as the tumour evolves, we now present an ODE model in which the three cell types are treated separately.

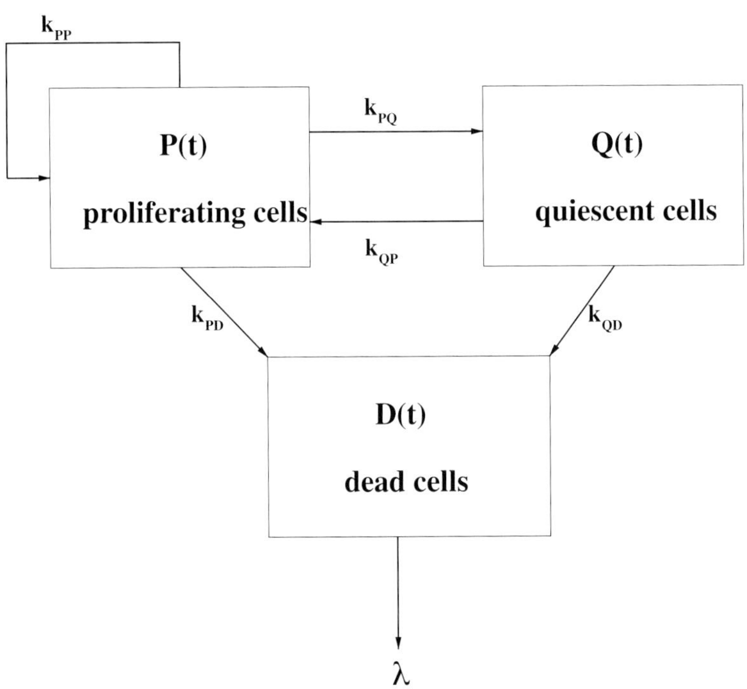

FIGURE 7. Schematic diagram of model of heterogeneous tumour growth (see equations (20)-(22)).

We denote by $P(t), Q(t)$ and $D(t)$ the numbers of proliferating, quiescent and dead cells at time t and by $N(t)$ their sum so that $N(t) = P(t) + Q(t) + D(t)$. We base our model on the schematic diagram presented in Figure 7, which shows how cells change from one state to another, how new cells are created and how dead cells are removed from the system. Using the figure, it is straightforward to arrive at the following evolution equations for P, Q and D:

$$\frac{dP}{dt} = (k_{PP} - k_{PQ} - k_{PD})P + k_{QP}Q, \tag{20}$$

$$\frac{dQ}{dt} = k_{PQ}P - (k_{QP} + k_{QD})Q, \tag{21}$$

$$\frac{dD}{dt} = k_{PD}P + k_{QD}Q - \lambda D, \tag{22}$$

with

$$P(0) = P_0, \quad Q(0) = Q_0, \quad D(0) = D_0. \tag{23}$$

In equations (20)-(22), k_{PP} and k_{PQ} represent, respectively, the rate at which proliferating cells divide (their daughter cells are also assumed to be in the proliferating state) and the rate at which proliferating cells become quiescent. The interpretation of the other transition rates follows similarly. In general, terms such as k_{PP} will be nonlinear to account for the fact that as the overall population size increases competition for space and resources will reduce k_{PP} and increase the rate at which cells become quiescent (k_{PQ}) and/or die (k_{PD} and k_{QD}). We assume, for simplicity, that the transition rates are constant. Equations (20)-(22) are then linear, the ODEs for P and Q decouple and the system admits general solutions of the form

$$(P, Q, D) = \Sigma_{i=1}^{3}(P_i, Q_i, D_i)e^{\sigma_i t}, \tag{24}$$

where $(P_i, Q_i, D_i)\,(i = 1, 2, 3)$ are constants and the growth rates σ_i are real. Without loss of generality, we suppose that $\sigma_1 > \sigma_2$ and fix $\sigma_3 = -\lambda$ (thus σ_1 is the fastest growing eigenvalue and we anticipate that it will dominate the tumour's dynamics at long times). It is possible to show that σ_1 and σ_2 solve

$$0 = (\sigma + k_{PQ} + k_{PD} - k_{PP})(\sigma + k_{QP} + k_{QD}) - k_{PQ}k_{QP} \equiv f(\sigma), \text{ say.} \tag{25}$$

By sketching $f(\sigma)$ or otherwise, it is also possible to show that if $k_{PP} \geq k_{PQ} + k_{PD}$ then $\sigma_1 > 0$. If $k_{PP} < k_{PQ} + k_{PD}$ then $\sigma_1 > 0$ if

$$(k_{PQ} + k_{PD} - k_{PP})(k_{QP} + k_{QD}) > k_{PQ}k_{QP}.$$

We deduce from (24) that when $\sigma_1 > 0$

$$(P, Q, D) \sim (P_1, Q_1, D_1)e^{\sigma_1 t} \quad \text{as } t \to \infty. \tag{26}$$

We explain below how this long time behaviour can be exploited to make predictions about the composition of the tumour at long times. First, however, we remark that the inequalities which guarantee $\sigma_1 > 0$ suggest that mutations which increase k_{PP} (and/or decrease k_{PQ} or k_{PD}) could cause σ_1 to increase through zero and, in so doing, stimulate a population of premalignant cells to divide uncontrollably.

Substituting with (26) in equations (20)-(22) we deduce that the coefficients P_1, Q_1 and D_1 satisfy

$$0 = \begin{pmatrix} \sigma_1 - k_{PP} + k_{PQ} + k_{PD} & -k_{QP} & 0 \\ -k_{PQ} & \sigma_1 + k_{QP} + k_{QD} & 0 \\ -k_{PD} & -k_{QD} & \sigma_1 + \lambda \end{pmatrix} \begin{pmatrix} P_1 \\ Q_1 \\ D_1 \end{pmatrix}. \tag{27}$$

By taking suitable combinations of (27), expressions for the proportion of the tumour occupied by proliferating, quiescent and dead cells at long times can be obtained. For example, we have that

$$\begin{aligned} \frac{P}{N} &\sim \frac{P_1}{P_1 + Q_1 + D_1} \\ &= \frac{k_{QP}\sigma_1(\sigma_1 + \lambda)}{k_{PP}p_{QP}(\sigma_1 + \lambda) - \lambda(k_{PD}k_{QP} + k_{QD}(\sigma + k_{PQ} + k_{PD} - k_{PP})}. \end{aligned} \tag{28}$$

We remark that equation (28) and its counterparts could be fit to experimental data to estimate rate parameters associated with a tumour's growth.

3.6. Discussion

In this section we have presented a range of mathematical models that may be used to describe the growth dynamics of solid tumours and their response to chemo- and radiotherapy.

There are many ways in which these models could be extended. For example, we could combine the model of heterogeneous tumour growth presented in Subsection 3.5 with the earlier models of chemo- and radiotherapy. Equally we could investigate how competition (for space and resources) with normal cells that are present in the tumour's environment affect its response to therapy. A natural way to do this would be to include additional ODEs to account for the number of normal cells present at time t and, possibly, the concentration of vital nutrients such as oxygen that affect the cells' growth rates and their response to therapy (for example, the potency of radiotherapy is known to increase with the local oxygen concentration [**134**]). Similarly, we could include additional ODEs to model changes in the vascular density of a vascularised tumour and its response to anti-angiogenic therapies such as angiostatin and endostatin which target the vasculature [**81**]. Additional equations could also be used to allow for the emergence of drug- or radio- resistant tumour subpopulations. Alternatively, as we explain in section 4, we could incorporate spatial effects.

Panetta and coworkers have developed a series of models, formulated as systems of ordinary differential equations and/or delay-differential equations, to investigate the impact of cell-cycle specific chemotherapy [**116**] and its optimal control [**59**]. They have also specialised their models to predict the response of breast and ovarian cancer to treatment with paclitaxel, a drug that acts at the G2/M phase of the cell cycle [**117**]. An alternative approach which enables the action of particular cell-cycle specific drugs, such as doxorubicin, to be investigated involves introducing a second independent variable $0 < a < T$, say. Then $n(t, a)$ denotes the number of cells at time t which are in position a of their cycle and $N(t)$, the total number of cells at time t, is related to $n(t, a)$ in the following way

$$N(t) = \int_0^T n(t, a)da$$

where, in the absence of therapy, $n(t,a)$ may satisfy the following partial differential equation

$$\frac{\partial n}{\partial t} + \frac{\partial n}{\partial a} = -\mu_0 n - \mu_1 n N,$$

with $n(0,a) = n_0(a)$ and $n(t,0) = \int_0^T \beta(a)n(t,a)da$. In the above equations, the constants μ_0 and μ_1 represent respectively rates of natural cell death and cell death due to overcrowding and $\beta(a)$ denotes the rate at which cells at position a of their cell cycle replicate. Models of this type are termed age-structured models and their analysis may be quite involved. Further details about age-structured tumour models can be found in [**21, 22, 80**]; for a more general treatment of age-structured models, see [**84**].

3.7. Exercises

1. Consider the ODE system for continuous infusion of a tumour with a chemotherapeutic drug:

$$\text{(29)} \qquad \frac{dN}{dt} = kN\left(1 - \frac{N}{\theta}\right) - \mu A N, \quad \frac{dA}{dt} = a_\infty - \lambda A - \gamma A N$$

   where $k, \theta, \mu, a_\infty, \lambda$ and γ are positive constants. Nondimensionalise equations (29), scaling N with θ, A with a_∞ and t with k^{-1}. Then derive expressions for the steady states of the resulting dimensionless equations. Show how the steady states vary with μ, the potency of the drug.

2. The fraction of tumour cells that survive n rounds of radiotherapy with dose D is

$$(\text{survival fraction}) = e^{n[g\Delta t - (\alpha D + \beta D^2)]},$$

   where g, α and β are tumour-specific parameters and Δt denotes the period between treatments. The associated damage to healthy tissue is

$$\text{Damage} = Dn^a(\Delta t)^b, \qquad (a, b \text{ positive constants}).$$

   When designing a treatment protocol, $n, \Delta t$ and D are chosen to minimise the survival fraction while ensuring that (Damage)$\leq R_{tol}$. Use calculus to show that the optimal protocol has

$$D = \frac{\alpha}{\beta}\frac{[1-(a+b)]}{[2(a+b)-1]}, \quad \Delta t = \frac{b(\alpha D + \beta D^2)}{g(a+b)}, \quad n = \left(\frac{R_{tol}}{D}\right)^{1/\alpha} (\Delta t)^{b/a}.$$

3. Consider the model of heterogeneous tumour growth defined by equations (20)-(22). Fix $k_{PQ} = k_{PD} = k_{QP} = k_{QD} = \lambda = 1$ and view k_{PP} as a parameter. Use equations (25) and (28) to determine how the maximum tumour growth rate and the proportion of the tumour occupied by proliferating cells (P/N) change as k_{PP} varies. Determine also how Q/N and D/N vary with k_{PP}. Comment upon your results.

4. The following system of differential equations describes how a tumour that contains a drug-sensitive and a drug-resistant population of cells responds to continuous infusion with a drug

$$\frac{dN_1}{dt} = k_1 N_1 \left(1 - \frac{N_1 + N_2}{\theta}\right) - \mu N_1 A, \tag{30}$$

$$\frac{dN_2}{dt} = k_2 N_2 \left(1 - \frac{N_1 + N_2}{\theta}\right), \tag{31}$$

$$\frac{dA}{dt} = a_\infty - \lambda A - \gamma N_1 A, \tag{32}$$

$$\text{with} \quad N_1(0) = N_{10},\ N_2(0) = N_{20},\ A(0) = A_0.$$

In the above equations, $N_1(t)$ and $N_2(t)$ denote the number of drug-sensitive and drug-resistant cells respectively and $A(t)$ the drug concentration. What is the physical interpretation of the parameters $k_1, k_2, \mu, \theta, a_\infty$, λ and γ? How do the number and nature of the equilibrium solutions change as μ varies?

4. Avascular tumours: radially-symmetric growth

4.1. Introduction

The earliest spatio-temporal models of avascular tumour growth [**32, 76, 131**] were developed to explain experimental results describing the dynamics and spatial structure of MCS culture *in vitro* [**63, 141**]. The experimentalists typically manipulated the environmental conditions in which the MCS were growing (by, for example, adding chemotherapeutic drugs or controlling the oxygen levels) and monitored the evolution of the radii of the tumours over time, supplementing their measurements, where possible, with information about the distribution within the tumours of diffusible chemicals such as oxygen and glucose and the volume fraction of the tumours that were necrotic and hypoxic [**78, 79**]. Faced with such data, it is not surprising that the key variables in the corresponding mathematical models are: the outer tumour radius of the assumed radially-symmetric tumour; the concentration within the tumour of a growth-rate limiting diffusible chemical (this may be a growth factor such as oxygen or glucose or a chemotherapeutic drug); and, the boundaries separating proliferating, quiescent and necrotic cells. Since the tumours change in size over time, the domains on which the models are formulated must be determined as part of the solution process. As such, the models can be termed moving boundary problems [**38, 51, 53**]. These models of early solid tumour growth have excited considerable interest in the mathematical community. In limited space we are unable to do justice to this increasing body of work. Fortunately there are several excellent reviews of the relevant literature to which we encourage the interested reader to refer [**3, 15, 123, 127**].

The remainder of this section is organised as follows. In Subsection 4.2 we present the moving boundary problems that we use to study avascular tumour growth. In Subsection 4.3 we show how, under certain conditions, the models may be reduced to systems of ODEs similar in form to those studied in Section 3. We indicate the sort of insight that these spatially-structured models can provide by presenting some analytical results in Subsection 4.4. The section then concludes with a critical review of these models.

4.2. Model development

The simplest models that describe the growth of a radially-symmetric, avascular tumour comprise equations governing the evolution of a single, growth rate limiting, diffusible chemical (henceforth oxygen) $c(r,t)$, the outer tumour radius $R(t)$ and the hypoxic and necrotic radii $R_H(t)$ and $R_N(t)$. The principle of mass balance is used to derive equations for $c(r,t)$ and $R(t)$ whereas $R_H(t)$ and $R_N(t)$ are defined implicitly, occurring when the chemical concentration passes through known threshold values.

The oxygen concentration, $c(r,t)$.

$$\frac{\partial c}{\partial t} = \underbrace{\frac{D}{r^2}\frac{\partial}{\partial r}\left(r^2\frac{\partial c}{\partial r}\right)}_{\text{flux due to diffusion}} - \underbrace{\Gamma(c,R,R_H,R_N)}_{\text{rate of chemical consumption}} . \tag{33}$$

In Equation (33) D denotes the assumed constant diffusion coefficient of the oxygen and $\Gamma(c,R,R_H,R_N)$ its rate of consumption. In practice $\Gamma(c,R,R_H,R_N)$ will be a non-linear function which depends on the tumour cell line being studied. For simplicity and to demonstrate the qualitative behaviour of the model, we suppose that oxygen is consumed at a constant rate Γ by proliferating and quiescent cells and fix

$$\Gamma(c,R,,R_H,R_N) = \Gamma H(r-R_N),$$

where $H(.)$ denotes the Heaviside step function ($H(x)=1$ if $x>0$ and $H(x)=0$ otherwise).

The outer tumour radius, $R(t)$.

$$\underbrace{\frac{1}{3}\frac{d}{dt}(R^3) = R^2\frac{dR}{dt}}_{\text{rate of change of tumour volume}} = \underbrace{\int_0^R S(c,R,R_H,R_N)r^2dr}_{\text{total rate of cell proliferation}} - \underbrace{\int_0^R N(c,R,R_H,R_N)r^2dr}_{\text{total rate of cell death}} . \tag{34}$$

In Equation (34), $S(c,R,R_H,R_N)$ and $N(c,R,R_H,R_N)$ denote respectively the rates of cell proliferation and death within the tumour. We assume that cell proliferation is localised in nutrient-rich regions (where $c_H < c$ or $R_H < r < R$) where it occurs at a rate which is proportional to the local nutrient concentration c. We suppose that apoptosis and necrosis contribute to the overall cell death rate, with apoptosis occurring at a constant rate throughout the tumour and necrosis being initiated in nutrient-poor regions (where $c < c_N$ or $r < R_N$). Thus, as simple representative examples, we write

$$S(c,R,R_H,R_N) = scH(r-R_H) \quad \text{and}$$
$$N(c,R,R_H,R_N) = s\lambda_A + s\lambda_N H(R_N - r),$$

where s, λ_A and λ_N are positive constants. Substitution with S and N in Equation (34) yields

$$\frac{1}{s}R^2\frac{dR}{dt} = \int_{R_H}^{R} cr^2 dr - \frac{1}{3}(\lambda_A R^3 + \lambda_N R_N^3). \tag{35}$$

The hypoxic and necrotic boundaries, $R_H(t)$ and $R_N(t)$.
The internal free boundaries $R_H(t)$ and $R_N(t)$ are defined implicitly in terms of threshold oxygen concentrations c_H and $c_N < c_H$. These constants denote respectively the minimum oxygen concentration at which cell proliferation can occur and the minimum concentration at which quiescent cells can remain alive. Three different cases arise according to whether the minimum oxygen concentration within the tumour falls below none, one or both these threshold values:

- $c(r,t) > c_H \ \forall\, r \in (0,R) \Rightarrow R_N = R_H = 0$;
- $\exists\, r \in (0,R)$ such that $c_N < c(r,t) \le c_H \Rightarrow R_N = 0 < R_H < R$ and $c(R_H,t) = c_H$;
- $\exists\, r \in (0,R(t))$ such that $c(r,t) \le c_N < c_H \Rightarrow 0 < R_N < R_H < R$ with $c(R_N,t) = c_N$ and $c(R_H,t) = c_H$.

In Figure 8 we provide a schematic diagram to illustrate the structure of a well-developed avascular tumour which possesses a central necrotic core and a hypoxic region.

FIGURE 8. Schematic diagram of a fully-developed avascular tumour. An outer proliferating tim (where $c > c_H$ and $R_H \le r \le R$) surrounds a hypoxic or quiescent annulus (where $c_N < c < c_H$ and $R_N \le r \le R_H$) and a central necrotic core (where $c(r,t) \le c_N$ and $o \le r \le R_N$).

Our model comprises equations (33)-(34) for $c(r,t)$ and $R(t)$ and supplementary equations for $R_H(t)$ and $R_N(t)$. It remains to specify appropriate boundary and initial conditions to close the model. We assume symmetry of the oxygen profile about $r = 0$, that on the outer tumour boundary it is maintained at a constant value, $c = c_\infty$ say. Finally we prescribe the initial tumour radius ($R(0) = R_0$) and the initial oxygen distribution ($c(r,0) = c_0(r)$). Thus we have

$$\left.\begin{array}{c} \dfrac{\partial c}{\partial r} = 0 \quad \text{at } r = 0, \\ c = c_\infty \quad \text{on } r = R(t), \\ c,\ \dfrac{\partial c}{\partial r} \text{ continuous across } r = R_H(t) \text{ and } r = R_N(t), \\ c(r,0) = c_0(r),\ R(t=0) = R_0. \end{array}\right\} \tag{36}$$

Nondimensionalisation

Before analysing the model equations, we recast them in dimensionless variables. Denoting typical lengthscales, timescales and oxygen concentrations by X, T and C, we introduce the following dimensionless variables

$$r^* = \frac{r}{X}, \quad t^* = \frac{t}{T}, \quad c^* = \frac{c}{C}, \quad R^* = \frac{R}{X}, \quad R_H^* = \frac{R_H}{X}, \quad R_N = \frac{R_N}{X}.$$

When written in terms of dimensionless variables, our model equations become

$$\frac{\partial c^*}{\partial t^*} = \left(\frac{DT}{X^2}\right)\frac{1}{{r^*}^2}\frac{\partial}{\partial r^*}\left({r^*}^2\frac{\partial c^*}{\partial r^*}\right) - \Gamma T H(r^* - R_N^*),$$

$${R^*}^2\frac{dR^*}{dt^*} = \int_0^{R^*}\left\{sTCc^*H(r^* - R_N^*) - \lambda_A^* - \lambda_N^* H(R_N^* - r^*)\right\}{r^*}^2 dr^*.$$

where $\lambda_A^* = sT\lambda_A$ and $\lambda_N^* = sT\lambda_N$.

At this stage, X, T and C are unspecified. Since we are concerned with simulating changes in the tumour's size and spatial structure, we focus on the tumour doubling timescale, choosing

$$T = \frac{1}{sC}.$$

Following [**1, 2, 32, 76, 103, 131**], we exploit the fact that the oxygen diffusion timescale ($\sim$ minutes) is much shorter than a typical tumour doubling time ($\sim$ weeks or months) and adopt a quasi-steady assumption in the nutrient equation so that

$$0 = \frac{1}{{r^*}^2}\frac{\partial}{\partial r^*}\left({r^*}^2\frac{\partial c^*}{\partial r^*}\right) - \Gamma^* H(r^* - R_N^*),$$

where $\Gamma^* = \frac{\Gamma X^2}{D} \sim O(1)$ (we remark that, as a result, prescription of the initial oxygen profile in (36) is redundant).

For completeness, we now state, in full, our nondimensionalised model equations, omitting the $*$s for clarity:

$$0 = \frac{1}{r^2}\frac{\partial}{\partial r}\left(r^2\frac{\partial c}{\partial r}\right) - \Gamma H(r - R_N), \tag{37}$$

$$R^2\frac{dR}{dt} = \int_0^R \left\{cH(r - R_N) - \lambda_A - \lambda_N H(R_N - r)\right\} r^2 dr, \tag{38}$$

$$R_H = 0 \text{ if } c > c_H \ \forall\, r \text{ and otherwise } c(R_H, t) = c_H, \tag{39}$$

$$R_N = 0 \text{ if } c > c_N \ \forall\, r \text{ and otherwise } c(R_N, t) = c_N, \tag{40}$$

$$\frac{\partial c}{\partial r} = 0 \quad \text{at } r = 0, \tag{41}$$

$$c = c_\infty \quad \text{on } r = R, \tag{42}$$

$$R(0) = R_0, \text{ prescribed.} \tag{43}$$

We remark that c_∞ could be eliminated from the model equations by fixing $C = c_\infty$. We want to investigate the effect of varying c_∞ and, hence, retain it as an explicit model parameter. For similar reasons, we choose not to scale lengths with R_0.

4.3. Model analysis

The simple functional forms used in our model equations enable us to solve for $c(r,t)$ and to derive algebraic equations for R_H and R_N in terms of R. The evolution of $R(t)$ is, in turn, governed by a non-linear differential equation. The form of these relations depends on whether the tumour contains regions of hypoxia and/or necrosis and, thus, on $R(t)$.

Small, uniformly proliferating tumour

When the tumour contains only proliferating cells, equations (37)-(43) reduce to give

$$c(r,t) = c_\infty - \frac{\Gamma}{6}(R^2 - r^2),$$

$$\frac{dR}{dt} = \frac{R}{3}\left(c_\infty - \frac{\Gamma R^2}{15} - \lambda_A\right), \tag{44}$$

with $R_H = R_N = 0$. This growth phase persists until either the tumour reaches a steady state for which $\frac{dR}{dt} = 0$ or quiescence is initiated and the model breaks down.

Since a spherically-symmetric tumour of radius $R(t)$ has volume $V(t) = 4\pi R^3(t)/3$, we deduce from (44) that

$$\frac{dV}{dt} = V\left[c_\infty - \lambda_A - \frac{\Gamma}{15}\left(\frac{3V}{4\pi}\right)^{2/3}\right],$$

which is of the same form as the growth law specified in equation (3), if we fix

$$\alpha = \frac{2}{3}, \quad k = \frac{2}{3}(c_\infty - \lambda_A), \quad \theta = \frac{4\pi}{3}\left[\frac{15}{\Gamma}(c_\infty - \lambda_A)\right]^{3/2}.$$

This shows that a more complex, spatially-structured model may produce the same dynamics as a simpler ODE model for a uniformly proliferating avascular tumour. By identifying the link between the two models we can see how local parameters associated with spatially-structured model may relate to parameters that appear in spatially-uniform models.

From (44) we deduce that $\frac{dR}{dt} = 0$ when $R = 0$ (the trivial solution) or, assuming $c_\infty > \lambda_A$, when $R = [15(c_\infty - \lambda_A)/\Gamma)]^{1/2}$. Linear stability analysis reveals that the trivial solution is unstable if $c_\infty > \lambda_A$ (*i.e.* where a nontrivial steady state exists) and stable otherwise.

It remains to determine whether the nontrivial steady state is physically realistic. This will be the case if $c(r,t) > c_H \; \forall \, r \in (0, R)$. Since c attains a minimum at $r = 0$ we deduce that $R = [15(c_\infty - \lambda_A)/\Gamma)]^{1/2}$ is a physically realistic steady state if the following inequality holds:

$$c_\infty < 5\lambda_A - 2c_H.$$

In practice, λ_A and c_H will be fixed for a given tumour cell line. If $\lambda_A < 2c_H/5$ then the above inequality can not hold, regardless of the choice of c_∞, and there is no feasible steady state solution. By contrast, if $\lambda_A > 2c_H/5$ then we predict that for $\lambda_A < c_\infty < 5\lambda_A - 2c_H$, there is a nontrivial steady state with $0 = R_H = R_N < R$ and which linear stability analysis shows to be stable. For larger values of c_∞, the model breaks down before the tumour reaches the steady state. In particular, a central quiescent region forms when $R = [6(c_\infty - c_H)/\Gamma]^{1/2}$.

Intermediate-sized tumour

When the tumour contains proliferating and quiescent cells, equations (37)-(43) supply

$$c(r,t) = c_\infty - \frac{\Gamma}{6}(R^2 - r^2),$$

$$\frac{dR}{dt} = \frac{R}{3}\left[\left(c_\infty - \frac{\Gamma R^2}{6}\right)\left(1 - \frac{R_H^3}{R^3}\right) + \frac{\Gamma R^2}{10}\left(1 - \frac{R_H^5}{R^5}\right) - \lambda_A\right], \tag{45}$$

$$\text{with} \quad R_N = 0 \quad \text{and} \quad R_H^2 = R^2 - \frac{6}{\Gamma}(c_\infty - c_H). \tag{46}$$

By differentiating with respect to time the algebraic expression for R_H our model may be reformulated as a pair of ODEs, with

$$\frac{dR_H}{dt} = \frac{R}{R_H}\frac{dR}{dt}.$$

Since $0 < R_H < R$ we deduce that $|\frac{dR_H}{dt}| > |\frac{dR}{dt}|$. Thus, if the tumour contains a quiescent region then $R_H(t)$ evolves more rapidly than $R(t)$. We note also that when the tumour mass is increasing (*i.e.* $\frac{dR}{dt} > 0$) then so too is the size of the quiescent region. Finally, instead of differentiating (46) with respect to time, we could use it to eliminate R_H from (45) and, in so doing, reduce the model to a single, nonlinear ODE for the tumour radius $R(t)$. In contrast to equation (44), the resulting ODE does not lend itself to physical interpretation. The absence here of a clear link between our spatially-structured model and a spatially-averaged ODE model underlines the difficulty associated with interpreting parameters in simple, spatially-homogeneous models.

Fully-developed tumour

In this case the tumour contains an outer proliferating rim ($R_H < r < R$), a quiescent region ($R_N < r < R_H$) and a central necrotic core ($0 < r < R_N$), and the model equations reduce to give

$$c(r,t) = \begin{cases} c_N & 0 < r < R_N \\ c_N + \Gamma(r - R_N)^2(r + 2R_N)/6r & R_N < r < R \end{cases}, \tag{47}$$

$$\begin{aligned} \frac{dR}{dt} &= \frac{R}{3}\left[c_N\left(1 - \frac{R_H^3}{R^3}\right) - \left(\lambda_A + \lambda_N\frac{R_N^3}{R^3}\right)\right] \\ &+ \frac{\Gamma R^3}{6}\left[\frac{1}{5}\left(1 - \frac{R_H^5}{R^5}\right) - \frac{R_N^2}{R^2}\left(1 - \frac{R_H^3}{R^3}\right) + \frac{R_N^3}{R^3}\left(1 - \frac{R_H^2}{R^2}\right)\right], \end{aligned} \tag{48}$$

$$\text{with} \quad \left(1 - \frac{R_N}{R}\right)^2\left(1 + \frac{2R_N}{R}\right) = \frac{6}{\Gamma R^2}(c_\infty - c_N), \tag{49}$$

$$\text{and} \quad \left(1 - \frac{R_N}{R_H}\right)^2 \left(1 + \frac{2R_N}{R_H}\right) = \frac{6}{\Gamma R_H^2}(c_H - c_N). \tag{50}$$

The kinetic terms that we used above are overly simplistic and should be replaced by experimentally-determined functions. In general, these functions will be non-linear and the resulting models must be solved numerically. Whilst numerical simulations are of tremendous value, they may obscure the manner in which the various mechanisms interact. Complementary insight into the system's behaviour can be gained by studying special cases for which the model equations simplify greatly. For example, in [**39**] asymptotic techniques were used to investigate the tumour's growth dynamics for several situations. Summarising, this analysis shows that when the tumour is small (so that $0 < R \ll 1$ and $R_H = R_N = 0$), it grows exponentially. Alternatively, if $c_\infty - c_N \ll 1$ and $\lambda_A + \lambda_N \ll 1$ then the tumour will evolve to a non-trivial equilibrium with a thin proliferating rim ($0 < R - R_N \ll 1$). This resembles the structure of MCS cultured *in vitro* where, typically, the outer proliferating rim is only three to seven cell layers thick [**63**].

4.4. Discussion

We summarise the insight gained from the simple, radially-symmetric models studied in this section. The models can be used to identify conditions under which a tumour will grow to an equilibrium configuration and to predict how its size, spatial structure and stability to time-dependent perturbations depend upon physically relevant parameters such as the externally-supplied nutrient concentration c_∞.

In addition to the deterministic models described in this section, several probabilistic models of avascular tumour growth that reproduce the same phenomena have been developed and shown to exhibit good qualitative and quantitative agreement with experimental data. Models of this type that focus on individual cells and their interactions with neighbouring cells use concepts ranging from Markov chain processes [**82**], through cellular automata [**56**] to stochastic energy minimisation techniques [**140**].

There are many ways in which the model presented above could be extended. For example, we could use experimentally-determined functions for the rates of cell proliferation, apoptosis and necrosis. Equally, we could include additional diffusion equations to describe the action of other diffusible growth factors which may be present in the tumour environment. These chemicals may be supplied externally or expressed by the tumour cells themselves and may promote or inhibit the tumour's growth. For example, bi-products of the cell degradation process that accompany necrosis are believed to inhibit cell proliferation (models of this type are studied in [**38**]). Equally, following Gatenby and coworkers [**74, 133**] and Webb *et al.* [**154**], we could introduce separate variables to describe oxygen, glucose, pH and lactate and, in so doing, not only include a more realistic description of cell metabolism but also account for the advantage over normal cells that tumour cells enjoy under acidosis. Alternatively, we could use the models developed in section 3 to investigate the impact of tumours to treatment with radiotherapy and/or chemotherapy [**58, 89, 86**]. A further model extension which provides a simplistic description of vascular tumour growth involves including a distributed source of nutrient (or a distributed sink of waste products) in the outer portion of the tumour so that, in place of equation (37), we use

$$0 = \frac{1}{r^2}\frac{\partial}{\partial r}\left(r^2\frac{\partial c}{\partial r}\right) + h(c_{vess} - c) - \Gamma H(r - R_N),$$

where h and c_{vess} denote respectively the rate at which oxygen exchanges with the vasculature and the oxygen concentration within the vessels. As a result, nutrients may be supplied to the tumour either by exchange with its vasculature or by diffusion across the tumour's outer boundary (for details, see [**36, 37, 87**]). The models presented in this section can be termed moving boundary problems since the domain on which they are formulated (*i.e.* the tumour size) must be determined as part of the solution procedure. As such, they have excited the interest of Friedman and coworkers who have constructed proofs of existence and uniqueness of the model solutions [**50, 69**], thereby placing the earlier analysis on a more rigorous footing.

Given that avascular tumours are usually radially-symmetric while vascular tumours possess highly irregular boundaries, it is natural to ask whether this change in morphology is due to the non-uniform distribution of blood vessels that accompanies vascularisation or whether the radially symmetric avascular tumours are intrinsically unstable to asymmetric perturbations. Although the spatially-structured models presented in this section are not amenable to such analysis, in the next section we show how they can be extended to address this important question.

4.5. Exercises

1. The following equations describe the growth of a one-dimensional tumour in response to an externally supplied nutrient:

$$0 = \frac{1}{r}\frac{\partial}{\partial r}\left(r\frac{\partial c}{\partial r}\right) - \lambda H(c - c_N),$$

$$\frac{dR}{dt} = \int_0^R s(c) r \, dr,$$

$$\text{where} \quad s(c) = \begin{cases} s(c - c^*) & c > c_N \\ -\mu & c \leq c_N \end{cases},$$

$$\text{with} \quad c = c_\infty \text{ at } r = R(t), \quad \frac{\partial c}{\partial x} = 0 \text{ at } x = 0,$$

$$c, \ \frac{\partial c}{\partial r} \text{ continuous across } r = R_N(t),$$

$$c = c_N \text{ when } x = R_N(t),$$

$$R = R_0 \text{ at } t = 0,$$

where $\lambda, c_N, c^*, c_\infty, s, \mu$ and R_0 are positive constants and $c_\infty > c_N$.

Suppose that the tumour does not possess a necrotic region so that $R_N = 0$. Solve for $c(r,t)$ in terms of $R(t)$ and use this expression to derive an ODE for the tumour radius, $R(t)$. Identify the steady state solutions of this ODE and determine the range of values of c_∞ for which these solutions exist.

Suppose now that $R_N > 0$. Determine the nutrient concentration in terms of $R(t)$ and $R_N(t)$ and derive an expression relating $R_N(t)$ to $R(t)$. By substituting with $c(r,t)$ in the equation for $R(t)$, obtain an ODE for $R(t)$.

2. The following equations describe the growth of a tumour in response to an externally-supplied nutrient such as oxygen:

$$\frac{\partial n}{\partial t} + \frac{\partial}{\partial x}(vn) = k(c)n \tag{51}$$

$$\frac{dR}{dt} = v(R,t), \tag{52}$$

$$0 = \frac{\partial^2 c}{\partial x^2} - \lambda n H(c - c_N),$$

$$\text{where} \quad k(c) = \begin{cases} k_+ & \text{if } c > c_N \\ -k_- & \text{if } c \leq c_N \end{cases},$$

$$\frac{\partial c}{\partial x}(R,t) = h(c_\infty - c(R,t)), \quad \frac{\partial c}{\partial r}(0,t) = 0 = v(0,t),$$

$$c, \ \frac{\partial c}{\partial r} \ \text{continuous across } r = R_N(t),$$

$$\text{and } R(0) = R_0.$$

In these equations, $n(x,t)$ denotes the tumour cell density, $v(x,t)$ the cell velocity, $c(x,t)$ the local oxygen concentration, $R(t)$ the position of the outer tumour radius and $R_N(t)$ the interface between proliferating and dead cells. The parameters $\lambda, c_N, c_\infty, h, R_0$ and $k_\pm$ are positive constants.

Assume that the tumour is densely packed with tumour cells, so that $n \equiv 1$ for $0 \leq x \leq R(t)$. Show that in this case equation (51) simplifies to give an expression for the cell velocity $v(x,t)$ in terms of $k(c)$. Use this result in (52) to show further that

$$\frac{dR}{dt} = \int_0^{R(t)} k(c)dx.$$

By solving for $c(x,t)$, derive an algebraic expression for R_N in terms of $R(t)$ and an ODE for R. Determine conditions under which the system evolves to a nontrivial steady state. Comment on how your results change as h varies and, in particular, as $h \to \infty$.

5. Symmetry breaking and invasion

5.1. Introduction

Here we extend the one-dimensional models of Section 4 to study avascular tumour growth in two and three dimensions, following an approach developed by Greenspan [**77**]. This involves introducing new dependent variables for the local cell velocity and pressure within the tumour and can be motivated in the following way. If the tumour is incompressible and contains no voids or holes, then cell proliferation and death will generate spatial variations in the pressure within the tumour which will drive cell motion, with cells moving down pressure gradients, away from regions of net cell proliferation and towards regions of net cell death. We also introduce surface tension effects into the model: such restraining forces maintain the tumour's compactness and counteract the expansive force caused by net cell proliferation.

The remainder of this section is organised as follows In Subsections 5.2 and 5.3 we develop the model and then show that it reduces to the one-dimensional models of Section 4 when growth is one-dimensional. The stability of steady, radially-symmetric solutions to symmetry-breaking perturbations is investigated in Subsection 5.4 before the section concludes, in Subsection 5.5, with a discussion of the models and suggestions for further research.

5.2. The model equations

The model that we study is presented below in dimensionless form (for details, see [**38, 77**]). For simplicity we assume that the tumour is small enough that it is nutrient-rich, with all cells proliferating and, hence, that there is no quiescence or necrosis. The key variables are the nutrient concentration $c(\boldsymbol{r}, t)$, the cell velocity $\boldsymbol{v}(\boldsymbol{r}, t)$ and the pressure $p(\boldsymbol{r}, t)$. The locus of the tumour boundary is represented by $\Gamma(\boldsymbol{r}, t) = 0 = r - R(\theta, \phi, t)$. The variables $c, \boldsymbol{v}$ and p satisfy the following system of partial differential equations

$$0 = \nabla^2 c - \Gamma, \tag{53}$$

$$\nabla \cdot \boldsymbol{v} = S(c) - N(c) \equiv c - \lambda_A, \tag{54}$$

$$\boldsymbol{v} = -\mu \nabla p. \tag{55}$$

Equation (53) is the natural extension in higher spatial dimensions of the diffusion equation that was used in section 4. Equation (54) expresses mass conservation within the (assumed incompressible) tumour. To reinforce the similarity with the models discussed in Section 4, we employ the same proliferation and death rates here. Following [**77**], we use Darcy's law to relate the velocity of the cells moving throughout the tumour to the pressure, the constant of proportionality μ denoting the sensitivity of the tumour cells to pressure gradients.

We combine equations (54) and (55) to eliminate $\boldsymbol{v}$ from the model equations, giving

$$0 = \mu \nabla^2 p + (c - \lambda_A). \tag{56}$$

We append to Equations (53) and (56) the following boundary conditions

$$c = c_\infty, \quad p = \gamma\kappa \quad \text{on } \Gamma(\boldsymbol{r}, t) = 0, \tag{57}$$

$$\frac{\partial c}{\partial r} = \frac{\partial p}{\partial r} = 0 \quad \text{at } \boldsymbol{r} = 0. \tag{58}$$

In (57) c_∞ and $p_\infty = 0$ are the assumed constant nutrient concentration and pressure outside the tumour, $\gamma \geq 0$ is the surface tension coefficient and κ the mean curvature of the tumour boundary. Thus equations (57) state that the nutrient concentration is continuous across the tumour boundary and that there is a jump discontinuity in the pressure, this jump being proportional to the curvature of the boundary (and playing the role of a surface tension force which maintains the tumour's compactness). Equations (58) guarantee that the nutrient and pressure profiles are bounded at the origin.

In order to complete our model, it remains to determine how the tumour boundary $\Gamma(\boldsymbol{r}, t) = 0$ evolves. For simplicity (and following [**77**]), we write $\boldsymbol{r} = (r, \theta)$ and parameterise the tumour boundary in the following way

$$\Gamma(\boldsymbol{r}, t) = 0 = r - R(\theta, t).$$

We assume further that cells located on the boundary move with the local velocity there so that

$$\frac{\partial R}{\partial t} = \boldsymbol{v} \cdot \boldsymbol{n}\big|_{r=R(\theta,t)} = -\left. \mu \nabla p \cdot \boldsymbol{n}\right|_{r=R(\theta,t)}, \qquad \text{with } R(\theta, 0) = R_0(\theta). \tag{59}$$

In Equation (59), $\boldsymbol{n}$ is the unit outward normal to the tumour boundary and $R_0(\theta) = r$ denotes the initial locus of the tumour boundary.

In summary, our model of solid tumour growth comprises equations (53) and (56)–(59).

5.3. Radially-symmetric model solutions

Under radial symmetry, $c = c(r, t), p = p(r, t)$ and $r = R(t)$ on the tumour boundary and the model equations reduce to give

$$0 = \frac{1}{r^2}\frac{\partial}{\partial r}\left(r^2 \frac{\partial c}{\partial r}\right) - \Gamma,$$

$$0 = \frac{\mu}{r^2}\frac{\partial}{\partial r}\left(r^2 \frac{\partial p}{\partial r}\right) + (c - \lambda_A),$$

$$\frac{dR}{dt} = -\left.\mu \frac{\partial p}{\partial r}\right|_{r=R(t)}.$$

Integrating the equation for p once with respect to r and imposing (58), we deduce that

$$-\mu \frac{\partial p}{\partial r} = \frac{1}{r^2}\int_0^r (c - \lambda_A)\rho^2 d\rho \Rightarrow R^2 \frac{dR}{dt} = \int_0^R (c - \lambda_A) r^2 dr. \tag{60}$$

This expression for $R(t)$ is equivalent to equation (34) of Section 4 when $R_H = R_N = 0$ and shows how the current model reduces to the simpler model under radial symmetry.

By integrating the equations for c and p we obtain the following expressions:

$$c(r, t) = c_\infty - \frac{\Gamma}{6}(R^2 - r^2),$$

$$p(r, t) = \frac{\gamma}{R} - \frac{\Gamma}{120\mu}(R^2 - r^2)^2 + \frac{1}{6\mu}\left(c_\infty - \lambda_A - \frac{\Gamma R^2}{15}\right)(R^2 - r^2).$$

Substituting with $c(r, t)$ in (60) we deduce that

$$\frac{dR}{dt} = \frac{R}{3}\left(c_\infty - \lambda_A - \frac{\Gamma R^2}{15}\right).$$

In the next section, we investigate what happens when the (non-trivial) steady radially-symmetric solutions are subjected to small, symmetry-breaking perturbations, so that, for example,

$$c \sim c_0(r) + \epsilon c_1(r, \theta, t), \qquad \text{where } \epsilon \ll 1.$$

Our aim is to identify conditions under which the perturbations grow over time and conditions under which they decay: in the former case the underlying, radially-symmetric solution is said to be unstable (to symmetry-breaking perturbations) and in the latter it is stable.

5.4. Linear stability analysis

For simplicity we consider the stability of the radially-symmetric solution that has reached steady-state configuration, so that

$$(c, p, R) \sim (c_0, p_0, R_0) = (c_0(r), p_0(r), R_0)$$

where, from section 5.3,

$$c_0(r) = c_\infty - \frac{\Gamma}{6}(R_0^2 - r^2), \quad p_0(r) = \frac{\gamma}{R_0} - \frac{s\Gamma}{120\mu}(R_0^2 - r^2)^2,$$

$$\text{and} \qquad R_0^2 = \frac{15}{\Gamma}(c_\infty - \lambda_A). \tag{61}$$

We determine the response to symmetry-breaking perturbations by seeking solutions to the governing equations of the form

$$\left.\begin{array}{lcl} c & \sim & c_0(r) + \epsilon c_1(r, \theta, t) \\ p & \sim & p_0(r) + \epsilon p_1(r, \theta, t) \\ R & \sim & R_0 + \epsilon R_1(\theta, t) \end{array}\right\} \quad \text{where } 0 < \epsilon \ll 1.$$

Substituting with the above trial solutions in equations (53), (56), (59), (57) and (58) and equating coefficients of $O(\epsilon)$ we deduce that (c_1, p_1, R_1) solve

$$0 = \nabla^2 c_1 = \nabla^2 p_1 + c_1, \tag{62}$$

$$\frac{\partial R_1}{\partial t} = -\mu \left[\frac{\partial p_1}{\partial r} + R_1 \frac{d^2 p_0}{dr^2}\right]_{r=R_0}, \tag{63}$$

$$\text{with} \quad \frac{\partial c_1}{\partial r} = \frac{\partial p_1}{\partial r} = 0 \quad \text{at } r = 0, \tag{64}$$

$$c_1 = -R_1 \left.\frac{dc_0}{dr}\right|_{r=R_0}, \tag{65}$$

$$p_1 = -R_1 \left.\frac{dp_0}{dr}\right|_{r=R_0} - \frac{\gamma}{R_0^2}\left(2R_1 + \mathcal{L}(R_1)\right)_{r=R_0}, \tag{66}$$

$$R_1(\theta, 0) = R_{10}(\theta), \ \text{prescribed.} \tag{67}$$

In (66),

$$\mathcal{L}(f) = \frac{1}{\sin\theta}\frac{\partial}{\partial\theta}\left(\sin\theta \frac{\partial f}{\partial\theta}\right)$$

so that

$$\nabla^2 f(r, \theta) = \frac{1}{r^2}\frac{\partial}{\partial r}\left(r^2 \frac{\partial f}{\partial r}\right) + \frac{\mathcal{L}(f)}{r^2}.$$

When deriving equations (66) we have used results presented in the appendix that follows this section to determine the $O(\epsilon)$ contributions to the curvature on the tumour boundary and the normal derivative of the pressure there (see Equations (71) and (74)).

Equations (62)-(67) are linear and, hence, admit separable solutions of the form

$$\left.\begin{aligned} c_1(r,\theta,t) &= \chi_k(t) r^k P_k(\cos\theta) \\ p_1(r,\theta,t) &= \left(\pi_k(t) - \tfrac{\chi_k r^2}{2\mu(2k+3)}\right) r^k P_k(\cos\theta) \\ R_1(\theta,t) &= \rho_k(t) P_k(\cos\theta) \end{aligned}\right\} \tag{68}$$

where the Legendre polynomials $P_k(\cos\theta)$ satisfy $\mathcal{L}(P_k) = -k(k+1)P_k$ so that $\nabla^2(r^k P_k) = 0$. It is easy to show that equations (62) and (64) are automatically satisfied by the above choice of c_1 and p_1.

The coefficients χ_k and π_k are determined by imposing conditions (66) and exploiting the orthogonality of the Legendre polynomials. This gives

$$\chi_k R_0^k = -\frac{\Gamma R_0}{3}\rho_k \quad \text{and} \quad \pi_k R_0^k = \frac{\gamma \rho_k}{2R_0^2}(k-1)(k+2) + \frac{\chi_k R_0^{k+2}}{2\mu(2k+3)}.$$

Substituting with these results in (63) we deduce that the growth rate of a perturbation to the tumour radius involving $P_k(\cos\theta)$ satisfies

$$\frac{1}{\rho_k}\frac{d\rho_k}{dt} = (k-1)\left(\frac{2\Gamma R_0^2}{15(2k+3)} - \frac{\gamma\mu}{2R_0^3}k(k+2)\right). \tag{69}$$

From (69) we note that all modes evolve independently. Also, the system is insensitive to perturbations involving the first Legendre polynomial. This is to be expected since $P_{k=1}(\cos\theta)$ corresponds simply to a translation of the co-ordinate axes. We note also that if surface tension effects are neglected ($\gamma = 0$) then the radially-symmetric steady state is unstable to all modes for which $k \geq 2$. More generally, if $\gamma > 0$ then the steady state is unstable to only a finite number of modes, those for which

$$k(k+2)(2k+3) < \frac{4\Gamma R_0^5}{15\mu\gamma}.$$

Since the steady state radius R_0 is defined in terms of the system parameters (see equation (61)), this result shows how the choice of parameter values influences the modes to which the steady state is unstable. We note also that as γ increases the number of unstable modes declines. In particular, if $2\Gamma R_0^5/15\mu\gamma < 15$ then there are no integers which satisfy the above inequality and we deduce that in this case the radially symmetric steady state is stable to perturbations involving Legendre polynomials of arbitrary order. Recalling that $\Gamma R_0^2 = 15(c_\infty - \lambda_A)$, we deduce that this will be the case if the external nutrient concentration c_∞ satisfies

$$c_\infty < \lambda + \frac{\Gamma}{15}\left(\frac{225\mu\gamma}{4\Gamma}\right)^{2/5}.$$

By differentiating (69) with respect to k we may determine the fastest growing mode for a given set of parameter values (and, hence, for a given value of R_0). After some manipulation, we deduce that the fastest growing mode satisfies

$$(2k+3)^2(3k^2+2k-2) = \frac{4\Gamma R_0^5}{3\mu\gamma}.$$

5.5. Discussion

The analysis presented above provides a mechanism which may explain how the irregular morphology that characterises certain solid tumours may be initiated. To understand this, consider a uniform cluster of tumour cells for which the surface tension coefficient γ is sufficiently large that the underlying radially symmetric solution is stable to all Legendre polynomials. Our analysis predicts that such a cluster will remain radially symmetric throughout the tumour's development. Suppose that there is a mutation which reduces γ *i.e.* the surface tension forces holding the tumour cells together are weakened. If this reduction is large enough then the tumour will become unstable to a finite range of asymmetric modes and develop an irregular morphology. We note that similar qualitative behaviour is obtained if, instead of invoking surface tension (and the associated jump in the pressure across the tumour boundary), we assume that the nutrient concentration is discontinuous across the tumour boundary, with a jump that is proportional to the local curvature [**33**]. The physical motivation for this boundary condition is that the associated nutrient (or its energy-equivalent) is utilised by cells on the tumour boundary to maintain its compactness.

We stress that the above mechanism is only one possible explanation for why solid tumours develop irregular, fractal-like morphologies. In practice, the acquisition of a blood supply, brought about by angiogenesis, may also break the tumour's radial symmetry, by making the supply of nutrients spatially heterogeneous. Whilst there are now a several realistic models of angiogenesis and vascular tumour growth (see Section 7), the development of mathematical models with the power to investigate the potential for vascularisation to stimulate asymmetric tumour growth remains to be developed.

There are many model modifications that may be studied. The exercises at the end of this section introduce several extensions which have been documented in the literature. These include a consideration of more general perturbations involving spherical harmonics $Y_{km}(\theta, \phi)$, other geometries and a study of the stability of more developed avascular tumours (*i.e.* those that contain quiescent and necrotic regions) [**34, 38**]. We note also that linear stability analysis predicts only local behaviour. Where instability is predicted it is natural to ask how the asymmetry develops. Two complementary approaches that have been used to investigate this issue. The first involves using weakly nonlinear analysis to extend the linear theory [**34, 41**]. Alternatively numerical methods can be used to solve the nonlinear governing equations [**52**].

The model that we have studied assumes that there is a single rate-limiting nutrient or chemical within the tumour environment. In practice, the tumour will be bathed in a variety of chemicals, some of which promote cell proliferation (*e.g.*, oxygen and glucose) and others which inhibit mitosis (*e.g.*, tumour necrosis factor, anti-cancer drugs). These growth factors may be supplied externally (*e.g.*, oxygen, drugs) or they may be produced as a bi-product of the cells' normal functions (*e.g.*, tumour necrosis factor is a bi-product of cell degradation). By extending our mathematical model to include additional diffusible chemicals, it should be possible to study the impact that the presence of multiple chemicals has on the tumour's development [**44**].

Two key features of the tumour growth model studied in this section concern the assumption of homogeneity of the tumour cell population and the validity of

Darcy's law to describe cell motion within the tumour. In the next section we explain how, by using a multiphase modelling framework, the current model can be extended to relax both of these assumptions.

5.6. Exercises

1. The following mathematical model describes the growth of an avascular tumour in three dimensions:

$$0 = \nabla^2 c - \Gamma,$$

$$\boldsymbol{v} = -\nabla p, \quad \nabla.\boldsymbol{v} = s(c - \tilde{c}),$$

$$\frac{\partial R}{\partial t} = \boldsymbol{v}.\boldsymbol{n} \quad \text{on } \Gamma(\boldsymbol{r}, t) = r - R(\theta, \phi, t) = 0,$$

$$\text{with} \quad c = 1, \quad p = \alpha\kappa \quad \text{on } \Gamma(\boldsymbol{r}, t) = 0,$$

$$\frac{\partial c}{\partial r} = \frac{\partial p}{\partial r} = 0 \quad \text{at } \boldsymbol{r} = 0$$

$$\text{and} \quad c \geq 0 \text{ throughout the tumour.}$$

Eliminate $\boldsymbol{v}$ from the governing equations. Suppose that the model admits solutions of the form

$$c = c_0(r) + \epsilon \chi_{lm}(t) r^l Y_{lm}(\theta, \phi),$$

$$p = p_0(r) + \epsilon \left(\pi_{lm}(t) - \frac{s\chi_{lm} r^2}{2(2l+3)} \right) r^l Y_{lm}(\theta, \phi),$$

$$R = R_0 + \epsilon \rho_{lm}(t) Y_{lm}(\theta, \phi),$$

where $\epsilon \ll 1$ and $Y_{lm}(\theta, \phi)$ $(-l \leq m \leq l)$ denote the spherical harmonics of order $l \geq 1$ so that $\nabla^2 Y_{lm} = -l(l+1)Y_{lm}$. Determine the leading order solutions (c_0, p_0, R_0) and then use linear stability analysis to derive differential equations for $\rho_{lm}(t)$, the growth rates of the asymmetric perturbations to the tumour boundary. Does your result vary for the $(2l+1)$ spherical harmonics of order l?

2. Consider a non-necrotic avascular tumour growing in a two-dimensional Cartesian domain, in response to a growth inhibitor β. The governing equations are

$$\frac{\partial^2 \beta}{\partial x^2} + \frac{\partial^2 \beta}{\partial y^2} - \Gamma = 0,$$

$$\frac{\partial^2 p}{\partial x^2} + \frac{\partial^2 p}{\partial y^2} + \frac{s}{\mu}(1 - \beta) = 0,$$

$$\frac{dX}{dt} = -\mu \left(\frac{\partial p}{\partial x} + \frac{\partial p}{\partial y} \right) \cdot \boldsymbol{n} \quad \text{on } x = X(y, t),$$

where $\boldsymbol{n}$ denotes the unit outward normal to the surface $x = X(y, t)$ and

$$\frac{\partial \beta}{\partial x} = \frac{\partial p}{\partial x} = 0 \quad \text{on } x = 0,$$

$$\frac{\partial \beta}{\partial y} = \frac{\partial p}{\partial y} = 0 \quad \text{on } y = 0, h,$$

$$\beta = \beta_\infty, \quad p = \alpha\kappa \quad \text{on } x = X(y,t),$$

and $\beta \geq 0$ throughout the tumour. Assume that, to leading order, the tumour grows uniformly in the x-direction and seek trial solutions of the form

$$\left.\begin{aligned} \beta(x,y,t) &\sim \beta_0(x,t) + \epsilon\beta_1(x,t)\cos(k\pi y/h) \\ p(x,y,t) &\sim p_0(x,t) + \epsilon p_1(x,t)\cos(k\pi y/h) \\ X(y,t) &\sim X_0(t) + \epsilon X_1(t)\cos(k\pi y/h) \end{aligned}\right\} \quad (\epsilon \ll 1, \quad k = 0,1,2,\ldots).$$

By substituting with these expressions in the model equations and equating coefficients of $O(1)$ and $O(\epsilon)$, determine (β_0, p_0, X_0) and state the partial differential equations that govern (β_1, p_1, X_1). Determine (β_1, p_1, X_1) for the special case when $X_0(t) = X_0$, constant.

Appendix

In this appendix we show how the $O(\epsilon)$ contributions to the normal derivative of the pressure and the curvature on the tumour boundary may be determined.

Mean curvature

The mean curvature κ of the surface $r = R(\theta,t)$ is given by

$$\kappa = -\frac{1}{2(R^2+R_\theta^2)^{3/2}}\left\{RR_{\theta\theta} - 3R_\theta^2 - 2R^2 + \frac{R_\theta}{R}(R^2+R_\theta^2)\frac{\cos\theta}{\sin\theta}\right\}, \tag{70}$$

where subscripts denote differentiation with respect to θ. Setting $R(\theta,t) = R_0 + \epsilon R_1(\theta,t)$ on the perturbed tumour boundary, viewing t as an independent parameter and equating coefficients of $O(\epsilon^n)$ we deduce that

$$\kappa = \frac{1}{R_0} - \frac{\epsilon}{2R_0^2}\left\{2R_1 + \frac{1}{\sin\theta}\frac{\partial}{\partial\theta}\left(\sin\theta\frac{\partial R_1}{\partial\theta}\right)\right\} + O(\epsilon^2). \tag{71}$$

Normal derivatives

The unit outward normal $\boldsymbol{n}$ to the surface $\Gamma(\boldsymbol{r},t) = 0 = r - R(\theta,t)$ is defined as follows

$$\boldsymbol{n} = \frac{\nabla\Gamma}{|\nabla\Gamma|}.$$

Thus, on the perturbed tumour boundary, where $R(\theta,t) = R_0 + \epsilon R_1(\theta,t)$,

$$\boldsymbol{n} = \left(1 + \frac{\epsilon^2 R_{1\theta}^2}{R_0^2 + \epsilon^2 R_{1\theta}^2}\right)^{-1/2}\left(\hat{\boldsymbol{r}} - \frac{\epsilon R_{1\theta}}{R_0 + \epsilon R_{1\theta}}\hat{\boldsymbol{\theta}}\right), \tag{72}$$

where $\hat{\boldsymbol{r}}$ and $\hat{\boldsymbol{\theta}}$ are unit vectors. Now

$$\nabla p = \left(\frac{\partial p}{\partial r}, \frac{1}{r}\frac{\partial p}{\partial\theta}\right) = \left(\frac{dp_0}{dr} + \epsilon\frac{\partial p_1}{\partial r}, \frac{\epsilon}{r}\frac{\partial p_1}{\partial\theta}\right) + O(\epsilon^2),$$

where $p(r,\theta,t) = p_0(r) + \epsilon p_1(r,\theta,t) + O(\epsilon^2)$. Evaluating ∇p on the perturbed tumour boundary we deduce

$$\nabla p|_{r=R_0+\epsilon R_1} = \left(\frac{dp_0}{dr} + \epsilon R_1\frac{d^2p_0}{dr^2} + \epsilon\frac{\partial p_1}{\partial r}\right)_{r=R_0} + O(\epsilon^2). \tag{73}$$

Combining (72) and (73) we deduce finally that on $\Gamma(\boldsymbol{r},t) = 0$

$$\nabla p\cdot\boldsymbol{n} = \left.\frac{dp_0}{dr}\right|_{r=R_0} + \epsilon\left(R_1\frac{d^2p_0}{dr^2} + \frac{\partial p_1}{\partial r}\right)_{r=R_0} + O(\epsilon^2). \tag{74}$$

6. Avascular tumours: multiphase models

6.1. Introduction

The spatially-structured models that we have studied thus far treat the tumour as a single, homogeneous population of cells, whose proliferation and death rates are regulated by a single, diffusible species. Additionally they assume that cell movement is governed by Darcy's law, an empirical constitutive law more usually associated with the flow of water through a porous medium! Several authors have developed spatially-structured models of avascular tumour growth that account for cellular heterogeneity. For example, Ward and King assume that all cellular species move with a common velocity and show how the cells naturally partition to reproduce the layered structure that characterises multicellular spheroids cultured *in vitro* [**151, 152, 153**]. An alternative approach is adopted by Thompson and Byrne [**145**] and Pettet and coworkers [**120**]. They allow for differential movement of distinct cell species but prescribe the velocities in a phenomenological manner. In this section we use a multiphase modelling framework to develop new models that can account for cellular heterogeneity and permit alternative descriptions for cell movement to be incorporated. The models can also be used to investigate the impact on cell proliferation and cell movement of mechanical effects due, for example, to interactions with the tissue in which the tumour is embedded.

The earliest multiphase models are due to Please and coworkers [**96, 121**] and based on an analogy between the layered structure of multicellular spheroids (MCS) and that of a compacting porous medium such as soil, the idea being that cells in the periphery are in direct physical contact (*i.e.* compacted) and therefore transmit any external mechanical forces that are acting whereas those in the necrotic core are no longer in contact (they are fluidised) and so any load is distributed between the cells and the fluid. In this section we follow an alternative approach developed by Byrne and coworkers [**30, 40, 43**] which is based on biomechanical models developed to study the response of cartilage to mechanical loading. An importance difference between the multiphase models we present here and those used to study the deformation of cartilage is that in the latter case no cell proliferation occurs: in our tumour models cell proliferation (and death) must be included, leading to mass conversion between the constituent phases.

The remainder of the section is organised as follows. In Subsection 6.2 we develop a two-phase model of avascular tumour growth *in vitro*, the two phases representing the tumour cells and the extracellular fluid in which they are bathed. In Subsection 6.3 we explain how the model equations can be reduced to a simpler system of PDEs. Subsection 6.3 also contains numerical and analytical results. We conclude in Subsection 6.4 by explaining how the two-phase model has been adapted to describe other aspects of solid tumour growth and outlining ideas for future work.[3]

6.2. Model development

Following [**30, 40, 43**], we view the tumour as a two-phase mixture of cells and extracellular fluid or water. We construct mass and momentum balances for each phase and close the model by making appropriate constitutive assumptions. For simplicity, we formulate the model in one-dimensional cartesian geometry.

[3]No exercises accompany this section: the complexity of the models makes it difficult to devise short, self-contained problems that are analytically tractable.

We denote by $n(x,t)$ and $w(x,t)$ the volume fractions occupied by tumour cells and extracellular fluid and by (v_n, σ_n, p_n) and (v_w, σ_w, p_w) their respective velocities, stress tensors and pressures.

The mass and momentum balance equations
Applying the principle of mass balance to n and w yields

$$\frac{\partial n}{\partial t} + \frac{\partial}{\partial x}(v_n n) = S_n, \tag{75}$$

$$\frac{\partial w}{\partial t} + \frac{\partial}{\partial x}(v_w w) = S_w \equiv -S_n. \tag{76}$$

In (75) and (76), S_n and S_w are the net rates at which cells and water are produced and we have assumed that mass is converted from one phase to the other (*i.e.* there is no local mass creation).

Applying the principle of momentum balance to n and w, and neglecting inertial effects, we have

$$0 = \frac{\partial}{\partial x}(n\sigma_n) + F_{nw} + p\frac{\partial n}{\partial x}, \tag{77}$$

$$0 = \frac{\partial}{\partial x}(w\sigma_w) - F_{nw} + p\frac{\partial w}{\partial x}. \tag{78}$$

In (77) the first term represents internal forces in each phase while F_{nw} denotes the force exerted by the water on the cells, with an equal and opposite force acting on the water. The third term is an interfacial term which arises when we average over discrete cells to obtain a continuum limit (for details, see [**55**]).

Constitutive assumptions
To close our model we must impose boundary and initial conditions and specify the functional forms for S_n, σ_n, σ_w and F_{nw}. We view the water as an inviscid fluid and the cell phase as a viscous fluid, with viscosity μ_n so that

$$\sigma_w = -p, \qquad \sigma_n = -p_n + 2\mu_n \frac{\partial v_n}{\partial x}, \qquad p_n = p + \Sigma_n(n). \tag{79}$$

We interpret the viscosity μ_n heuristically as a measure of the cells' affinity for each other: as the cells become more pathological, and less well differentiated, they will be less likely to maintain contact with each other and/or to shear and so their viscosity will decrease. To account further for the way in which cells differ from bags of water and following [**43**], we introduce in (79) an additional cell pressure $\Sigma_n(n)$ where

$$\Sigma(n) = \begin{cases} 0 & 0 \le n < n_0, \\ \alpha \dfrac{(n-n_0)(n-n_2)}{(1-n)^\beta} & n_0 \le n < 1, \end{cases}, \tag{80}$$

and

$$\beta = \frac{(1-n_1)(3n_1 - 2n_2 - n_0)}{(n_1 - n_0)(n_2 - n_1)}, \qquad \alpha = -\frac{\Sigma_1 (1-n_1)^\beta}{(n_1 - n_0)^2 (n_2 - n_1)}.$$

Thus if $n < n_0$ the cells exert no influence on each other. By contrast, when $n_0 < n < n_2$ the cells are attracted to each other. Finally, when $n_2 < n < 1$ the cells repel each other, the repulsive force becoming infinite as $n \to 1$. We illustrate these effects in Figure 9 where we sketch $\Sigma(n)$.

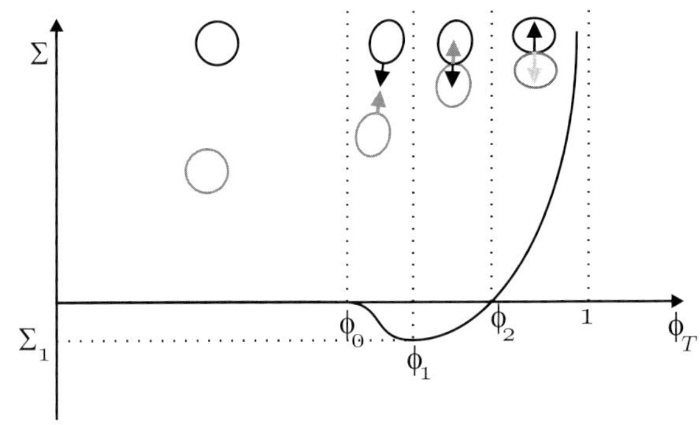

FIGURE 9. Schematic diagram showing how the function $\Sigma(n)$ varies with n. The black arrows represent the forces experienced by the black cells due to the presence of the grey cells (and conversely).

We assume that the interaction force F_{nw} is due to relative motion of the two phases and accordingly specify

$$F_{nw} = knw(v_w - v_n),$$

assuming, for simplicity, that the drag coefficient is proportional to the volume fractions of each phase, with constant of proportionality k (for other choices, see [**64**]).

With σ_n, σ_w and F_{nw} defined above, equations (77)-(78) specify v_n and v_w. An additional equation is needed to determine the pressure p. We assume that there are no voids in the tumour mass and accordingly prescribe

$$n + w = 1. \tag{81}$$

We now consider the cell proliferation rate S_n that appears in (75). Again, following [**43**], we assume that $S_n = S_n(n, c)$ where $c(x, t)$ denotes an externally-supplied nutrient and

$$S_n(n,c) = \begin{cases} \dfrac{S_0(c - c_N)n}{1 + S_1 c} - \delta n, & \text{if } c \geq c_N, \\ -\left(\dfrac{1 + S_2 \tilde{c}}{1 + S_2 c}\right) n, & \text{if } c \leq c_N, \end{cases} \tag{82}$$

where $S_0, S_1, S_2, c_N, \tilde{c}$ and μ are positive constants. The first term models cell proliferation as an increasing, saturating function of nutrient concetration (in practice, there is a physical limit to how rapidly cells can divide), the second models apoptosis and the third necrosis. As in earlier sections, we assume that $c(x,t)$ satisfies a quasi-steady reaction diffusion equation and introduce positive constants Q_0 and Q_1 such that

$$0 = \frac{\partial^2 c}{\partial x^2} - \frac{Q_0 c n}{1 + Q_1 c}. \tag{83}$$

It remains to specify the tumour's growth rate. As in Section 5 we assume that the tumour boundary $x = R(t)$ moves with the local cell velocity there so that

$$\frac{dR}{dt} = v_n(R,t). \tag{84}$$

Boundary and initial conditions
Our two-phase model of avascular tumour growth comprises equations (75)-(84) which we close by specifying the following boundary and initial conditions:

$$v_n = 0 = v_w = \frac{\partial c}{\partial x} \quad \text{at } x = 0, \tag{85}$$

$$\sigma_w = -p = 0, \ \sigma_n = \mu_n \frac{\partial v_n}{\partial x} - \Sigma = 0, \ c = c_\infty \quad \text{at } x = R(t), \tag{86}$$

$$n(x,0) = n_0(x), \ w(x,0) = 1 - n(x,0) = 1 - n_0(x), \ R(0) = R_0. \tag{87}$$

Equations (85) ensure symmetry about $x = 0$ while the first two of equations (86) guarantee continuity of (the normal component of) the cell and water stress tensors across $x = R(t)$, with the pressure outside the tumour normalised so that $p = 0$ there. Additionally we assume that c is continuous across $x = R$ and denote by c_∞ the nutrient concentration in the surrounding medium. Finally, equations (87) specify the initial cell and water distributions within the tumour and its initial size.

6.3. Model simplification

Using the constitutive assumptions specified above, it is straightforward to show that equations (75)-(87) can be written

$$\frac{\partial n}{\partial t} + \frac{\partial}{\partial x}(n v_n) = S_n, \quad \frac{\partial}{\partial x}(n v_n + (1-n) v_w) = 0, \tag{88}$$

$$\frac{\partial}{\partial x}\left(-p - n\Sigma + \mu_n n \frac{\partial v_n}{\partial x}\right) = 0, \quad -\frac{\partial p}{\partial x} = kn(v_w - v_n), \tag{89}$$

$$0 = \frac{\partial^2 c}{\partial x^2} - \frac{Q_0 c n}{1 + Q_1 c}, \tag{90}$$

$$\frac{dR}{dt} = v_n(R,t). \tag{91}$$

Integrating the second of equations (88) and the first of (89), subject to (85) and (86), we deduce that

$$v_w = -\frac{n v_n}{1-n} \quad \text{and} \quad p = \mu_n n \frac{\partial v_n}{\partial x} - n\Sigma.$$

Using these relations to eliminate p and v_w, our model then reduces to the first of equations (88), together with equations (90), (91) and

$$\frac{\partial}{\partial x}\left(\mu_n n \frac{\partial v_n}{\partial x} - n\Sigma\right) = \frac{kn}{1-n} v_n. \tag{92}$$

These equations for n, v_n, c and R are solved subject to conditions (85)-(87). A typical numerical simulation taken from [**43**] is presented in Figure 10 below. It shows how the cell volume fraction $n(x,t) \equiv \phi(x,t)$ and tumour radius $R(t) = L(t)$ evolve over time towards a steady state. For further details of numerical simulations and model analysis, see [**30, 43**]: henceforth we focus on showing how existing models of solid tumour growth may be recovered from our model when appropriate asymptotic limits are taken.

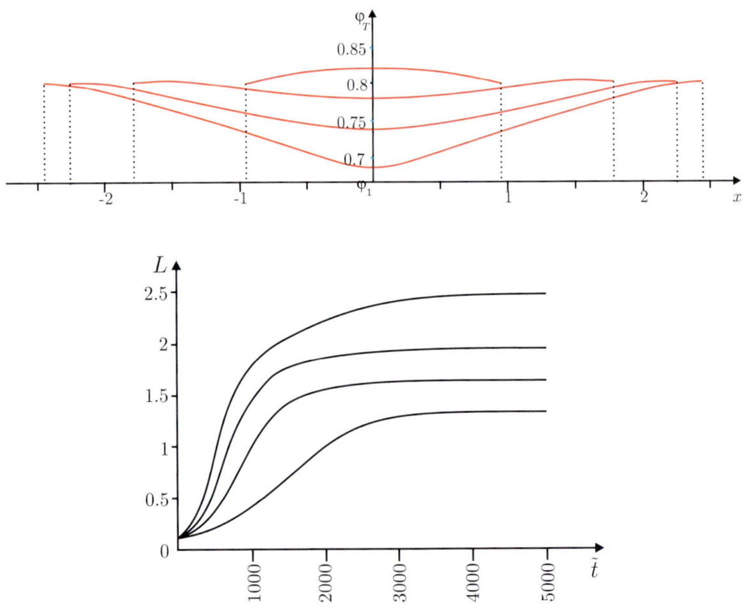

FIGURE 10. (a) Diagram showing the evolution of the cell volume fraction $n(x,t) = \phi_T(x,t)$ towards a steady state at times $t = 500, 1000, 2000, 4000$; (b) Diagram showing how the evolution of the tumor radius, $R(t) = L(t)$, to its equilibrium value varies as the maximum tumour cell proliferation rate is increased. As S_0 increases the tumour's equilibrium size increases while the time taken to reach the steady state decreases. From the lower to the upper curve S_0 increases from $S_0 = 0.005$ to $S_0 = 0.0125$.

Viscosity negligible $\mu_n \to 0$

In this limit equation (92) may be used to eliminate v_n from the first of (88), yielding

$$\frac{\partial n}{\partial t} = \frac{\partial}{\partial x}\left(\frac{1-n}{k}\frac{\partial}{\partial x}(n\Sigma)\right) + S_n.$$

Thus when $\mu_n \to 0$ our model comprises a quasi-steady diffusion equations for $c(x,t)$, a nonlinear diffusion equation for n and associated boundary and initial conditions. This model is similar to existing reaction diffusion models of tumour growth [**73, 130**]. The key differences here are our formulation of the problem on a moving domain and the physical mechanisms driving cell diffusion: here they are drag and cell-cell interactions rather than random cellular motion.

It is also possible to show that when viscous effects are neglected the pressures and velocities in each phase are such that

$$v_w = -\frac{1}{k}\frac{\partial p_w}{\partial x} \quad \text{and} \quad v_n = -\frac{(1-n)}{kn}\frac{\partial}{\partial x}\left(\frac{n}{1-n}p_n\right).$$

Thus for one-dimensional growth, a Darcy-type law governs the velocities of each phase, providing justification for the models of tumour growth originally developed by Greenspan [**77**] and discussed in Section 5.

Cell viscosity and drag negligible, $\mu_n, k/(n\Sigma)' \to 0$
With viscosity negligible, v_n and v_w are given by

$$v_w = \frac{1}{k(1-n)/(n\Sigma(n))'}\frac{\partial n}{\partial x} \qquad \text{and} \qquad v_n = -\frac{(1-n)}{nk(\alpha)/(n\Sigma(n))'}\frac{\partial n}{\partial x}. \tag{93}$$

Assuming that the velocities are finite and $n \geq n_0$, in the further limit $k/(n\Sigma)' \to 0$, we have $\partial n/\partial x \to 0$. Hence the low drag (or high stiffness) of the tumour causes n to adjust rapidly to the constant value $n = n_2$ (see equation (80)). Equations (88) and (91) then supply

$$\frac{dR}{dt} = \frac{1}{n_2}\int_0^R S_n(n_2, c)dx. \tag{94}$$

Finally, we deduce that in this limit our model reduces to give (94), with $R = R_0$ at $t = 0$, and

$$0 = \frac{\partial^2 c}{\partial x^2} - \frac{Q_0^* c}{1 + Q_1 c},$$

$$\text{with } \frac{\partial c}{\partial x} = 0 \text{ at } x = 0 \quad \text{and } c = c_\infty \text{ when } x = R(t),$$

where $Q_0^* = Q_0 n_2$. In particular, the nutrient distribution within the tumour, its rate of consumption by the cells and its impact on the cell proliferation rate control the tumour's evolution. This model is qualitatively similar to the models of avascular tumour growth discussed in Section 4 and, as a result, places them on a stronger, theoretical foundation.

6.4. Discussion

In this section we have used a multiphase approach to develop a model of avascular tumour growth that accounts for cellular heterogeneity and allows for a more general description of cell movement than was possible with the simpler models of Sections 4 and 5. We have also shown how diffusion models can be recovered when appropriate asymptotic limits are taken and provided stronger justification for the use of Darcy's law to describe cell movement [**77**].

There are many ways in which this two-phase model could be extended. Motivated to a large extent by experimental work which suggests that mechanical effects can influence cell proliferation, in [**43**] Byrne and Preziosi assumed $S_n = S_n(c, p)$ in equation (82) and showed that a proliferation rate that was a decreasing function of the pressure p could limit cell growth as an alternative to nutrient limitation. A different approach was adopted in [**49**]. There the growth of a two-phase tumour embedded in a poroelastic gel was studied, the gel compressing in response to the expansive force exerted on it by the growing tumour mass, and exerting an equal and opposite force on the tumour. The model reproduced experimental results which showed that the equilibrium size attained by a tumour spheroid cultured in such a gel decreases as its stiffness increases [**83**]. Recently, Roose *et al.* revised this model, treating the tumour mass as a two-phase poroelastic solid [**126**].

Other authors have developed multiphase models involving three or more phases to describe vascular tumour growth [**31**], the response of tumours to treatment involving macrophages that have been genetically engineered to release anti-tumour chemicals when they localise in low oxygen tumour regions [**115**], and tumour encapsulation [**88**]. Tumour encapsulation refers to the process by which a dense collagenous rim forms around a tumour, keeping it localised and preventing it from invading the surrounding tissue. Using a three-phase model, Jackson and Byrne showed that the capsule was more likely to form as a result of compression of the existing extracellular matrix (caused by expansion of the tumour mass) than as a result of the immune response stimulating the deposition of new collagenous material.

Several other authors have used solid mechanics to develop models of solid tumour growth. In [**45**], Chaplain and Sleeman viewed the proliferating rim of a radially-symmetric expanding multicellular spheroid (MCS) as an elastic shell, enclosing a fluid-like necrotic core. A weakness of this model is that the volume of the proliferating rim does not change. Consequently, as the tumour grows, the rim getting progressively thinner: in practice, however, when MCS are culture *in vitro* the width of the proliferating rim settles to a constant value and, so, the volume of the proliferating rim increases as the tumour radius increases. Jones *et al.* formulated a single phase mechanical model in which the usual constitutive law for linear elasticity was reformulated in terms of rates of strain and stress in order to account for the continuous (and nonuniform) cell growth and death within the tumour mass [**90**]. A weakness of this model is that the stress becomes unbounded (and, hence, physically unrealistic) when the tumour reaches an equilibrium configuration. Several authors have since suggested modifications that resolve this deficiency. Araujo and McElwain introduce anisotropy, assuming that cell proliferation increases the stress in the tangential direction whereas cell death relieves stress in the radial direction [**17**]. By contrast, MacArthur and Please show that the introduction of viscosity into the material constitutive law relieves the excessive stresses that develop in Jones *et al.*'s purely elastic tumour [**100**], while Roose *et al.* decompose the tumour into a two-phase mixture [**126**].

Other areas that remain to be investigated include studying the multiphase models in two and three spatial dimensions and investigating the impact on the tumour's growth dynamics of using different constitutive laws to describe its mechanical properties. For example, we might view the cell phase as an elastic [**126**] or viscoelastic material. Finally, when developing mechanical models of the type discussed in this section, it is important to verify that the constitutive laws used to close the models do not violate the laws of thermodynamics (for details, see [**9**] and references therein).

7. Angiogenesis and vascular tumour growth

7.1. Introduction

Until recently, difficulties associated with performing suitable studies and collecting relevant data meant that the mechanisms that control angiogenesis and vascular tumour growth were less well understood than those for avascular tumour growth. As a result, there are fewer models of these stages of tumour growth. In this section we illustrate the types of models that have been developed and the insight that they yield by focussing on a deterministic model of tumour angiogenesis and a hybrid model of vascular tumour growth.

The remainder of the section is organised as follows. In Subsection 7.2 we describe a model of tumour angiogenesis [**35**]. It is based on a model originally developed by Balding and McElwain [**20**] and subsequently extended by Chaplain and coworkers [**35, 46, 47, 113**]. In Subsection 7.3 we derive a caricature of the full model for which explicit analytical solutions can be found. In Subsection 7.4 we present a multiscale model of vascular tumour growth developed by Alarcon and coworkers [**6, 7**] and some numerical results. The section concludes in Subsection 7.6 with a summary and suggestions for future work.

7.2. Angiogenesis: model development

The one-dimensional model of tumour angiogenesis that we study derives from a model which was developed by Balding and McElwain [**20**] and is based on an analogy with fungal growth, both processes involving the formation of interconnected, branched structures in response to environmental cues [**57**]. There are three physical variables of interest: capillary tips $n(x,t)$, vessels $b(x,t)$ and a generic, diffusible chemical $a(x,t)$ that is produced by the tumour and termed a tumour angiogenesis factor (TAF: in practice, there are many such chemicals and they include the different isoforms of vascular endothelial growth factor and transforming growth factor beta). The model describes how new vessels emerge from existing vessels and then migrate via random motion and chemotaxis towards the tumour. The governing equations are derived by applying the principle of mass balance to each species and are stated below in dimensionless form:

$$\frac{\partial a}{\partial t} = \frac{\partial^2 a}{\partial x^2} - \lambda a, \tag{95}$$

$$\frac{\partial n}{\partial t} = \mu \frac{\partial^2 n}{\partial x^2} - \chi \frac{\partial}{\partial x}\left(n \frac{\partial a}{\partial x}\right) + \alpha_0 ab + \alpha_1 H(a-\hat{a})n - \alpha_2 nb, \tag{96}$$

$$\frac{\partial b}{\partial t} = -\mu \frac{\partial n}{\partial x} + \chi n \frac{\partial a}{\partial c} - \gamma b. \tag{97}$$

In (95) we assume that the TAF undergoes natural decay (with half-life λ^{-1}) as it diffuses away from its source, the tumour (which is located at $x = 1$). In equation (96), $\mu, \chi, \alpha_0, \alpha_1, \alpha_2$ and $\hat{a}$ are nonnegative constants and $H(.)$ denotes the Heaviside step function ($H(x) = 1$ if $x > 0$, $H(x) = 0$ otherwise). We suppose that the capillary tips migrate via random motion and chemotaxis towards the tumour, so that μ and χ denote the random motility and chemotaxis coefficients respectively. We assume that new capillary tips are emanate from pre-existing vessels and tips at rates that increase with a (rate constants α_0 and α_1 respectively) and that they are removed as a result of fusion or anastomosis with existing vessels (rate constant α_2) The production of vessels is driven by migration of the capillary tips, the vessel density increasing at a rate that matches the flux of capillary tips via a mechanism termed the 'snail trail'. The vessels also undergo natural death, with half-life γ^{-1}.

We close equations (95)-(97) by imposing the following boundary and initial conditions:

$$a = 0, \quad n = e^{-kt} \quad \text{at } x = 0, \tag{98}$$

$$a = 1, \quad n = 0 \quad \text{at } x = 1, \tag{99}$$

$$a(x,0) = a_0(x), \quad n(x,0) = n_0(x), \quad b(x,0) = b_0(x). \tag{100}$$

Equations (98) specify the concentration of TAF and capillary tips at the parent vessels. We assume that any TAF reaching $x = 0$ is rapidly removed by the vasculature. We also suppose that the density of capillary tips there decays exponentially, following an initial surge in their density. At the tumour boundary (see equations (99)), the TAF is maintained at a constant level (here scaled to unity) by the tumour cells and the concentration of capillary tips is zero (as a result, the model ceases to be valid when $n(1,t) > 0$: this is to be expected since at this time the tumour will be vascularised and there will blood flow will have been established to through the new vasculature). Finally, equations (100) specify the initial distributions of a, n and b: we assume, for simplicity, that the region is initially devoid of TAF, capillary tips and vessels.

1D numerical results

Equations (95)-(100) may be solved using standard numerical packages for systems of partial differential equations such as the NAG library routine D03PCF. Repeated simulations show that depending on the parameter values used the model reproduces successful and failed angiogenesis. A typical simulation for the case of successful angiogenesis is presented in Figure 11. It exhibits many of the characteristic features of angiogenesis that have been observed *in vivo* [**111**]. These include acceleration of the capillary tips and vessels towards the tumour, with the peak vessel density lagging behind the peak capillary tip density (for details, see [**35, 113**]).

7.3. Angiogenesis: caricature model

Numerical simulations of equations (95)-(100) suggest that the TAF profile rapidly evolves to a steady configuration and so we fix $\frac{\partial}{\partial t} = 0$ in (95). To render the analysis tractable we make the additional assumptions that $\frac{\partial b}{\partial t} \sim 0$ and $\mu \to 0$. Then equations (95) and (97) supply

$$a(x,t) = \frac{\sinh \sqrt{\lambda} x}{\sinh \sqrt{\lambda}} \quad \text{and} \quad b(x,t) = \frac{\chi}{\gamma} n \frac{\partial a}{\partial x},$$

while (96) becomes

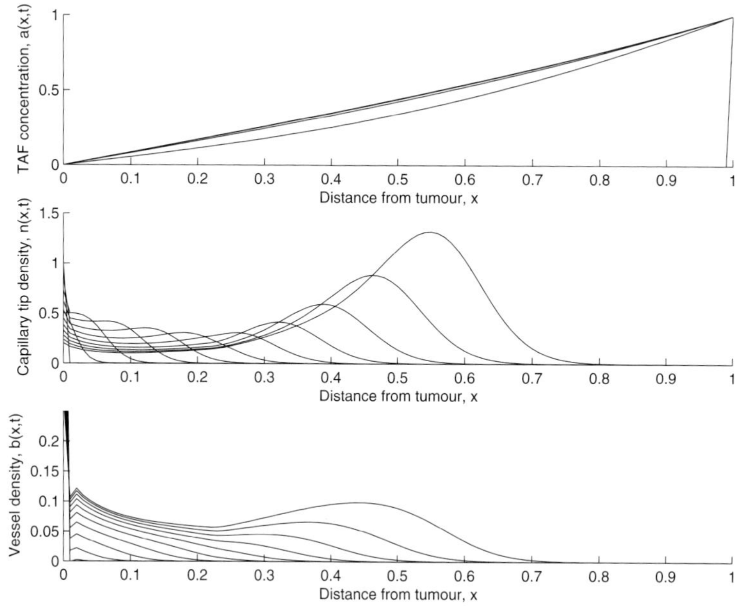

FIGURE 11. Profiles of the TAF concentration $a(x,t)$, the capillary tip density $n(x,t)$ and the vessel density $b(x,t)$ during successful angiogenesis. The TAF profile rapidly evolves to a steady state while the tips and vessels migrate towards the tumour at $x = 1$. Profiles are plotted at times $t = 0, 0.16, 0.32, 0.48, ..., 1.6$. Parameter values: $\mu = 0.001, \chi = 0.4, \alpha_0 = 50, \alpha_1 = 10, \alpha_2 = 50, \lambda = 10, \gamma = 0.25, \hat{a} = 0.2$.

(101)

$$\frac{\partial n}{\partial t} + \chi \frac{\partial a}{\partial x}\frac{\partial n}{\partial x} = an\left(\frac{\alpha_0 \chi}{\gamma}\frac{\partial a}{\partial x} + \alpha_1 H(a-\hat{a}) - \chi\lambda\right) - \frac{\alpha_2\chi}{\gamma}\frac{\partial a}{\partial x}n^2 \equiv F(n,a), \quad \text{say.}$$

Equation (101) is a one-dimensional nonlinear wave equation which can be solved using the method of characteristics. We note that the wavespeed is given by

$$\frac{dx}{dt} = \chi\frac{\partial a}{\partial x} = \frac{\chi \cosh\sqrt{\lambda}x}{\sinh\sqrt{\lambda}} > 0.$$

Thus the capillary tips migrate with increasing speed towards the tumour. This result agrees with the numerical simulations and is consistent with experimental measurements which show capillary tip velocities increasing from 0.2mms^{-1} to 0.8mms^{-1} as the new vasculature approaches the tumour [**60, 61**].

We now explain how to construct explicit solutions to equation (101). We transform the independent variables from (x,t) to (s,r) and assume that s and r parameterise arc length along characteristic curves and the initial data respectively. Then we deduce that

$$\frac{\partial n}{\partial s} = \frac{\partial n}{\partial t} + \chi \frac{\partial a}{\partial x}\frac{\partial n}{\partial x} = F(n, a).$$

Recalling that $b = (\frac{\chi}{\gamma})n\frac{\partial a}{\partial x}$, we deduce further that along characteristic curves

$$\frac{\partial b}{\partial s} = \frac{\partial b}{\partial t} + \chi \frac{\partial a}{\partial x}\frac{\partial b}{\partial x} = \frac{\chi}{\gamma}\frac{\partial a}{\partial x}\left(\frac{\partial n}{\partial s} + \chi n \frac{\partial^2 a}{\partial x^2}\right).$$

With $a = \sinh\sqrt{\lambda}x/\sinh\sqrt{\lambda}$, we conclude that at a stationary point for the capillary tips

$$\frac{\partial n}{\partial s} = 0 \Rightarrow \frac{\partial b}{\partial s} = \frac{\chi^2 n}{\gamma}\frac{\partial^2 a}{\partial x^2} = \frac{\lambda\chi^2}{\gamma} na > 0,$$

so that when n attains its maximum value b is still increasing. Similarly

$$\frac{\partial b}{\partial s} = 0 \Rightarrow \frac{\partial n}{\partial s} = -\chi\lambda na < 0,$$

so that where b attains its maximum n is decreasing. Noting now, that for a fixed value of x, s may be regarded as a time-like variable, these results prove, for the simplified model, that as the vascular front migrates towards the tumour the maximum capillary tip density precedes the maximum vessel density, a phenomenon that is observed experimentally [**111**].

To simplify the analysis we now assume that $0 < \lambda \ll 1$ and fix $\alpha_1 = 0$ in which case $c(x,t) \sim x$, $b = \chi n/\gamma$ and (101) simplifies to give (to leading order in λ):

(102)
$$\frac{\partial n}{\partial t} + \chi\frac{\partial n}{\partial x} = \left(\frac{\alpha_0\chi}{\gamma}\right) n \left(x - \frac{\alpha_2}{\alpha_0} n\right).$$

Since the characteristic curves associated with (102) are the straight lines $x - \chi t =$ constant, the solution structure for $n(x,t)$ differs depending on whether $x > \chi t$. In particular, it is possible to show that $n(x,t) = 0$ if $x > \chi t$ and that otherwise

(103)
$$n(x,t) = \exp\left[\frac{\alpha_0}{2\gamma}x^2 - k(t - x/\chi)\right]\left(1 + \frac{\alpha_2}{\chi\alpha_0}e^{-k(t-x/\chi)}\int_0^x e^{\alpha_0\xi^2/2\gamma}d\xi\right)^{-1}.$$

We note also that the tips reach the tumour at time $t = 1/\chi$, this being when the characteristic curve through the origin reaches $x = 1$. From (103) we have that

$$n(1, 1/\chi) = e^{\alpha_0/2\gamma}\left(1 + \frac{\alpha_2}{\chi\alpha_0}\int_0^1 e^{\alpha_0\xi^2/2\gamma}d\xi\right)^{-1}.$$

Further, since

$$1 < \int_0^1 e^{\alpha_0\xi^2/2\gamma}d\xi < e^{\alpha_0/2\gamma},$$

we deduce that

$$n_L = \frac{e^{\alpha_0/2\gamma}}{1 + \frac{\alpha_2}{\chi\alpha_0}e^{\alpha_0/2\gamma}} < n(1, 1/\chi) < \frac{e^{\alpha_0/2\gamma}}{1 + \frac{\alpha_2}{\chi\alpha_0}} = n_U.$$

We now say that angiogenesis fails if $n(1, 1/\chi) < 1$. Using the above inequalities, we infer that angiogenesis fails if $n_U < 1$ or, equivalently, if

$$e^{\alpha_0/2\gamma} < 1 + \frac{\alpha_2}{\chi\alpha_0}$$

and that it is successful if $n_L > 1$ or

$$\frac{1}{1 - \alpha_2/\chi\alpha_0} < e^{\alpha_0/2\gamma}.$$

In conclusion, using our model we predict that the success, or failure, of angiogenesis is controlled by the balance between two parameter groupings: $\alpha_0/2\gamma$ and $\alpha_2/\chi\alpha_0$. As $\alpha_0/2\gamma$ increases (*i.e.* as the proliferation rate of TAF increases or the rate of vessel death decreases), angiogenesis is more likely to occur. Equally, as $\alpha_2/\chi\alpha_0$ increases (due to $\alpha_0\chi$ decreasing or α_2 increasing), angiogenesis is more likely to fail.

7.4. Multiscale modelling of vascular tumour growth

Thus far we have studied mathematical models that are deterministic and focus on a single spatial scale (usually the macro- of tissue scale). However, tumour growth, particularly that of vascularised tumours, is a complex, multiscale process in which a variety of inter-related phenomena act on widely varying spatial and temporal scales. Several authors have developed PDE models of vascular tumour growth. For example, in [**31**] the vasculature is treated as a distinct phase in a multiphase modelling framework. In [**16, 29**] models are developed to predict those tumour regions in which the immature angiogenic vessels will collapse. Other authors have developed models which account for the multiscale nature of solid tumour growth by coupling phenomena that act on different scales. For example, Gatenby *et al.* [**118**] modelled vascular tumour growth by coupling a cellular automata for cell-cell interactions with reaction-diffusion equations for chemicals such as oxygen and glucose which act on a macroscale (a fixed, non-adaptive vasculature acts as a distributed source of oxygen and glucose).

Recently Arakelyan *et al.* [**13, 14**] have developed a complex, multiscale model of vascular tumour growth that accounts for vessel maturation and the regression of immature and mature vessels, phenomena that depend on growth factors, such as VEGF and the angiopoietins, and pericytes (smooth muscle cells that attach to the outer wall of vessels, enabling them to alter blood flow by contracting and dilating). Model simulations suggest that rapid tumour growth will occur when the vessel maturation rate is high, due to the rapid increase in vessel density that this causes. By contrast, the model also predicts that lower vessel maturation rates in tumours with lower background vessel densities will create slower growing tumours that possess larger proportions of immature vessels and may exhibit oscillatory growth dynamics (when the immature vessels regress, perfusion is limited and a new cycle of angiogenesis is stimulated). Experimental results confirm the model predictions and, used in combination with the mathematical model, strengthen the hypothesis that variations in tumour growth rates may be strongly influenced by the strength of the angiogenic response elicited in the tumour environment.

Anderson and coworkers [**101, 102, 135, 136**] have developed an alternative hybrid model that incorporates blood flow and vascular remodelling. In this section we describe a multiscale model of vascular tumour growth which generalises the models in [**101, 102, 118, 135, 136**] by linking together phenomena that act on the subcellular, cellular and macro- or tissue scale. After describing the model, we present a series of numerical simulations and discuss the predictions they generate before outlining how the model may be extended in the future. Since space is limited, our discussion will be brief (see [**6, 7**] for further details).

Model development
Our multiscale model describes the growth of a tumour located in a region of healthy vascularised tissue. Physical mechanisms of interest include cell division, the expression of angiogenic factors such as VEGF, competition between normal and cancer cells, vascular remodelling, blood flow and oxygen transport. The model is formulated as a hybrid cellular automaton, with different submodels describing phenomena at the subcellular, cellular and macro- (or vascular) scales (see Figure 12). We summarise below the submodels that we use and, simultaneously, explain how phenomena on different scales interact.

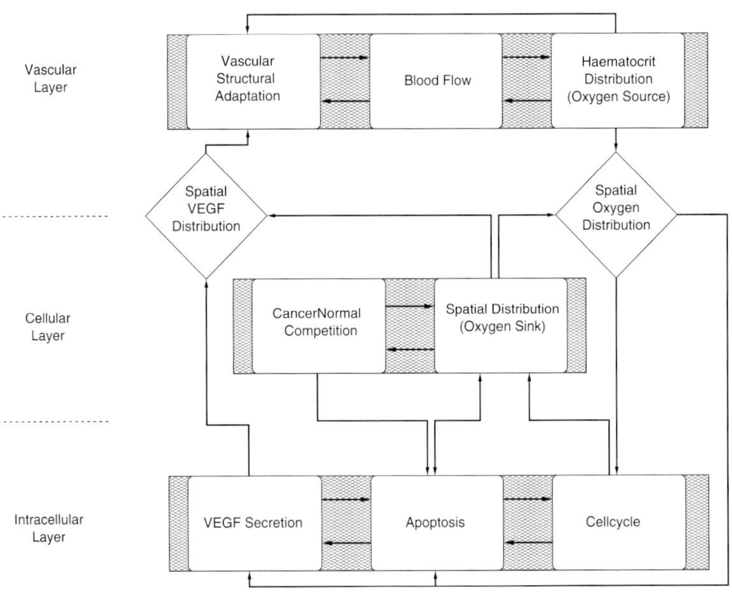

FIGURE 12. Schematic representation of our multiscale model of vascular tumour growth, reproduced from [**7**]. Copyright 2005, Society for Industrial and Applied Mathematics. Reprinted with permission. All right reserved.

The macroscale: vascular adaptation and blood flow
We consider a vascular structure composed of a regular hexagonal network and specify the initial network configuration (*i.e.* vessel radii and lengths). We impose a pressure drop across the vasculature, assuming that blood flows into (out of) the region through a single inlet (outlet) vessel. We use the Poiseuille approximation and Kirchoff's laws to determine the flow rates through, and pressure drops across, each vessel. Following Pries *et al.* [**124**] we assume that the radius $R(t)$ of a vessel evolves over a time period Δt as follows:

$$(104)\quad R(t+\Delta t) = R(t) + R(t)\Delta t\Bigg(\underbrace{\log\left(\frac{\tau_w}{\tau(\pi)}\right)}_{\substack{\text{haemodynamic}\\ \text{stimulus}}} + \underbrace{k_m(V)\log\left(\frac{\dot{Q}_{ref}}{\dot{Q}H}+1\right)}_{\substack{\text{metabolic}\\ \text{stimulus}}} - \underbrace{k_s}_{\substack{\text{shrinking}\\ \text{tendency}}} \Bigg).$$

In equation (104) $\dot{Q}$ is the flow rate, $\dot{Q}_{ref}$, k_m and k_s are constants, H is the haematocrit (red blood cell volume), $\tau_w = R\Delta\pi/L$ is the wall shear stress acting on a vessel of length L and π is the transmural pressure. The magnitude of the corresponding "set point" value of the wall shear stress, $\tau(\pi)$, is obtained from an empirical fit to experimental data. We assume that the metabolic stimulus increases as the haematocrit decreases. Coupling to the subcellular level is achieved by assuming that $k_m = k_m(V)$ where V denotes the local, extracellular VEGF concentration (see below). We account for the pro-angiogenic impact of VEGF by assuming that $k_m(V)$ is an increasing, saturating function, with

$$k_m(V) = k_m^0\left(1 + k_m^V\frac{V}{V_0+V}\right),$$

for constants k_m^0, k_m^V and V_0. Finally, the shrinking tendency is included so that, in the absence of mechanical and metabolic stimuli, the vessels shrink.

Once equation (104) has been used to update each vessel radius, the associated blood flow and haematocrit can be determined. When computing the blood flow we use a complex function of H and R which was determined empirically in [**124**] to describe its viscosity. When updating H, we assume that at branch points the red blood cells split in a manner which is proportional to the flow velocity in each daughter vessel [**70**].

The macroscale: the diffusible chemicals
Coupling between the vascular, cellular and subcellular scales is effected by the diffusive transport of oxygen and VEGF, their respective distributions within the tissue being governed by reaction-diffusion equations (RDEs) with zero-flux conditions on the boundaries of the computational domain. As in earlier sections, we make a quasi-steady approximation in the RDEs, the timescales for oxygen and VEGF diffusion being much shorter than the tumour doubling time which is the timescale of interest.

We denote by P the oxygen concentration and treat the cells as sinks and the vessels as sources of oxygen so that the relevant RDE can be written

$$(105)\qquad 0 = D_P\nabla^2 P + h_P(P_{vess} - P) - \lambda_{cell}P.$$

In (105), D_P denotes the assumed constant oxygen diffusion coefficient, h_P is the rate at which oxygen is transported across the vessel wall (h_P is only nonzero where vessels are located), $P_{vess} = P_{vess}(H)$ is the oxygen concentration associated with a haematocrit H (this is determined from the vascular problem). Finally, λ_{cell}

denotes the assumed constant rate at which cells consume oxygen, with different values for normal and cancerous cells.

The local VEGF profile is calculated in a similar manner, except that cells now act as sources of VEGF and vessels as sinks. If we further assume that VEGF is rapidly eliminated from the vasculature (so that its concentration in the vasculature V_{vess} is zero) and denote by λ_V its degradation rate, then it is straightforward to show that V satisfies

$$0 = D_V \nabla^2 V - h_V V + \gamma_{cell} - \lambda_V V. \tag{106}$$

In (106), D_V denotes the assumed constant diffusion coefficient of VEGF and h_V the rate at which VEGF crosses vessel walls. Finally, γ_{cell} represents the rate at which sites occupied by cells release their intracellular stores of VEGF to the extracellular environment (this occurs only when the internal levels of VEGF exceed a threshold value). As with λ_{cell} in equation (105), γ_{cell} differs between normal and cancerous cells.

The cellular level

The cell colony is viewed as a two-dimensional cellular automaton [**4, 110**]. Each element is characterised by a state vector, whose components correspond to features of interest. These include: (i) occupation status (whether an element is occupied by a normal cell, a cancer cell, an empty space or a vessel), (ii) cell status (whether the cell is in a proliferative or quiescent state), (iii) the local oxygen concentration, and (iv) the intracellular levels of VEGF, p53 and cell-cycle proteins (see below). The state vector evolves according to prescribed rules that update a given element on the basis of its status and those of its neighbours at the previous time step [**4**].

On each timestep elements of the cellular automaton are selected randomly and updated in turn. If a particular element is occupied by a cell, we update its internal dynamics (cell cycle proteins, VEGF and p53 levels – see below). If a cell is ready to divide, the daughter cell is placed in the (empty) adjacent cell with the largest oxygen concentration. If there is no empty site, the cell dies, without dividing. The mechanism for cell death differs between normal and cancer cells. In normal cells, if the intracellular level of p53 exceeds a threshold value (which depends on the cells in its nearest neighbourhood), then the cell dies. By contrast, under low oxygen, cancer cells become quiescent, suspending most cell functions, including proliferation. On entering this state, a clock is started and incremented each timestep that the cell remains quiescent. If the clock reaches a threshold value then the cell dies. However, if the oxygen level increases above the threshold for cancer cell quiescence before this time then the cell recommences cycling and its clock is reset to zero.

The subcellular level

We use ODEs to describe the evolution of the proteins that control the subcellular processes of interest. These include progress through the cell cycle and expression of VEGF and p53 [**5, 149**]. These variables are strongly coupled to behaviour at the cellular and macroscopic levels. For example, under low oxygen concentration p53 expression and, hence, rates of apoptosis are upregulated while progress through the cell cycle may be retarded or arrested. Equally, VEGF expression at the subcellular level impacts on vascular remodelling at the macroscale (see equation (104)).

Numerical algorithm
The following algorithm is used on each timestep to construct model solutions (for details, see [**7**]).

(1) Update the vascular network using the structural adaptation rules;
(2) Calculate the blood flow and haematocrit at all points within the vascular network;
(3) Update the oxygen and VEGF profiles by solving the relevant boundary value problems;
(4) Update the elements of the cellular automaton, and associated intracellular variables associated with each non-empty element according to whether it is occupied by a normal or cancerous cell.

7.5. Vascular tumour growth: numerical simulations and model predictions

Figures 13 to 16 show how, for a typical simulation, the spatial composition of the tumour and surrounding tissue changes over time and Figure 17 summarises the overall dynamics of the tumour cells. Since there is a single inlet and outlet to the vasculature, the incoming blood flow and haematocrit peak near these points and become diluted as they pass through the hexagonal lattice. Over time, the tumour cells proliferate and spread towards oxygen-rich regions. In regions where the number of tumour cells increases, the demand for oxygen exceeds the supply from the vasculature, and the cells become quiescent. These cells produce VEGF which diffuses through the tissue (see equation (106) and Figures 14, 15 and 16), stimulating vessel adaptation (see equation (104)) and biasing blood flow towards low oxygen regions. If the vasculature does not adapt to the VEGF stimulus quickly enough then the quiescent cells die. VEGF levels also decline and blood flow to the remaining tumour cells rises, enabling them to increase in number until their demand for oxygen exceeds that being supplied, and so the cycle repeats, with pronounced oscillations in the number of quiescent cells (see Figure 17). Also presented in Figure 17 are the results of a simulation which was identical in all respects except that vascular adaptation was independent of VEGF ($k_m(V) = k_m^0$, constant, in equation (104)). In both cases, the tumours grow to similar sizes. However, when $k_m(V) = k_m^0$ the approach to equilibrium is monotonic, the oscillations in the cell populations disappear and the number of quiescent cells is consistently much lower. These results show how coupling between the different spatial scales influences the tumour's growth dynamics and its spatial composition. In particular, the pronounced oscillations in the number of quiescent cells suggest that the time at which chemotherapy is administered could dramatically influence the tumour's response to treatment.

The simulations presented in Figure 18 illustrate further the sensitivity of the model predictions to the underlying assumptions. In Figure 18 we show how the growth dynamics of a tumour are influenced by the way in which the haematocrit is distributed in the vasculature. If it is heterogeneously distributed (as described above) then the tumour's growth is highly irregular and becomes localised in well-oxygenated regions. By contrast if the haematocrit is distributed uniformly through the vasculature then the tumour grows as a compact mass, expanding radially outwards.

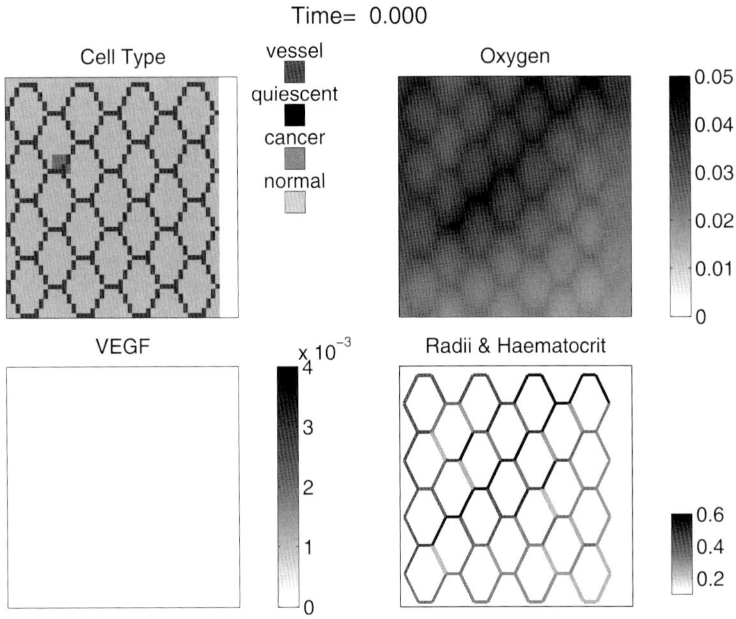

FIGURE 13. Series of panels showing the initial spatial distribution of the cells, the oxygen concentration, the VEGF concentration, and the radii and haematocrit associated with each branch of the vascular network, the width of the lines being proportional to the vessel radii and the colour intensity being indicative of the haematocrit (dark=high levels of H). From the plots on the right-hand side we observe that the oxygen is distributed nonuniformly across the domain, with the highest levels occurring near the vessel inlet (bottom left corner) and the outlet (top right corner) and corresponding to regions where the vessels are widest and the haematocrit largest. Since there are no quiescent cells, there is no VEGF present. The vessel radii, haematocrit and oxygen are maximal along the diagonal connecting the vessel inlet and outlet.

Taken together, the results presented above highlight the need for accurate data and information about the behaviour of tumours for model building in order that the resulting models generate accurate predictions.

7.6. Conclusions

In this section we have presented and analysed a one-dimensional model of tumour angiogenesis and a hybrid model of vascular tumour growth and shown how each model can be used to provide insight into these crucial phases of tumour growth. We now explain how the models have been developed and discuss their strengths and weaknesses, introducing alternative modelling approaches where appropriate.

Since Balding and McElwain's model of tumour angiogenesis was developed, more detailed knowledge of the specific chemicals and biochemistry involved has

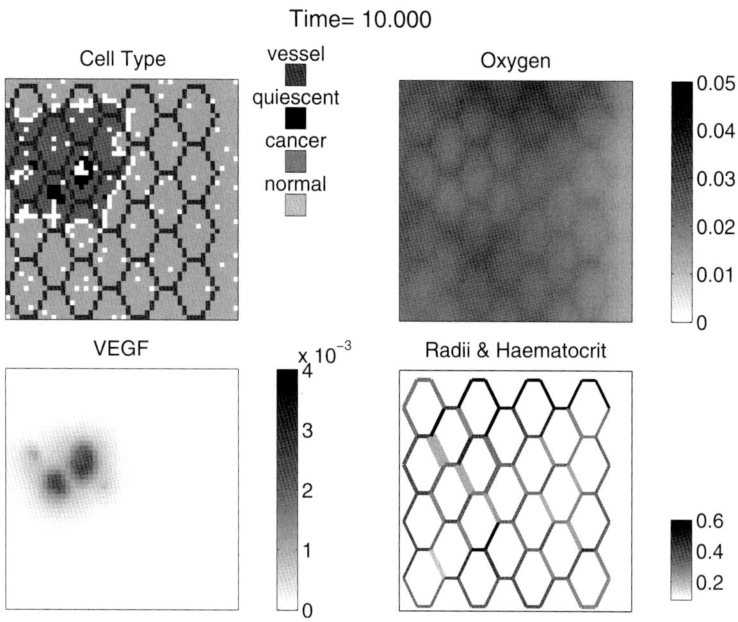

FIGURE 14. Series of plots showing how the initial conditions presented in Figure 13 have developed at $t = 10$ (dimensionless time unit). The tumour has increased in size and now contains quiescent cells which produce small amounts of VEGF. The oxygen and vessel profiles remain unchanged.

been elucidated. Following Orme and Chaplain [**113**], Levine and coworkers [**99**] developed complex models that account for much of this detail and, as such, these represent an important step towards understanding angiogenesis using biochemically-based rather than phenomenological arguments. For example, Levine and coworkers account for interactions between endothelial cells, angiogenic factors and cells such as pericytes and macrophages that are also involved in angiogenesis. One important difference between the models of angiogenesis in [**20, 99**] is the treatment of the endothelial cells: in [**99**] endothelial cell movement is based on the theory of reinforced random walks developed in [**114**], rather than diffusive movement. Preziosi and coworkers [**71, 129**] focus on the formation of vascular networks *in vitro* and, using Burger's Equation to describe the motion of individual endothelial cells, are able to reproduce experimental observations in which the cells migrate and aggregate on a two-dimensional gel to form a branched network.

The models of angiogenesis we have studied do not account for nutrient delivery to the tissue by the developing vasculature and cease to apply once the capillary tips reach the tumour. Recently, Breward *et al.* [**31**] have used a multiphase modelling framework (see Section 6) to develop continuum models of vascular tumour growth that address these shortcomings.

Whilst it is straightforward to extend Balding and McElwain's model to two and three space dimensions [**113**], the resulting models highlight some of the shortcomings of using a continuum framework to study angiogenesis. This is primarily

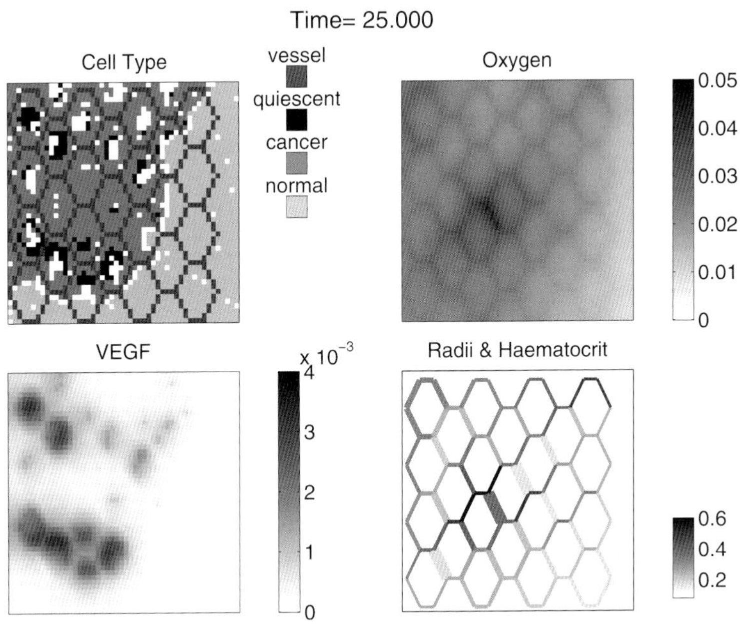

FIGURE 15. Series of plots showing how the initial conditions presented in Figure 13 and 14 have developed at $t = 25$ (dimensionless time unit).

because angiogenesis is a multi-dimensional process, with tips sprouting in directions other than that of the propagating vascular front. Further, it is not obvious how the snail trail should be generalised in higher space dimensions. In addition, because the dependent variables are densities, the models do not discriminate between different vascular morphologies (e.g. a region perfused by one, large vessel would be identical to a region perfused by many small vessels, even though their surface areas and, hence, nutrient delivery rates may differ markedly). Moreover the models do not distinguish between anastomosis and capillary tip death, even though the former will increase nutrient supply to the tissue while the latter will not. Finally, no account is taken of vascular remodelling and, in particular, the impact of blood flow and haematocrit on the evolving vasculature. In spite of these weaknesses, the models provide useful insight into the ways in which different physical mechanisms (e.g. the strength of the chemotactic response, the rate of TAF production, the rate of capillary tip formation) influence angiogenesis. In particular, the success of angiogenesis is tightly controlled by the balance between endothelial cell proliferation and migration: as the strength of the chemotactic response increases, the tips migrate towards the tumour more rapidly, thereby reducing both the time available for tip proliferation and the density of the vasculature when the tumour is reached.

Possibly the main criticism of the above models of angiogenesis stems from their inability to track individual capillary tips and, hence, to reproduce accurately the patterns of vascular growth observed during tumour angiogenesis. Anderson and Chaplain [**10, 11**] developed a continuum-discrete model in which the movement

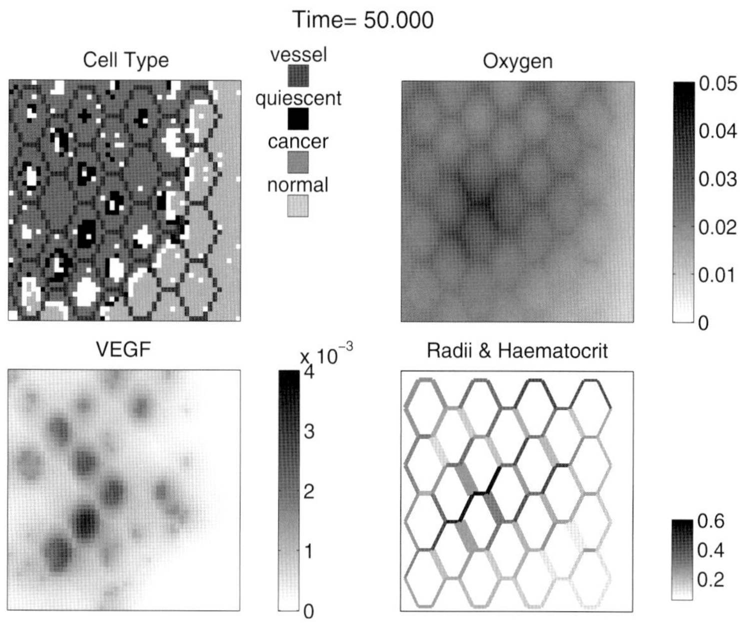

FIGURE 16. Series of plots showing how the simulation presented in Figures 13, 14 and 15 has developed at $t = 50$ (dimensionless time unit). The tumour continues to expand into the tissue region. The number of quiescent cells is now large enough to elicit an angiogenic response, with VEGF localised in a neighbourhood of the quiescent cells (compare the cell and VEGF profiles on the left hand side). The profiles on the right hand side show that the VEGF has caused remodelling of the vasculature, with blood flow and oxygen supply (haematocrit) being directed primarily towards the tumour mass.

of individual vessel tips was modelled by means of a biased random walk whose transition probabilities were derived by discretisation of a PDE model previously developed. In this work the role of haptotaxis (migration upwards gradients of adhesive molecules, in particular fibronectin) was examined and shown to be a key factor in successful angiogenesis. In an alternative approach, Stokes and Lauffenburger modelled the velocity of individual endothelial cells as a persistent random walk, using the Ornstein-Uhlenbeck process and including a directional bias term to account for chemotactic movement [**138**].

We turn now to the multiscale model of vascular tumour growth. In comparison with the earlier model of angiogenesis, this model is undoubtedly complex and, while the underlying submodels may be amenable to analytical investigations, solution of the full model must be numerical. Even so, we have shown how such a model can be used to generate experimental predictions and how these predictions depend on the underlying modelling assumptions (*e.g.* whether the haematocrit is uniformly distributed through the vasculature and whether vascular adaptation is sensitive to VEGF). While there are many ways in which the hybrid model can be extended, one of the most important involves introducing cell movement, of tumour

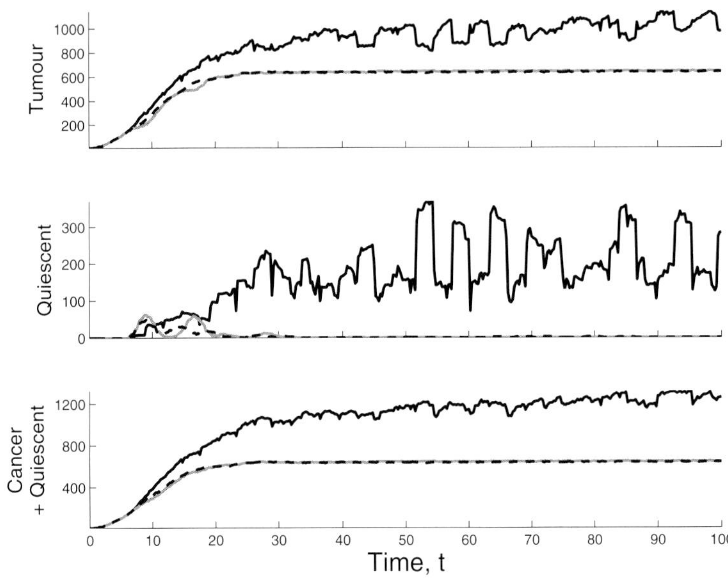

FIGURE 17. Series of curves showing how, for the simulation presented in Figures 13 to 16 the numbers of proliferating (upper panel), quiescent tumour (middle panel) and total number of tumour cells (lower panel) change over time. While the number of proliferating cells increases fairly steadily, the dynamics of the quiescent cells are more complex, initially undergoing oscillations of increasing amplitude. For $t > 30$, the tumour is sufficiently large that the quiescent cells are never eliminated (quiescent cells that die are replaced by proliferating cells that become quiescent). The dashed and grey lines show the evolution of tumours which are identical in all respects except for the extent to which their vasculature is regulated by VEGF: the grey line corresponds to a case in which the response to VEGF is 10 times smaller than that for the solid line; the dashed line to a case in which the response to VEGF is absent, so that $k_m^V = 0$ in equation (104). In both of these cases the size to which the tumour evolves is similar and smaller than that obtained when vascular adaptation is strong (solid line); the oscillations in the cell populations disappear and the number of quiescent cells is negligible.

cells (to account for invasion) and endothelial cells (to account more realistically for angiogenesis). Regarding the former, Anderson has developed hybrid models in which the tumour cells produce enzymes that degrade the host's extracellular matrix, liberating space into which the tumour cells can invade, showing how heterogeneity of the underlying extracellular matrix can influence the tumour's pattern of invasion [**12**]. In more recent work, Anderson and coworkers have embedded this invasion model within a multiscale framework that includes componenets at the molecular, cellular and tissue levels [**19**]. The inclusion of a detailed and realistic description of angiogenesis is, perhaps, more challenging: in practice, capillary tips

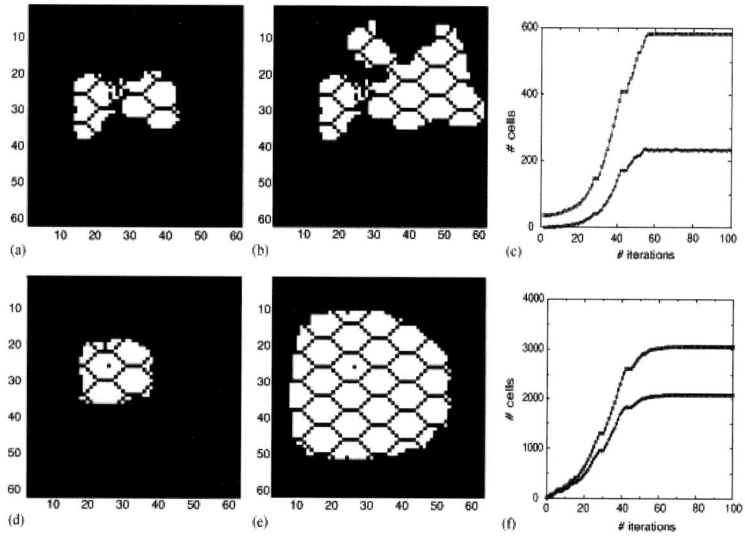

FIGURE 18. Series of profiles showing how the pattern of tumour invasion depends on the way in which the vessels deliver oxygen to the tissue. When the haematocrit is distributed uniformly through the vasculature (lower figures), the tumour grows as a compact, almost circular mass and eventually colonises the entire tissue. When the haematocrit is distributed heterogeneously (upper figures), the tumour's growth is more irregular and confined to those regions of the tissue where oxygen levels are highest. In panels (a), (b), (d) and (e), white spaces represent tumour cells and black spaces are either empty or occupied by vessels. Panels (c) and (f) show the time evolution of the number of cancer cells for nonuniform and uniform oxygen distribution respectively. The upper curves (squares) represent the total number of cancer cells (proliferating and quiescent) and the lower curves (diamonds) represent the number of quiescent tumour cells. In both simulations the normal cells are rapidly eliminated by the tumour population and therefore do not appear. Reproduced, with permission, from [**5**].

are blunt-ended and do not permit blood flow until the capillaries have fused, or anastomosed, with other vessels and/or capillary tips. Further, such blunt-ended capillary tips will regress if they do not establish a circulating flow within a fixed time period. An additional mechanism that we may wish to include is vasculogenesis. A process more usually associated with embryonic development, it involves the formation of new blood vessels from (stem-cell-like) endothelial progenitor cells which are recruited from the bone marrow, via the blood stream, to the tumour. On extravasation, the endothelial progenitor cells proliferate, producing endothelial cells that align to form new vessels [**94, 139**]. There is currently considerable debate about the relative importance of vasculogenesis as a mechanism for new blood vessel formation and its implications for treating and managing tumours with anti-angiogenic and/or anti-vascular agents. Thus, a related, and equally important, area involves using the multiscale model first to predict a patient's response to a course of chemotherapy [**42, 125**] (or radiotherapy) and then to determine whether better protocols exist. The model could also be used to predict a patient's response to treatments involving combinations of drugs and, in consequence, to address questions such as whether it is better to administer an anti-angiogenic drug before a chemotherapeutic agent.

7.7. Exercises

1. Chemicals produced in the centre of a one-dimensional wound that occupies the region $0 \le x \le L$ stimulate endothelial cells to proliferate and migrate towards the wound centre. We model the early stages of this process by writing equations for the chemical attractant $a(x,t)$ and the endothelial cell density, $n(x,t)$:

$$\frac{\partial a}{\partial t} = D\frac{\partial^2 a}{\partial x^2} - \lambda a, \qquad \frac{\partial n}{\partial t} = -\frac{\partial}{\partial x}\left(\chi(a)n\frac{\partial a}{\partial x}\right).$$

We close these equations by imposing the following boundary and initial conditions:

$$a = 1 \text{ at } x = 0, \qquad a = 0,\ n = 1 \text{ at } x = L,$$

$$a(x,0) = a_{in}(x),\ \ n(x,0) = 0 \text{ for } x \in [0,L).$$

We suppose that the chemical diffuses very rapidly compared to endothelial cell proliferation and migration and that chemical decay is extremely slow. Accordingly we set $\frac{\partial a}{\partial t} = 0 = \lambda$ in the PDE for $a(x,t)$. Determine the steady state profile for a and show that, in this case, if $\chi =$ constant then the PDE for $n(x,t)$ becomes

$$\frac{\partial n}{\partial t} - \frac{\chi}{L}\frac{\partial n}{\partial x} = 0.$$

Show further (by direct substitution or by using the method of characteristics) that $n(x,t) = N(z)$ where $z = x + \chi t/L$ is a general solution of the reduced PDE for n.

Sketch the solution profiles for $n(x,t)$, distinguishing between times $0 < t < L^2/\chi$ and $t > L^2/\chi$. What is the effect of increasing χ or decreasing L on the wound healing time?

Remarks:

- The cells migrate with constant velocity $v = -\chi/L$ into the wound space, providing the region with new blood vessels.
- The cells reach $x = 0$ at $t = L^2/\chi$ (=time at which vessels restored to wounded region)
- For $t > L^2/\chi$, $n(x,t) = 0\ \forall\ x \in [0,L]$ i.e. the cells evolve to a spatially uniform steady state in finite time $t = L^2/\chi$. Hence the profile for $n(x,t)$ is different when $0 < t < L^2/\chi$ and $t \ge L^2/\chi$.
- Physically speaking, the model breaks down when $t = L^2/\chi$: once there is a connected blood supply, oxygen levels will be restored to normal levels,

chemical production will cease, and cell migration and proliferation will halt.

2. We now consider the effect of endothelial cell migration on wound healing. We proceed as in the previous question but consider instead the following equations:

$$0 = \frac{\partial^2 a}{\partial x^2}, \qquad \frac{\partial n}{\partial t} = -\frac{\partial}{\partial x}\left(\chi n \frac{\partial a}{\partial x}\right) + \alpha a n,$$

$$a = 1 \text{ at } x = 0, \qquad a = 0,\ n = 1 \text{ at } x = L,$$

$$a(x,0) = a_{in}(x),\ \ n(x,0) = 0 \text{ for } x \in [0, L),$$

where α denotes the rate of endothelial cell proliferation. By solving for $a(x,t)$ and then eliminating it from the PDE for $n(x,t)$, show that the problem reduces to the following PDE for $n(x,t)$:

$$\frac{\partial n}{\partial t} - \frac{\chi}{L}\frac{\partial n}{\partial x} = \alpha\left(1 - \frac{x}{L}\right) n.$$

Use the method of characteristics to show that

$$n(x,t) = \begin{cases} 0 & \text{for } 0 < x + \chi t/L < L, \\ \exp\left\{\frac{\alpha L^2}{2\chi}\left(1 - \frac{x}{L}\right)^2\right\} & \text{for } L < x + \chi t/L. \end{cases}$$

3. We now consider the combined effects of cell proliferation and enhanced chemotaxis on wound healing. The equations of interest are

$$0 = D\frac{\partial^2 a}{\partial x^2}, \qquad \frac{\partial n}{\partial t} = -\frac{\partial}{\partial x}\left(\chi a n \frac{\partial a}{\partial x}\right) + \alpha a n,$$

$$a = 1 \text{ at } x = 0, \qquad a = 0,\ n = 1 \text{ at } x = L,$$

$$a(x,0) = a_{in}(x),\ \ n(x,0) = n_{in}(x) \text{ for } x \in [0, L).$$

By solving for and then eliminating $a(x)$, show that the model reduces to solving the following PDE for $n(x,t)$:

$$\frac{\partial n}{\partial t} - \frac{\chi}{L}\left(1 - \frac{x}{L}\right)\frac{\partial n}{\partial x} = \left[\alpha\left(1 - \frac{x}{L}\right) - \frac{\chi}{L^2}\right] n.$$

where we now specify

$$n(x,0) = \begin{cases} 0 & \text{for } 0 \leq x < x^* \\ \dfrac{x - x^*}{L - x^*} & \text{for } x^* \leq x \leq L. \end{cases}$$

Use the method of characteristics to show that where $n(x,t) \neq 0$ we have

$$n(x,t) = \left(1 - \frac{(L-x)}{(L-x^*)}\, e^{-\chi t/L^2}\right) \exp\left\{\frac{\alpha L^2}{\chi}\left(1 - \frac{x}{L}\right)\left(1 - e^{-\chi t/L^2}\right) - \frac{\chi t}{L^2}\right\}.$$

Comment on the form of the steady state solutions and the time taken to attain them.

4. Let $c(x)$ and $m(x,t)$ denotes the concentration of a chemical attractant and the density of infiltrating macrophages at position x and time t. You are given that

$$0 = \frac{d^2c}{dx^2} + 1 - x, \qquad \frac{\partial m}{\partial t} = -\chi \frac{\partial}{\partial x}\left(m\frac{dc}{dx}\right),$$

$$\frac{dc}{dx}(0) = 0, \quad c(1) = 0, m(x,0) = 0 \text{ for all } x \in [0,1), \quad m(1,t) = 1.$$

Determine the steady state profile for the chemical attractant $c(x)$.

By substituting for $c(x)$ in the PDE for $m(x,t)$, deduce that $m(x,t)$ solves

$$\frac{\partial m}{\partial t} - \chi x\left(1 - \frac{x}{2}\right)\frac{\partial m}{\partial x} = \chi(1-x)m.$$

Use the method of characteristics to show that

$$\frac{x}{1 - x/2}e^{\chi t}$$

is constant along characteristic curves and construct explicit solutions for $m(x,t)$, stating clearly in which regions of (x,t) space your solutions apply. Use your solutions to describe the long time behaviour of the cells.

8. Summary and future directions

There is now an extensive (and rapidly growing) theoretical literature devoted to modelling solid tumour growth. It embraces so many branches of applied mathematics that it is not possible to do justice to all of them in limited space. In this lecture series we have, therefore, focussed on continuum and deterministic models, selecting for discussion those which we believe best illustrate the type of insight that mathematical modelling of this type can provide while also showing how the field has developed over the past twenty years and as our understanding of the underlying biology has progressed. We reviewed a variety of models, ranging from spatially-averaged ODE models that can be used to describe the response of tumours to chemo- and radiotherapy (Section 3) to spatially-structured moving boundary problems that can be used to describe the radially-symmetric development of early avascular tumours in response to a single, growth-rate limiting diffusible species (Section 4) and their stability to symmetry-breaking perturbations (Section 5). We also explained how a multiphase framework can be used to account for cellular heterogeneity and to permit a fuller investigation of the way in which mechanical effects (*e.g.* pressure and stress) may influence the tumour's development (Section 6). In Section 7 we presented a deterministic model of tumour angiogenesis, before outlining a hybrid model of vascular tumour growth.

An important area of theoretical research that we have not reviewed is tumourigenesis, the process by which cells acquire the genetic mutations that are needed for a tumour to develop, the number and type of genes required remaining open questions [**85**]. The earliest models of tumourigenesis were stochastic and developed to investigate the age-specific incidence rates for certain adult cancers [**18**]. These multistage models were extended to distinguish between inherited and spontaneous cancers [**93, 105**] and adapted to investigate the effects of apoptosis [**146**] and variable mutation rates [**147**]. More recently, Plotkin and Nowak [**122**] developed a stochastic model that investigates the extent to which loss of DNA repair genes and tumour suppressor genes contribute to tumourigenesis. For a comprehensive review of models that have been used to describe these early stages of tumour initiation we refer to the recent book by Wodarz and Komarova [**157**] and the article by van Leeuwen and coworkers [**98**] in which existing models of colorectal cancer initiation and growth are reviewed. Another modelling approach that may be particularly relevant to early tumour growth, when cell numbers are so small that a continuum model may no longer be valid, involves treating each cell as a separate and discrete entity. Probabilistic models have been used to describe the behaviour of individual cells and their interactions with neighbouring cells use concepts ranging from Markov chain processes [**82**], through cellular automata [**4, 56, 110, 118**] to stochastic energy minimisation techniques [**140, 150**].

Two other important areas of research that we are unable to discuss in detail are tumour invasion and interactions between the tumour and the immune system. In addition to the discrete hybrid models mentioned in Section 7 [**12, 19, 118**], a range of deterministic models have been developed [**48, 49, 119, 130, 148, 154**]. For example, in a series of papers in which reaction-diffusion models are used to describe competition between tumour cells and their normal counterparts, Gatenby and coworkers show how tumour cells obtain survival and invasive advantages. In particular, under low oxygen the cancer cells switch to anaerobic respiration, producing lactate which reduces the local pH to levels that the normal cells cannot tolerate. The normal cells die, liberating space into which the tumour cells can invade [**73, 74, 133**]. Indeed, these models predicted the existence of a hypocellar gap between the tumour and normal cells which was subsequently identified experimentally. The complexity of some of the above models of invasion mean that they are not typically amenable to analysis. By contrast, Marchant and coworkers have developed highly idealised models of invasion which admit biologically relevant, stable steady travelling wave solutions characterised by a jump discontinuity in the cancer cell concentration. These models can be analysed with non-standard phase plane analysis [**106**] and used to explain experimental observations in which tumour cells invade dense tissues at the same speed as they invade less dense tissue, with maximal invasion seen for intermediate tissue densities [**107, 119**].

Turning now to the role of interactions between the immune system and the tumour, in partnership with experimental colleagues [**112**], Byrne and coworkers have developed PDE models to investigate the feasibility of genetically engineering macrophages to target hypoxic tumour regions in which macrophages are known to localise [**91, 115**]. The aim is to manipulate the macrophages so that they are activated to release anti-tumour agents (either cytotoxic compounds or anti-angiogenic drugs) when they experience the low oxygen that is unique to tumours growing *in vivo*. In order to account in more detail for tumour-immune system interactions which take place at the cellular level, Bellomo and colleagues have adopted a different approach, using ideas from kinetic theory, originally developed in statistical mechanics, to construct evolution equations for tumour cells, host and immune cells which are based on mass conservation principles. The resulting models are typically formulated as systems of integro-differential equations which, under appropriate closure assumptions, reduce to ODE models similar in form to those studied in Section 3. This approach was initially proposed by Bellomo and Forni [**27**] and was subsequently developed in a range of articles that include [**26, 28**].

For further details, we refer the interested reader to the following pair of review articles [**23, 25**].

We anticipate that in the future, multiscale models, like the hybrid model of vascular tumour growth presented in Section 7 will become more widespread, such models being needed to explore the ways in which phenomena acting on different scales interact and to integrate experimental data collected at these widely varying spatial (and temporal) scales [**24**]. The development of these models poses many new theoretical challenges, important issues being to establish the level of detail that is needed at a particular spatial scale. For example, in order to predict accurately the macroscale dynamics of an avascular tumour cultured *in vitro* is it necessary to include a detailed model of the cell cycle or would similar results be obtained when a simple model is used? Another open problem concerns the development of appropriate (and robust) numerical and analytical techniques to investigate the behaviour of the resulting multiscale models.

An obvious criticism of the models that we have reviewed is that they are all generic in nature: the geometry is highly idealised, the models are not specialised to a particular cell line or tissue and the identity of the growth factors is not defined. As the field matures and a better understanding of the fundamental mechanisms that regulate tumour growth is gained, the scope for developing more specific models will increase. With this in mind, Swanson and coworkers have developed three dimensional models that describe the spread of gliomas through the brain [**142, 143, 144**]. While the mathematical model that they use is highly idealised (Fisher's equation describes how the tumour cells migrate and proliferate), they have gone to considerable efforts to replicate the detailed anatomical structure of the brain and to validate/calibrate their models against MRI scans. In particular, the random motility and proliferation rates of the tumours are assumed to differ between grey and white matter. As a result, their models show excellent agreement with clinical data and are being used to predict the response of patients to different treatment protocols.

On a different note, Franks and coworkers have developed a series of models to describe ductal carcinoma in situ (DCIS). DCIS represents the early, non-invasive stage of breast cancer during which the tumour cells are confined by an intact but compliant duct wall [**66, 67, 68**]. The resulting models involve coupling existing models of avascular tumour growth in a cylindrically-symmetric tube with mechanical models for the finite deformation of the compliant membrane that confines the tumour, the coupling being mediated by interactions between the expansive forces created by tumour cell proliferation and the stresses that develop in the compliant membrane. The models provide insight into DCIS progression and also generate several hypotheses that could be tested experimentally. For example, the pressure on the duct wall is likely to be greatest at the centre of the tumour and, hence, this is where the duct wall is likely to be breached [**67**]. The model is also used to test two hypotheses for the localisation, near the duct wall, of proteases that compromise its integrity. Model simulations and asymptotic analysis suggest that the elevated pressure that accompanies tumour growth is more likely to stimulate protease production and localisation near the duct wall than hypoxia [**68**].

In conclusion, we believe that theoretical modelling of solid tumour growth is an exciting and challenging field of mathematical biology. Numerous phenomena are of interest, from stochastic effects relating to the mutation or survival of individual cells to mechanical effects which may not only influence how a growing tumour interacts with its surroundings [**49, 83**] but may also create difficulties in, say, delivering drugs against an adverse pressure gradient [**159**]. While most of the existing models are generic in nature, they are able to provide useful insight into the mechanisms that control the different phases of tumour growth and also to generate new models which are of independent mathematical interest. We believe that if such mathematical investigations are to be of clinical value the underlying theoretical models must become highly specialised by, for example, focussing on particular tumours, signalling pathways and treatment protocols. Equally advances in biotechnology mean that parts of the cancer jigsaw detailing subcellular, cellular and macroscopic phenomena are now being identified. While it is not currently clear how to integrate all the available data, we believe that mathematics represents an ideal framework within which to establish how the pieces of the cancer jigsaw interlock. Additionally, we believe that formulating such models will, in itself, stimulate new directions for theoretical research in applied mathematics, in particular with regard to the development and analysis of multiscale and hybrid models and the design of efficient, accurate and robust numerical methods for performing simulations [**52, 75, 158**].

Bibliography

[1] J.A. Adam (1986). A simplified mathematical model of tumour growth, *Math. Biosci.* 81: 229-242.

[2] J.A. Adam (1987). A mathematical model of tumour growth. II Effects of geometry and spatial uniformity on stability. *Math. Biosci.* 86: 183-211.

[3] J.A. Adam and N. Bellomo (1997). A survey of models for tumour immune system dynamics. Birkhauser Press.

[4] T. Alarcón, H.M. Byrne and P.K. Maini (2003). A cellular automaton model for tumour growth in inhomogeneous environment. *J.theor.Biol.* 225: 257-274.

[5] T. Alarcón, H.M. Byrne and P.K. Maini (2003). A mathematical model of the effects of hypoxia on the cell-cycle of normal and cancer cells. *J. Theor. Biol.* 229: 395-411.

[6] T. Alarcón, H.M. Byrne and P.K. Maini (2004). Towards whole-organ modelling of tumour growth. *Prog. Biophys. Mol. Biol.* 85: 451-472.

[7] T. Alarcón, H.M. Byrne and P.K. Maini (2005). A multiple scale model for tumour growth. *SIAM J. Multiscale Mod. & Sim.* 3: 440-475.

[8] T. Alarcón, H.M. Byrne and P.K. Maini (2005). A design principle for vascular beds: the effects of complex blood rheology. *Microvasc. Res.* **69**: 156-172.

[9] D. Ambrosi and L. Preziosi (2002). On the closure of mass balance models for tumor growth, *Math. Models Methods Appl. Sci.* 12: 737-754.

[10] A.R.A. Anderson and M.A.J. Chaplain (1998). A mathematical model for capillary network formation in the absence of endothelial cell proliferation. *Appl. Math. Lett.* **11**: 109-114.

[11] A.R.A. Anderson and M.A.J. Chaplain (1998). Continuous and discrete mathematical models of tumour-induced angiogenesis. *Bull. Math. Biol* **60**: 857-899

[12] A.R. Anderson (2005). A hybrid mathematical model of solid tumour invasion: the importance of cell adhesion. *Math. Med. Biol.* 22: 163-86.

[13] L. Arakelyan, Y. Merbl and Z. Agur (2005). Vessel maturation effects on tumour growth: validation of a computer model in implanted human ovarian carcinoma spheroids. *Eur. J. Cancer* 41: 159-167.

[14] L. Arakelyan, V. Vainstein and Z. Agur (2002). A computer algorithm describing the process of vessel formation and maturation, and its use for predicting the effects of anti-angiogenic and anti-maturation therapy on vascular tumor growth. *Angiogenesis* 5: 203-14.

[15] R.P. Araujo and D.L.S. McElwain (2004). A history of the study of solid tumor growth: the contribution of mathematical modelling. *Bull. Math. Biol.* 66: 1039-1091.

[16] R.P. Araujo and D.L.S. McElwain (2004). New insights into vascular collapse and growth dynamics in solid tumours. *J. Theor. Biol.* 228: 335-346.

[17] R.P. Araujo and D.L.S. McElwain (2004). A linear-elastic model of anisotropic tumour growth. *Eur. J. Appl. Math.* 15: 365-384.

[18] P. Armitage and R. Doll (1954). The age distribution of cancer and multistage theory of carcinogenesis. *Br. J. Cancer* 8: 1-12.

[19] B.P. Ayati, G.F. Webb and A.R.A. Anderson (2006). Computational methods and results for structured multiscale models of tumor invasion. *Multiscale Model. Simul.* 5: 1-20.

[20] D. Balding and D.L.S. McElwain (1985). A mathematical model of tumour-induced capillary growth. *J. theor. Biol.* 114: 53-73

[21] B. Basse, B.C. Baguley, E.S. Marshall, W.R. Joseph, B. van Brunt, G. Wake and D.J. Wall (2003). A mathematical model for analysis of the cell cycle in cell lines derived from human tumors. *J. Math. Biol.* 47: 295-312.

[22] B. Basse, B.C. Baguley, E.S. Marshall, G. Wake and D.J. Wall (2005). Modelling the flow [corrected] cytometric data obtained from unperturbed human tumour cell lines: parameter fitting and comparison. *Bull. Math. Biol.* 67: 815-30.

[23] N. Bellomo and E. de Angelis (1998). Strategies of applied mathematics towards an immuno mathematical theory on tumours and immune system interactions. *Math. Mod. Methods Appl. Sci.* 8: 1403-1429.

[24] N. Bellomo, E. de Angelis and L. Preziosi (2003). Multiscale modelling and mathematical problems related to tumor evolution and medical therapy. *J. Theor. Med.* 5: 111–136.

[25] N. Bellomo, A. Bellouquid and E. de Angelis (2003). The modelling of the immune competition by generalized kinetic (Boltzmann) models: review and research perspectives. *Math. Comp. Modelling* 37: 65–86.

[26] N. Bellomo, B. Firmani and L. Guerri (1999). Bifurcation analysis for a nonlinear system of integro-differential equations modelling tumor immune cells competition. *Appl. Math. Letters* 12: 39–44.

[27] N. Bellomo and G. Forni (1994). Dynamics of tumour interaction with the host immune system. *Math. Comp. Modelling* 20: 107–122.

[28] N. Bellomo, L. Preziosi and G. Forni (1996). On a kinetic cellular theory for competition between tumours and the host immune system. *J. Biol. Systems* 4: 476-502.

[29] Y. Boucher and R.K. Jain (1992). Microvascular pressure is the principal driving force for interstitial hypertension in solid tumours: implications for vessel collapse. *Cancer Res.* 52: 5110-5114.

[30] C.J.W. Breward, H.M. Byrne and C.E. Lewis (2002). The role of cell-cell interactions in a two-phase of solid tumor growth, *J. Math. Biol.* 45: 125-152.

[31] C.J.W. Breward, H.M. Byrne and C.E. Lewis (2003). A multiphase model describing vascular tumour growth. J. Math. Biol. 65: 609-640.

[32] A.C. Burton (1996). Rate of growth of solid tumours as a problem of diffusion, *Growth* 30: 157-176.

[33] H.M. Byrne (1997). The importance of intercellular adhesion in the development of carcinomas, *IMA J. Math. Appl. Med. Biol.* 14: 305-323.

[34] H.M. Byrne (1999). A weakly nonlinear analysis of a model of avascular solid tumour growth, *J. Math. Biol.* 33: 59-89.

[35] H.M. Byrne and M.A.J. Chaplain (1995). Mathematical models for tumour angiogenesis: numerical simulations and nonlinear wave solutions. *Bull. Math. Biol.* 57: 461-486 .

[36] H.M. Byrne and M.A.J. Chaplain (1995). Growth of non-necrotic tumours in the presence and absence of inhibitors, *Math. Biosci.* 130: 151-181.

[37] H.M. Byrne and M.A.J. Chaplain (1995). Growth of necrotic tumours in the presence and absence of inhibitors, *Math. Biosci.* 131: 187-216.

[38] H.M. Byrne and M.A.J. Chaplain (1997). Free boundary value problem associated with the growth and development of multicellular spheroids, *Eur. J. Appl. Math.* 8: 639-658.

[39] H.M. Byrne and M.A.J. Chaplain (1998). Necrosis and apoptosis: Distinct cell loss mechanisms in a mathematical model of solid tumour growth, *J. Theor. Med.* 1: 223-236.

[40] H.M. Byrne, J.R. King, D.L.S. McElwain and L. Preziosi (2003). A two-phase model of solid tumor growth, *Appl. Math. Lett.* 16: 567-573.

[41] H.M. Byrne and P.C. Matthews (2002). Asymmetric growth of avascular tumours: exploiting symmetries, *IMA J. Math. Appl. Med. Biol.* 19: 1-29.

[42] H.M. Byrne, M.R. Owen, T. Alarcón, J. Murphy and P.K. Maini (2006). Modelling the response of vascular tumours to chemotherapy. *Math. Mod. Meth. Appl. Sci.* 16: 1219-1241.
[43] H.M. Byrne and L. Preziosi (2003). Modelling solid tumor growth using the theory of mixtures, *IMA J. Math. Appl. Med. Biol.* 20: 341-366.
[44] M.A.J. Chaplain, M. Ganesh and I. Graham (2001). Spatio-temporal pattern formation on spherical surfaces: numerical simulation and application to solid tumour growth, *J. Math. Biol.* 42: 387-423.
[45] M.A.J. Chaplain and B.D. Sleeman (1993). Modelling the growth of solid tumours and incorporating a method for their classification using nonlinear elasticity theory. *J. Math. Biol.* 31: 431-473.
[46] M.A.J. Chaplain and A.M. Stuart (1993). A model mechanism for the chemotactic response of endothelial cells to tumour angiogenesis factor. *IMA J. Math. Appl. Med. Biol.*, 10: 149-168.
[47] M.A.J. Chaplain, S.M. Giles, B.D. Sleeman and R.J. Jarvis (1995). A mathematical analysis of a model for tumour angiogenesis. *J. Math. Biol.* 33: 744-770.
[48] M.A.J. Chaplain and G. Lolas (2005). Mathematical modelling of cancer cell invasion of tissue: The role of the urokinase plasminogen activation system. *Math. Modell. Methods. Appl. Sci.* 15: 1685-1734.
[49] C.Y. Chen, H.M. Byrne and J.R. King (2001). The influence of growth-induced stress from the surrounding medium on the development of multicell spheroids, *J. Math. Biol.* 43: 191-220.
[50] S.B. Cui and A. Friedman (2000). Analysis of a mathematical model of the effect of inhibitors on the growth of tumours, *Math. Biosci.* 164: 103-137.
[51] J. Crank, *Free and Moving Boundary Problems.* Clarendon Press, Oxford, 1984.
[52] V. Cristini, J. Lowengrub and Q. Nie (2003). Nonlinear simulation of tumour growth. *J. Math. Biol.* 46: 191-224.
[53] E. De Angelis and L. Preziosi (2000). Advection-diffusion models for solid tumour evolution *in vivo* and related free boundary problems, *Math. Models Methods App. Sci.*, 10: 379-407.
[54] M. Dobbelstein (2004). Replicating adenoviruses cancer therapy. *Curr. Top. Microbiol. Immunol.* 273: 291-334.
[55] D.A. Drew and L.A. Segel (1971). Averaged equations for two-phase flows. *Stud. Appl. Math.* 50: 205-231.
[56] W. Duchting (1996). Cancer: a challenge for control theory and computer modelling. *Eur. J. Cancer* 32: 1283-1292.
[57] L. Edelstein (1982). The propagation of fungal colonies: a model of tissue growth. J. theor. Biol. 98: 679-701.
[58] H. Enderling, A.R.A. Anderson, M.A.J. Chaplain, A.J. Munro, J.S. Vaidya (2006). Mathematical Modelling of Radiotherapy Strategies for Early Breast Cancer *J. theor. Biol.* (in press).
[59] K.R. Fister and J.C. Panetta (2000). Optimal control applied to cell-cycle-specific cancer chemotherapy *SIAM J. Appl. Math.* 60: 1059-1072.
[60] J. Folkman (1974). Tumour angiogenesis. *Adv. Cancer Res.* 19: 331-358.
[61] J. Folkman (2003). Fundamental concepts of the angiogenic process. *Curr. Mol. Med.* 3: 643-651.
[62] J. Folkman (2003). Antiangiogenic activity of a matrix protein. *Cancer Biol. Ther.* 2: 53-54.
[63] J. Folkman and M. Hochberg (1973). Self-regulation of growth in three-dimensions, *J. Exp. Med.* 138: 745-753.
[64] A.C. Fowler (1997). *Mathematical Models in the Applied Sciences.* Cambridge University Press, Cambridge.
[65] L.M. Franks and N. Teich (1988). *Introduction to the cellular and molecular biology of cancer.* Oxford Science Publications.
[66] S.J. Franks, H.M. Byrne, J.R. King and C.E. Lewis (2003). Modelling the growth of ductal carcinoma in situ. *J. Math. Biol.*. 47: 424-452.
[67] S.J. Franks, H.M. Byrne, H.S. Mudhar, J.C.E. Underwood and C.E. Lewis (2003). Mathematical modelling of comedo ductal carcinoma in situ of the breast, *Math. Med. Biol.* 20: 277-308.
[68] S.J. Franks, H.M. Byrne, J.C.E. Underwood and C.E. Lewis (2005). Biological inferences from a mathematical model of comedo duct carcinoma in situ of the breast, *J. theor. Biol.* 232: 523-543.

[69] A. Friedman and F. Reitich (1999). Analysis of a mathematical model for the growth of tumours, *J. Math. Biol.* 38: 262-284.

[70] Y.C. Fung (1993). Biomechanics. *Springer, New York.*

[71] A. Gamba, D. Ambrosi, A. Coniglio, A. de Candia, S. Di Talia, E. Giraudo, G. Serini, L. Preziosi and F. Bussolino (2003). Percolation, morphogenesis, and Burgers dynamics in blood vessels formation. *Phys. Rev. Lett.* 90: 118101.

[72] D. Gammack, H.M. Byrne and C.E. Lewis (2001). Estimating the selective advantage of mutant p53 tumour cells to repeated rounds of hypoxia. *Bull. Math. Biol.* 63: 135-166.

[73] R.A. Gatenby and E.T. Gawlinski (1996). A reaction-diffusion model of cancer invasion. *Cancer Res.* 56: 5745-5753.

[74] R.A. Gatenby and E.T. Gawlinski (2003). The glycolytic phenotype in carcinogenesis and tumor invasion: insights through mathematical models. *Cancer Res.* 63: 3847-3854.

[75] A. Gerisch and M.A.J. Chaplain (2006). Robust numerical methods for taxis–diffusion–reaction systems: applications to biomedical problems. *Math. Comput. Modelling* 43: 49-75.

[76] H.P. Greenspan (1972). Models for the growth of a solid tumour by diffusion. *Stud. Appl. Math.* 52: 317-340.

[77] H.P Greenspan (1976). On the growth and stability of cell cultures and solid tumours. *J. theor. Biol.* 56: 229-242.

[78] K. Groebe and W. Mueller-Klieser (1991). Distributions of oxygen, nutrient and metabolic waste concentration in multicellular spheroids and their dependence on spheroid parameters. *Eur. Biophys. J.* 19: 169-181.

[79] K. Groebe and W. Mueller-Klieser (1996). On the relation between size of necrosis and diameter of tumour spheroids. *Int. J. Radiat. Oncol.* 34: 395-401.

[80] M. Gyllenberg and G. Webb (1990). A nonlinear structured population model of tumour growth with quiescence, *J. Math. Biol.* 28: 671-684.

[81] P. Hahnfeldt, D. Panigraphy, J. Folkman and L. Hlatky (1999). Tumour development under angiogenic signalling: a dynamical theory of tumour growth, treatment response and postvascular dormancy, *Cancer Research* 59: 4770-4775.

[82] L.G. Hanin, S.T. Rachev, A.D. Tsodikov and A.Y. Yakovlev (1997). A stochastic model of carcinogenesis and tumour size at detection. *Adv. Appl. Probab.* (1997) 29: 607-628.

[83] G. Helmlinger, P.A. Netti, H.C. Lichtenbeld, R.J. Melder and R.K. Jain (1997). Solid stress inhibits the growth of multicellular tumour spheroids. *Nature Biotech.* 15: 778-783.

[84] F. Hoppensteadt, *Mathematical Theories of Populations: Demographics, Genetics and Epidemics*, SIAM, Philadelphia, 1975.

[85] M. Ilyas, J. Straub, I.P. Tomlinson and W.F. Bodmer (1999). Genetic pathways in colorectal and other cancers. *Eur. J. Cancer* 35: 1986-2002.

[86] T.L. Jackson (2003). Intracellular accumulation and mechanism of action of doxorubicin in a spatio-temporal tumor model. *J. Theor. Biol.* 220: 201-13.

[87] T.L. Jackson and H.M. Byrne (2000). A mathematical model to study the effects of drug resistance and vasculature on the response of solid tumors to chemotherapy *Math. Biosci.* 164: 17-38.

[88] T.L. Jackson and H.M. Byrne (2002). A mechanical model of tumor encapsulation and transcapsular spread. *Math. Biosci.* 180: 307-328.

[89] T.L. Jackson, S.R. Lubkin and J.D. Murray (1999). Theoretical analysis of conjugate localization in two-step cancer chemotherapy *J. Math. Biol.* 39: 353-376.

[90] A.S. Jones, H.M. Byrne, J.W. Dold and J. Gibson (2000). A mathematical model of the stress induced during solid tumour growth. *J. Math. Biol.* 40: 473-499.

[91] C.E. Kelly, R.D. Leek, H.M. Byrne, S.M. Cox, A.L. Harris and C.E. Lewis (2002). Modelling macrophage infiltration into avascular tumours, *J. Theor. Med.* 4: 21-38.

[92] J.F.R. Kerr (1971). Shrinkage necrosis: a distinct mode of cell death, *J. Path.* 105: 13-20.

[93] A.G. Knudson (1971). Mutation and cancer: statistical study of retinoblastoma. *Proc. Natl. Acad. Sci.* 68: 820-823.

[94] N.L. Komarova and V. Mirinov (2005). On the role of endothelial progenitor cells in tumour neovascularisation. *J. theor. Biol.* 235: 338-349.

[95] G. Landini and J.W. Rippin (1996). How important is tumour shape? *J. Pathol.* 179: 210-217.

[96] K.A. Landman and C.P. Please (2001). Tumour dynamics and necrosis: Surface tension and stability, *IMA. J. Math. Appl. Med.* 18, 131-158.

[97] V. Larreta-Garde and H. Berry (2002). Modelling extracellular matrix degradation balance with proteinase/transglutamine cycle. *J. theor. Biol.* 217: 105-124.

[98] I. van Leeuwen, H.M Byrne, O.E. Jensen and J.R. King(2005). Crypt dynamics and colorectal cancer: advances in mathematical modelling. *Cell Prolif.* (in press).

[99] H. Levine, S. Pamuk, B.D. Sleeman and M. Nilsen-Hamilton (2001). Mathematical modelling of capillary formation and development in tumour angiogenesis: penetration into the stroma. *Bull. Math. Biol.* 63: 801-863.

[100] B.D. MacArthur and C.P. Please (2004). Residual stress generation and necrosis formation in multicell tumour spheroids. *J. Math. Biol.* 49: 537-552.

[101] S.R. McDougall, A.R.A. Anderson, M.A.J. Chaplain and J.A. Sherratt (2002). Mathematical modelling of flow through vascular networks: implications for tumour-induced angiogenesis and chemotherapy strategies. *Bull. Math. Biol.* 64: 673-702.

[102] S. McDougall, A.R.A. Anderson and M.A.J. Chaplain (2006). Mathematical Modelling of Dynamic Adaptive Tumour-Induced Angiogenesis: Clinical Implications and Therapeutic Targeting Strategies. *J. theor. Biol.* (to appear)

[103] D.L.S. McElwain and L.E. Morris (1978). Apoptosis as a volume loss mechanism in mathematical models of solid tumour growth. *Math. Biosci.* 39: 147-157.

[104] N. Mantzaris, S. Webb, S. and H.G. Othmer (2004). Mathematical modelling of tumour-induced angiogenesis. *J. Math. Biol.* 95: 111-187.

[105] J.H. Mao, K.A. Lindsay, A. Balmain and T.E. Wheldon (1998). Stochastic modelling of tumorigenesis in p53 deficient mice. *Br. J. Cancer* 77: 243-252.

[106] B. Marchant, A.J. Perumpanani and J. Norbury (2000). Travelling shock waves arising in a model of malignant invasion. *SIAM J. Appl. Math.* 60: 463-476.

[107] B.P. Marchant, J. Norbury and H.M. Byrne (2006). Biphasic behaviour in malignant invasion. *Math. Med. Biol.* (in press).

[108] M. Marusic, Z. Bajzer, S. Vukpavlovic and J.P. Freyer (1994). Tumour-growth in-vivo and as multicellular spheroids compared by mathematical models. *Bull. Math. Biol.* 56: 617-631.

[109] M. Marusic, Z. Bajzer, J.P. Freyer and S. Vukpavlovic (1994). Analysis of growth of multicellular tumour spheroids by mathematical models. *Cell Prolif.* 27: 73-94.

[110] J. Moreira and A. Deutsch (2002). Cellular automaton models of tumor development: A critical review. *Adv. in Complex Systems* 5: 247-267.

[111] V.R. Muthukkaruppan, L. Kubai and R. Auerbach (1982). Tumour-induced neovascularisation in the mouse eye. *J. Natl. Cancer Inst.* 69: 699-705.

[112] C. Murdoch and C.E. Lewis (2005). Macrophage migration and gene expression in response to tumour hypoxia. *Int. J. Cancer* 117: 701-708.

[113] M.E. Orme and M.A.J. Chaplain (1996). A mathematical model of vascular tumour growth and invasion. *Math. Comp. Mod.* 23: 43-60.

[114] H.G. Othmer and A. Stevens (1997). Aggregation, Blowup and Collapse: The ABC's of generalized taxis. *SIAM J. Applied Math.* 57 1044-1081.

[115] M.R. Owen, H.M. Byrne and C.E. Lewis (2004). Mathematical modelling of the use of macrophages as vehicles for drug delivery to hypoxic tumour sites. *J. theor. Biol.* 226: 377-391.

[116] J.C. Panetta and J.A. Adam (1995). A mathematical model of cycle-specific chemotherapy. *Math. Comput. Modell.* 22: 67-82.

[117] J.C. Panetta (1997). A mathematical model of breast and ovarian cancer treated with paclitaxel. *Math. Biosci.* 146: 89-113.

[118] A.A. Patel, E.T. Gawlinsky, S.K. Lemieux and R.A. Gatenby (2001). Cellular automaton model of early tumour growth and invasion: the effects of native tissue vascularity and increased anaerobic tumour metabolism. *J. Theor. Biol.* 213: 315-331.

[119] A.J. Perumpanani and H.M. Byrne (1999). Extracellular matrix concentration exerts selective pressure on invasive cells. *Eur. J. Cancer* 35: 1274-1280.

[120] G.J. Pettet, C.P. Please, M.J. Tindall and D.L.S. McElwain (2001). The migration of cells in multicell tumour spheroids. *Bull. Math. Biol.* 63: 231-257.

[121] C.P. Please, G.J. Pettet and D.L.S. McElwain (1998). A new approach to modelling the formation of necrotic regions in tumours. *Appl. Math. Letters* 11: 89-94.

[122] J.B. Plotkin and M.A. Nowak (2002). The different effects of apoptosis and DNA repair on tumorigenesis. *J. theor. Biol.* 214: 453-467.

[123] L. Preziosi (2003). *Cancer modelling and simulation.* Chapman and Hall/CRC.

[124] A.R. Pries, T.W. Secomb and P. Gaehtgens (1998). Structural adaptation and stability of microvascular networks: theory and simulations. *Am. J. Physiol.* 275: H349-H360.

[125] B. Ribba, K. Marron, Z. Agur, T. Alarcón, T. and P.K. Maini (2005). A mathematical model of doxorubicin treatment efficacy for non-Hodgkin's lymphoma: investigation of the current protocol through theoretical modelling results. *Bull. Math. Biol.* 67: 79-99.

[126] T. Roose, P.A. Netti, L.L. Munn, Y. Boucher and R.K. Jain (2003). Solid stress generated by spheroid growth estimated using a linear poroelasticity model. *Microvasc. Res.* 66: 204-212.

[127] T. Roose, S.J. Chapman and P.K. Maini (2006). Mathematical models of avascular tumour growth: a review. *SIAM Review* (in press).

[128] J.A. Royds, S.K. Dower, E.E. Qwarstrom and C.E. Lewis (1998). Resonse of tumour cells to hypoxia: role of p53 and NF$\kappa\beta$. *J. Clin. Path.: Mol. Pathol.* 51: 55-61.

[129] G. Serini, D. Ambrosi, E. Giraudo, A. Gamba, L. Preziosi and F. Bussolino (2003). Modeling the early stages of vascular network assembly. *EMBO J.* 22: 1771-9.

[130] J.A. Sherratt (1993). Cellular growth and travelling waves of cancer. *SIAM Appl. Math.* 53: 1713-1730.

[131] R.M. Shymko and L. Glass (1976). Cellular and geometric control of tissue growth and mitotic instability. *J. Theor. Biol.* 63: 355-374.

[132] Z.T. Sinik, T. Albibay, O. Ataoglu, H. Biri, S. Sozen, N. Deniz, U. Karaoglan and I. Bozkirli (1997). Nuclear p53 overexpression in bladder, prostate and renal carcinomas. *Int. J. Urol.* 4: 546-551.

[133] K. Smallbone, D.J. Gavaghan, R.A. Gatenby and P.K. Maini (2005). The role of acidity in solid tumour growth and invasion. *J. Theor. Biol.* 235: 476-484.

[134] P. Sminia, K.A. Hartmann, C.J. Schneider, U.M. Carl, A.J. van der Kleij, B.J. Slotman and M.C. Huishof (2001). Radiosensitizing effect of carbogen breathing during pulsed irradiation of the rat R1H tumour. *Acta Oncol.* 40: 870-874.

[135] A. Stephanou, S.R. McDougall, A.R.A. Anderson and M.A.J. Chaplain (2005). Mathematical Modelling of Flow in 2D and 3D Vascular Networks: Applications to Anti-angiogenic and Chemotherapeutic Drug Strategies. *Math. Comput. Modell.* 41: 1137-1156.

[136] A. Stephanou, S. McDougall, A. Anderson and M.A.J. Chaplain (2006). Mathematical modelling of the influence of blood rheological properties upon adaptive tumour-induced angiogenesis. *J. Math. Comp. Mod.* (submitted).

[137] W.G. Stetler-Stevenson, S. Aznavoorian and L.A. Liotta (1993). Tumour cell interactions with the extracellular matrix during invasion and metastasis. *Ann. Rev. Cell. Biol.* 9: 541-573.

[138] C.L. Stokes and D.A. Lauffenburger (1991). Analysis of the roles of microvessel endothelial cell random motility and chemotaxis in angiogenesis. *J. theor. Biol.* 152: 377-403.

[139] B.R. Stoll, C. Migliorini, A. Kadambi, L.L. Munn and R.K. Jain (2003). A mathematical model of the contribution of endothelial progenitor cells to angiogenesis in tumors: implications for antiangiogenic therapy. *Blood* 102: 2555-2561.

[140] E.L. Stott, N.F. Britton, J.A. Glazier and M. Zajac (1999). Stochastic simulation of benign avascular tumour growth using the Potts model. *Math. Comp. Mod.* 30: 183-198.

[141] R.M. Sutherland and R.E. Durand (1984). Growth and cellular characteristics of multicell spheroids. *Recent Results in Cancer Research* 95: 24-49.

[142] K.R. Swanson, E.C. Alvord and J.D. Murray (2000). A quantitative model for differential motility of gliomas in grey and white matter. *Cell. Prolif.* 33: 317-329, 2000.

[143] K.R. Swanson, E.C. Alvord and J.D. Murray (2002). Quantifying efficacy of chemotherapy of brain tumors with homogeneous and heterogeneous drug delivery. *Acta Biotheoretica* 50: 223-237.

[144] K.R. Swanson, C. Bridge, J.D. Murray and E.C. Alvord, E.C. (2003). Virtual and real brain tumors: using mathematical modeling to quantify glioma growth and invasion. *J. Neur. Sci.* 216: 1-10.

[145] K.E. Thompson and H.M. Byrne (1999). Modelling the internalisation of labelled cells in tumour spheroids. *Bull. Math. Biol.* 61: 601-623.

[146] I. Tomlinson, P.M. Novelli and W.F. Bodmer (1995). Failure of programmed cell-death and differentiation as causes of tumours – some simple mathematical-models. *Proc. Nat. Acad. Sci.* 92: 11130-11134.

[147] I. Tomlinson, P.M. Novelli and W.F. Bodmer (1995). The mutation rate and cancer. *Proc. Nat. Acad. Sci.* 93: 14800-14803.

[148] P. Tracqui (1995). From passive diffusion to active cellular migration in mathematical models of tumour invasion. *Acta Biotheoretica* 43: 443-464.

[149] J.J. Tyson and B. Novak (2001). Regulation of the eukariotic cell-cycle: molecular anatagonism, hysteresis, and irreversible transitions. *J. Theor. Biol.* 210: 249-263.

[150] S. Turner and J.A. Sherratt (2002). Intercellular adhesion and cancer invasion: a discrete simulation using the extended Potts model. *J. Theor. Biol.* 216: 85-100.

[151] J.P. Ward and J.R. King (1997). Mathematical modelling of avascular-tumour growth. *IMA. J. Math. Appl. Med.* 14: 39-69.

[152] J.P. Ward and J.R. King (1998). Mathematical modelling of avascular-tumour growth II: Modelling growth saturation. *IMA. J. Math. Appl. Med.*, 15: 1-42.

[153] J.P. Ward and J.R. King (1999). Mathematical modelling of the effects of mitotic inhibitors on avascular tumor growth. *J. Theor. Med.*, 1: 171-211.

[154] S.D. Webb, J.A. Sherratt and R.G. Fish (1999). Alterations in proteolytic activity at low pH and its association with invasion: a theoretical model. *Clin. Exp. Metastasis* 17: 397-407.

[155] L.M. Wein, J.T. Wu and D.H. Kirn (2003). Validation and analysis of a mathematical model of a replication-competent oncolytic virus for cancer treatment: implications for virus design and therapy. *Cancer Res.* 15: 1317-1324.

[156] T.E. Wheldon, J. Kirk and J.S. Orr (1977). Optimal radiotherapy of tumour cells following exponential-quadratic survival curves and exponential repopulation kinetics. *Brit. J. Radiol.* 50: 681-682.

[157] D. Wodarz and N. Komarova (2004). *Computational biology of cancer: lecture notes and mathematical modelling.* World Scientific Publishing Company.

[158] X. Zheng, S.M. Wise and V. Cristini (2005). Nonlinear simulation of tumour necrosis, neovascularisation and tissue invasion via an adaptive finite-element/level-set method. *Bull. Math. Biol.* 67: 211-260.

[159] R.A. Zlotecki, L.T. Baxter, Y. Boucher and R.K. Jain (1995). Pharmacological modification of tumour blood flow and interstitial fluid pressure in a human tumour xenograft – network analysis and mechanistic interpretation. *Microvascular Res.* 50: 429-443.

Lectures in Mathematical Neuroscience

Paul C. Bressloff

IAS/Park City Mathematics Series
Volume 14, 2005

Lectures in Mathematical Neuroscience

Paul C. Bressloff

Neurons in the brain communicate with each other by transmitting electrical spikes known as action potentials (see Figure 1). An action potential propagates along the axon of a neuron until it reaches a terminal that forms the upstream or presynaptic component of the synaptic connection to a downstream or postsynaptic neuron. The arrival of the action potential induces the release of chemical transmitters into the synapse. These subsequently bind to receptors in the postsynaptic membrane resulting in the opening of various ion channels. This generates a synaptic current that flows along the dendritic tree of the postsynaptic neuron and combines with currents from other activated synapses. If the total synaptic current forces the membrane potential at a certain location within the cell body to cross some threshold, then the postsynaptic neuron fires an action potential and the process continues. One can thus view the brain as a vast collection of synaptically–coupled networks of spiking neurons. Each network is specialized for a particular range of tasks and receives inputs from other networks and possibly from external sources (eg. sensory stimuli). Some of the networks generate motor commands that control the movement of the organism in its environment. Many of the synaptic connections within and between networks are modifiable by experience (synaptic plasticity).

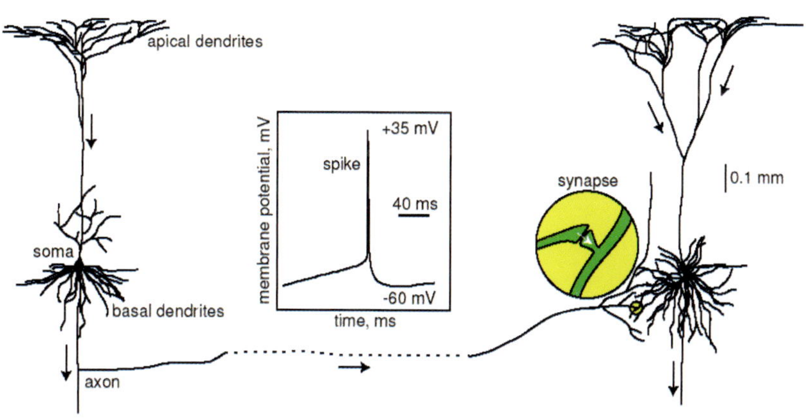

FIGURE 1. Communication with spikes. [Reprinted from [**1**], figure 1.1, by permission of the MIT Press.]

Department of Mathematics, Salt Lake City, Utah, USA 84112
E-mail address: bressloff@math.utah.edu
The author was supported in part by NSF grant DMS 0515725

One of the ultimate challenges in neuroscience is to understand how networks of spiking neurons with modifiable connections allow an organism to carry out goal–directed behavior within a changing environment, based on current sensory input and previous experience? In addition to understanding computation and information processing during normal brain function, there is also the important clinical goal of finding the causes of dysfunctional brain states and behavior. In these lectures we will carry out the much more modest goal of unraveling some of the molecular, cellular and network mechanisms underlying the dynamics of synaptically–coupled spiking neurons. We will not focus on any particular specialized region of the brain, but consider some general principles of neuronal dynamics together with illustrative examples. The basic topics covered are single neuron models (Lecture 1), synaptic and dendritic processing (Lecture 2), firing rates, spike train statistics and the neural code (Lecture 3), network oscillations and synchrony (Lecture 4), and neural pattern formation (Lecture 5). In each lecture we describe in some detail a particular mathematical or modeling approach. For example, in Lecture 1 we consider phase reduction methods for analyzing single neural oscillators, which are extended to the case of synaptically coupled oscillators in Lecture 4. In Lecture 2 we emphasize the biophysical modeling of synapses at the molecular level, paying particular attention to the important topic of synaptic plasticity. In Lecture 3 we focus on the use of stochastic methods for studying noise in spiking neuron models. Finally, in Lecture 5 we consider the analysis of integrodifferential equations describing the large-scale dynamics of cortical tissue.

LECTURE 1

Single neuron models

We begin the first lecture by considering conductance–based models of a single neuron, in which the spatial structure of the neuron is neglected [**2, 3**]. Such models can reproduce electrophysiological measurements to a high degree of accuracy, but their intrinsic complexity makes them difficult to analyze, particularly at the network level. One way around this is to simplify the single neuron dynamics by carrying out some form of dimensional reduction. Reduction schemes typically exploit the presence of some small parameter in the system such as the strength of input, the degree of nonlinearity (as determined by distance from a bifurcation point, for example), or the rate of change of one or more slow variables. Regular or singular perturbation methods can then be combined with techniques from low–dimensional dynamical systems theory such as phase–plane analysis and bifurcation theory in order to analyze neural excitability, spiking and bursting [**4, 5, 1**]. Here we focus on the so–called phase reduction method [**6**], which is used to analyze the dynamics of a regular spiking neural oscillator under weak periodic forcing. One of the important applications of this method is analyzing conditions for synchrony in networks of synaptically coupled neural oscillators (see §4). We end the lecture by describing some formal threshold models of neuronal firing that are highly popular in studies of neural information processing, large–scale network dynamics, and memory [**7**].

1.1. Conductance–based models

The standard biophysical model for describing a neuron with spatially constant membrane potential u is based upon conservation of electric charge, so that

$$C\frac{\mathrm{d}u}{\mathrm{d}t} = -F + I_{\text{syn}} + I_{\text{ext}}, \tag{1.1}$$

where C is the cell capacitance, F the membrane current, I_{syn} the sum of synaptic currents entering the cell (see §2) and I_{ext} describes any externally injected currents. Ions can diffuse in and out of the cell through ion specific channels embedded in the cell membrane. Ion pumps within the cell membrane maintain concentration gradients, such that there is a higher concentration of Na^+ and Ca^{2+} outside the cell and a higher concentration of K^+ inside the cell. The membrane current through a specific channel varies approximately linearly with changes in the potential u relative to some equilibrium or reversal potential, which is the potential at which there is a balance between the opposing effects of diffusion and electrical forces. Summing over all channel types, the total membrane current (flow of positive ions) leaving the cell through the cell membrane is

$$F = \sum_i g_i(u - u_i), \tag{1.2}$$

where g_i is the conductance due to channels of type i and u_i is the corresponding reversal potential. In the case of a channel selective to a single ion, u_i satisfies the Nernst equation

$$u_i = \frac{k_B T}{q} \ln\left(\frac{[outside]_i}{[inside]_i}\right), \tag{1.3}$$

where q is the charge of the ion, k_B is the Boltzman constant, T is temperature (in degrees Kelvin) and $[outside]_i, [inside]_i$ denote the extracellular and intracellular concentrations of the given ion. Typical values for the common ion species are $u_K \approx -75mV$, $u_{Na} \approx 50mV$, $u_{Ca} \approx 150mV$ and $u_{Cl} \approx -60mV$ (which is close to the resting potential of the cell).

The generation and propagation of an action potential arises from nonlinearities associated with active membrane conductances. Recordings of the current flowing through single channels indicate that channels fluctuate rapidly between open and closed states in a stochastic fashion. Nevertheless, most models of a neuron use deterministic descriptions of conductance changes, under the assumption that there are a large number of approximately independent channels of each type. It then follows from the law of large numbers that the fraction of channels open at any given time is approximately equal to the probability that any one channel is in an open state. The conductance g_i for ion channels of type i is thus taken to be the product $g_i = \bar{g}_i P_i$ where $\bar{g}_i$ is equal to the density of channels in the membrane multiplied by the conductance of a single channel and P_i is the fraction of open channels. The voltage-dependence of the probabilities P_i in the case of a delayed-rectifier K^+ current and a fast Na^+ current were originally obtained by Hodgkin and Huxley [**8**] as part of their Nobel prize winning work on the generation of action potentials in the squid giant axon. The delayed-rectifier K^+ current is responsible for repolarizing a neuron after an action potential. One finds that opening of the K^+ channel requires structural changes in 4 identical and independent subunits so that $P_K = n^4$ where n is the probability that any one gate subunit has opened. In the case of the fast Na^+ current, which is responsible for the rapid depolarization of a cell leading to action potential generation, the probability of an open channel takes the form $P_{Na} = m^3 h$ where m^3 is the probability that an activating gate is open and h is the probability that an inactivating gate is open. Depolarization causes m to increase and h to decrease, whereas hyperpolarization has the opposite effect.

The dynamics of the gating variables m, n, h are usually formulated in terms of a simple kinetic scheme that describes voltage-dependent transitions of each gating subunit between open and closed states. More specifically, for each $X \in \{m, n, h\}$

$$\frac{dX}{dt} = \alpha_X(u)(1 - X) - \beta_X(u)X, \tag{1.4}$$

where $\alpha_X(u)$ is the rate of the transition *closed* $\rightarrow$ *open* and $\beta_X(u)$ is the rate of the reverse transition *open* $\rightarrow$ *closed*. Equation (1.4) can be rewritten in the alternative form

$$\tau_X(u)\frac{\mathrm{d}X}{\mathrm{d}t} = X_\infty(u) - X, \qquad \text{with } X \in \{m, n, h\}, \tag{1.5}$$

where

$$\tau_X(u) = \frac{1}{\alpha_X(u) + \beta_X(u)}, \qquad X_\infty(u) = \alpha_X(u)\tau_X(u).$$

It follows that the conductance variables m, n and h approach the asymptotic values $m_\infty(u)$, $n_\infty(u)$ and $h_\infty(u)$ exponentially with time constants $\tau_m(u)$, $\tau_n(u)$ and $\tau_h(u)$ respectively. The following voltage-dependent rates $\alpha_X(u)$ and $\beta_X(u)$, $X \in \{m, n, h\}$ were obtained by Hodgkin and Huxley [**8**] using fits with experimental data:

$$\alpha_m = \frac{0.1(u+40)}{1-\exp[-0.1(u+40)]} \qquad \alpha_h = 0.07\exp[-0.05(u+65)],$$
$$\alpha_n = \frac{0.01(u+55)}{1-\exp[-0.1(u+55)]} \qquad \beta_m = 4.0\exp[-0.556(u+65)],$$
$$\beta_h = \frac{1}{1+\exp[-0.1(u+35)]} \qquad \beta_n = 0.125\exp[-0.125(u+65)].$$

All potentials are measured in mV, all times in ms and all currents in μA per cm^2. The corresponding asymptotic functions $X_\infty(u)$ and time constants $\tau_X(u)$ are plotted in Figure 1.

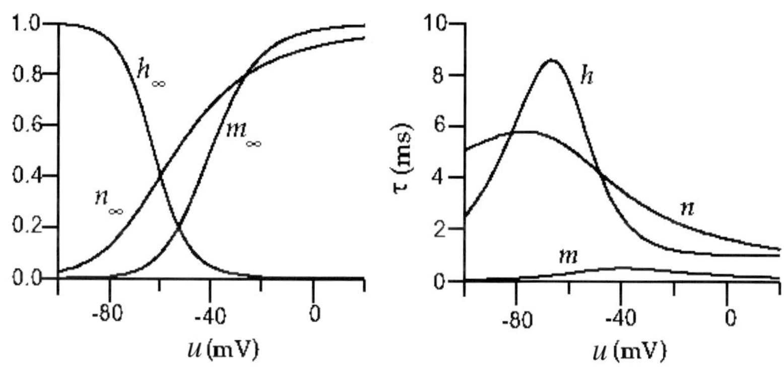

FIGURE 1. Voltage-dependent steady-state levels of activation and inactivation (left panel) and voltage-dependent time constants (right panel) for the Hodgkin-Huxley model.

We can now write down the Hodgkin-Huxley model for the generation of an action potential, which takes the membrane current to be the sum of a leakage current, a delayed-rectifier K$^+$ current and a fast Na$^+$ current,

$$F(u,m,n,h) = g_L(u-u_L) + g_K n^4(u-u_K) + g_{Na}hm^3(u-u_{Na}). \tag{1.6}$$

The maximal conductances and reversal potentials used in the model are $g_L = 0.003\ ms/mm^2$, $g_K = 0.36\ mS/mm^2$, $g_{Na} = 1.2\ mS/mm^2$, $u_L = -54.387\ mV$, $u_K = -77\ mV$ and $u_{Na} = 50\ mV$. Note that the leakage current groups together various voltage-independent processes such as the currents carried by ion pumps that maintain the concentration gradients. The variables m, n, h evolve according to equation (1.4). The temporal evolution of the variables u, m, n, h during a single action potential is shown in Figure 2. Injection of a depolarizing current induces a rapid increase in the m variable describing activation of the Na$^+$ current. Since the slower h variable is initially around 0.6, there is a large influx of Na$^+$ ions, producing a sharp downward spike in the membrane current and a rapid depolarization through positive feedback. However, the rise in the membrane potential causes the Na^+ conductance to inactivate by driving h towards zero. In addition, the depolarization activates the K^+ conductance by driving n towards a, resulting in a subsequent hyperpolarization.

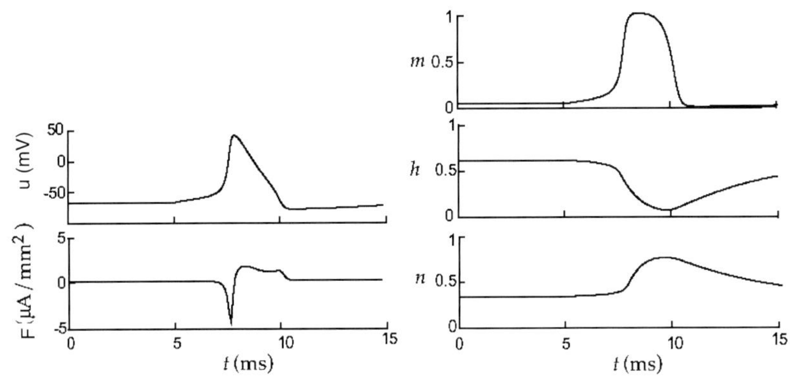

FIGURE 2. The dynamics of u, n, m, h in the Hodgkin-Huxley model during the firing of an action potential induced by a current injection at $t = 5\ ms$.

Since the introduction of the Hodgkin-Huxley model, there has been considerable experimental progress in identifying the molecular mechanisms underlying the opening and closing of ion channels, and many other types of ion channels and associated currents have been discovered [**3**]. This has led to the development of more detailed conductance–based models that can account for a much wider variety of complex spiking behavior including bursting. In these lectures we mainly focus on regular spiking (§1.2 and §4) and irregular spiking behavior (§3). However, we do briefly describe some reduced integrate–and–fire models of bursting in §1.3. For extensive reviews of conductance–based models of bursting see [**1**] and the collection of articles in [**9**].

1.2. Periodically forced neural oscillator

We can generally formulate a conductance-based model of a neuron with constant input current as an M-dimensional ($M \geq 2$) system of ODEs

$$\frac{d\mathbf{x}}{dt} = \mathbf{f}(\mathbf{x}), \qquad \mathbf{x} = (x_1, \ldots x_M). \tag{1.7}$$

Here x_1, say, represents the membrane potential of the neuron (treated as a point processor) and x_m, $m > 1$, represent various ionic channel gating variables. Suppose that the neuron has a stable periodic solution $\mathbf{x}(t) = \mathbf{x}(t + \Delta_0)$ where $\omega_0 = 2\pi/\Delta_0$ is the *natural frequency* of the oscillator. In *phase space* the solution is an isolated attractive trajectory called a *limit cycle*. The dynamics on the limit cycle can be described by a uniformly rotating phase such that

$$\frac{d\phi}{dt} = \omega_0 \tag{1.8}$$

and $\mathbf{x}(t) = \mathbf{g}(\phi(t))$ with $\mathbf{g}$ a 2π-periodic function. Note that the phase is *neutrally stable* with respect to perturbations along the limit cycle – this reflects invariance of an autonomous dynamical system with respect to time shifts. Now suppose that a small external periodic input is applied to the oscillator such that

$$\frac{d\mathbf{x}}{dt} = \mathbf{f}(\mathbf{x}) + \varepsilon \mathbf{p}(\mathbf{x}, t), \tag{1.9}$$

where $\mathbf{p}(\mathbf{x}, t) = \mathbf{p}(\mathbf{x}, t+\Delta)$ and $\omega = 2\pi/\Delta$ is the forcing frequency. If the amplitude ε is sufficiently small and the cycle is stable then deviations transverse to the limit cycle are small so that the main effect of the perturbation is to induce shifts in the phase. Therefore, we need to extend the definition of phase to a neighborhood of the limit cycle. We accomplish this using the notion of an *isochrone* [**10, 11, 12**].

Isochrones

Suppose that we observe the unperturbed system stroboscopically at time intervals of length Δ_0. This leads to a Poincare mapping

$$\mathbf{x}(t) \to \mathbf{x}(t + \Delta_0) \equiv \Phi(\mathbf{x}(t)).$$

This mapping has all points on limit cycle as fixed points. Choose a point $\mathbf{x}^*$ on the cycle and consider all points in the vicinity of $\mathbf{x}^*$ that are attracted to it under the action of Φ. They form an $(M-1)$-dimensional hypersurface I, called an isochrone, crossing the limit cycle at $\mathbf{x}^*$ (see Figure 3). A unique isochrone can be drawn through each point on the limit cycle so we can parametrize the isochrones by the phase $I = I(\phi)$. Finally, we extend the definition of phase by taking all points on $I(\phi)$ to have the same phase ϕ, which then rotates at the natural frequency ω_0.

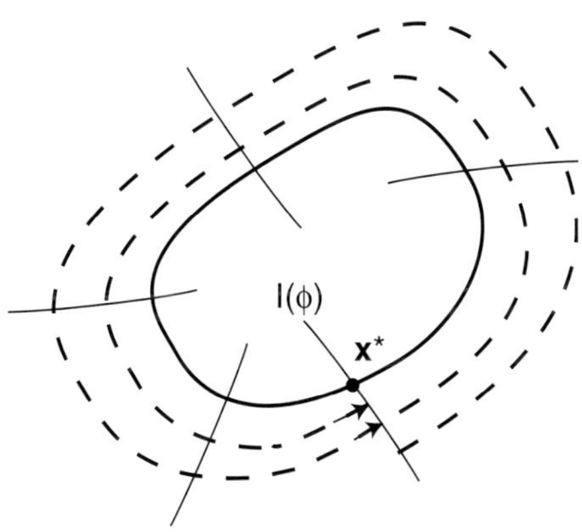

FIGURE 3. Isochrones in the neighborhood of a stable limit cycle

Example: Limit cycle oscillator Consider as an example the amplitude equation that arises for a limit cycle oscillator close to a Hopf bifurcation [**13**]:

$$\frac{dA}{dt} = (1 + i\eta)A - (1 + i\alpha)|A|^2 A. \tag{1.10}$$

In polar coordinates $A = Re^{i\theta}$,

$$\frac{dR}{dt} = R(1 - R^2), \quad \frac{d\theta}{dt} = \eta - \alpha R^2.$$

The solution for arbitrary initial data $R(0) = R_0$, $\theta(0) = \theta_0$ is

$$
\begin{aligned}
R(t) &= \left[1 + \frac{1-R_0^2}{R_0^2}\mathrm{e}^{-2t}\right]^{-1/2}, \\
\theta(t) &= \theta_0 + \omega_0 t - \frac{\alpha}{2}\log(R_0^2 + (1-R_0^2)\mathrm{e}^{-2t}),
\end{aligned}
\tag{1.11}
$$

where $\omega_0 = \eta - \alpha$ is the natural frequency of the stable limit cycle at $R = 1$. Strobing the solution at times $t = n\Delta_0$, we see that

$$\lim_{n\to\infty} \theta(n\Delta_0) = \theta_0 - \alpha \ln R_0.$$

Hence, we can define a phase on the whole plane

$$\phi(R, \theta) = \theta - \alpha \ln R. \tag{1.12}$$

It follows that the isochrones are logarithmic spirals with $\theta - \alpha \ln R = constant$.

Phase equation

For an unperturbed oscillator in the vicinity of the limit cycle we have

$$\omega_0 = \frac{d\phi(\mathbf{x})}{dt} = \sum_k \frac{\partial \phi}{\partial x_k}\frac{dx_k}{dt} = \sum_k \frac{\partial \phi}{\partial x_k} f_k(\mathbf{x}).$$

Now consider the perturbed system but with the unperturbed definition of the phase:

$$\frac{d\phi(\mathbf{x})}{dt} = \sum_k \frac{\partial \phi}{\partial x_k}(f_k(\mathbf{x}) + \varepsilon p_k(\mathbf{x}, t)) = \omega_0 + \varepsilon \sum_k \frac{\partial \phi}{\partial x_k} p_k(\mathbf{x}, t).$$

To a first approximation we can neglect deviations of $\mathbf{x}$ from the limit cycle which we denote by $\mathbf{x}^*$:

$$\frac{d\phi(\mathbf{x})}{dt} = \omega_0 + \varepsilon \sum_k \frac{\partial \phi(\mathbf{x}^*)}{\partial x_k} p_k(\mathbf{x}^*, t).$$

Finally, since points on the limit cycle are in 1:1 correspondence with the phase ϕ we obtain the closed phase equation

$$\frac{d\phi}{dt} = \omega_0 + \varepsilon Q(\phi, t), \tag{1.13}$$

where

$$Q(\phi, t) = \sum_k \frac{\partial \phi(\mathbf{x}^*(\phi))}{\partial x_k} p_k(\mathbf{x}^*(\phi), t) \tag{1.14}$$

is a 2π-periodic function of ϕ and a Δ-periodic function of t.

Example: forced limit cycle oscillator. Rewrite equation (1.10) in Cartesian coordinates

$$
\begin{aligned}
\frac{dx}{dt} &= x - \eta y - (x^2 + y^2)(x - \alpha y) + \varepsilon \cos \omega t, \\
\frac{dy}{dt} &= y + \eta y - (x^2 + y^2)(y + \alpha x),
\end{aligned}
$$

where we have added a periodic modulation in the x-direction. Rewrite the phase (1.12) as

$$\phi = \tan^{-1}\frac{y}{x} - \frac{\alpha}{2}\log(x^2+y^2)$$

so that

$$\frac{\partial\phi}{\partial x} = -\frac{y}{x^2+y^2} - \alpha\frac{x}{x^2+y^2}.$$

On the limit cycle $\mathbf{x}_0(\phi) = (\cos\phi, \sin\phi)$ and hence

$$\frac{\partial\phi(\mathbf{x}_0)}{\partial x} = -\sin\phi - \alpha\cos\phi.$$

It follows that the corresponding phase equation is

$$\frac{d\phi}{dt} = \omega_0 - \varepsilon(\alpha\cos\phi + \sin\phi)\cos\omega t.$$

Phase resetting curves for conductance-based models

The phase reduction method is particularly useful because the function $Q(\phi, t)$ can be related to an easily measurable property of a neural oscillator, namely its *phase resetting curve* (PRC), which we denote by the 2π-periodic function $R(\phi)$. The PRC is found experimentally (or numerically) by perturbing the oscillator with a brief depolarizing voltage stimulus at different times in its cycle and measuring the resulting phase-shift from the unperturbed system [**10, 12**]. Taking the coordinate x_1 as the membrane potential, it follows from equation (1.13) that

$$\frac{d\phi}{dt} = \omega_0 + \varepsilon\frac{\partial\phi(\mathbf{x}^*(\phi))}{\partial x_1}\delta(t-t_0). \tag{1.15}$$

Integrating this equation over a small interval around t_0, we see that the impulse induces a phase-shift $\Delta\phi = \varepsilon R(\phi_0)$ where $R(\phi) = \partial\phi(\mathbf{x}^*(\phi))/\partial x_1$. Thus comparing the phase at large times for the unperturbed and perturbed cases generates the phase resetting curve. Given the phase resetting curve $R(\phi)$, the response of the neuron to a more general time-dependent voltage perturbation $\varepsilon P(t)$ is determined by the phase equation

$$\frac{d\phi}{dt} = \omega_0 + \varepsilon R(\phi)P(t). \tag{1.16}$$

We can also express the PRC in terms of the firing times of a neuron (assuming fast reconvergence to the limit cycle). Suppose that there exists a well-defined threshold κ signalling the onset of fast somatic membrane depolarization and the subsequent firing of an action potential spike. If T^n denotes the nth firing time of the neuron then

$$u(T^n) = \kappa, \qquad \frac{du}{dt}(T^n) > 0, \tag{1.17}$$

where $u(t)$ is the membrane potential at time t. (Alternatively we could define the firing times in terms of when the neuron reaches its maximum depolarization). Since $u(t) = x_1(\phi(t))$, the threshold corresponds to a particular phase of the limit cycle, which we choose to be $\phi = 0$. Since $\phi(t) = 2\pi t/\Delta_0$ in the absence of perturbations, the firing times are $T^n = n\Delta_0$ where Δ_0 is the natural period of oscillation. On the other hand, a small perturbation applied at the point ϕ on the limit cycle at time $t, T^n < t < T^{n+1}$, induces a phase-shift that changes the next time of firing according to (see Figure 4)

$$\frac{T^{n+1} - T^n}{\Delta_0} = 1 - \frac{R(\phi)}{2\pi\Delta_0}. \tag{1.18}$$

For certain types of neuron a depolarizing stimulus always advances the onset of

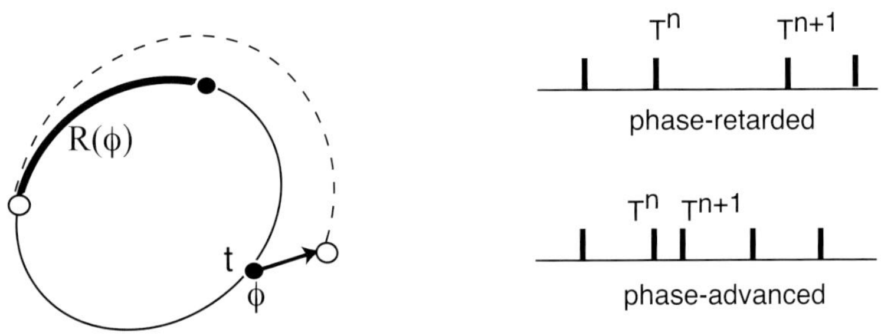

FIGURE 4. Phase-shift $R(\phi)$ induced by a small perturbation of the membrane potential applied at time $t = 0$ when the phase of the limit cycle is ϕ

the next spike, that is, the PRC is always positive, whereas for others the stimulus may also delay the next spike. We refer to models with a strictly positive PRC as Type I and those for which the PRC has a negative regime as Type II.

Example: Morris-Lecar model. A numerical example illustrating both types of PRC is shown in Figure 5(a) for the Morris–Lecar model of a neuron, which was originally introduced to describe how under constant current injection barnacle muscle fibers

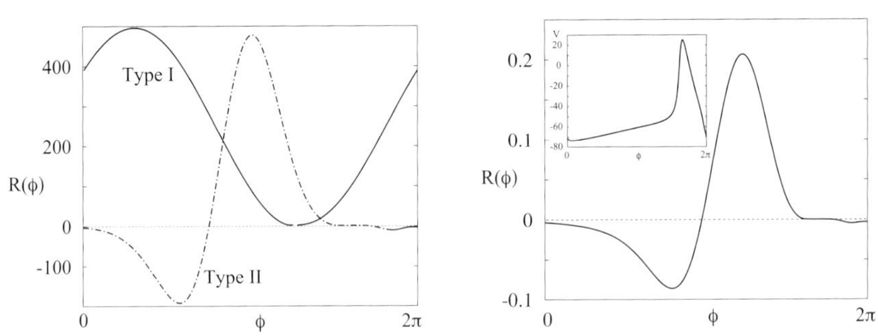

FIGURE 5. (a) Morris-Lecar model showing two different response types. In both cases $u_k = -0.7$, $u_L = -0.5$, $u_{Ca} = 1$, $g_K = 2$, $g_L = 0.5$, $u_1 = -0.01$, $u_2 = 0.15$. For a type I response, $g_{Ca} = 1.33$, $u_3 = 0.1$, $u_4 = 0.145$, $\phi = 1/3$ and $I = 0.0695$. For a type II response, $g_{Ca} = 1.1$, $u_3 = 0$, $u_4 = 0.3$, $\phi = 0.2$ and $I = 0.25$. Responses have been scaled to the same ranges. (b) Hodgkin-Huxley model with external drive $I = 10$ showing Type II phase resetting curve. Corresponding orbit over a single cycle is shown in the inset.

respond with a host of oscillatory voltage waveforms [**14**]. It takes the form

$$\frac{\mathrm{d}u}{\mathrm{d}t} = I - g_L(u-u_L) - g_K w(u-u_K) - g_{Ca} m_\infty(u)(u-u_{Ca}),$$

$$\frac{\mathrm{d}w}{\mathrm{d}t} = \lambda(u)(w_\infty(u) - U), \tag{1.19}$$

with

$$m_\infty(u) = 0.5(1+\tanh[(u-u_1)/u_2]),$$
$$w_\infty(u) = 0.5(1+\tanh[(u-u_3)/u_4]),$$
$$\lambda(u) = \phi\cosh[(u-u_3)/(2u_4)].$$

Here g_L is the leakage conductance, g_K, g_{Ca} are potassium and calcium conductances, u_L, u_K, u_{Ca} are corresponding reversal potentials, $m_\infty(u)$, $w_\infty(u)$ are voltage-dependent gating functions and $\lambda(u)$ is a voltage-dependent rate. The Type II PRC for the Hodgkin-Huxley model is shown in Figure 5(b)

Averaging theorem

Expand $Q(\phi, t)$ in the phase equation (1.13) as a double Fourier series

$$Q(\phi,t) = \sum_{l,k} a_{l,k} \mathrm{e}^{ik\phi + il\omega t}.$$

Substitute for ϕ using the zero-order approximation $\phi = \omega_0 t + \phi_0$:

$$Q(\phi,t) = \sum_{l,k} a_{l,k} \mathrm{e}^{ik\phi_0 + i(k\omega_0 + l\omega)t}.$$

It follows that Q contains fast oscillating terms (compared to the time-scale $1/\varepsilon$) together with slowly varying terms that satisfy the *resonance condition*

$$k\omega_0 + l\omega \approx 0. \tag{1.20}$$

Only the latter will lead to large variations in the phase, so we can average the forcing term Q keeping only the resonant terms. The simplest case is $\omega \approx \omega_0$ for which the resonant terms satisfy $l = -k$ and

$$Q(\phi,t) \to \sum_k a_{-k,k} \mathrm{e}^{ik(\phi - \omega t)} = q(\phi - \omega t). \tag{1.21}$$

The phase equation then becomes

$$\frac{d\phi}{dt} = \omega_0 + \varepsilon q(\phi - \omega t).$$

The phase difference between the oscillator and external drive, $\psi = \phi - \omega t$, then satisfies the equation

$$\frac{d\psi}{dt} = -\Delta\omega + \varepsilon q(\psi), \tag{1.22}$$

where $\Delta\omega = \omega - \omega_0$ is the degree of *frequency detuning*. Similarly, if $\omega \approx m\omega_0/n$ then

$$Q(\phi,t) \to \sum_k a_{-nj,mj} \mathrm{e}^{ij(m\phi - n\omega t)} = \hat{q}(m\phi - n\omega t) \tag{1.23}$$

and

$$\frac{d\psi}{dt} = m\omega_0 - n\omega + \varepsilon m \hat{q}(\psi), \tag{1.24}$$

where $\psi = m\phi - n\omega t$.

The above is an example of an application of the averaging theorem [**13**]. Assuming that $\Delta\omega = \omega - \omega_0 = \mathcal{O}(\varepsilon)$ and defining $\psi = \phi - \omega t$ we have

$$\frac{d\psi}{dt} = -\Delta\omega + \varepsilon Q(\psi + \omega t, t) = \mathcal{O}(\varepsilon). \tag{1.25}$$

Define

$$q(\psi) = \lim_{T\to\infty} \frac{1}{T}\int_0^T Q(\psi + \omega t, t)dt, \tag{1.26}$$

and consider the averaged equation

$$\frac{d\psi}{dt} = -\Delta\omega + \varepsilon q(\psi). \tag{1.27}$$

It is easy to establish that q only contains the resonant terms of Q as above. The averaging theorem states that there exists a change of variables that maps solutions of the full equation to those of the averaged equation. The question then remains as to what extent solutions of the averaged equations are a good approximation to the solutions of the full equation. In general, one can only establish that a solution of the full equation is ε-close to a corresponding solution of the averaged system for times of $\mathcal{O}(\varepsilon^{-1})$. No such problem occurs however for hyperbolic fixed points corresponding to phase-locked states.

Phase-locking and synchronization

Suppose that the 2π-periodic function $q(\psi)$ has a unique maximum q_{max} and a unique minimum q_{min} in the interval $[0, 2\pi)$. We can then distinguish between two regimes:

Synchronization regime: If the degree of detuning for a given drive amplitude is sufficiently small,

$$\varepsilon q_{min} < \Delta\omega < \varepsilon q_{\max},$$

then there exists at least one pair of stable/unstable fixed points (ψ_s, ψ_u). (This follows from the fact that $q(\psi)$ is 2π-periodic and thus has to cross any horizontal line an even number of times). The system evolves to the synchronized state

$$\phi(t) = \omega t + \psi_s,$$

in which the oscillator is *phase-locked* to the external drive and is *frequency entrained*. Note that the stability of a phase-locked state is determined by the sign of $q'(\psi)$ with $q'(\psi_s) < 0$ and $q'(\psi_u) > 0$ (see Figure 6(a)).

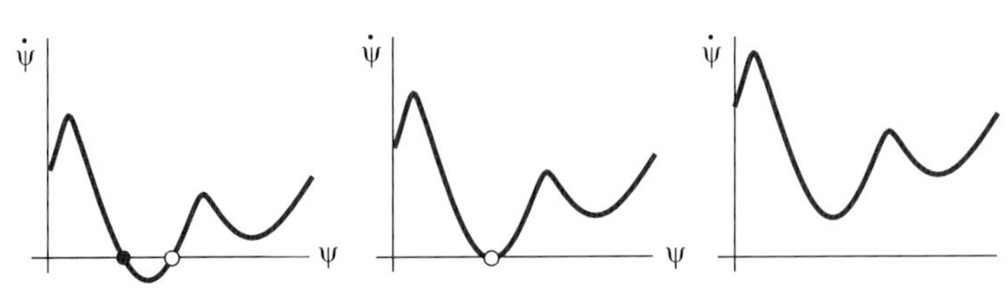

FIGURE 6. Saddle-node bifurcation signalling transition from a synchronized to a drifting state as size of frequency detuning $|\Delta\omega|$ increases.

Drift regime: As $|\Delta\omega|$ increases it approaches one of the critical values $\varepsilon q_{min,max}$ where the two fixed points coalesce in a saddle-node bifurcation and phase-locking disappears, see Figure 6(b,c). Hence, if the degree of tuning is large then $d\psi/dt$ never changes sign and the oscillation frequency differs from the drive frequency ω. The phase $\psi(t)$ rotates through 2π with period

$$T_\psi = \left| \int_0^{2\pi} \frac{d\psi}{\varepsilon q(\psi) - \Delta\omega} \right|. \tag{1.28}$$

The mean frequency of rotation is thus $\Omega = \omega + \Omega_\psi$ where $\Omega_\psi = 2\pi/T_\psi$ is known as the *beat frequency*. One is often interested in how the behavior varies in the $(\Delta\omega, \varepsilon)$-plane (see Figure 7). First the boundary between the two regimes consists of the two straight lines $\Delta\omega = \varepsilon q_{max,min}$. Second, close to the boundary Ω_ψ has a characteristic form. Suppose, for example, that $\Delta\omega - \Delta\omega_{max}$ is small for fixed ε with $\Delta\omega_{max} = \varepsilon q_{max}$. The integral in equation (1.28) is then dominated by a small region around ψ_{max}. Expanding $q(\psi)$ as a Taylor series,

$$\begin{aligned} \Omega_\psi = \frac{2\pi}{T_\psi} &\approx 2\pi \left| \int_{-\infty}^{\infty} \frac{d\psi}{\varepsilon q''(\psi_{max})\psi^2 - (\Delta\omega - \Delta\omega_{max})} \right|^{-1} \\ &= \sqrt{\varepsilon |q''(\psi_{max})| (\Delta\omega - \Delta\omega_{max})}. \end{aligned} \tag{1.29}$$

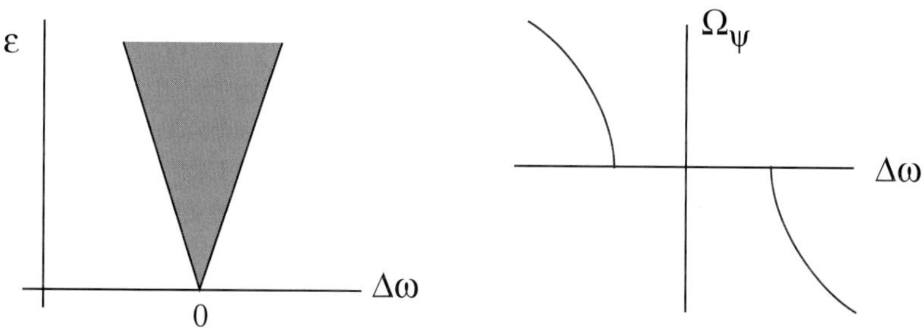

FIGURE 7. (a) Synchronization regime (shaded) in $(\Delta\omega, \varepsilon)$-plane. (b) Variation of beat frequency with $\Delta\omega$ for fixed ε.

1.3. Integrate–and–fire models

Integrate–and–fire (IF) neuron models neglect details regarding the spike generation process by reducing the latter to an all–or–nothing threshold event. That is, whenever the membrane potential crosses a firing threshold, the neuron fires a spike, typically modeled as a Dirac delta function, and the membrane potential is reset to some subthreshold value. The output of a spiking neuron can thus be characterized in terms of the sequence of threshold–crossing or firing times. Although they are less realistic than conductance–based models, they have provided a very useful platform for exploring computational issues such as neural coding and spike train statistics (see §3).

Leaky integrate–and–fire model

The simplest example of a spiking neuron is the so–called leaky integrate–and–fire (LIF) model [**15**]. The basic circuit of the LIF model consists of a capacitor C in parallel with a resistor R driven by a total current $I(t) = I_{\text{ext}}(t) + I_{\text{syn}}(t)$. The voltage $u(t)$ across the resistor then satisfies the differential equation

$$\frac{du}{dt} = -\frac{u(t)}{\tau} + \frac{I_{\text{ext}}(t) + I_{\text{syn}}(t)}{C}, \tag{1.30}$$

where $\tau = RC$ is the membrane time constant of the cell. The form of the action potential is not described explicitly. Spikes are formal events characterized by the ordered sequence of firing times $\{T^m, m \in \mathbf{Z}\}$ determined by the threshold crossing conditions

$$T^m = \inf\{t|u(t) = u_\kappa, t > T^{m-1}\}, \tag{1.31}$$

where u_κ is the firing threshold. Immediately after firing, the potential is reset to a value $u_r < u_\kappa$,

$$\lim_{t \to T^m_+} u(t) = u_r. \tag{1.32}$$

For simplicity we set $u_\kappa = 1$, $u_r = 0$ and $C = 1$.

One of the useful features of the LIF model is that one can analyze the response of a neuron to external and synaptic input currents without requiring that the inputs are sufficiently small as in phase reduction methods. We will illustrate this in the case of periodic forcing. Suppose that $I_{\text{syn}}(t) = 0$ and decompose the external drive as $I_{\text{ext}}(t) = I_0 + \varepsilon I_1(t)$ with $I_1(t)$ assumed to be Δ-periodic. Integrate the LIF equation between successive firing times T^n and T^{n+1} using $u(T^n_+) = 0$ and $u(T^{n+1}_-) = 1$:

$$\mathrm{e}^{T^{n+1}/\tau} = \tau I_0 \left[\mathrm{e}^{T^{n+1}/\tau} - \mathrm{e}^{T^n/\tau}\right] + \varepsilon \int_{T^n}^{T^{n+1}} \mathrm{e}^{t/\tau} I_1(t) dt. \tag{1.33}$$

When $\varepsilon = 0$ this equation can be solved explicitly to give

$$T^{n+1} - T^n = \tau \log \frac{\tau I_0}{\tau I_0 - 1} \equiv \Delta_0, \tag{1.34}$$

which describes an oscillator that fires at uniform time intervals of length Δ_0 (provided that $\tau I_0 > 1$), see Figure 8(a). We interpret $\omega_0 = 2\pi/\Delta_0$ as the natural

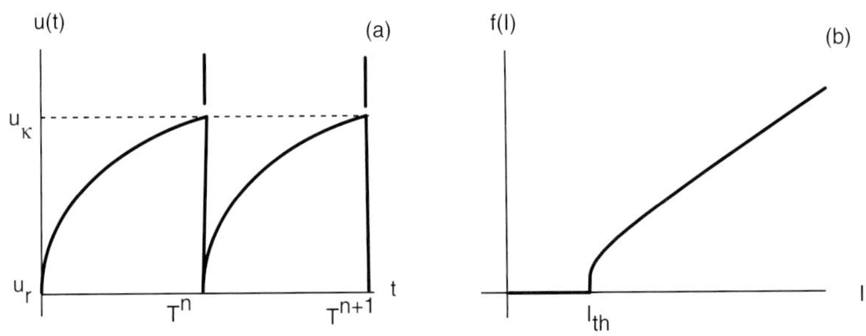

FIGURE 8. Integrate-and-fire oscillator. (a) Plot of $u(t)$ as a function of time t for constant current I. (b) Firing rate r as a function of current injection I with $I_{th} = 1/\tau$ when $u_\kappa = 1, u_r = 0$.

frequency of the LIF oscillator. The so–called $f - I$ curve relating the steady–state firing rate as a function of input current I_0 is then given by the firing rate function

$$f(I_0) = \frac{1}{\tau}\left[\frac{1}{\log(\tau I_0/[\tau I_0 - 1])}\right]_+ . \tag{1.35}$$

where $[x]_+ = 1$ if $x > 0$ and zero otherwise. The firing rate function f is plotted in Figure 8(b).

In order to analyze the case $\varepsilon \neq 0$, it is useful to introduce the function

$$F(t) = [\tau I_0 - 1]e^{t/\tau} + \varepsilon \int_{-\infty}^{t} e^{s/\tau} I_1(s)ds \tag{1.36}$$

such that the iterative firing time map can be written in the form

$$F(T^{n+1}) = F(T^n) + e^{T^n/\tau}. \tag{1.37}$$

This equation has been studied in great detail by Keener *et al* [**15**] for the particular case of the sinusoidal forcing $I(t) = I_0 + \epsilon \sin(2\pi t)$. Here the time-scale has been chosen so that the forcing period $\Delta = 1$. The function F then has the explicit form

$$F(t) = e^{t/\tau}\left(\tau I_0 - 1 + \frac{\epsilon\tau}{\sqrt{1 + 4\pi^2\tau^2}}\sin(2\pi t - \eta)\right), \tag{1.38}$$

where $\tan\eta = 2\pi\tau$. Keener *et al* [**15**] showed that major differences in behavior occur depending upon whether or not F is invertible. If F is invertible then we have an explicit first–order map

$$T^{n+1} = \Psi(T^n) \equiv F^{-1}\left[F(T^n) + e^{T^n/\tau}\right]. \tag{1.39}$$

The condition for invertibility is that $F'(t) \neq 0$ for all t, which implies that

$$I_0 - \frac{1}{\tau} > \varepsilon \max_t\{-I_1(t)\}. \tag{1.40}$$

The invertibility condition reduces to $I_0 - \tau^{-1} > \varepsilon$ in the case of sinusoidal forcing. Since (for $\Delta = 1$)

$$F(\Psi(t+1)) = e^{1/\tau}[F(t) + e^{t/\tau}] = e^{1/\tau}F(\Psi(t)) = F(\Psi(t) + 1),$$

it follows that $\Psi(t+1) = \Psi(t) + 1$. By identifying the interval $0 \leq t < 1$ with the unit circle we thus obtain a smooth invertible circle map. When F is not invertible, the mapping $T^n \to T^{n+1}$ of equation (1.37) can still be defined provided that F^{-1} is taken to be the smallest branch of the inverse of F, that is, T^{n+1} is taken to be the smallest value of $t, t > T^n$, for which equation (1.37) is satisfied. The dynamics is now described by a discontinuous, piecewise monotonic circle map.

The qualitative dynamics of the equivalent circle map can be characterized in terms of its rotation number ρ. This is defined according to [**15**]

$$\rho = \lim_{n\to\infty}\frac{T^n - T^0}{n}, \tag{1.41}$$

and measures the mean inter-spike interval (time between successive firings). It can be shown that ρ is independent of the initial state T^0 and is a continuous function of parameters in both the invertible and non–invertible regimes. If ρ is rational , that is, $\rho = p/q$ for integers p, q, then every sequence of firing times converges to a periodic orbit on which $T^{n+p} = T^n + q$, and $p : q$ phase–locking is said to occur. In the invertible regime, ρ is irrational on a parameter set of positive Lebesgue measure and the resulting quasiperiodic solutions are ergodic, that is, the sequence

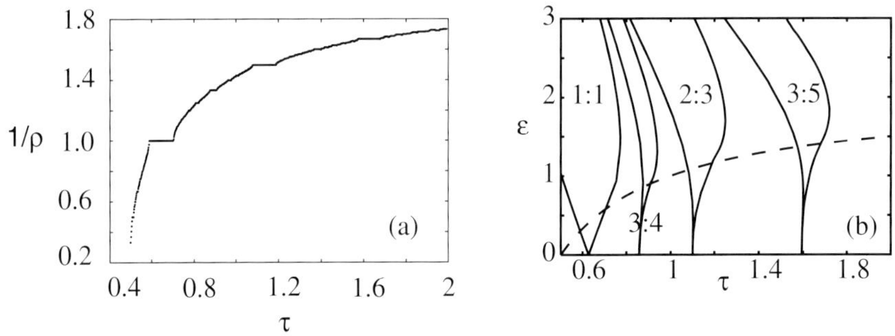

FIGURE 9. (a) Plot of the inverse mean firing period as a function of the membrane time constant τ for $I_0 - \epsilon = \tau^{-1}$ with $I_0 = 2$. Note that the dominant solutions are $1:1$, $3:4$, $2:3$ and $3:5$ (with increasing τ). (b) The Arnold tongue structure for the dominant modes. The dashed curve shows the boundary between invertible and non-invertible regimes of the firing time map

T_n mod 1 is dense in the interval $[0,1)$. On the other hand, in the noninvertible regime, ρ is irrational on a set of Lebesgue measure zero and the quasiperiodic orbits are no longer ergodic in $[0,1)$. In Figure 9(a) we plot $1/\rho$ as a function of τ for $\varepsilon = I_0 - \tau^{-1}$. The resulting devil's staircase structure shows that the preferred mode-locked solutions are those with low ratios of q to p. In Figure 9(b) we plot the borders of the Arnold tongues where these dominant modes become unstable in the (τ, ϵ)-plane. In the invertible regime both boundaries are associated with the disappearance of a phase–locked state via a smooth saddle-node bifurcation, whereas in the noninvertible regime the left–hand boundary signals the non-smooth disappearance of a stable phase-locked state [**15**].

Weak periodic forcing and phase averaging

For sufficiently small ε, the conditions for mode–locking reduce to those obtained using the phase reduction method of §1.2. Consider an LIF neuron with a weak periodic input $\varepsilon I(t)$ evolving as

$$\frac{du}{dt} = f(u) + \varepsilon I(t),$$

where $f(u) = I_0 - u$ and we have set $\tau = 1$. Taking $u_r = 0$ and $u_\kappa = 1$ the period of oscillation for $\varepsilon = 0$ is

$$T_0 = \int_0^1 \frac{dx'}{f(x')} = \log[I_0/(I_0 - 1)].$$

Introducing the phase variable

$$\theta(t) = \frac{2\pi}{T_0} \int_0^u \frac{dx'}{f(x')}, \tag{1.42}$$

we can transform the LIF equation to a phase equation identical in form to (1.16)

$$\frac{d\theta(t)}{dt} = \omega_0 + \varepsilon R(\theta) I(t),$$

where $\omega_0 = 2\pi/T_0$ and

$$R(\theta) = \frac{2\pi}{I_0 T_0} e^{T_0\theta/2\pi}, \quad 0 \leq \theta < 2\pi. \tag{1.43}$$

We identify the 2π–periodic function $R(\theta)$ as the PRC of an LIF neuron (see §1.2). Suppose that the input $I(t) = \cos \omega t$ with $\omega - \omega_0 = \mathcal{O}(\varepsilon)$. Introduce the slowly varying phase variable $\psi(t) = \theta(t) - \omega t$ such that

$$\frac{d\psi(t)}{dt} = \varepsilon\left[\Delta\omega + R(\psi + \omega t)\cos(\omega t)\right],$$

with $\omega - \omega_0 = \varepsilon\Delta\omega$. From the averaging theorem [**13**], solutions of this equation can be approximated by solutions of the reduced equation (at least up to times of order $1/\varepsilon$)

$$\frac{d\psi(t)}{dt} = \varepsilon\left[\Delta\omega + q(\psi))\right],$$

where

$$q(\psi) = \frac{1}{T}\int_0^T R(\psi + \omega t)\cos(\omega t)dt$$

and $\omega = 2\pi/T$. A fixed point of the reduced phase equation corresponds to a $1:1$ mode–locked solution, and from this the boundaries of the $1:1$ Arnold tongue can be deduced for small ε.

Spike response model

The LIF model assumes that it is the capacitative nature of the cell that in conjunction with a simple thresholding process dominates the production of spikes. The spike response (SR) model is a more general framework that can accommodate the apparent reduced excitability (or increased threshold) of a neuron after the emission of a spike [**16, 17, 7**]. Spike reception and spike generation are combined with the use of two separate response functions. The first, $\eta(t)$, describes the postsynaptic response to an incoming spike in a similar fashion to the LIF model, whereas the second, $\eta(t)$, mimics the effect of refractoriness. The refractory function $\eta(t)$ can in principle be related to the detailed dynamics underlying the description of ionic channels. In practice an idealized functional form is often used, although numerical fits to the Hodgkin-Huxley equations during the spiking process are also possible [**18**]. In more detail, a sequence of incoming spikes $\{T^m\}$ evokes a postsynaptic potential in the neuron of the form $u_s(t) = \sum_m \epsilon(t - T^m)$ where $\epsilon(t)$ incorporates details of axonal, synaptic and dendritic processing (see §2). The total membrane potential of the neuron is then taken to be

$$u(t) = \sum_p \eta(t - \widehat{T}^p) + \sum_m \epsilon(t - T^m), \tag{1.44}$$

where $\{\widehat{T}^p\}$ is the sequence of output firing times and $\eta(t)$ is the the refractory function reflecting the reduced excitability seen after firing. Since the reset condition of the LIF model is equivalent to a sequence of current pulses, $-(u_\kappa - u_r)\sum_p \delta(\widehat{T}^p)$, the LIF model is a special case of the SR model. More specifically, integrating the LIF equation (1.30) in the case of a synaptic input of the form $I(t) = \sum_m g_{\text{syn}}(t - T^m)$, we obtain the equivalent formulation (1.44) with

$$(1.45)\ \eta(t) = -(u_\kappa - u_r)\mathrm{e}^{-t/\tau}, \quad \epsilon(t) = \int_0^t \mathrm{e}^{-(t-t')/\tau} g_{\mathrm{syn}}(t')\mathrm{d}t', \quad t > 0.$$

Nonlinear IF model and parabolic bursting

In a general nonlinear IF model, equation (1.30) is replaced by [**19**]

$$\tau \frac{du}{dt} = F(u) + G(u)I. \tag{1.46}$$

(Note that one can always eliminate the nonlinear factor $G(u)$ by a change of variables). A specific example of a nonlinear IF model is the quadratic model

$$\tau \frac{du}{dt} = a_0(u - u_{\mathrm{rest}})(u - u_c) + RI, \tag{1.47}$$

with $a_0 > 0$ and $u_c > u_{\mathrm{rest}}$. For $I = 0$ and initial condition $u < u_c$, the voltage decays to the resting potential u_{rest}. For $u > u_c$ it increases to infinity in a finite time due to the quadratic nonlinearity, which defines when a spike is emitted. Immediately after firing, the neuron is reset to $-\infty$. Again the quadratic nonlinearity ensures that the membrane potential returns to a finite value in a finite time.

After shifting the voltage and current variables, the quadratic IF model can be rewritten as

$$\frac{\mathrm{d}u}{\mathrm{d}t} = \left(p + qu^2\right), \tag{1.48}$$

which represents the normal form of any neuron exhibiting a saddle–node on a limit cycle bifurcation [**20, 21**], see Figure 10. Such a neuron displays Class I excitability, which is characterized by the appearance of oscillations with arbitrary low frequency; the frequency increases with increasing injected current. This should be contrasted with Class II excitability for which the onset of repetitive firing occurs at a non-zero frequency that is relatively insensitive to changes in the applied current. Class II excitability is usually associated with a Hopf bifurcation. The

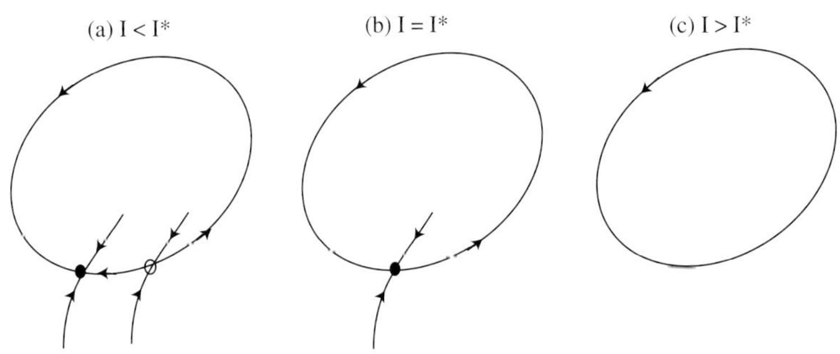

FIGURE 10. Schematic diagram of a saddle-node bifurcation on a limit cycle occurring as the applied current I is increased. (a) For $I < I^*$ there is a unique asymptotically stable fixed point and a saddle-point with a one-dimensional unstable manifold whose branches form a closed loop. (b) At criticality, $I = I^*$, the node and saddle coalesce. (c) For $I > I^*$ all that remains is a limit cycle

behavior of the model becomes more transparent by performing the change of variables $u = \tan(\theta/2)$ which gives the so–called theta model [**22**]

$$\frac{d\theta}{dt} = q(1 - \cos(\theta)) + p(1 + \cos(\theta)). \tag{1.49}$$

If q and p have opposite signs then equation (1.49) has two fixed points, one stable and the other unstable, and the neuron is in an excitable state. On the other hand, if q and p have the same sign (which we take to be positive) then there are no fixed points and the phase monotonically increases with time. In fact, one can solve equation (1.49) explicitly to obtain the solution

$$\theta(t) = 2\tan^{-1}\left[\sqrt{\frac{p}{q}}\tan\left(\sqrt{pq}t + \phi\right)\right] \tag{1.50}$$

for some arbitrary phase ϕ. It follows that the frequency of rotation is $\omega = 2\sqrt{pq}$. Each time $\theta(t)$ passes through π the potential $u(t)$ 'blows up' thus signalling the firing of a single spike in the full system.

A simple extension of the above canonical model of a Class I excitable neuron can be used to describe what is known as a parabolic burster [**20**]. *Bursting* is the rhythmic generation of several action potentials during a short time, followed by a period of inactivity. There are a wide variety of burst phenomena, but it appears that many are due to a similar underlying mechanism. We first note that the various chemical and electrical dynamics of the neuron operate on many time-scales, and for some neurons one can *dissect* their full dynamics into a fast system coupled to a slowly oscillating sub-system [**23, 1, 4, 5**]. Typically the fast system has a time-scale of milliseconds and models the membrane potential, and hence spike generation. The slow sub-system operates on a time-scale that varies from hundreds of milliseconds to tens of seconds and typically models trans-membrane ionic currents. For many types of bursting, the fast subsystem exhibits bistability between a quiescent fixed point and an active oscillatory state. At least one slow variable is then required to alternate back and forth through the region of bistability such that the neuron periodically jumps between the active and quiescent phases along a hysteresis loop. It is also possible to generate bursting without bistability in the fast subsystem, as in the case of parabolic bursting, provided there are at least two variables in the slow subsystem that periodically sweep the fast subsystem through a sequence of bifurcations. Both cases are illustrated in Figure 11.

Ermentrout and Kopell [**20**] showed how parabolic bursting models can be reduced to the following canonical form

$$\dot{\theta} = [1 - \cos(\theta)] + [1 + \cos(\theta)]\, g(\phi), \quad \dot{\phi} = \omega \tag{1.51}$$

where $(\phi, \theta) \in S^1 \times S^1$ is on a torus, the frequency of the slow oscillations is $\omega \ll 1$, and $g : S^1 \to \mathbf{R}$ is a continuous function describing the influence of ϕ on the fast variable θ. If g is negative for all ϕ then the fast variable is quiescent, and there is no bursting. If $g(\phi) > 0$ the fast variable fires all the time and is never quiescent. Bursting occurs when g changes sign over the interval $\phi \in [0, 2\pi)$. A typical choice of g is $g(\phi) = a + b\cos\phi$ resulting in the so called atoll model. Typical activity of this system in the (θ, ϕ) plane is shown in Figure 12 for $(a, b) = (0, 1)$.

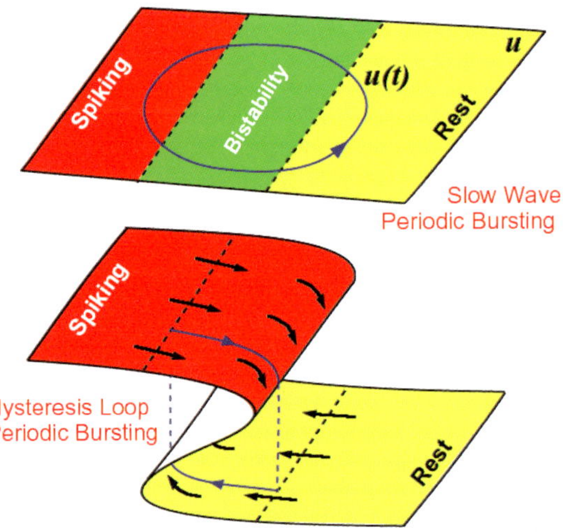

FIGURE 11. Slow variation in the parameter s switches the neuron between quiescent and spiking states resulting in bursting. In the case of slow wave bursting the periodic variation of s is independent of the fast system, whereas in hysteresis type bursting the slow and fast systems are mutually coupled. [Reprinted from [**1**], figure 9.17, by permission of the MIT Press.]

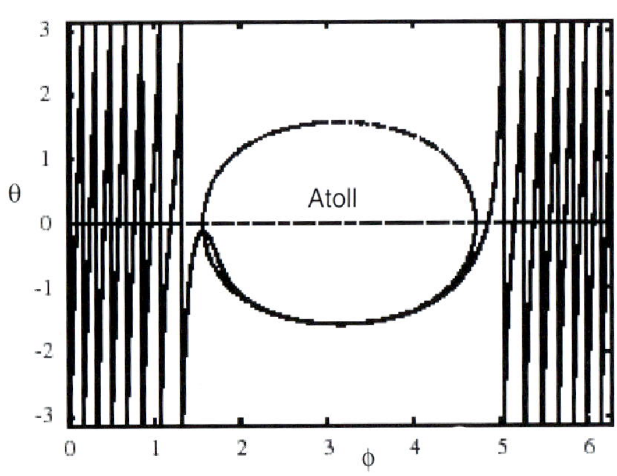

FIGURE 12. Activity of the atoll model for $\omega = 0.05, a = 0, b = 1$. The dashed circle is the set of equilibria for the fast variable θ. Outside the atoll region, θ completes full 2π rotations, so that the neuron fires action potentials whereas within the atoll θ is attracted to a fixed point and the neuron is quiescent. The system periodically enters and leaves the atoll due to the periodic rotation of the slow phase variable ϕ.

Integrate-and-fire-or-burst model

Several models of bursting neurons include a low-threshold calcium T-current [**24**]. Such a current is slowly de–inactivated when the neuron is sufficiently hyperpolarized so that it can cause the neuron to fire a burst of action potentials when released from sustained inhibition, so-called *post-inhibitory* rebound. Recently the integrate-and-fire model has been extended to incorporate this slow acting T-current leading to the aptly named *integrate-and-fire-or-burst* (IFB) model [**25**]. The current balance equation for the model is

$$C\frac{du}{dt} = -I - I_L - I_T, \tag{1.52}$$

where C is a membrane capacitance, I represents external or synaptic current, $g_L(u-u_L)$ is a leakage current with constant conductance g_L and reversal potential u_L. The low-threshold Ca^{2+} current is given by

$$I_T = g_T h(t)(u - u_T)H(u - u_h), \tag{1.53}$$

where H is a Heaviside function and the slow variable h has the dynamics

$$\frac{dh}{dt} = \begin{cases} -h/\tau_-, & u \geq u_h \\ (1-h)/\tau_+, & u < u_h \end{cases}. \tag{1.54}$$

The variable h, which represents slow de-inactivation of the low-threshold Ca^{2+} conductance, relaxes to zero with time constant τ_- when $u \geq u_h$ and relaxes to unity with time constant τ_+ when $u < u_h$. Assuming $u_h < u_r < u_\kappa$, where u_κ is the firing threshold, it follows that sustained hyperpolarization is required in order that the potential crosses the u_h boundary from above resulting in an increase in h. If the cell is then released from inhibition so that u recrosses the threshold u_h from below, a non-zero I_T current is generated that can depolarize the cell and produce a burst of spikes. In the absence of such hyperpolarization, the system acts as a regular integrate-and-fire neuron. Also note that if $u_r < u_h < u_\kappa$ then bursting can be induced by direct excitation, as occurs for inhibitory reticular cells (see below). In the case of a constant external input, it is possible to obtain an exact solution of the IFB model in terms of characteristic curves in the (u, h)–plane [**26**]. The Arnold tongue structure for the IFB model under periodic forcing has also been determined [**27**].

One particular structure where calcium T–currents are thought to be important is the thalamus. The thalamus is the primary relay station for most of the sensory information on route to the cerebral cortex. Oscillations at 7-14 Hz (spindles) and 4-7 Hz (delta) are common in cortex and thalamus during various stages of non-REM sleep [**28, 29**]. The EEG spindles are typical of brain electrical synchronization at the onset of sleep associated with the loss of perceptual awareness. It has been established that these oscillations are generated as the result of synaptic interactions between excitatory thalamocortical cells and GABA inhibitory neurons of the thalamic reticular nucleus. The latter is a thin sheet of inhibitory cells between the thalamus and cortex that is uniquely placed to play a regulatory role in levels of awareness. From a computational perspective, one can basically reduce sleep into two phases that correspond roughly speaking to two modes of firing of thalamic cells. During the slow oscillations of non-REM sleep, thalamic cells are in a bursting mode in which there is sufficient hyperpolarization to de-inactivate the low-threshold calcium current. Rather than communicating information about the sensory world, the thalamocortical neurons participate in a globally synchronized bursting state mediated by the corticothalamic loop. Such a coherent brain state may perform the function of consolidating memory traces acquired during wakefulness [**28**]. However, it can also lead to hypersynchronous episodes similar to those observed in some epileptic seizures. During REM sleep, the thalamic neurons are more depolarized such that the calcium current is inactivated. The neurons now fire in a tonic (regular spiking) rather than a bursting mode and may exhibit high frequency γ oscillations similar to the awake state. Transitions between non-REM sleep, REM sleep, and arousal are controlled by a variety of neurotransmitters released from brain stem nuclei including acetylcholine, norepinephrine and serotonin [**31**]. Until recently, it was thought that during wakefulness the thalamic cells only operate in tonic mode, which relays sensory information to the cortex. However, recent experimental studies indicate that burst firing can also serve as an effective relay mode in the awake state [**30**], see Figure 13. In particular burst firing is more effective than tonic firing in detecting weak signals in background noise. It

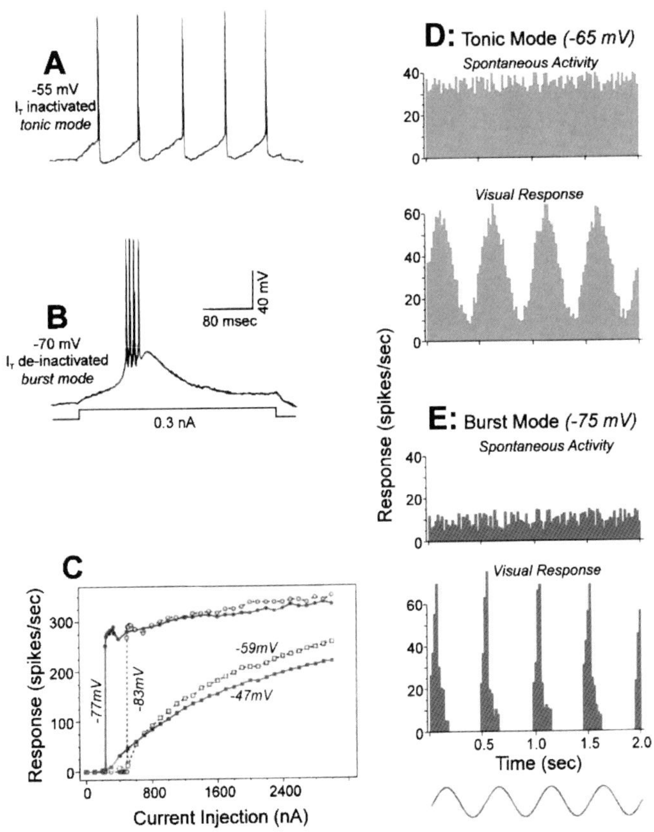

FIGURE 13. Two modes of firing in thalamic cells during the wake state. (A) Tonic firing with T–current inactivated. (B) Burst firing when the T–current is de-inactivated. (C) Firing rate as a function of current injection in the tonic mode (at -47 and -59 mV) and in the burst mode (-77 and -83 mV). The former is approximately linear whereas the latter is approximately a step function. (D) Peri-stimulus time histograms of a neuron in tonic mode showing response to a periodic stimuli and spontaneous activity in the absence of the stimulus. (E) Corresponding histograms in burst mode. [Reprinted from [**30**], with permission of Elsevier.]

is also possible that bursting is more effective in stimulating neurons downstream. This has led to the suggestion that the burst mode is used to detect novel stimuli. Once the change is detected, the neuron switches to tonic mode so that the new object can be analyzed more faithfully. Recently, IFB models have been used to investigate both bursting and tonic firing in models of thalamic tissue [**32, 33**].

LECTURE 2

Synaptic and dendritic processing

In the single neuron models discussed in §1, we decomposed the total input current into an external part $I_{\text{ext}}(t)$ and a synaptic part $I_{\text{syn}}(t)$. In order to develop recurrent network models, it is necessary to consider the detailed form of the synaptic current $I_{\text{syn}}(t)$ that mediates neuronal interactions. In this lecture we begin by considering a single synapse and the sequence of events underlying conductance changes in the postsynaptic membrane due to the arrival of an action potential at the presynaptic terminal. We then show how these conductance changes can be modeled in terms of a kinetic scheme describing the opening and closing of ion channels in the postsynaptic membrane. Typically, a single neuron in cerebral cortex has up to 10,000 synapses, which are spatially distributed along the dendritic tree (and perhaps on the cell body and proximal part of the axon). In order to find the total synaptic current $I_{\text{syn}}(t)$ entering the cell body, it is necessary to determine how the various local currents flow along the dendritic tree and combine at the soma. We show that if the dendrites are modeled as passive electrical cables, then the dendritic tree acts as a linear spatio-temporal filter of synaptic currents. We end the lecture by describing some biophysical models of synaptic plasticity, in which the strength or weight of a synapse is modifiable by experience. (The synaptic weight can be identified as the maximal conductance change induced by a single action potential). In particular, we consider the phenomenon of spike timing–dependent plasticity (STDP), where changes in synaptic weights depend on the precise timing of pre– and post–synaptic spikes at the millisecond level. STDP is emerging as an important paradigm in computational approaches to adaptive behavior and learning.

2.1. Excitatory and inhibitory synapses

The basic stages of synaptic processing induced by the arrival of an action potential at an axon terminal are shown in Figure 1. An action potential arriving at the terminal of a presynaptic axon causes voltage-gated Ca^2+ channels within an active zone to open. The influx of Ca^2+ produces a high concentration of Ca^2+ near the active zone [**34, 35**], which in turn causes vesicles containing neurotransmitter to fuse with the presynaptic cell membrane and release their contents into the synaptic cleft (a process known as exocytosis). The released neurotransmitter molecules then diffuse across the synaptic cleft and bind to specific receptors on the postsynaptic membrane. These receptors cause ion channels to open, thereby changing the membrane conductance and membrane potential of the postsynaptic cell. A single synaptic event due to the arrival of an action potential at time T induces a synaptic current of the form

$$I_{\text{syn}}(t) = g_{\text{syn}}(t - T)(u_{\text{syn}} - u(t)), \tag{2.1}$$

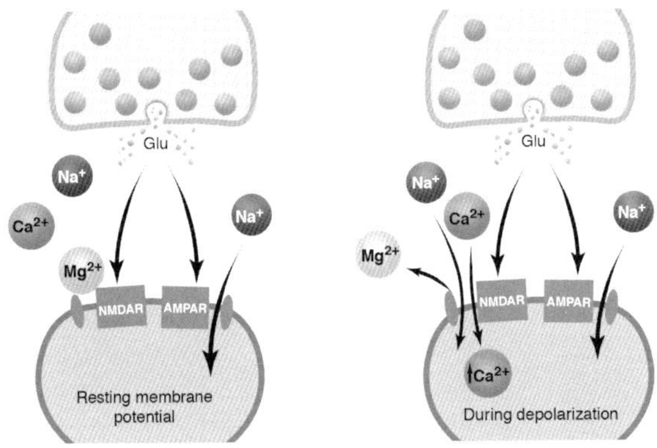

FIGURE 1. An NMDA synapse. Arrival of an action potential at the presynaptic terminal induces vesicles containing glutamate to merge with the cell membrane and release the neurotransmitter into the synaptic cleft. This binds to NMDA and AMPA receptors on the postsynaptic cell membrane. The AMPA associated channel opens but the NMDA channel is blocked by magnesium. Depolarization of the postsynaptic membrane unblocks magnesium allowing the influx of calcium. Depolarization may be caused by a back propagating action potential (BPAP) that travels up the dendrites. [Reprinted from [**38**], with permission from AAAS.]

where u is the voltage of the postsynaptic neuron, u_{syn} is the synaptic reversal potential and $g_{\text{syn}}(t)$ is the change in synaptic conductance with $g_{\text{syn}}(t) = 0$ for $t < 0$. The sign of u_{syn} relative to the resting potential u_{rest} (typically $u_{\text{rest}} \approx -65mV$) determines whether the synapse is excitatory ($u_{\text{syn}} > u_{\text{rest}}$) or inhibitory ($u_{\text{syn}} < u_{\text{rest}}$). For simplicity, it is often assumed that a neuron spends most of its time close to rest such that $u_{\text{syn}} - u \approx u_{\text{syn}} - u_{\text{rest}}$, with the factor $u_{\text{syn}} - u_{\text{rest}}$ absorbed into g_{syn} (see also §3.3). One is then effectively taking the arrival of a spike as generating a synaptic current rather than a change in conductance.

The predominant fast, excitatory neurotransmitter of the vertebrate central nervous system is the amino acid *glutamate*, whereas in the peripheral nervous system it is *acetylcholine*. Glutamate-sensitive receptors in the post-synaptic membrane can be subdivided into two major types, namely, NMDA and AMPA [**3**]. At an AMPA receptor the postsynaptic channels open very rapidly. The resulting increase in conductance peaks within a few hundred microseconds, with an exponential decay of around 1 msec. The time course of the synaptic conductance change can be modeled in terms of an nth state Markov process [**37**] (see §2.2). Usually a simplified representation of $g_{\text{syn}}(t)$ is used that is given by the difference of exponentials

$$g_{\text{syn}}(t) = \bar{g}\left(\frac{1}{\tau_2} - \frac{1}{\tau_1}\right)(\mathrm{e}^{-t/\tau_1} - \mathrm{e}^{-t/\tau_2})H(t), \tag{2.2}$$

with $H(t)$ the Heaviside function. In the limit $\tau_2 \to \tau_1 = \alpha^{-1}$, equation (2.2) reduces to the well known α function

$$g_{\text{syn}}(t) = \bar{g}\alpha^2 t\mathrm{e}^{-\alpha t}H(t). \tag{2.3}$$

These expressions for the conductance are also used for GABA inhibitory synapses (see below). In contrast to an AMPA receptor, the NMDA receptor operates about 10 times slower and the amplitude of the conductance change depends on the post-synaptic membrane potential, see Figure 1. If the postsynaptic potential is at rest and glutamate is bound to the NMDA receptor then the channel opens but it is physically obstructed by Mg^{2+} ions. As the membrane is depolarized, the Mg^{2+} ions move out and the channel becomes permeable to Na^+ and Ca^{2+} ions. The rapid influx of calcium ions due to the opening NMDA channels is thought to be the critical trigger for the onset of *long term potentiation* or LTP, a major component of synaptic plasticity (see §2.4).

The most common inhibitory neurotransmitter in the central nervous system of both vertebrates and invertebrates appears to be GABA. There are two major forms of postsynaptic receptors termed A and B. The GABA_A receptors opens channels selective to chloride ions, whose reversal potential $u_{\text{syn}} = -70mV$ is close to that of the resting potential. The postsynaptic conductance change is quite fast, rising within 1 msec and decaying within 10-20 msec. GABA_B receptors are at least 10 times slower and open ion channels selective for K^+ ions. Thus they tend to be considerably more hyperpolarizing with $u_{\text{syn}} \approx -100mV$. The two receptor classes tend to be segregated with GABA_A occurring at or close to the soma and GABA_B further out on the dendrites. Another way to distinguish between GABA_A and GABA_B receptors is that the former is ionotropic (as are NMDA and AMPA) whilst the latter is metabotropic [**3**]. Neurotransmitter binding to an ionotropic receptor directly opens an ion channel through a series of conformational changes of the receptor. On the other hand, neurotransmitter binding to a metabotropic receptor indirectly opens an ion channel elsewhere in the membrane through a sequence of biochemical steps.

Synaptic facilitation and depression

A single synaptic event due to the arrival of an action potential at time T induces a synaptic current of the form (2.1). As a crude approximation we might try summing individual responses to model the synaptic current arising from a train of action potentials arriving at times T^m, integer m:

$$I_{\text{syn}}(t) = \sum_m g_{\text{syn}}(t - T^m)(u_{\text{syn}} - u(t)). \tag{2.4}$$

Note that this sum only includes spikes for which $T^m < t$ since $g_{\text{syn}}(t) = 0$ for $t < 0$ (causality condition). For many synapses such a simple ansatz does not hold, since processes such as synaptic facilitation and short-term depression cause the amplitude of the response to depend on the previous history of presynaptic firing [**39, 40**]. One way to incorporate these history–dependent effects is to take [**41**]

$$I_{\text{syn}}(t) = \left[\sum_m A(T^m) g_{\text{syn}}(t - T^m)\right](u_{\text{syn}} - u(t)), \tag{2.5}$$

where the factor $A(T^m)$ scales the response evoked by an action potential by an amount that depends upon the details of the previous spike train data. One interpretation of the factor A is that it represents short-term (reversible) changes in the release probability for synaptic transmission. One possible mechanism for synaptic depression is a depletion in the number of vesicles that can readily fuse with the cell membrane, whereas facilitation is thought to be due to residual calcium remaining in active fusion regions [**42**].

A common phenomenological model of facilitation or depression is to assume that between spikes $A(t)$ relaxes at a rate τ_A to its steady state value of one, but that directly after the arrival of a spike it changes discontinuously. τ_A can vary between around 100msecs and a few seconds [**40**]. For facilitation A is increased, whilst for depression it is decreased. It is mathematically convenient to model the former as an additive process and the latter as a multiplicative process in order to avoid possible divergences. That is, $A \to A + \gamma - 1$ with $\gamma > 1$ for facilitation, and $A \to \gamma A$ with $\gamma < 1$ for depression. The two models may be written succinctly as

$$\tau_A \frac{dA}{dt} = (1 - A) + (\gamma - 1) \sum_n A^\beta \delta(t - T^n), \quad A(0) = 1 \tag{2.6}$$

with $\beta = 1$ for the multiplicative model and $\beta = 0$ for the additive model. They have solutions of the form

$$A(T^m) = 1 + (\gamma - 1) \sum_{n<m} \hat{\gamma}^{[m-n-1]\beta} \mathrm{e}^{-(T^m - T^n)/\tau_A}, \tag{2.7}$$

where $\hat{\gamma} = \gamma$ (depression) and $\hat{\gamma} = 1$ (facilitation). Assuming a regular sequence of incoming spikes $T^n - T^{n-1} = \Delta$ for all n we find that the asymptotic amplitude $A_\infty(\Delta) \equiv \lim_{m\to\infty} A(T^m)$ is given by

$$A_\infty(\Delta) = 1 + \frac{(\gamma - 1)}{(\mathrm{e}^{\Delta/\tau_A} - 1)} \qquad \text{(facilitation)}, \tag{2.8}$$

and

$$A_\infty(\Delta) = \frac{1 - \mathrm{e}^{-\Delta/\tau_A}}{1 - \gamma \mathrm{e}^{-\Delta/\tau_A}} \qquad \text{(depression)}. \tag{2.9}$$

One possible computational role for synaptic depression is as a mechanism for cortical gain control [**40**]. The basic idea can be understood from the dependence of the asymptotic amplitude $A_\infty(\Delta)$ on the stimulus frequency $f = \Delta^{-1}$. Assuming that $\tau_A \gg \Delta$, we can Taylor expand A_∞ in equation (2.9) to find that

$$A_\infty(f) \approx \frac{\Gamma}{f}, \tag{2.10}$$

where $\Gamma = \tau_A/(1-\gamma)$. The main point to note is that the postsynaptic response per unit time is approximately independent of f (assuming that each spike elicits the same response in the steady-state). This means that the synapse is very sensitive to changes in the stimulus frequency. The instantaneous response to a rapid increase Δf in the stimulus rate is given by $\Gamma \Delta f / f$. In other words, the synapse responds to relative rather than absolute changes in the rate of input stimulation.

2.2. Kinetic model of a synapse

Let $g_{\mathrm{syn}}(t) \sim s(t)$ where $s(t)$ is the fraction of synaptic receptor channels that are in an open conducting state. The probability of being in an open state depends on the presence and concentration $\mathcal{T}$ of neurotransmitter released by the presynaptic neuron. Assuming a first order kinetic scheme, in which a closed receptor in the presence of a concentration of neurotransmitter $\mathcal{T}$ equilibrates with the open receptor state, we have

$$C \overset{r_1(u,\mathcal{T})}{\underset{r_2(u)}{\rightleftharpoons}} O,$$

where C and O represent the closed and open states of the channel and $r_1(u,\mathcal{T})$ and $r_2(u)$ are the associated rate constants. However, in many cases synaptic channels are found to have time dependent properties that are more accurately modeled with a second order kinetic scheme. In fact the presence of one or more receptor sites on a channel allows the possibility of transitions to *desensitized states*. Such states are equivalent to the inactivated states of voltage-dependent channels. The addition of such a desensitized state to the first order process generates a second order scheme:

$$\begin{aligned} \frac{\mathrm{d}s}{\mathrm{d}t} &= r_1(u,\mathcal{T})(1-s-z) - [r_2(u)+r_3(u)]s + r_4(u)z, \\ \frac{\mathrm{d}z}{\mathrm{d}t} &= r_6(u,\mathcal{T})(1-s-z) - [r_4(u)+r_5(u)]z + r_3(u)s, \end{aligned} \tag{2.11}$$

where z is the fraction of channels in the desensitized state. All neurotransmitter dependent rate constants have the form $r_i(u,\mathcal{T}) = r_i(u)\mathcal{T}$. It is common for detailed Markov models of voltage-gated channels to assume that the voltage dependence of all rates takes a simple exponential form. However, it has been shown that the number of states needed by a model to more accurately reproduce the behavior of a channel may be reduced by adopting sigmoidal functions for the voltage-dependent transition rates (see Destexhe *et al* [**37**] for a discussion), so that we write

$$r_i(u) = \frac{a_i}{1+\exp\left[-(u-c_i)/b_i\right]}. \tag{2.12}$$

The a_i set the maximum transition rate, b_i the steepness of the voltage dependence and c_i the voltage at which the half-maximal rate is reached. Furthermore, the concentration of neurotransmitter can also often be successfully approximated by a sigmoidal function of the presynaptic potential u_{pre}:

$$\mathcal{T}(u_{\text{pre}}) = \frac{\mathcal{T}_{\max}}{1+\exp[-(u_{\text{pre}}-u_\Delta)/\Delta]}. \tag{2.13}$$

Here $\mathcal{T}_{\max}$ is the maximal concentration of transmitter in the synaptic cleft, u_{pre} is the presynaptic voltage, Δ gives the steepness and u_Δ sets the value at which the function is half activated. It is common to take $\Delta = 5$mV and $u_\Delta = 2$mV. One of the main advantages of using an expression such as (2.13) is that it provides a smooth transformation between presynaptic voltage and transmitter concentration from which postsynaptic currents can be easily calculated from (2.1), (2.11) (2.12) and (2.13).

Now consider the following second-order gating scheme

$$\begin{array}{ccc} C & \overset{r_1(\mathcal{T})}{\underset{r_2}{\rightleftharpoons}} & C_1 \\ \nwarrow & & \swarrow \\ r_4 & & r_3 \\ & O & \end{array}$$

where C and C_1 are the closed forms of the receptor, O is the open (conducting) form, and the r_i are voltage independent transition rates. Under certain assumptions it may be shown that this particular second order scheme describes the so-called alpha function response commonly used in synaptic modeling. The following approximations are required: (i) The transmitter concentration ($\mathcal{T}$) occurs as a pulse $\delta(t-t_0)$ for a release event occurring at time $t = t_0$. Then $r_1(\mathcal{T}) = r_1\delta(t-t_0)$ and (ii) The fraction of channels in C is considered constant and ~ 1. Then the kinetic equations (2.11) reduce to

$$\frac{\mathrm{d}\mathbf{p}(t)}{\mathrm{d}t} = \mathbf{Q}\mathbf{p}(t) + \mathbf{I}(t)$$

(assuming $\mathbf{p}(0) = \mathbf{0}$) where

$$\mathbf{Q} = \begin{pmatrix} -\frac{1}{\tau_1} & 0 \\ r_3 & -\frac{1}{\tau_2} \end{pmatrix}, \quad \mathbf{I}(t) = \begin{pmatrix} r_1\delta(t-t_0) \\ 0 \end{pmatrix}, \quad \mathbf{p} = \begin{pmatrix} z \\ s \end{pmatrix}$$

and $\tau_1 = 1/(r_2 + r_3)$, $\tau_2 = 1/r_4$ and z and s represent the fraction of receptors in the forms C_1 and O respectively. This Markov chain system has a solution of the form

$$\mathbf{p}(t) = \int_0^t \mathbf{G}(t-s)\mathbf{I}(s)\mathrm{d}s, \quad \mathbf{G}(t) = \mathrm{e}^{t\mathbf{Q}}.$$

The eigenvectors of $\mathbf{Q}$ are $(1, r_3/(\tau_2^{-1} - \tau_1^{-1}))$ and $(0, 1)$ with associated eigenvalues $-1/\tau_1$ and $-1/\tau_2$ respectively. Hence, one finds that

$$s(t) = r_1 r_3 \left(\frac{1}{\tau_2} - \frac{1}{\tau_1}\right)^{-1} (\mathrm{e}^{-(t-t_0)/\tau_1} - \mathrm{e}^{-(t-t_0)/\tau_2}), \quad t > t_0.$$

In the limit $\tau_2 \to \tau_1 \to \tau_s$ this reduces to an alpha function

$$s(t) = r_1 r_3 (t - t_0)\mathrm{e}^{-(t-t_0)/\tau_s}, \quad t > t_0.$$

This kinetic derivation of the alpha function only holds for $s \ll 1$ in order to remain consistent with condition (ii).

The time course of some ion channel open and closed states seem to follow a power law rather than multiexponential law at large times [**43**]. In order to understand such power law behavior, consider an ion channel with N closed states such that the transition to an open state can only take place from state 1 at one end of a chain

$$0 \overset{\alpha}{\leftarrow} 1 \underset{\beta_1}{\overset{\gamma_1}{\rightleftharpoons}} 2 \ldots \underset{\gamma_{N-1}}{\overset{\beta_{N-1}}{\rightleftharpoons}} N.$$

The corresponding kinetic equations are

$$\begin{aligned}
\frac{dp_1}{dt} &= \beta_1 p_2 - (\gamma_1 + \alpha)p_1, \\
\frac{dp_n}{dt} &= \gamma_{n-1}p_{n-1} + \beta_n p_{n+1} - (\gamma_n + \beta_{n-1})p_n, \quad 1 < n < N, \\
\frac{dp_N}{dt} &= \gamma_{N-1}p_{N-1} - \beta_{N-1}p_N.
\end{aligned}$$

In the following we take $\gamma_n = \beta_n = 1$ for all n and $\alpha = 1$, so that the system of equations describes a discrete diffusion process along a chain with a reflecting boundary at $n = N$ and an absorbing boundary at $n = 0$. In the large N limit, it can be shown that given the initial condition $p_n(0) = \delta_{n,1}$, the exact solution is

$$p_n(t) = \mathrm{e}^{-2t}[I_{n-1}(t) - I_{n+1}(t)], \tag{2.14}$$

where $I_n(t)$ is the modified Bessel function of integer order:

$$I_n(t) = \int_{-\pi}^{\pi} \mathrm{e}^{ink} \mathrm{e}^{2t\cos(k)} \frac{dk}{2\pi}.$$

By carrying out an asymptotic expansion for large t, it can be shown that

$$p_n(t) \approx \frac{n}{2\pi^{1/2}t^{3/2}}.$$

Define $F(t)$ to be the total probability of finding the system in a closed state:

$$F(t) = \sum_{n=1}^{N} p_n(t).$$

It follows that $dF/dt = -\alpha p_1$ and, hence, $F(t) \approx (\pi t)^{-1/2}$ for large N, t and $\alpha = 1$. More recently, it has been suggested that synapses with multiple states, which exhibit dynamics over a wide-range of time-scales and show power–law like behavior, could have some interesting computational properties [**44, 45**]. For example, it has been suggested that such synapses could provide a way of combining high levels of memory storage with long retention times [**44**].

Stochastic model of receptor binding

In the above kinetic scheme, the stochastic opening and closing of ion channels was modeled in terms of deterministic rate equations that keep track of the mean number of channels in each available state. An implicit assumption of such a model is that the number of ion channels is sufficiently large so that one can ignore fluctuations. In order take into account such fluctuations, it is necessary to consider the associated master equation. We will illustrate this by considering a simple first order kinetic scheme, in which the opening or closing of a channel is equivalent to the binding or unbinding a neurotransmitter to a postsynaptic receptor. For simplicity, we will assume that there is a fixed, uniform concentration $\mathcal{T}$ of neurotransmitter in the synaptic cleft. Consider a cell surface with a fixed number N of receptors. Let k_+ and k_- denote the binding and disassociation rates of neurotransmitter molecules. If B denotes the mean fraction of bound receptors then the standard kinetic binding equations is (from the law of mass action)

$$\frac{dB}{dt} = k_+\mathcal{T}(1-B) - k_-B.$$

However, this equation ignores fluctuations in the number of bound receptors, which become significant when N is sufficiently small. A stochastic version of the binding problem is described by an associated master equation for transitions between bound and unbound states. More specifically, if $P(n,t)$ denotes the probability that n out of N receptors are bound at time t, then the master equation takes the form

$$\frac{\partial P}{\partial t} = \lambda_+(n-1)P(n-1,t) + \lambda_-(n+1)P(n+1,t) - [\lambda_+(n) + \lambda_-(n)]P(n,t),$$

with

$$\lambda_+(n) = (N-n)k_+\mathcal{T}, \quad \lambda_-(n) = nk_-.$$

Define the mean number of bound receptors at time t by

$$\overline{n}(t) = \sum_{n=0}^{N} np(n,t).$$

By differentiating both sides with respect to t and using the master equation, it can be shown that in the limit of large N (where the upper limit in the sum can be taken to be ∞) we recover the kinetic equation with $B = \overline{n}/N$.

The steady-state solution $P_s(n)$ satisfies $J(n) = J(n+1)$ with

$$J(n) = \lambda_-(n)P_s(n) - \lambda_+(n-1)P_s(n-1).$$

Using the fact that n is a nonnegative integer, that is, $P_s(n) = 0$ for $n < 0$, it follows that $J(n) = 0$ for all n. Hence, by iteration,

$$P_s(n) = P_s(0)\prod_{m=1}^{n}\frac{\lambda_+(m-1)}{\lambda_-(m)} = P_s(0)\left[\frac{k_+\mathcal{T}}{k_-}\right]^n \frac{N!}{n!(N-n)!}. \tag{2.15}$$

Taking logs of both sides of this equation and use Stirling's formula $\log(n!) \approx n\log n - n$ it can be shown that for large n, N, $P_s(n) \approx p(\phi)$ where $\phi = n/N$,

$$p(\phi) = p(0)\mathrm{e}^{NA(\phi)} \tag{2.16}$$

and

$$A(\phi) = \phi\log(k_+\mathcal{T}/k_-) - \phi\log(\phi) - (1-\phi)\log(1-\phi).$$

Let ϕ^* be the (unique) stationary solution, that is, $A'(\phi^*) = 0$. Since N is large, the method of steepest descent leads to the Gaussian approximation

$$p(\phi) \approx p(0)\exp\left[NA(\phi^*) + NA''(\phi^*)(\phi-\phi^*)^2/2\right]. \tag{2.17}$$

Under this approximation, the mean and variance of the fraction of bound receptors are given by

$$\langle\phi\rangle = \phi^* = \frac{k_+\mathcal{T}}{k_- + k_+\mathcal{T}}, \quad \langle(\phi-\phi^*)^2\rangle = \frac{\phi^*(1-\phi^*)}{N}. \tag{2.18}$$

It is clear that fluctuations become negligible in the large–N limit.

2.3. Dendritic filtering of synaptic inputs

So far we have neglected the fact that the synapses of a neuron are spatially distributed across the neuron's dendritic tree, see Figure 2(a). At the simplest level, the dendritic tree can be treated as a passive electrical medium that filters incoming synaptic stimuli in a diffusive manner. The current flow and potential changes along a branch of the tree may be described with a second–order, linear partial differential equation commonly known as the *cable equation* [**46, 2**]. The cable equation is based on a number of approximations: (i) magnetic fields due to the movement of electric charge can be neglected, (ii) changes in ionic concentrations are sufficiently small so that Ohm's law holds, (iii) radial and angular components of voltage can be ignored so that the cable can be treated as one-dimensional medium, and (iv) dendritic membrane properties are voltage-independent, that is, there are no active elements. Given a distribution of synaptic inputs innervating the dendritic tree, what is the net synaptic current I_{syn} entering the soma or cell body of a neuron? In order to address this problem we consider, for simplicity, a semi-infinite uniform dendritic cable, $0 \leq x < \infty$, with the soma located at the end $x = 0$, see

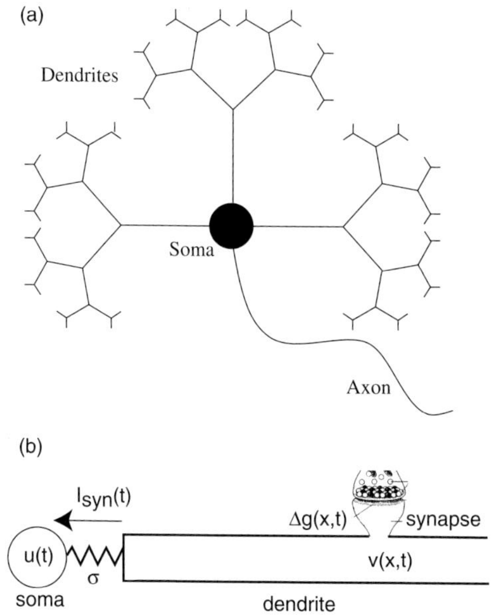

FIGURE 2. (a) Branching dendritic tree of an idealized single neuron. (b) Schematic diagram of a neuron consisting of a soma resistively coupled to one end of a dendritic cable. A synaptic conductance change $\Delta g(x,t)$ at position x on the cable induces a synaptic current into the soma at $x = 0$.

Figure 2(b). We model the soma as a LIF neuron that is passively coupled to the dendritic cable via a resistor with conductance σ,

$$\tau \frac{du}{dt} = -u + RI_{\text{ext}} + \sigma[v(0,t) - u(t)], \tag{2.19}$$

where $\sigma[v(0,t) - u(t)]$ is the net current density flowing into the soma from the dendritic cable at $x = 0$ and $v(x,t)$ is the membrane potential of the cable at position x. The dendritic potential $v(x,t)$ evolves according to the cable equation

$$\widehat{\tau}\frac{\partial v(x,t)}{\partial t} = -v(x,t) + \lambda^2 \frac{\partial^2 v(x,t)}{\partial x^2} + I(x,t), \tag{2.20}$$

where $\widehat{\tau}$ is the dendritic membrane time constant and λ is the corresponding space constant, both of which are determined by the passive electrical properties of the cable. Here $I(x,t)$ is the synaptic input at location x at time t:

$$I(x,t) = g(x)[V_{\text{syn}} - v(x,t)] \sum_m \alpha(t - T^m(x)), \tag{2.21}$$

where $\alpha(t)$ is an α function, say, $g(x)$ determines the maximal conductance and $\{T^m(x)\}$ is the sequence of spikes arriving into the synapses located at x. It follows from current conservation that we also have the boundary condition

$$-\frac{1}{r}\frac{\partial v}{\partial x}(0,t) = \sigma[v(0,t) - u(t)], \tag{2.22}$$

where r is the intracellular resistance per unit length of cable. Since we can eliminate the term $-\sigma u(t)$ in equation (2.19) by shifting the somatic membrane time constant τ, it follows that the total synaptic current into the soma is $I_{\rm syn}(t) = \sigma v(0,t)$.

Under the approximations $V_{\rm syn} \gg v$, we can formally solve the inhomogeneous boundary value problem for $v(0,t)$ using the Green's function G for the semi-infinite cable with a closed boundary [**47**]:

$$\begin{aligned} v(0,t) &= V_{\rm syn}\int_{-\infty}^{t}\int_{0}^{\infty} G(0,x',t-t')I(x',t')dx'dt' \\ &\quad -\sigma\int_{-\infty}^{t} G(0,0,t-t')[v(0,t')-u(t')]dt', \end{aligned} \tag{2.23}$$

with

$$G(x,y,t) = G_0(x-y,t) + G_0(x+y,t), \tag{2.24}$$

and

$$G_0(x,t) = \frac{1}{2\lambda\sqrt{\pi t/\hat{\tau}}}\mathrm{e}^{-t/\hat{\tau}}\mathrm{e}^{-\hat{\tau}x^2/4\lambda^2 t} \tag{2.25}$$

is the Green's function of the cable equation on an infinite domain. We see that the effective synaptic current $I_{\rm syn}$ flowing into the soma will itself be affected by the cell firing an action potential, due to the dependence of $v(0,t)$ on $u(t)$. This could be taken into account within the spike response model framework by using an effective reset function [**7**]. Here we will assume for simplicity that the second term on the r.h.s. of equation (2.23) is small compared to the first term arising from synaptic inputs. This approximation corresponds to imposing the homogeneous boundary condition $\partial v/\partial x|_{x=0} = 0$. It then follows that the total synaptic input into the soma is

$$I_{\rm syn}(t) = \sigma V_{\rm syn}\int_{-\infty}^{t}\int_{0}^{\infty} G(0,x',t-t')I(x',t')dx'dt'. \tag{2.26}$$

A similar analysis can also be carried out for more general dendritic topologies with the soma coupled to one of the terminals of the tree. We conclude that under the given approximations, the passive dendritic tree acts like a spatio-temporal linear filter on incoming spike trains, whose properties are determined by the underlying Green's function on the tree. The effects of the dendritic filtering of synaptic inputs on network dynamics is reviewed in some detail by Bressloff and Coombes [**48**].

Dendritic spines

The majority of excitatory synapses that occur in the cortex are located on tiny specialized protoplasmic protuberances called dendritic *spines* [**49**], see Figure 3. They typically occupy 20–70% of the total dendritic membrane. Since the input impedance of a spine head is typically large, a small excitatory synaptic current can produce a large local depolarization. Moreover, the thin stem neck connecting the spine to the main body of the dendrite provides an axial resistance that partially decouples the spine-head dynamics from the dendritic tree. Hence, it has long been theorized that the dendritic spine is a favorable site for the initiation of an action potential [**50, 51**], and is thus a possible substrate for local computations [**52**]. Modeling studies also suggest that if the heads of dendritic spines have excitable membrane properties, then the spread of current from one spine along the dendrites could bring adjacent spines to their thresholds for impulse generation. The result

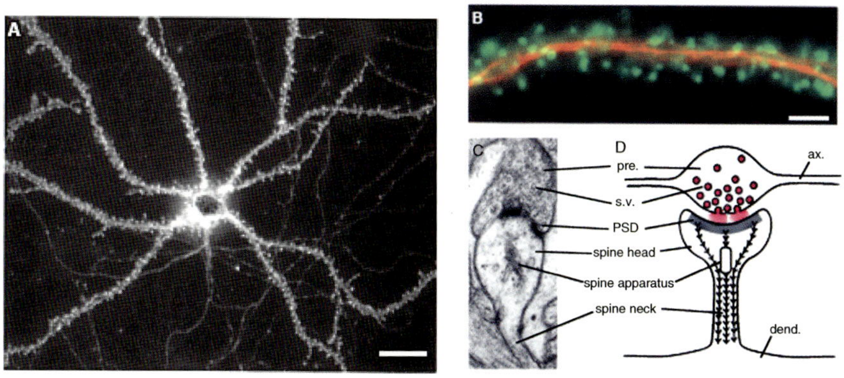

FIGURE 3. Dendritic spines. (A,B) Dendritic spines of a cultured hippocampal neuron revealed by green fluorescent protein (GFP) tagged actin. (C) A single spine revealed by electron microscopy. (D) Schematic diagram of a spine with barbed lines representing actin filaments.[Reprinted from Matus [**56**], with permission of AAAS.]

would be a sequence of spine-head action potentials, representing a saltatory propagating wave in the distal dendritic branches [**53, 54**]. Calcium imaging experiments provide strong evidence that the spine heads are endowed with voltage-dependent Ca^{2+} channels that can indeed support an all-or-nothing response to an excitatory synaptic input [**55**].

Early theoretical studies of spines also considered their potential role in synaptic plasticity and Hebbian learning. This was motivated by the idea that small changes in spine morphology, such as changes in the width of the spine neck, could lead to large changes in the amplitude of response to excitatory synaptic inputs on to the spine. It is now known that spines are rich in actin filaments, which have the capacity to drive such changes in spine shape [**56**]. Moreover, there is increasing experimental evidence that the growth and removal of spines provides an important substrate for structural changes during brain development [**57, 58, 59**]. It is less clear whether changes in spine morphology play a significant role in adult plasticity. Nevertheless, the basic geometry of a spine does provide an isolated biochemical microenvironment for Ca^{2+} to accumulate, and Ca^{2+} is thought to be a major chemical signal for the induction of synaptic plasticity [**60, 61**], see §2.4. The dynamics of calcium diffusion in dendritic spines has been explored in a number of computational models [**62, 63**].

2.4. Synaptic plasticity

An important property of a synapse is its effective strength, which can be identified as the maximal conductance change induced by the arrival of a single action potential eg. the factor $\bar{g}$ in equation (2.3). A great deal of experimental evidence has accumulated that these synaptic weights can undergo experience–dependent persistent changes over time. Such changes are thought to be the neural correlates of learning and memory. Most experimental studies of synaptic plasticity have focused on refining the original theoretical postulate due to Hebb [**64**]: *When an axon of cell A is near enough to excite cell B or repeatedly or persistently takes* *part in firing it, some growth process or metabolic change takes place in one or both cells such that A's efficiency, as one of the cells firing B, is increased.* A more modern interpretation of this postulate is that synaptic modification is driven by correlations in the firing activity of presynaptic and postsynaptic neurons.

In many regions of the brain, neurons are found to exhibit bidirectional plasticity in which the strength of a synapse can increase or decrease depending on the stimulus protocol [**65, 66, 67, 68, 69**]. Long term potentiation (LTP) is a persistent increase in synaptic efficacy produced by high-frequency stimulation of presynaptic afferents or by the pairing of low frequency presynaptic stimulation with robust postsynaptic depolarization. Long–term synaptic depression (LTD) is a long-lasting decrease in synaptic strength induced by low-frequency stimulation of presynaptic afferents. More recent experimental studies suggest that both the sign and degree of synaptic modification arising from repeated pairing of pre- and postsynaptic action potentials depend on their relative timing [**70, 71, 72**]. Long-term strengthening of synapses occurs if presynaptic action potentials precede postsynaptic firing by no more than about 50 ms. Presynaptic action potentials that follow postsynaptic spikes produce long-term weakening of synapses. The largest changes in synaptic efficacy occur when the time difference between pre- and postsynaptic action potentials is small, and there is a sharp transition from strengthening to weakening. This phenomenon of spike timing–dependent plasticity (STDP) is illustrated in Figure 4. In the following, we consider some of the molecular mechanisms

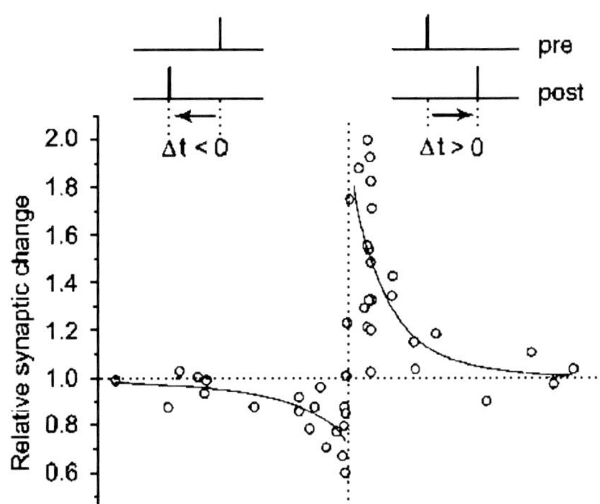

FIGURE 4. Spike-timing-dependent synaptic plasticity observed in hippocampal neurons. Each data point represents the relative change in the amplitude of evoked postsynaptic current after repetitive application of pre- and postsynaptic spiking pairs (1 Hz for 60 s) with fixed spike timing Δt, which is defined as the time interval between pre- and postsynaptic spiking within each pair. Long-term potentiation (LTP) and depression (LTD) windows are each fitted with an exponential function. [Reprinted from Bi and Poo [**73**], with permission from J. Neurosci.]

underlying bidirectional synaptic plasticity and how they can be modeled. Before proceeding, however, it is useful to distinguish between the induction, expression and maintenance of LTP and other forms of synaptic plasticity. LTP induction refers to the early steps that trigger the modification process, LTP expression is the proximal cause of synaptic modifications that may last hours or days, and LTP maintenance concerns the more general problem of how a synaptic change can endure over longer periods in the face of constant molecular turnover. The maintenance of synaptic plasticity, which requires modifications in gene expression and protein synthesis, is still poorly understood and will not be discussed further here.

NMDA receptors and calcium influx

Calcium has long been suggested as a major signaling agent for the induction of LTP and LTD [**74, 75, 66, 76**], and more recently for STDP [**72**]. However, the details of the induction mechanism depend on the type of synapse and the stimulus protocol. We will focus on NMDA–dependent mechanisms, which arise in the most commonly studied synapses, namely those connecting the axons of the Schaffer collateral/commissural system to pyramidal neurons in the CA1 region of the hippocampus. As illustrated in Figure 1, calcium can enter these cells through channels controlled by NMDA receptors. Two conditions are necessary for the opening of such channels: (i) arrival of an action potential at the presynaptic terminal releases the excitatory neurotransmitter glutamate which then binds to NMDA receptors and (ii) the postsynaptic cell is sufficiently depolarized so that magnesium ions are unblocked from the NMDA channels allowing the influx of calcium. One source of strong depolarization is the back propagation of an action potential into the dendritic tree, which is a signature that the postsynaptic neuron has fired a spike. Given that calcium ions can enter the cell only if glutamate has been released by presynaptic activity *and* if the postsynaptic membrane is sufficiently depolarized, it follows that the NMDA receptor is a natural candidate for the biophysical implementation of a Hebbian–like mechanism for detecting correlations between presynaptic and postsynaptic activity.

Recently Shouval *et al* [**77**] have proposed a mechanism for calcium–induced bidirectional synaptic plasticity, based on the idea that the maximum level of postsynaptic calcium concentration determines whether LTP or LTD occurs. That is, at intermediate calcium levels LTD occurs whereas at high calcium levels LTP occurs. Suppose that a presynaptic spike arrives at an NMDA synapse at time T and the postsynaptic membrane potential is $u(t)$. The calcium current through an NMDA receptor–controlled channel is taken to be of the form [**77**]

$$I_{\rm Ca}(t) = \bar{g}_{\rm Ca}\alpha(t-T)[u(t) - u_{\rm Ca}]B(u(t)), \tag{2.27}$$

where $\bar{g}_{\rm Ca}$ is the maximal conductance of the channel and $u_{\rm Ca}$ is the reversal potential of calcium ($130mV$). The function $\alpha(t-T)$ describes the time course of binding to NMDA receptors and is typically taken to be the sum of a fast ($\tau_f = 50ms$) and a slow ($\tau_s = 200ms$) exponential:

$$\alpha(t) = I_f \mathrm{e}^{-t/\tau_f} + I_s \mathrm{e}^{-t/\tau_s}, \quad t > 0. \tag{2.28}$$

The experimentally fitted function

$$B(u) = \frac{1}{1 + \eta[Mg^{2+}]\mathrm{e}^{-\gamma u}}, \tag{2.29}$$

with $\gamma = 0.062, \eta = 1/3.57$, describes the unblocking of the channel at depolarized levels of membrane potential [**78**]. The extracellular magnesium concentration is approximately $1mM$ (10^{-3} molars or moles per liter). If there are several presynaptic spikes within 100ms then calcium accumulates inside the cell. The change in the calcium concentration $[\mathrm{Ca}^{2+}]$ can be described by the simple first order kinetic equation

$$\frac{d[\mathrm{Ca}^{2+}]}{dt} = I_{\mathrm{Ca}}(t) - \frac{[\mathrm{Ca}^{2+}]}{\tau_{\mathrm{Ca}}} \tag{2.30}$$

with $\tau_{\mathrm{Ca}} = 50ms$.

In order to completely determine the calcium current, it is necessary to specify the dynamics of the postsynaptic membrane potential $u(t)$. In the case of STDP, the main source of depolarization is thought to arise from a back propagating action potential (although this might not be the whole story [**79**]). This can be modeled by taking [**77**]

$$u(t) = \eta(t - \widehat{T}), \quad \eta(t) = U\left(\widehat{I}_s \mathrm{e}^{-t/\hat{\tau}_s} + \widehat{I}_f \mathrm{e}^{-t/\hat{\tau}_f}\right) \tag{2.31}$$

where $\widehat{T}$ is the last firing time of the postsynaptic cell and $\eta(t)$ is the profile of the BPAP, which is taken to be the sum of a fast ($\hat{\tau}_s = 5ms$) and a slow ($\hat{\tau}_f = 25ms$) exponential. Equations (2.27), (2.30) and (2.31) determine how the maximum level of calcium influx depends on the relative timing of presynaptic and postsynaptic spikes, and hence whether LTP or LTD occurs (see Figure 5) [**77**]. One potential

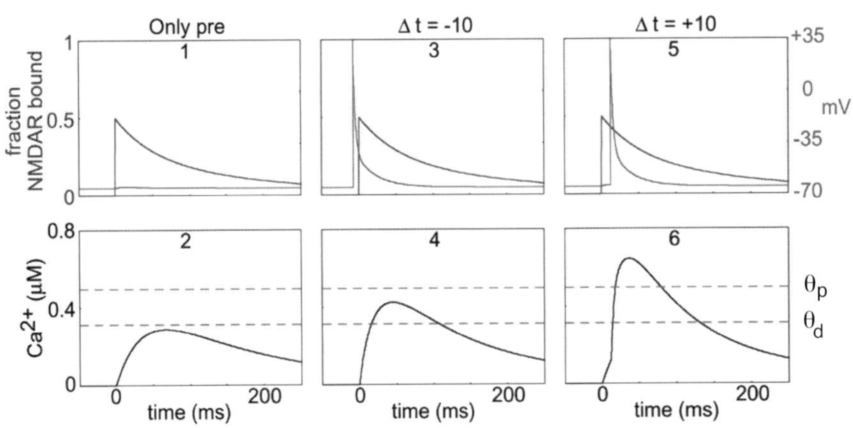

FIGURE 5. (1,2) Presynaptic stimulation alone results in the binding of glutamate to NMDA receptors but only a small depolarization so that there is only moderate calcium influx. (3,4) Post–pre stimulation ($\Delta t = -10ms$) results in a large brief depolarization due to a BPAP, which partially overlaps with the onset of glutamate binding to NMDA receptors. The calcium influx is increased so that it crosses the threshold θ_d for LTD and there is a negative weight change. (5,6) Pre–post stimulation ($\Delta t = +10ms$) results in a large brief depolarization that completely overlaps with the period of significant glutamate binding to NMDA receptors. The resulting calcium influx is well above the threshold θ_p for LTP and there is a positive weight change. [Reprinted from Shouval *et al.* [**77**], with permission of the National Academy of Sciences.]

problem with this particular calcium control mechanism for STDP is that it predicts LTD can also be induced by a sequence of pre–post spikes if the separation between spikes is larger than about $\Delta t = 40ms$. The reason is that in this case the removal of the magnesium block (induced by the BPAP) occurs at a time when the probability of glutamate binding is reduced. As a consequence less calcium enters the cell so that only the LTD threshold is reached. There is currently no convincing experimental evidence for such a form of LTD. This has motivated a more detailed biophysical model in which the time course of the calcium current rather than just its level acts as signal for STDP [**80**].

Regulation of AMPA receptors

There is growing experimental evidence that a major component of the postsynaptic expression of LTP and LTD is the regulation of AMPA receptor trafficking, which results in changes in receptor number at the postsynaptic membrane, and hence modifications in synaptic strength [**81, 82, 83**]. The total number of synaptic AMPA receptors is determined by a nonequilibrium steady–state, in which the flux of receptors between the synaptic and extrasynaptic regions of the postsynaptic membrane is balanced by the rate of exchange of receptors with intracellular recycling pools (see Figure 6).

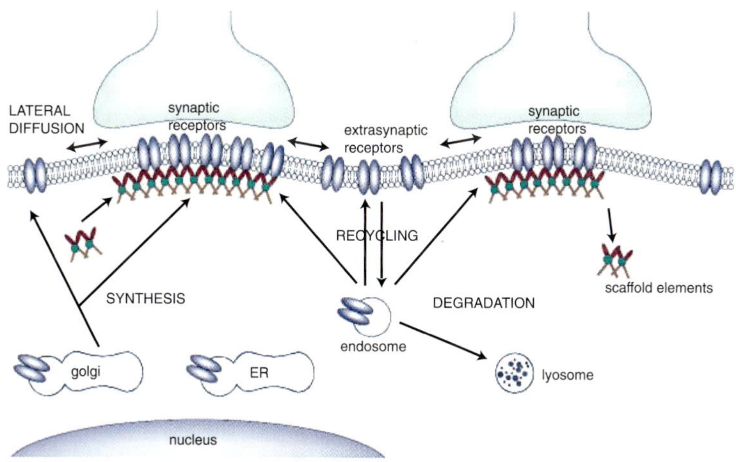

FIGURE 6. AMPA receptor trafficking involves exchanges between synaptic, extrasynaptic and intracellular compartments. The extrasynaptic diffusive receptors are immobilized by the scaffold-cytoskeleton complex at postsynaptic densities (PSDs). The extrasynaptic receptors are also exchanged with the intracellular recycling pool. Increasing or diminishing the number (or affinity) of scaffold molecules at the synapse changes proportionally the confinement capabilities, and subsequently the number of receptors at steady state. [Reprinted from Choquet and Triller [**82**], with permission from Macmillan Publishing Ltd: *Nature Reviews Neuroscience*, copyright 2003.]

A mathematical model of AMPA receptor trafficking has recently been constructed [**84**] in terms of a simplified two-compartment model of a dendritic spine's surface membrane (see Figure 7). The first compartment represents the postsynaptic density (PSD) of the spine head, which is the protein-rich domain in the

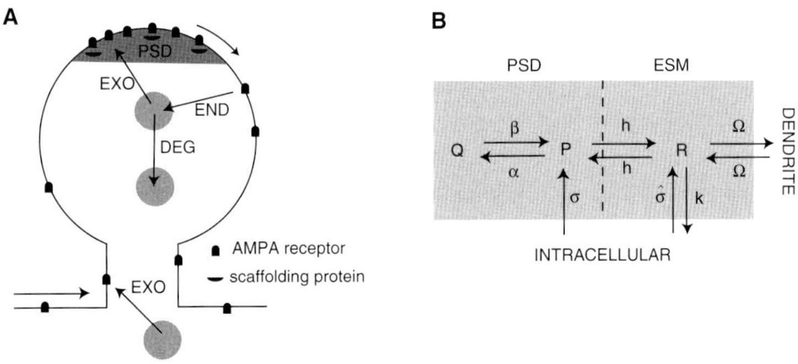

FIGURE 7. Compartmental model of AMPA receptor trafficking. (A) Schematic of AMPA receptor trafficking at a dendritic spine. Receptors stored in intracellular pools are continually exchanged with surface receptors through exo/endocytosis (EXO/END) and sorted for degradation (DEG). Surface receptors diffuse in the dendritic membrane and can be immobilized at the PSD through interactions with scaffolding proteins. (B) Simplified two-compartment model of a dendritic spine. Free receptors (P) bind to scaffolding proteins within the PSD to form bound receptors (Q) at a rate α and unbind at a rate β. Free receptors flow between the the PSD and ESM at a hopping rate h, and flow between the ESM and surface of the dendritic cable at a rate Ω. Free receptors (R) within the ESM are internalized at a rate k. Receptors are inserted into the PSD at a rate σ.

postsynaptic membrane of a dendritic spine that is directly apposed to the presynaptic active zone, and the second compartment represents the extrasynaptic membrane (ESM) of the remaining spine head and neck. AMPA receptors are assumed to diffuse freely in the extrasynaptic membrane. Within and near the PSD, however, AMPA receptor diffusion proceeds in a highly obstructed fashion. This is probably due to a number of factors including the attraction and binding of receptors to scaffolding proteins, the transient corralling of receptors to restricted domains by the underlying actin cytoskeleton, and the repulsion of receptors by picket-like transmembrane proteins. Single-particle tracking data suggests that the PSD acts as a confinement domain for diffusing receptors, and that about half of the AMPA receptors within the PSD are mobile [**82**]. Experimental estimates for the diffusivity D in the PSD and ESM range from 0.01 to 0.5 $\mu m^2 s^{-1}$. Given that the typical length–scale of a spine is $\ell_{sp} = 1\mu m$, it follows that the time–scale $\sqrt{\ell_{sp}^2/D}$ for diffusion within each compartment is 1-10 sec, which is faster than the typical time–scale (minutes) for the expression of LTP/LTD. Thus, assuming that each compartment is homogeneous, the corresponding free receptor concentration can be taken to be spatially uniform. In order to model the confinement of diffusing particles within the PSD, the boundary between the PSD and ESM compartments is treated as a potential barrier over which receptors must hop in order to enter or exit the PSD. For simplicity, the net flux across the boundary is taken to be proportional to the difference in concentrations on either side of the barrier, with the hopping rate dependent on the barrier height. Receptors can also hop between the ESM and the surrounding membrane of the dendritic cable, with the net flux proportional to the difference between the free receptor concentrations in the ESM and dendrite. The latter is taken to have a fixed background receptor density. The hopping rate between the ESM and dendrite incorporates the effects of spine neck geometry on restricting the flow of receptors. In addition to the lateral movement of AMPA receptors within the plasma membrane, there is a continual exchange of surface receptors with pools of intracellular receptors through exo/endocytosis. Experimentally it is found that under basal conditions, the rates of receptor exo/endocytosis are of the order 10-30 minutes. (In fact there are at least two separate sources of intracellular AMPA receptors, one consisting of GluR1/2 heteromers that are inserted into the ESM during LTP and the other consisting of GluR2/3 heteromers that are inserted into the PSD during constitutive recycling. Here we do not distinguish between the two receptor types.)

Let P, Q denote the concentration of free and bound receptors in the PSD and R denote the corresponding free receptor concentration in the ESM. It is also assumed that there is a fixed concentration L of active binding sites within the PSD. Receptor trafficking in the PSD is then described in terms of the kinetic equations [**84**]

$$\frac{dP}{dt} = \sigma - \alpha[L-Q]P + \beta Q - h[P-R] \tag{2.32}$$

$$\frac{dQ}{dt} = \alpha[L-Q]P - \beta Q \tag{2.33}$$

where σ is the rate of exocytosis of receptors into the PSD, and α, β denote the rates at which receptors bind and unbind to scaffolding proteins. The corresponding kinetic equation for the receptor concentration in the ESM is given by

$$\frac{dR}{dt} = \hat{\sigma} - kR + h[P-R] - \Omega[R-\overline{R}] \tag{2.34}$$

where $\hat{\sigma}$ is the rate of exocytosis into the ESM and k is the rate of endocytosis. The last term on the right–hand side of equation (2.34) represents the hopping of receptors between the ESM and surrounding dendritic cable with hopping rate Ω. The fixed background receptor concentration in the dendritic membrane is denoted by $\overline{R}$. Two basic assumptions of the model are (1) the strength of a synapse can be identified with the number of synaptic receptors $(P+Q)$ and (2) there is a separation of time-scales between the activation of the signaling pathways by the postsynaptic calcium signal during the induction of LTP/LTD (seconds) and the subsequent expression of LTP/LTD via receptor trafficking (minutes). One can then determine the time–dependent variation in receptor number induced by rapid modifications in one or more model parameters and compare these with experimentally measured time courses for LTP/LTD. Good agreement is obtained for physiologically reasonable choices of parameters provided that changes in receptor trafficking are correlated with changes in the concentration L of binding sites (see [**84**] for more details)

Model of a CaMKII switch

One of the major challenges concerning synaptic plasticity is unraveling the signaling pathways that convert a transient elevation in Ca^{2+} concentration, which can last for just a few hundred milliseconds, to a persistent change in synaptic strength, which can last for hours or even days. One molecule that appears to play a critical role in LTP is calcium/calmodulin–dependent protein kinase II (CaMKII). It has been shown that CaMKII activation is persistent, is required for LTP and can by itself produce potentiation (see [**85**] for a review). Further evidence that CaMKII may serve as a memory molecule is provided by the finding that it has autocatalytic properties that would allow it to act as a bistable molecular switch. The basic biochemical autocatalytic mechanism is as follows (see Figure 8(a,b)). CaMKII is composed of two rings, each with six kinase subunits. Each subunit has a single

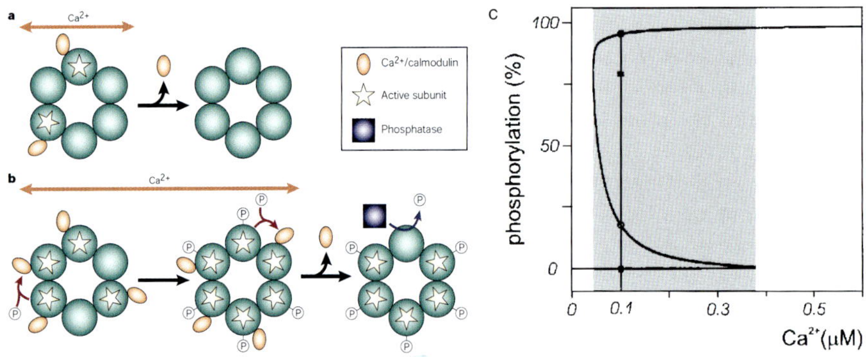

FIGURE 8. (a) At low Ca^{2+} concentrations there is little autophosphorylation, so that subunits of CaMKII are only active if bound by Ca^{2+}/calmodulin. (b) A transient increase in Ca^{2+} concentration causes an increase in the rate of phosphorylation leading to persistent activity after Ca^{2+} falls. [Reprinted from Lisman *et al* [**85**], with permission from Macmillan Ltd: *Nature Reviews Neuroscience*, copyright 2002.] (c) Steady–state levels of phosphorylation as a function of Ca^{2+} concentration calculated using a biophysical model [**86**]. A pair of stable states exists over a range of concentrations.

phosphorylation site that, when activated, makes the subunit active independently of Ca^{2+}. In the absence of phosphorylation, a subunit is only active when bound by a calcium binding protein known as Ca^{2+}/calmodulin. Initial autophosphorylation of a CaMKII ring requires the binding of two Ca^{2+}/calmodulin molecules, one to activate a catalytic subunit and another to expose the phosphorylation site of a neighboring substrate subunit. At the resting Ca^{2+} concentration, this process is slow with an average rate of one initial autophosphorylation every few hours. Once one or more sites of a ring are phosphorylated, subsequent phosphorylation is much faster since only a single Ca^{2+}/calmodulin molecule is required, namely, to expose the phosphorylation site on a substrate subunit adjacent to an already activated subunit.

A recent biophysical model [**86, 87, 88**] has shown how the interplay between autophosphorylation and dephosphorylation by phosphotase-1 (PP1) enzymes can generate two stable states of phosphorylation at resting levels of Ca^{2+}, see Figure 8(c). A transient increase of Ca^{2+}, induced for example by an LTP stimulus protocol, can switch the system from an unphosphorylated (DOWN) state to a persistent, highly phosphorylated (UP) state. Such a persistent change in the activation of CaMKII following LTP induction could then mediate persistent changes in synaptic strength. The system of equations describing the reaction kinetics takes the following form [**86, 87**]:

$$\frac{dP_0}{dt} = -v_1 M P_0 + v_3 P_1$$
$$\frac{dP_1}{dt} = v_1 M P_0 - v_3 P_1 - v_2 P_1 + v_3 2 P_2$$
$$\frac{dP_j}{dt} = v_2 w_{j-1} P_{j-1} - v_3 j P_j - v_2 w_j P_j + v_3 (j+1) P_{j+1}$$
$$\frac{dP_M}{dt} = v_2 P_{M-1} - v_3 M P_M$$

where P_j denotes the concentration of j–fold phosphorylated rings, M is the number of subunits on a ring, v_1 is the initial per site rate of autophosphorylation, v_2 is the per site rate of autophosphorylation of partially phosphorylated CaMKII, and v_3 is the rate of dephosphorylation by PP1. The factors w_j count the number of ring configurations that can undergo the transition $P_j \rightarrow P_{j+1}$. The autophosphorylation rates v_1 and v_2 are taken to depend on calcium concentration according to

$$v_1 = \frac{k_1([Ca^{2+}]/K_{H1})^8}{(1 + ([Ca^{2+}]/K_{H1})^4)^2}, \quad v_2 = \frac{k_1([Ca^{2+}]/K_H 1)^4}{1 + ([Ca^{2+}]/K_{H1})^4} \tag{2.35}$$

where k_1 is a rate constant, and the resting Ca^{2+} concentration $(100nM)$ is well below the constant K_{H1} $(0.7\mu M)$. Dephosphorylation proceeds according to a Michaelis-Menten scheme:

$$SP + E \rightleftharpoons SPE \rightarrow E + S,$$

where S is an unphosphorylated subunit, SP is a phosphorylated subunit, and E denotes the enzyme PP1. Thus

$$v_3 = \frac{k_2[E]}{K_M + \sum_{j=1}^{M} j P_j} \tag{2.36}$$

for constants k_2, K_M. The concentration $[E]$ of PP1 is itself modulated by calcium concentration as detailed elsewhere [**86**]. The resulting full system of equations can be solved numerically to determine the steady state levels of phosphorylation as a function of Ca^{2+} concentration, and hence establish bistability.

Recall from §2.2 that certain care has to be taken in using rate equations when the number of reacting molecules is small, since stochastic fluctuations become important. Recent measurements indicate that for a typical postsynaptic density the number of CaMKII molecules is only around 30 [**89**]. In order to determine the stability of the CaMKII switch with respect to fluctuations, Monte Carlo simulations of the full stochastic chemical equations have recently been carried out. These simulations suggest that even a small number of molecular switches can maintain stable persistent activation [**88**].

Computational role of STDP

One of the problems with standard formulations of Hebbian learning is that without additional constraints, the level of activity in a network can grow or shrink in an uncontrolled manner. The observation that LTP and LTD can occur at the same synapse depending on the pre-post correlations suggests that STDP can provide a mechanism for self-stabilization of weights and firing rates in a network. A second important requirement of any learning scheme is that networks need to operate competitively. Thus, normally it does not make sense to drive all synapses into similar states through learning, because the network will then no longer contain inner structure. Instead, most real networks in the brain are highly structured, forming maps or other subdivisions. There have been a number of recent theoretical studies exploring the possible role of STDP in stability and competition during learning (see [**7**] for a review). We end this lecture by briefly describing another potential application of STDP, namely to prediction and reward learning [**90**]. Animals in the environment often have to react quickly to the earliest signs of harmful stimuli or potential prey. That is, they have to predict or anticipate future events. STDP provides a hint at how a simple form of predictive coding could be implemented at the cellular level. Consider as an example a single neuron receiving inputs from a set of N presynaptic cells that fire sequentially (see Figure 9). This could be the result of a moving stimulus. Initially all synapses have the same weight $w_j = w_0$

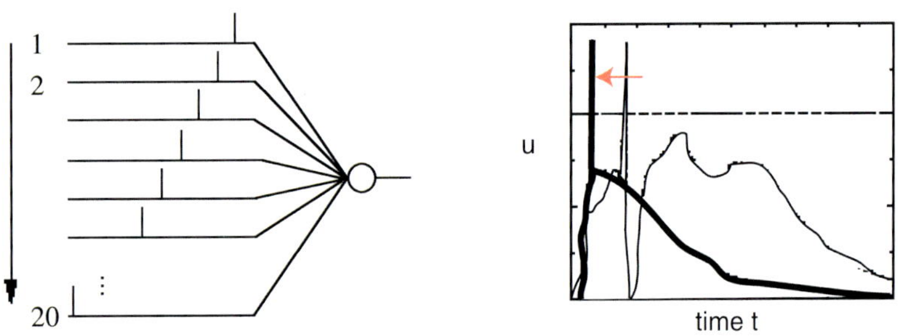

FIGURE 9. A) A sequence of presynaptic spikes is presented to a single neuron. Such a sequence could be generated by a moving stimulus. B) After many trials the synapses in the early part of the sequence have been strengthened so that the neuron fires earlier by an amount Δt. Initial response is shown by thin lines and adapted response by thick lines

for all $j = 1, \ldots N$. Suppose that the postsynaptic neuron initially fires at time T during the presentation of the sequence. All synapses that have been activated prior to the postsynaptic spike are strengthened while synapses that have been activated immediately afterward are depressed. After many trials, the firing time of the postsynaptic neuron within the sequence has shifted by an amount $-\Delta t$. Interestingly, this effective shift in the distribution of synaptic connections is consistent with experimental observations of a shift in the place fields of hippocampal neurons [**91**].

The shift of responses towards earlier predictors plays a central role in classical conditioning as exemplified by Pavlov's dog. An unconditioned stimulus (US) – food – is preceded by a conditioned stimulus (CS) – a bell – at a fixed time interval ΔT. Before learning the US evokes an immediate response – salivation – wheres the CS evokes no response. After learning, the dog starts to salivate in response to the CS. Thus the reaction has moved from the US to the CS, which reliably predicts the US. STDP can replicate this result if the time difference ΔT between two stimuli is less than the width of the learning window. The mechanism is identical to the previous example illustrated in Figure 9, except that the inputs are clustered into two groups corresponding to US and CS. A shift to early predictors is also observed experimentally in recordings from dopamine neurons in the basal ganglia concerned with reward processing [**92**]. Before learning some dopamine neurons respond to an unpredicted reward. During learning this response decreases and the neurons now increase their firing to a reward-predicting stimulus. The more predictable the reward becomes, the less strongly the neurons fire when it appears, and if it does not appear the neurons are inhibited. There are in fact a number of different types of reward–predicting dopamine neurons with distinct responses. For example, some neurons fire a prolonged discharge following a trigger stimulus until a reward is delivered. Dopamine neurons have a central role in guiding our behavior and thoughts [**93**]. Their performance is impaired by every addictive drug and in mental illness. They are also lost in dramatically impairing illnesses such as Parkinson's disease. If dopamine systems are overstimulated, we may hear voices, experience elaborate bizarre cognitive distortions, or engage excessively in dangerous goal-directed behavior. Dopamine function is also central to the way that we value our world, including the way that we value money and other human beings.

LECTURE 3

Firing rates, spike statistics and the neural code

A major debate in the neuroscience literature is whether neurons use a rate code or a spike code to carry out information processing [**94, 95**]. The former is some form of smoothed version of the sequence of spikes obtained by averaging with respect to time, experimental trials or populations of neurons. The latter on the other hand preserves temporal information at the level of individual spikes, that is, on a millisecond time scale. Since generating a rate code requires averaging over a distribution of spikes, it is necessary to specify the associated spike statistics. It turns out that in many cases the output spike train of a single neuron can be modeled quite well by an inhomogeneous Poisson process. Such a spike train generates a stochastic input current to other neurons via synaptic coupling. Hence, it is important to understand how neurons respond to stochastic inputs at both the single neuron and network levels. In the latter case, the degree of cross– correlation between spike trains of distinct neurons is an important factor in determining the appropriate description of network dynamics. In cases where the spike trains are incoherent, it is often possible to carry out some form of population averaging, leading to a rate–based description of network dynamics (see §5). On the other hand, when there is a high level of coherence across spike trains the rate reduction procedure breaks down. One situation where this occurs is when the individual neurons act as intrinsic oscillators along the lines of §1.2. One can then use phase reduction methods to investigate network oscillations and synchrony as outlined in §4. In this third lecture, we focus on some of the basic issues regarding neural coding and noise.

3.1. The neural code

Rate codes

The most commonly used definition of a firing rate $\nu(t)$ refers to a *temporal average*. Suppose that we represent a sequence of spikes at times $\{T^1, \dots, T^n\}$ in terms of the sum of delta functions

$$\rho(t) = \sum_m \delta(t - T^m). \tag{3.1}$$

Introducing a temporal filter function $K(t)$ with $K(t) = 0$ for $t < 0$, we set

$$\nu(t) = \int_{-\infty}^{\infty} K(t-\tau)\rho(\tau)d\tau = \sum_m K(t - T^m). \tag{3.2}$$

The degree of smoothing will depend on the choice of K as illustrated in Figure 1. Note that if $K(t) = T^{-1}$ for $0 \le t < T$ and is zero otherwise, then $\nu(t)$ is simply the mean number of spikes within the time interval $[t - T, t]$. There are a number of problems with the idea that neurons use a time–averaged rate code [**97**]. From

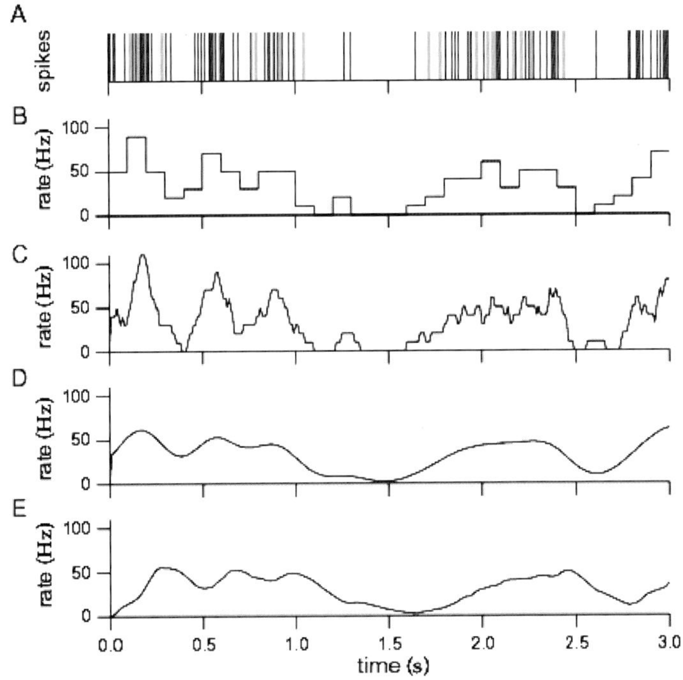

FIGURE 1. Firing rates obtained by filtering a spike train with various convolution functions $K(t)$. (A) Spike train recorded from a neuron in the inferotemporal lobe of a monkey. (B) Discrete-time firing rate obtained by binning time and counting spikes with $\Delta t = 100$ ms. (C) Filtering with a step function of width 100 ms. (D) Filtering with a Gaussian $K(t) = (2\pi\sigma^2)^{-1/2}e^{-t^2/2\sigma^2}$ with $\sigma^{-1} = 100$ ms. (E) Filtering with an alpha function having $\alpha^{-1} = 100ms$. [Reprinted from Baddeley, et al. [**96**], with permission from the Royal Society.]

behavioral experiments it is known that reaction times tend to be too short for neurons to sample over a sufficient number of spikes. For example, a fly can react to new stimuli and change the direction of flight within $30 - 40$ ms [**94**]. Moreover, humans can recognize certain aspects of visual scenes in just a few hundred milliseconds [**98**], even though recognition is believed to involve several processing stages. (See, however, Rossum *et al.* [**99**] who show how fast computation through a number of processing levels can be achieved using a population rate code). Finally, given that real spike trains are highly irregular, a reliable estimate of the mean firing rate requires averaging over a larger number of spikes.

There is a second definition of rate, which is more applicable to time-dependent stimuli. This is constructed by averaging over *experimental trials* and representing the data in a Peri-Stimulus-Time-Histogram (PSTH). For each trial the time interval is divided into bins of size Δt and the number of spikes appearing in each bin is counted. Summing over the N trials, the total number of spikes $n(t, t + \Delta)$ in the interval $[t, t+\Delta t)$ is

$$n(t, t+\Delta t) = \sum_{j=1}^{N} \int_{t}^{t+\Delta t} \rho_j(t)dt, \tag{3.3}$$

where j, $j = 1, \ldots, N$, labels trials. That is, $\rho_j(t) = \sum_m \delta(t - T_j^m)$ where T_j^m is the mth firing time of the jth trial. We then identify the rate with the *spike density*

$$\nu(t) = \frac{1}{\Delta t} \frac{n(t, t+\Delta t)}{N} = \langle \rho(t) \rangle, \tag{3.4}$$

where $\langle\rangle$ denotes trial averaging. Sometimes this is combined with time averaging to obtain a smooth rate variable. An obvious problem with this approach is that it cannot be used by neurons in the brain – a frog does not catch a fly by waiting for the insect to fly repeatedly along exactly the same trajectory; it has to act on a single trial! Nevertheless, the spike density can make sense experimentally if there is a large population of independent neurons that respond to the same stimulus. Instead of recording from the population of N neurons in a single run, it is experimentally easier to record from a single neuron and average over many trials. This leads to the third definition of rate based on *population averaging*. Given a homogeneous population of N identical neurons, the rate is defined by equations (3.3) and (3.4) except that now $j = 1, \ldots N$ labels the neurons within the population and T_j^m is the mth firing time of the jth neuron. One of the useful properties of the population activity is that it can respond very quickly to changes in inputs [**100**], see §3.4. However, it does require that the population is sufficiently homogeneous. Population rate codes are used extensively in modeling the large–scale spatio-temporal dynamics of cortex [**101, 102**], see §5.

Spike codes

Three possible coding strategies based on spike timing are shown in Figure 2:

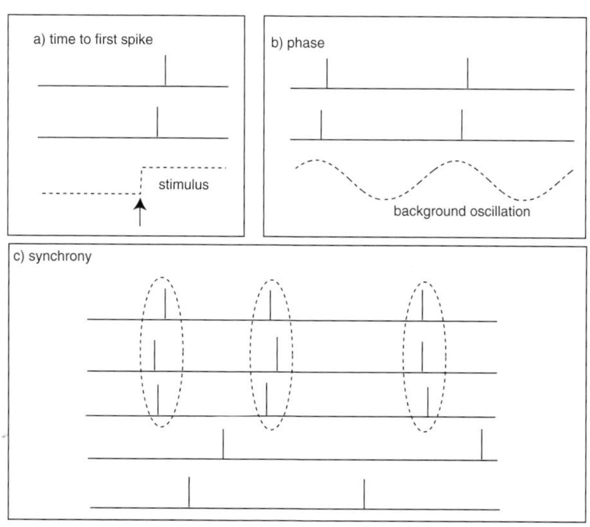

FIGURE 2. Three examples of spike codes

Time-to-first-spike: Suppose that there is a rapid change of input at a time t_0. Such an abrupt change could be due to a saccade, for example, in which an animal's gaze jumps to a new location. One can then imagine a code based on how quickly a neuron fires after t_0, with a shorter latency signalling a stronger stimulus. This form of coding scheme has been advocated as a means of carrying out fast computations [**98**]. However, as we have already mentioned, it is also possible to achieve fast computations using a population code. Moreover, one expects there to be a correlation between the time-to-first-spike and the firing rate of a neuron.

Phase: A second coding strategy is based on the timing of spikes relative to some reference background oscillation. For example, extensive studies of rodents have demonstrated direct behavioral correlates of hippocampal neuronal activity, the most robust of which is the selective activation of certain hippocampal cells at particular locations in space called place fields [**103**]. When a rodent is exploring its environment (or is in REM sleep) cells in the hippocampus participate in 4-10 Hz theta rhythms. A key observation is that the spike activity of a hippocampal place cell advances to earlier phases of the theta cycle as the animal passes through the cell's place field [**103**]. One possible interpretation of this phase advance is that it reflects a prediction of the sequence of upcoming locations based on the rat's current position [**104**], that is, it is part of a path integration system. Theta oscillations are superimposed upon faster gamma (40 Hz) oscillations, and these may organize the readout of sequential locations during a theta cycle.

Correlations/Synchrony: Various experimental studies have shown that correlations in the firing patterns of groups of neurons are stimulus-dependent. For example, neurons in the antennal lobe of insects are typically synchronized by 20 Hz oscillations – when such synchronization is artificially disrupted the specificity of downstream responses to different odors decreases [**105**]. Stimulus-induced oscillations have also been observed in the olfactory lobe of mollusks and rabbits [**106**], and the visual cortex of cats [**107**]. Correlated activity between neurons can also be considered without reference to intrinsic oscillations. For example, analysis of the activity of visual neurons in the lateral geniculate nucleus has shown that significantly more information about the stimulus can be extracted from their spike trains if coincident spikes are taken into account separately from the non-coincident ones [**108**]. Similarly, recordings from primary auditory cortex indicate that when a stimulus is turned on, neurons respond by changing their firing rates and their correlations; in many cases, the firing rate modulations are transient whereas the evoked changes in correlations can be sustained [**109**]. The possible role of stimulus-induced correlated or synchronous activity is still unclear. One suggestion is that synchrony between pairs of neurons in the visual cortex provides a mechanism for signalling whether or not such neurons are responding to a single object or two distinct objects that induce the same firing rates [**110**]. Indeed, stimulus-induced phase synchronization has been proposed as a solution to the so-called binding problem, that is, how to bind together the multitude of different representations (color, form, location) of an object, although it remains a controversial issue [**111, 112**]. An alternative suggestion is that correlations do not contain information about the nature of a stimulus, but provide an attentional gating mechanism that controls the flow of information [**113**]. The issue of network oscillations and synchrony will be considered in §4.

Reverse correlation methods

One important question concerns how a spike or rate code is used to represent information about a stimulus. Reverse correlation methods provide a powerful tool for investigating stimulus encoding and decoding [**94, 95**]. Consider a neuron that is driven by a time dependent stimulus $s(t)$. Suppose that every time a spike occurs, we record the time course of the stimulus in a time window of about 100 msec immediately before the spike. Averaging the results for several spikes yields the typical time course of the stimulus just before a spike. Such a procedure is called a *reverse correlation* method. In contrast to the Peri-Stimulus-Time-Histogram (PSTH), which is constructed by averaging the neuron's response over several trials with the same stimulus, reverse correlation involves averaging the stimulus under the condition of the same response, that is, the firing of a single spike. The so-called spike-triggered average stimulus $C(\tau)$ is defined according to

$$C(\tau) = \left\langle \frac{1}{M}\sum_{m=1}^{M} s(T^m-\tau)\right\rangle \approx \frac{1}{\langle M\rangle}\left\langle \sum_{m=1}^{M} s(T^m-\tau)\right\rangle \tag{3.5}$$

where the stimulus is first averaged over all M spikes $T^1,\ldots,T^M$ in a given trial of length T and then it is averaged over trials. It is also assumed that for large N the total number of spikes on each trial is well approximated by the average number of spikes per trial, $M \approx \langle M\rangle$. Under such an approximation, we can rewrite $C(\tau)$ in the form

$$C(\tau) = \frac{1}{\langle M\rangle}\int_0^T \langle \rho(t)s(t-\tau)\rangle dt = \frac{1}{\bar{\nu}T}\int_0^T \nu(t)s(t-\tau)dt \tag{3.6}$$

where $\rho(t) = \sum_m \delta(t-T^m)$, $\bar{\nu} = \langle M\rangle/T$ is the mean firing rate and

$$\nu(t) = \langle \rho(t)\rangle = \frac{1}{N}\sum_{j=1}^{N}\sum_{m=1}^{M}\delta(t-T_j^m) \tag{3.7}$$

with j labeling the trial. A schematic illustration of the spike-triggered average is shown in Figure 3. One common application of spike–triggered averaging is to determine the linear response properties of a neuron. That is, using a white noise stimulus one can identify the resulting spike–triggered average as the optimal linear filter for modeling the relationship between an input stimulus and the output firing rate of a neuron [**95**]. Another interesting application has been to decoding the spike trains of the H1 neuron of the fly [**94**]. The H1 neuron detects the motion of visual images during flight in order to generate and guide stabilizing motor corrections. In experiments, the fly is usually held fixed while a visual image with a time-varying velocity is presented. The basic idea of spike decoding is to generate a linear estimate of a stimulus based on the neuron's firing times.

3.2. Spike statistics and the Poisson process

In vivo recordings of neuronal activity are characterized by a high degree of irregularity. One often finds that the statistics of neuronal spike trains can be modeled quite well by a Poisson process. Suppose that a neuron emits a sequence of spikes at times $T^1, T^2, \ldots, T^n$. One way to characterize this sequence statistically is in terms of the probability density $\rho(T^1,\ldots T^n)$ of finding such a sequence over many experimental trials. In other words, the probability of having a sequence of n spikes in the interval $[0,T]$ with the ith spike falling between the times T^i and $T^i+\Delta t$ is

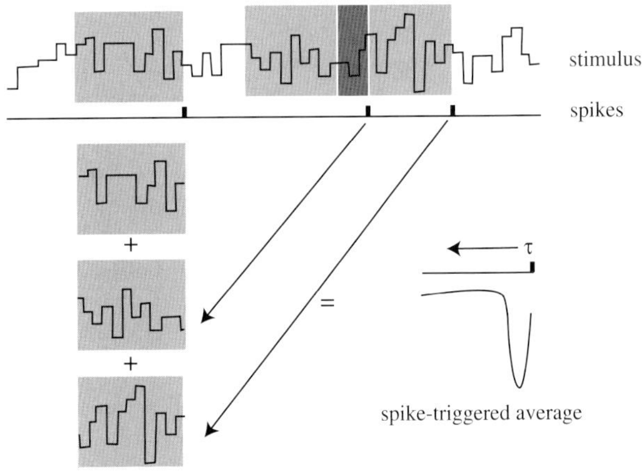

FIGURE 3. Schematic diagram illustrating the procedure for computing the spike-triggered average. Each gray rectangle contains the stimulus prior to one of the spikes shown along the time axis. These are averaged over spikes and trials to produce the waveform shown in the lower right.

$P[T^1, \ldots T^n] = \rho(T^1, \ldots T^n)(\Delta t)^n$. In principle, the probability of an event occurring, namely the firing of the next spike, could depend on the entire history of past spikes. If this dependence only extends to the previous spike so that the inter-spike intervals are statistically independent, then the stochastic process is said to be a *renewal process*. If there is no dependence at all on preceding spikes so that the firing times are themselves independent then we have a *Poisson process*.

Homogeneous Poisson process

Consider a sequence of spikes generated by a homogeneous Poisson process, that is, one with a time-independent mean firing rate r. Divide a given time interval T into M bins of size $\Delta t = T/M$ and assume that Δt is small enough so that the probability of finding two spikes within any one bin can be neglected. Then the probability $P_T[n]$ of finding n spikes over the interval T is given by

$$P_T[n] = \lim_{\Delta t \to 0} \frac{M!}{(M-n)!n!}(r\Delta t)^n(1-r\Delta t)^{M-n}.$$

This consists of the probability $(r\Delta t)^n$ of finding n spikes in n specific bins multiplied by the probability $(1-r\Delta t)^{M-n}$ of not finding spikes in the remaining bins. The binomial factor is the number of ways of choosing n out of M bins with spikes. Using the approximation $M - n \approx M = T/\Delta t$ and defining $\varepsilon = -r\Delta t$, we have that

$$\lim_{\Delta t \to 0}(1-r\Delta t)^{M-n} = \lim_{\varepsilon \to 0}\left((1+\varepsilon)^{1/\varepsilon}\right)^{-rT} = \mathrm{e}^{-rT}.$$

For large M, $M!/(M-n)! \approx M^n = (T/\Delta t)^n$, so that we obtain the Poisson distribution

$$P_T[n] = \frac{(rT)^n}{n!}\mathrm{e}^{-rT}. \tag{3.8}$$

Given that there are n independent spikes over the interval $[0, T]$, the probability that these spikes lie within specified bins of size Δt is $n!(\Delta t/T)^n$. Hence

$$\rho(T^1, \dots T^n) = n! \left(\frac{1}{T}\right)^n P_T[n] = r^n \mathrm{e}^{-rT}. \tag{3.9}$$

Also note that the fastest way to generate a sequence of Poisson spikes for constant r is to iterate the firing times $T^{n+1} = T^n - \log(x_{rand})/r$ with x_{rand} uniformly distributed over $[0, 1]$. A simple method for calculating the moments of the Poisson distribution is to introduce the moment generating function

$$G(s) = \sum_{n=0}^{\infty} P_T[n] \mathrm{e}^{sn}. \tag{3.10}$$

Differentiating with respect to s shows that

$$\left.\frac{d^k G(s)}{ds^k}\right|_{s=0} = \langle n^k \rangle. \tag{3.11}$$

The generating function for the Poisson process can be evaluated explicitly as

$$G(s) = \exp(-rT) \exp(rT\mathrm{e}^s), \tag{3.12}$$

from which we deduce that

$$\langle n \rangle = rT, \quad \sigma_n^2 = rT. \tag{3.13}$$

Another useful quantity is the interspike-interval (ISI) distribution. Suppose that a spike was last fired at time T^n. The probability of a homogeneous Poisson process generating the next spike in the interval $T^n + \tau \le T^{n+1} \le T^n + \tau + \Delta\tau$ is equal to the probability that no spike is fired for a time τ, which is $P_\tau[0] = \mathrm{e}^{-r\tau}$ multiplied by the probability $r\Delta\tau$ of generating a spike within the following interval $\Delta\tau$:

$$\mathbf{Pr}[\tau \le T^{n+1} - T^n \le \tau + \Delta\tau] = r\Delta\tau \mathrm{e}^{-r\tau}.$$

The ISI probability density is thus an exponential, $\rho(\tau) = r\mathrm{e}^{-r\tau}$. It follows that the mean interspike interval is

$$\langle \tau \rangle = \int_0^\infty r\mathrm{e}^{-r\tau} \tau d\tau = \frac{1}{r}$$

and the variance is

$$\sigma_\tau^2 = \int_0^\infty r\mathrm{e}^{-r\tau} \tau^2 d\tau - \langle \tau \rangle^2 = \frac{1}{r^2}.$$

The ratio of the standard deviation to the mean is called the *coefficient of variation* $C_V = \sigma_\tau / \langle \tau \rangle$. It follows that for a homogeneous Poisson process $C_V = 1$.

How well do Poisson statistics describe experimentally measured spike trains? One often finds that for ISIs longer than about $10msec$ the ISI distribution is indeed approximately exponential. However, for shorter intervals there is a rapid decrease in the distribution reflecting the fact that neurons are refractory immediately after firing. This is illustrated in Figure 4. The data can be fitted more accurately by a gamma distribution $\rho[\tau] = r(r\tau)^k \mathrm{e}^{-t\tau}/k!$. Alternatively, one can introduce a refractory period into a standard Poisson model. Note that C_V values extracted from the spike trains of cortical neurons also take values around unity, provided that the mean interspike interval is not too small [**115**].

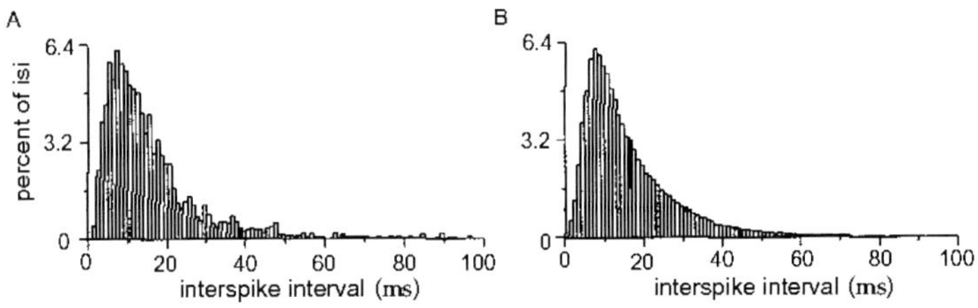

FIGURE 4. (A) Interspike interval distribution from a neuron from the MT cortical visual area of a macaque monkey responding to a moving image. (B) Interspike interval generated with a Poisson model with a stochastic refractory period. [Reprinted from [**114**], with permission from J. Neurosci.]

Inhomogeneous Poisson process

It is possible to generalize the above Poisson model to the case of a time-dependent rate $r(t)$. The simplest way to analyze this inhomogeneous Poisson process is to consider the probability distribution $\rho(T^1, \ldots T^n)$. This is given by the product of the probabilities $r(T^i)\Delta t$ that the neuron fires within the time intervals $T^i \leq t \leq T^i + \Delta t$ and the probabilities of not firing during the interspike intervals. The latter is given by

$$\mathbf{Pr}[\text{no spikes in } (T^i, T^{i+1})] = \prod_{m=1}^{M} (1 - r(T^i + m\Delta t)\Delta t),$$

where we have partitioned the interval (T^i, T^{i+1}) into M bins of size Δt. Taking the logarithm,

$$\begin{aligned}\log \mathbf{Pr}[\text{no spikes in } (T^i, T^{i+1})] &= \sum_{m=1}^{M} \log(1 - r(T^i + m\Delta t)\Delta t) \\ &\approx -\sum_{m=1}^{M} r(T^i + m\Delta t)\Delta t.\end{aligned}$$

Taking the limit $\Delta t \to 0$ and exponentiating again shows that

$$\mathbf{Pr}[\text{no spikes in } (T^i, T^{i+1})] = \exp\left(-\int_{T^i}^{T^{i+1}} r(t)dt\right).$$

Hence

$$\rho(T^1, \ldots T^n) = \prod_{i=1}^{n} r(t_i) \exp\left(-\int_0^T r(t)dt\right). \tag{3.14}$$

In the case of a time-dependent rate, one generates x_{rand} at each time step and a spike is fired if $r(t)\Delta t > x_{rand}$.

3.3. Stochastically driven IF neuron

The origin of neural irregularity is still not very well understood, and whether the variability is simply noise or some intricate neural encoding of information remains unclear. Nevertheless, one can distinguish between intrinsic sources of noise that generate stochastic behavior at the level of single neuron dynamics, and extrinsic sources that arise from network effects. Examples of the former include thermal noise inducing membrane fluctuations, fluctuations in the opening and closing of ion channels in the postsynaptic cell membrane (see §2.2), and synaptic transmission failure. One way to model the effects of intrinsic noise is to define a noisy threshold in terms of an instantaneous escape rate, which depends on the distance of the membrane potential from the threshold [**7**]. Extrinsic noise reflects the fact that a typical neuron such as a pyramidal cell in the cerebral cortex receives input spikes from thousands of other neurons, which in turn receive input from their own presynaptic neurons and so forth. Rather than trying to model the brain as a single huge network, one typically focuses on a specific subset of neurons and treat inputs from other parts of the brain as a stochastic background. The stochastic arrival of excitatory and inhibitory pulses leads to a diffusion of the membrane potential that can be analyzed using Fokker–Planck equations [**116, 117, 118**]. We shall consider the effects of such diffusive noise on the dynamics of IF neurons.

Suppose that an IF neuron receives an external input $I_{\rm ext}(t)$ together with a set of N spike trains generated by N background neurons. Denote the arrival times of the spikes from the kth presynaptic neuron by $\{T_k^n, n \in \mathbf{Z}\}$ for $k = 1, \ldots, N$. The membrane potential evolves according to the equation

$$\tau \frac{du}{dt} = -u + RI_{\rm ext} + \tau \sum_{n\in\mathbf{Z}} \sum_{k=1}^{N} w_k g_{\rm syn}(t - T_k^n), \tag{3.15}$$

together with the reset condition that $u(t^+) = 0$ whenever $u(t) = 1$. Here $g_{\rm syn}(t)$ represents a normalized synaptic conductance and w_k is the strength or efficacy of the kth synapse. Note that this equation treats input spikes as generating current rather than conductance fluctuations (ie. we have dropped the factor $u_{\rm syn} - u$). The latter can be accounted for heuristically by introducing an input–dependent effective membrane time constant [**119**]. As a further simplification, we will ignore the shape of each postsynaptic potential by taking $g_{\rm syn}(t) = \delta(t)$, where $\delta(t)$ is a Dirac delta function. Each input spike from the kth neuron then generates a change in the postsynaptic potential of the form

$$\Delta u(t) = w_k h(t - T_k^n), \quad h(t) = \mathrm{e}^{-t/\tau} H(t),$$

where $H(t)$ is the Heaviside function. Thus each spike induces a jump of size w_k, which represents the strength of the connection or synapse from the kth presynaptic neuron, and then decays exponentially. Suppose that the spikes at synapse k are generated by an inhomogeneous Poisson process with arrival rate $\nu_k(t)$. This means that in each small time interval $[t, t + \Delta t]$ the probability that a spike arrives on the kth synapse is $\nu_k(t)\Delta t$, and each spike is uncorrelated with any other. We will derive a Fokker–Planck equation for the probability density $p(u, t)$ for u evolving according to the stochastic ODE (3.15), assuming that the neuron last fired at $t = 0$. We also set $I_{\rm ext} = 0$ for simplicity.

The probability that no spike arrives in a short time interval Δt is

$$\mathrm{Prob}\{\text{no spike in } [t, t + \Delta t]\} = 1 - \sum_k \nu_k(t)\Delta t. \tag{3.16}$$

If no spike arrives then the membrane potential changes from $u(t) = u'$ to $u(t + \Delta t) = u'\mathrm{e}^{-\Delta t/\tau}$. On the other hand, if a spike arrives at synapse k, the membrane changes from u' to $u'\mathrm{e}^{-t/\tau}+w_k$. Therefore, given a value u' at time t, the probability density of finding a membrane potential u at time $t+\Delta t$ is

$$
\begin{aligned}
P(u,t+\Delta t|u',t) &= \left[1-\Delta t\sum_k \nu_k(t)\right]\delta\left(u-u'\mathrm{e}^{-\Delta t/\tau}\right) \\
&\quad +\Delta t\sum_k \nu_k(t)\delta\left(u-u'\mathrm{e}^{-\Delta t/\tau}-w_k\right).
\end{aligned} \tag{3.17}
$$

Since the input spikes are generated by a Poisson process, it follows that the random variable $u(t)$ evolves according to a Markov process:

$$p(u,t+\Delta t)=\int P(u,t+\Delta t|u',t)p(u',t)du'. \tag{3.18}$$

Substituting for P using equation (3.17) shows that

$$
\begin{aligned}
p(u,t+\Delta) &= \left[1-\Delta t\sum_k \nu_k(t)\right]\mathrm{e}^{\Delta t/\tau}p(\mathrm{e}^{\Delta t/\tau}u,t) \\
&\quad +\Delta t\sum_k \nu_k(t)\mathrm{e}^{\Delta t/\tau}p(\mathrm{e}^{\Delta t/\tau}u-w_k,t).
\end{aligned} \tag{3.19}
$$

Rearranging and taking the limit $\Delta t\to 0$ leads to the master equation

$$\frac{\partial p}{\partial t} = \frac{1}{\tau}\frac{\partial}{\partial u}[up(u,t)]+\sum_k \nu_k(t)[p(u-w_k,t)-p(u,t)]. \tag{3.20}$$

If the jump amplitudes w_k are sufficiently small, then we can formally Taylor expand the right–hand side of the master equation as a series in w_k corresponding to the so–called Kramers–Moyall expansion [**117, 118**]. Neglecting terms of order w_k^3 then leads to the Fokker–Planck equation

$$\frac{\partial p}{\partial t} = -\frac{1}{\tau}\frac{\partial}{\partial u}\left[-u+\tau\sum_k \nu_k(t)w_k\right]p(u,t)+\frac{1}{2}\left[\sum_k \nu_k(t)w_k^2\right]\frac{\partial^2}{\partial u^2}p(u,t). \tag{3.21}$$

The Fokker-Planck equation determines the time evolution of the probability density of a membrane potential evolving according to the equivalent stochastic differential equation (Langevin equation)

$$\tau du=-u(t)dt+\mu(t)dt+d\xi(t). \tag{3.22}$$

Here $\mu(t)$ is the mean background synaptic input

$$\mu(t)=\tau\sum_k \nu_k(t)w_k \tag{3.23}$$

and $\xi(t)$ is a Gaussian white noise process,

$$\langle\xi(t)\rangle=0,\quad \langle\xi(t)\xi(t')\rangle=\tau\sigma^2(t)\min(t,t'), \tag{3.24}$$

with the size of the membrane fluctuations given by

$$\sigma^2(t)=\tau\sum_k \nu_k(t)w_k^2. \tag{3.25}$$

In the case of constant noise, the resulting Langevin equation describes the well known Ornstein–Uhlenbeck process. Note that in the derivation of the Fokker–Planck equation we have suppressed higher–order terms of the form

$$\sum_{n=3}^{\infty} \frac{(-1)^n}{n!} A_n(t) \frac{\partial^n}{\partial u^n} p(u,t),$$

with $A_n = \sum_k \nu_k(t) w_k^n$. This becomes exact in the so–called diffusion limit $w_k \to 0$ such that $\mu(t), \sigma^2(t)$ are unchanged and $A_n \to 0$ for $n \geq 3$.

First passage times

In our derivation of the Fokker–Planck equation we neglected the threshold u_κ. This can be incorporated as an absorbing boundary condition

$$p(u_\kappa, t) \equiv 0 \text{ for all } t. \tag{3.26}$$

We can then look for the solution $p = p(u,t|U,0)$ of the Fokker–Planck equation assuming the initial condition $p(u,0|U,0) = \delta(u-U)$. At any time $t > 0$, the probability that the neuron has not reached threshold is

$$S(U,t) = \int_{-\infty}^{u_\kappa} p(u,t|U,0)dt. \tag{3.27}$$

Let $\psi(U,t)\Delta t$ be the probability that the neuron fires its next spike between t and $t+\Delta t$. It follows that $\psi(U,t)\Delta t = S(U,t) - S(U,t+\Delta t)$ so that in the limit $\Delta t \to 0$,

$$\psi(U,t) = -\frac{d}{dt} \int_{-\infty}^{u_\kappa} p(u,t|U,0)du. \tag{3.28}$$

The density $\psi(U,t)$ determines the distribution of first passage times. When this is combined with the reset condition $U = u_r$, we see that it also determines the distribution of interspike intervals $\Delta^n = T^{n+1} - T^n$. Unfortunately, no general solution is known for the first passage time problem of the Ornstein–Uhlenbeck process. However, in the case of constant inputs such that $\mu(t) = \mu_0$ and $\sigma(t) = \sigma$, one can carry out a moment expansion of the first passage time distribution. In particular, a closed form expression for the mean first passage time can be obtained.

In the case of constant inputs, $p(u,t|U,0)$ evolves according to the Fokker–Planck equation

$$\tau \frac{\partial p}{\partial t} = -\frac{\partial}{\partial u}[-u + \mu_0]p + \frac{\sigma^2}{2} \frac{\partial^2 p}{\partial u^2},$$

with $p(u,0|U,0) = \delta(u-U)$ and absorbing boundary condition $p(u_\kappa, t|U,0) = 0$. It can then be shown that $S(U,t)$ satisfies the backwards FP equation [**116, 117**]

$$\tau \frac{\partial S}{\partial t} = [-U + \mu_0] \frac{\partial S}{\partial U} + \frac{\sigma^2}{2} \frac{\partial^2 S}{\partial U^2}, \tag{3.29}$$

with $S(U,0) = 1$ for $U \leq u_\kappa$ and zero otherwise. Defining the mean first passage time (MFPT) by

$$T(U) = \int_0^\infty t\psi(U,t)dt,$$

it follows that $T(U)$ satisfies the ODE

$$-\tau = [-U + \mu_0] \frac{dT}{dU} + \frac{\sigma^2}{2} \frac{d^2 T}{dU^2}. \tag{3.30}$$

The solution for $T(U)$ is then

$$T(U) = \frac{2\tau}{\sigma^2}\int_U^{u_\kappa} du e^{2\Phi(u)/\sigma^2}\int_{-\infty}^{u} e^{-2\Phi(u')/\sigma^2}du', \tag{3.31}$$

where $\Phi(u) = u^2/2 - \mu_0 u$. Hence, we obtain the following expression for the MFTP of an IF neuron whose initial state is the reset potential, $U = u_r$:

$$T(u_r) = \int_0^\infty s\psi(s)ds = \tau\sqrt{\pi}\int_{(u_r-\mu_0)/\sigma}^{(u_\kappa-\mu_0)/\sigma} e^{u^2}(1+\mathrm{erf}(u))du. \tag{3.32}$$

Subthreshold regime

A major aim of noisy neuron models is to try to understand the large variability of interspike intervals found *in vivo* [**97, 120**]. Given that a neuron receives a large number of inputs at any one time, one might expect the law of large numbers to apply so that the neuron's output reflects the average input activity and thus exhibits small fluctuations about a mean firing rate. This is certainly true if the neuron operates in the superthreshold regime, in which the total input to the neuron in the absence of noise drives it to fire. Diffusive noise has little influence other than slightly broadening the interspike interval distribution. On the other hand, if the noise–free input is subthreshold, that is, it generates a membrane potential that remains below the firing threshold, then diffusive noise has a significant effect. In the subthreshold regime spikes are generated by the fluctuations of the membrane potential, rather than by its mean, and the interspike interval distribution is broad. The differences between the two regimes are illustrated in Figure 5. Another interesting subthreshold effect is *stochastic resonance*. This refers to a range of phenomena in which noise improves the signal transmission properties of a system and this effect is optimal at a particular noise level [**121**]. Within the neural context, a subthreshold input stimulus without noise does not generate action potentials, so no information regarding the stimulus is transmitted. However, if noise is added then spikes occur and there is an optimal noise level for signal transmission [**122**]. A related phenomenon is *coherence resonance* [**123**]. Here it is assumed that an excitable neuron is at rest in the absence of noise, but fires a sequence of spikes when noise is added. The optimal noise level now corresponds to the condition under which the resulting spike train is most regular.

3.4. Homogeneous population of IF neurons

Let us now consider a large population of identical IF neurons each being driven by a set of Poisson distributed spike trains. We assume that the kth synapse of every neuron receives a Poisson spike train with the same instantaneous rate $\nu_k(t)$, but that the spike trains across the population are statistically independent. The derivation of the Fokker–Planck equation proceeds along very similar lines to the single neuron case, and takes the form

$$\frac{\partial p}{\partial t} = -\frac{\partial}{\partial u}\left[\frac{-u+\mu(t)+RI_{\mathrm{ext}}(t)}{\tau}\right]p(u,t) + \frac{\sigma^2(t)}{2\tau}\frac{\partial^2}{\partial u^2}p(u,t), \tag{3.33}$$

with $\mu(t), \sigma(t)$ given by equations (3.23) and (3.25) and we now include an external input. However, there are a number of important differences between the single neuron and population cases. First, $p(u,t)$ is now interpreted as the probability

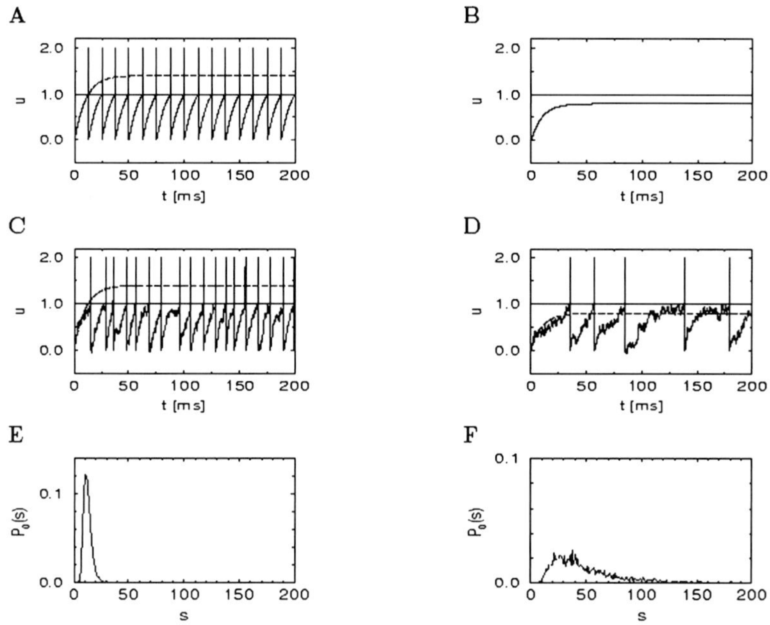

FIGURE 5. IF neuron with superthreshold (left column) and subthreshold (right column) inputs. (A,B) Noise–free case showing either regular firing or zero firing. (C,D) Spike trains in the presence of diffusive noise. (E,F) Corresponding interspike intervals. [Reprinted from [**7**], with permission of Cambridge University Press.]

density of membrane potentials across the population of neurons. Consequently, the normalization is different. In the case of a single neuron, the integrated density

$$\int_{-\infty}^{u_\kappa} p(u,t)du \leq 1$$

was interpreted as the probability that the neuron under consideration has not yet fired, which changes over time. On the other hand, if a neuron in the population fires, it remains part of the population so that we have the constant normalization

$$\int_{-\infty}^{u_\kappa} p(u,t)du = 1. \tag{3.34}$$

The second major difference is that we now have to incorporate the reset condition explicitly in the evolution of the probability density. First, note that the Fokker–Planck equation can be rewritten as a continuity equation reflecting conservation of probability:

$$\frac{\partial}{\partial t}p(u,t) = -\frac{\partial}{\partial u}J(u,t), \quad \text{for } u \neq u_\kappa, u_r, \tag{3.35}$$

where

$$J(u,t) = \frac{1}{\tau}\left[-u + \mu(t) + RI_{\text{ext}}(t)\right]p(u,t) - \frac{\sigma^2(t)}{2\tau}\frac{\partial}{\partial u}p(u,t). \tag{3.36}$$

In a population of N neurons, the fraction of active neurons is calculated by counting the number of output spikes $n(t;t+\Delta t)$ in a small time interval Δt and dividing by N. Further division by Δt yields the population activity

$$\nu(t) = \lim_{\Delta t \to 0} \frac{1}{\Delta t} \frac{n(t,t+\Delta t)}{N} = \frac{1}{N} \sum_{j=1}^{N} \sum_{m} \delta(t - T_j^m). \tag{3.37}$$

The double sum runs over all firing times T_j^m of all neurons in the population. The fraction of neurons that flow across the threshold per unit time should then be equal to the population averaged activity $\nu(t)$, that is, $\nu(t) = J(u_\kappa, t)$. Equation (3.36) together with the absorbing boundary condition

$$p(u_\kappa, t) = 0 \tag{3.38}$$

implies that

$$\frac{\partial}{\partial u} p(u_\kappa, t) = -\frac{2\nu(t)\tau}{\sigma^2(t)}. \tag{3.39}$$

Due to the reset condition, the neurons that "disappear" across threshold are reinjected at the reset potential u_r, which implies that there is a discontinuity in the flux at u_r, $J(u_r^+, t) - J(u_r^-, t) = \nu(t)$. Continuity of p,

$$p(u_r^+, t) - p(u_r^-, t) = 0, \tag{3.40}$$

together with equation (3.36) then shows that there is a discontinuity in the first derivative of $p(u,t)$ at $u = u_r$:

$$\frac{\partial}{\partial u} p(u_r^+, t) - \frac{\partial}{\partial u} p(u_r^-, t) = -\frac{2\nu(t)\tau}{\sigma^2(t)}. \tag{3.41}$$

In summary, one has to solve the Fokker–Planck equation (3.33) together with the boundary conditions (3.38), (3.39), (3.40) and (3.41).

Steady–state solution

Suppose that the background rates ν_k and external input I_{ext} are time–independent so that the total mean input

$$h_0 = RI_0 + \tau \sum_k \nu_k w_k \tag{3.42}$$

is a constant. The steady–state Fokker–Planck equation implies that the flux

$$J(u) = \frac{-u + h_0}{\tau} p(u) - \frac{1}{2} \frac{\sigma^2}{\tau} \frac{\partial}{\partial u} p(u) \tag{3.43}$$

is constant except at $u = u_r$ where it jumps by an amount ν_0, which is the steady–state population activity. Taking $J(u) = 0$ for $u < u_r$ we can solve equation (3.43) to obtain the Gaussian distribution

$$p_0(u) = \frac{c_1}{\sigma} \exp\left[-\frac{(u - h_0)^2}{\sigma^2} \right], \quad \text{for } u \le u_r \tag{3.44}$$

for some constant c_1. However, such a solution cannot be valid for $u > u_r$, since it does not satisfy the absorbing boundary condition $p_0(u_\kappa) = 0$. It turns out that in this domain the solution is of the form [**124**]

$$p_0(u) = \frac{c_2}{\sigma^2} \exp\left[-\frac{(u - h_0)^2}{\sigma^2} \right] \int_u^{u_\kappa} \exp\left[\frac{(x - h_0)^2}{\sigma^2} \right] dx, \quad \text{for } u_r < u \le u_\kappa \tag{3.45}$$

for some constant c_2. Equation (3.43) shows that $c_2 = 2\tau J(u)$ for $u_r < u \le u_\kappa$ with $J(u) = \nu_0$. Continuity of the solution at $u = u_r$ implies that

$$c_1 = \frac{c_2}{\sigma} \int_u^{u_\kappa} \exp\left[\frac{(x-h_0)^2}{\sigma^2}\right] dx. \tag{3.46}$$

Finally, the constant c_2 is determined by the normalization condition (3.34). On setting $\nu_0 = c_2/2\tau$, one finds a firing rate that is consistent with the MFPT of equation (3.32):

$$\nu_0 = \left[\tau\sqrt{\pi} \int_{(u_r-h_0)/\sigma}^{(u_\kappa-h_0)/\sigma} \mathrm{e}^{u^2}(1+\mathrm{erf}(u))du\right]^{-1} \equiv f(h_0), \tag{3.47}$$

where f is the so–called gain function for the population.

Asynchronous firing in recurrent networks

The above analysis assumed that the neurons were independent of each other so that the only synaptic inputs were from some stochastic background. Now suppose that we have a fully connected network such that there is an additional contribution to the synaptic current into each neuron of the form

$$I_{\text{rec}}(t) = \frac{\tau}{R}\frac{\Gamma_0}{N}\sum_{j=1}^{N}\sum_m \delta(t-T_j^m) = \tau\Gamma_0\nu(t)/R, \tag{3.48}$$

where Γ_0/N is the strength of connection between any pair of neurons within the population, and we have used the definition (3.37) of the population activity $\nu(t)$. Suppose that the neuronal population is in a macroscopic state with constant activity $\nu(t) = \nu_0$, which is referred to as a state of asynchronous firing. The steady–state activity can then be determined self–consistently from equation (3.47) by setting

$$h_0 = RI_0 + \tau\left[\sum_k \nu_k w_k + \Gamma_0\nu_0\right]. \tag{3.49}$$

This is illustrated in Figure 6(a). One can also determine the stability of the asynchronous state by considering small perturbations of the steady–state probability distribution. One finds that in the limit of low noise, the asynchronous state is unstable and the neurons tend to split up into several synchronized groups that fire alternately. The overall activity then oscillates several times faster than the individual neurons [**125, 126, 124**], see §4.3. One of the interesting properties of the asynchronous state from a computational perspective is that the population activity can respond rapidly to a step input [**7**], as illustrated in Figure 6(b). The basic intuition behind this is that in the asynchronous state there will always be a significant fraction of neurons that are sitting close to the firing threshold so that as soon as a step increase in input current occurs they can respond immediately. However, the size of the step has to be at least as large as the noise amplitude σ, since the threshold acts as an absorbing boundary, that is, the density of neurons vanishes as $u \to u_\kappa$.

In the above example noise is added explicitly in the form of stochastic background activity. It is also possible for a network of deterministic neurons with fixed random connections to generate its own noise [**124, 127, 128, 129**]. In particular, suppose that each neuron in the population of N neurons receives input from C randomly selected neurons in the population with $C \ll N$. The assumption of

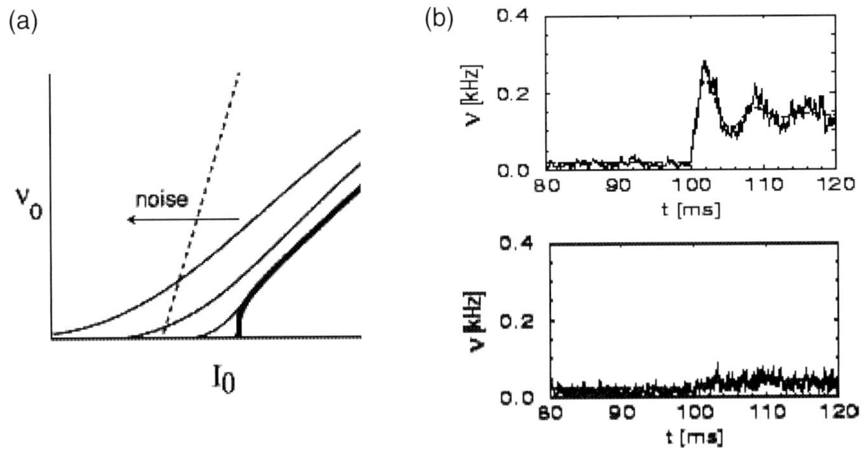

FIGURE 6. (a) Steady–state population activity ν_0 as a function of external input I_0 for different levels of noise (increasing from right to left). Thick curve shows firing rate function for a single noiseless IF neuron. Intersection with the dashed line of slope $1/\Gamma_0$ gives possible solutions for the stationary activity in a population with excitatory coupling Γ_0. (b) Response of a network of 1000 IF neurons with diffusive noise to a step current input. Rapid response to a large current step for low noise (upper figure). Slow response to a small current step for high noise (lower figure). [Reprinted from [**7**], with permission of Cambridge University Press.]

sparse connectivity means that two neurons share only a small number of common inputs. Hence, if the presynaptic neurons fire stochastically then the input spike trains that arrive at distinct postsynaptic neurons can be treated as statistically independent. Since the presynaptic neurons belong to the same population, it follows that each neuron's output should itself be stochastic in the sense that it should have a sufficiently broad distribution of interspike intervals. This will tend to occur if the neurons operate in a subthreshold regime, that is, the mean total input is below threshold so that threshold crossings are fluctuation driven. Finally, note that there have been a number of studies of recurrent network dynamics based on extensions of the single neuron master equation (3.20) to neural populations. In the case of large populations, numerical solutions of the resulting master equation compare very well with direct simulations of the full network equations, capturing fine details not possible under the diffusion approximation [**130**]. In certain special cases, it is also possible to obtain analytical solutions [**131**]. One recent application of the master equation approach has been to the study of fluctuation–driven dynamics in recurrent network models of primary visual cortex [**132**].

LECTURE 4

Network oscillations and synchrony

Synchronized neural activity appears to be prevalent in many areas of the brain including sensory cortices, thalamus and hippocampus. Synchrony can occur on the spike-to-spike level or on the bursting level, and is usually associated with oscillatory activity. Oscillations may be generated by intrinsic properties of single cells or may arise through excitatory and inhibitory synaptic interactions within a local population of cells. In this lecture we explore some of the mechanisms underlying oscillatory activity and synchronization at the network level by considering populations of synaptically coupled spiking neurons. From a theoretical perspective there are a number of issues that need to be addressed: Is the basic oscillatory unit a single neuron or a local population of neurons? What are the roles of recurrent excitatory and inhibitory connections in determining the patterns of oscillations? How are the patterns affected by the intrinsic ionic properties of the neurons? Which conditions determine whether synchronous activity is at the level of spikes, bursts or firing rates, and in which frequency range? How are synchronous states affected by external inputs? Under what conditions does one expect to observe traveling (phase) waves rather than synchronous oscillations?

One popular approach to tackling some of these issues is to assume that each neuron (or group of neurons) is intrinsically oscillatory, and that interactions among the neurons, or groups of neurons that comprise an oscillator unit, are weak. The phase reduction method outlined in §1 can then be applied at the network level. This leads to a system of coupled equations for the phases of the oscillators in the network, in which the effective phase interaction function H depends on both the nature of the synaptic interactions and the intrinsic properties of each oscillator as expressed through its phase resetting curve. For simplicity, we shall take the fundamental oscillatory unit to be a single neuron that transmits a regular sequence of spikes to other neurons in the network (see Figure 1). The relative shift in the spike trains of two oscillating neurons will then depend on the relative phases of the two oscillators. Note, however, that the phase reduction techniques outlined below can also be applied to models in which the basic oscillator consists of more than one neuron. For example, when modeling spindle oscillations in thalamus, the minimal oscillator might consist of an excitatory thalamocortical neuron and an inhibitory reticular neuron. As another example, we note that stimulus-induced oscillations in visual cortex are likely to arise at the level of a cortical column, involving several thousand excitatory and inhibitory neurons that form the basic oscillatory unit. In this case it is more appropriate to express the output of the oscillator in terms of a population-averaged firing rate (see §3). Networks of weakly interacting cortical columns can then be described in terms of a set of coupled phase equations, whose phase interaction function is stimulus–dependent [**133, 134**]. Note that oscillations at the population level can occur, even when the individual neurons are not themselves intrinsic oscillators, due to the combined effects of external stimuli and recurrent synaptic interactions. One particular network mechanism for stimulus–induced oscillations is through a Hopf bifurcation of a stationary activity bump, leading to the formation of a spatially–localized oscillation or breather that is aligned with the stimulus [**135, 136**]. Activity bumps are discussed in §5.3.

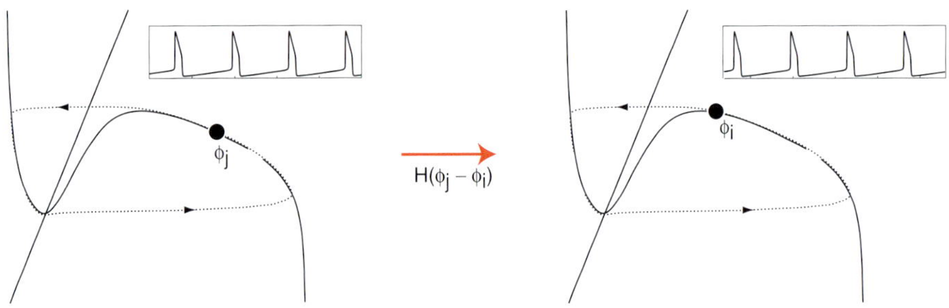

FIGURE 1. A system of two synaptically coupled limit cycle oscillators can be reduced to a pair of phase equations involving an effective phase interaction function H. See text for details.

4.1. Phase reduction for synaptically coupled neural oscillators

Let us consider a network of N neural oscillators labeled $i = 1, \ldots, N$. Denoting the state of the ith neuron by $\mathbf{x}_i \in \mathbf{R}^m$, the general form of the network dynamics is the coupled system of ODEs

$$\frac{d\mathbf{x}_i}{dt} = \mathbf{f}(\mathbf{x}_i) + \varepsilon \sum_{j=1}^{N} w_{ij}\mathbf{P}(\mathbf{x}_i, \mathbf{x}_j), \tag{4.1}$$

where $w_{ij}\mathbf{P}(\mathbf{x}_i, \mathbf{x}_j)$ represents the synaptic input from the jth to the ith neuron. We assume that the synaptic interactions are identical in form but are allowed to have different relative strengths or weights w_{ij}. (It is straightforward to extend the analysis to non-identical oscillators with different intrinsic frequencies for example). If the coupling between the oscillators is weak then one can derive closed equations for the phases of the oscillators. We shall establish this using the method of isochrones [**11**], following along similar lines to the analysis of a periodically forced oscillator described in §1. An alternative derivation based on the Fredholm alternative is described at the end of the section.

Method of isochrones

Suppose that in the absence of interactions each neuron has a stable limit cycle $\mathbf{X}$ with natural frequency ω_0. For each oscillator we introduce the phase variable ϕ_i such that

$$\frac{d\phi_i}{dt} = \omega_0, \tag{4.2}$$

and there is a one-to-one mapping $\phi_i \to \mathbf{X}(\phi_i)$. The definition of the phase is then extended into a neighborhood of the limit cycle using isochrones. That is, we associate with any point $\mathbf{x}_i$ in the vicinity of the ith limit cycle a phase $\phi_i(\mathbf{x}_i)$ such that

$$\frac{d\phi_i(\mathbf{x}_i)}{dt} = \omega_0 + \varepsilon \frac{\partial \phi_i}{\partial \mathbf{x}_i} . \sum_{j=1}^{N} w_{ij} \mathbf{P}(\mathbf{x}_i, \mathbf{x}_j).$$

To a first approximation we can evaluate the r.h.s on the unperturbed limit cycles $\mathbf{X}$. Since these are defined uniquely by the phase we obtain the coupled phase equations

$$\frac{d\phi_i}{dt} = \omega_0 + \varepsilon \sum_{j=1}^{N} w_{ij} Q(\phi_i, \phi_j), \tag{4.3}$$

where

$$Q(\phi_i, \phi_j) = \frac{\partial \phi_i}{\partial \mathbf{x}_i}(\mathbf{X}(\phi_i)).\mathbf{P}(\mathbf{X}(\phi_i), \mathbf{X}(\phi_j)). \tag{4.4}$$

Q is a 2π-periodic function of the phases.

It remains the specify the structure of $\mathbf{P}$. The first point to note is that the coupling term only contributes to the equation for the membrane potential $u_i = [\mathbf{x}_i]_1$ so that $P_k = \delta_{1,k} P$ and we can write

$$Q(\phi_i, \phi_j) = R(\phi_i) P(\mathbf{X}(\phi_i), \mathbf{X}(\phi_j)), \tag{4.5}$$

where R is the phase response curve of the oscillator, see §1.2. If we neglect dendritic processes and synaptic depression then the synaptic input from the jth to the ith neuron is of the form

$$I_{syn}(t) = w_{ij} \sum_m g_{\rm syn}(t - T_j^m - \tau_a)[u_{syn} - u_i(t)], \tag{4.6}$$

where the sum is taken over all the spikes emitted by neuron j at the firing times T_j^m, $g_{\rm syn}(t)$ is the alpha function (2.3) and τ_a is a discrete axonal delay. The firing times can be related to the phase of the oscillators as follows. Assume that the neuron fires whenever the phase ϕ_j crosses zero. The firing times then satisfy the condition

$$\phi_j(T_j^m) = 2\pi m, \quad \dot{\phi}_j(T_j^m) > 0. \tag{4.7}$$

If we now set $\psi_j = \phi_j - \omega_0 t$ then ψ_i is a slowly varying function of time so that to a first approximation

$$T_j^m = (m - \psi_j/2\pi)\Delta_0 = t + (m - \phi_j(t)/2\pi)\Delta_0. \tag{4.8}$$

Hence

(4.9)

$$P(\phi_i, \phi_j) \equiv P(\mathbf{X}(\phi_i), \mathbf{X}(\phi_j)) = \sum_m g_{\rm syn}((\phi_j/2\pi - m)\Delta_0 - \tau_a)[u_{syn} - u^*(\phi_i)].$$

Thus we can rewrite Q as

$$Q(\phi_i, \phi_j) = \widehat{R}(\phi_i) \widehat{P}(\phi_j - \omega_0 \tau_a), \tag{4.10}$$

where

$$\widehat{R}(\phi_i) = [u_{syn} - u^*(\phi_i)] R(\phi_i), \quad \widehat{P}(\phi_j) = \sum_m g_{\rm syn}((\phi_j/2\pi - m)\Delta_0). \tag{4.11}$$

The final step of the analysis is to use the method of averaging to obtain effective interactions that depend only on phase differences. Substituting $\psi_i = \phi_i - \omega_0 t$ into the phase equation (4.3) gives

$$\frac{d\psi_i}{dt} = \varepsilon \sum_{j=1}^{N} w_{ij} Q(\psi_i + \omega_0 t, \psi_j + \omega_0 t). \tag{4.12}$$

The corresponding averaged equation is then

$$\frac{d\psi_i}{dt} = \varepsilon \sum_{j=1}^{N} w_{ij} H(\psi_j - \psi_i - \omega_0 \tau_a), \tag{4.13}$$

where

$$\begin{aligned} H(\psi_j - \psi_i) &= \frac{1}{\Delta_0}\int_0^{\Delta_0} \widehat{R}(\psi_i + \omega_0 t)\widehat{P}(\psi_j + \omega_0 t])dt \\ &= \frac{1}{2\pi}\int_0^{2\pi} \widehat{R}(\phi + \psi_i - \psi_j)\widehat{P}(\phi)d\phi, \end{aligned} \tag{4.14}$$

and we have exploited the periodicity of the functions $\widehat{R}, \widehat{P}$. A useful alternative expression for H is obtained by substituting for $\widehat{P}$:

$$\begin{aligned} H(\psi) &= \frac{1}{2\pi}\int_0^{2\pi} \widehat{R}(\phi - \psi)\sum_m g_{\text{syn}}([\phi/2\pi - m]\Delta_0)d\phi \\ &= \frac{1}{2\pi}\int_0^{\infty} \widehat{R}(\phi - \psi) g_{\text{syn}}(\phi\Delta_0/2\pi)d\phi. \end{aligned} \tag{4.15}$$

Our analysis also establishes that averaging in the weak coupling regime reduces the effect of an axonal delay to that of an additional phase-shift.

Fredholm alternative

An alternative derivation of the phase equation (4.13) involves the Fredholm alternative and perturbation theory [**137, 138**]. To simplify the notation we set $\Delta_0 = 2\pi$. The basic idea is to seek solutions of equation (4.1) of the form

$$\mathbf{x}_i = \mathbf{X}(t + \psi_i(\tau)) + \varepsilon \mathbf{U}_i(t, \tau, \varepsilon), \tag{4.16}$$

where $\mathbf{X}(t)$ represents the 2π-periodic solution of the uncoupled equation. Substituting into equation (4.1) yields at order ε

$$\begin{aligned} &\mathbf{f}(\mathbf{X}(t+\psi_i))\frac{d\psi_i}{d\tau} + \frac{d\mathbf{U}_i(t,\tau,0)}{dt} \\ &\quad = \mathcal{D}\mathbf{f}(\mathbf{X}(t+\psi_i))\mathbf{U}_i(t,\tau,0) + \sum_{j=1}^{N} w_{ij}\mathbf{P}(\mathbf{X}(t+\psi_i), \mathbf{X}(t+\psi_j)), \end{aligned} \tag{4.17}$$

where $\mathcal{D}\mathbf{f}$ is the Jacobian of $\mathbf{f}$. This equation may be rewritten in the more suggestive form

$$\mathcal{L}(t+\psi_i)\mathbf{U}_i(t,\tau,0) \equiv \left[\frac{d}{dt} - \mathcal{D}\mathbf{f}(\mathbf{X}(t+\psi_i))\right]\mathbf{U}_i(t,\tau,0) = \mathbf{b}_i(t,\tau), \tag{4.18}$$

where

$$\mathbf{b}_i(t,\tau) = -\mathbf{f}(\mathbf{X}(t+\psi_i))\frac{d\psi_i}{d\tau} + \sum_{j=1}^{N} w_{ij}\mathbf{P}(\mathbf{X}(t+\psi_i), \mathbf{X}(t+\psi_j)). \tag{4.19}$$

Using the Fredholm alternative, the inhomogeneous linear equation (4.18) has a periodic solution if and only if

$$\frac{1}{\Delta_0}\int_0^{\Delta_0} \overline{\mathbf{Z}}(t+\psi_i).\mathbf{b}_i(\omega_0 t,\tau)dt = 0, \tag{4.20}$$

where $\mathbf{Z}(t)$ is the unique 2π-periodic solution of the adjoint homogeneous equation

$$\left[\frac{d}{dt} + \mathcal{D}\mathbf{f}(\mathbf{X}(t))^\dagger\right]\mathbf{Z}(t) = 0 \tag{4.21}$$

with chosen normalization

$$\frac{1}{2\pi}\int_0^{2\pi} \overline{\mathbf{Z}}(t).\mathbf{f}(\mathbf{X}(t))dt = 1. \tag{4.22}$$

Using this normalization condition and the expression for $\mathbf{b}_i$, we conclude that equation (4.18) has a solution if and only if

$$\frac{d\psi_i}{d\tau} = \frac{1}{2\pi}\int_0^{2\pi} \overline{\mathbf{Z}}(t+\psi_i).\sum_{j=1}^N w_{ij}\mathbf{P}(\mathbf{X}(t+\psi_i),\mathbf{X}(t+\psi_j))dt. \tag{4.23}$$

Exploiting the periodicity of the functions on the r.h.s, we can shift t to obtain the phase-difference equation (4.13) with H given by

$$H(\psi_j-\psi_i) = \frac{1}{2\pi}\int_0^{2\pi} \overline{\mathbf{Z}}(t).\mathbf{P}(\mathbf{X}(t),\mathbf{X}(t+\psi_j-\psi_i))dt. \tag{4.24}$$

Equation (4.24) can be further reduced by substituting the explicit form for the synaptic interactions $\mathbf{P}$ as given by equation (4.9). One then recovers equation (4.14) with $\overline{Z}_1$ replacing the phase resetting curve R. (The two forms of H are equivalent to lowest order in ε). Note that the adjoint solution can be determined using standard numerical methods. On the other hand, the phase resetting curve has a more direct physical interpretation, and hence we will adopt the version given by equation (4.14) in our subsequent analysis.

4.2. Phase-locked solutions

We define a $1:1$ phase-locked solution of the phase equation (4.13) to be of the form

$$\psi_i(t) = \Delta\omega\, t + \bar{\psi}_i, \tag{4.25}$$

where $\omega = \omega_0 + \Delta\omega$ is the *collective frequency* of the coupled oscillators and $\Psi = (\bar{\psi}_1, \dots, \bar{\psi}_n)$ is a set of constant phases. Substitution into equation (4.13) leads to the fixed point equations

$$\Delta\omega = \varepsilon\sum_j w_{ij}H(\bar{\psi}_j - \bar{\psi}_i), \quad j = 1,\dots,N. \tag{4.26}$$

After choosing some reference oscillator, the N phase equations determine the collective period ω and $N-1$ relative phases with the latter independent of ϵ. In order to analyze the local stability of a phase-locked solution Ψ, we linearize equation (4.13) by setting $\psi_i(t) = \bar{\psi}_i + \Delta\omega t + \Delta\psi_i(t)$ and expanding to first-order in

$\widetilde{\psi}_i$:

$$\frac{\mathrm{d}\Delta\psi_i}{\mathrm{d}t} = \epsilon \sum_{j=1}^{N} \widehat{\mathcal{H}}_{ij}(\Phi)\Delta\psi_j, \tag{4.27}$$

where

$$\widehat{\mathcal{H}}_{ij}(\Phi) = w_{ij}H'(\bar{\psi}_j - \bar{\psi}_i) - \delta_{i,j}\sum_k w_{ik}H'(\bar{\psi}_k - \bar{\psi}_i), \tag{4.28}$$

and $H'(\psi) = dH(\psi)/\mathrm{d}\psi$. One of the eigenvalues of the Jacobian $\widehat{\mathcal{H}}$ is always zero, and the corresponding eigenvector points in the direction of the flow, that is $(1, 1, \ldots, 1)$. The phase-locked solution will be stable provided that all other eigenvalues have a negative real part [**139**].

Pair of identical neural oscillators

Consider the case of two identical oscillators ($N = 2$) that are symmetrically coupled, in other words $w_{12} = w_{21} = 1$ and $w_{11} = w_{22} = 0$. The phase difference $\psi = \psi_2 - \psi_1$ satisfies the simple equation

$$\frac{d\psi}{dt} = \varepsilon H^-(\psi), \tag{4.29}$$

where $H^-(\psi) = H(-\psi) - H(\psi)$. Phase-locked states are then given by the zeros of the odd function H^-:

$$H^-(\psi) = 0. \tag{4.30}$$

A given phase-locked state is stable provided that

$$\epsilon\frac{\mathrm{d}H^-(\psi)}{\mathrm{d}\psi} < 0. \tag{4.31}$$

Note that by symmetry both the in-phase ($\psi = 0$) and anti-phase ($\psi = \pi$) states are guaranteed to exist [**140**]. However, there may be additional fixed points depending on the parameters. The collective frequency of oscillation is determined by the even part of H, $H^+(\psi) = H(\psi) + H(-\psi)$:

$$\omega = \omega_0 + \frac{\varepsilon}{2}H^+(\psi). \tag{4.32}$$

It turns out that the stability of the in-phase state for a pair of neurons depends on a number of factors [**141, 142**]: whether the synapses are excitatory ($u_{syn} > 0$) or inhibitory ($u_{syn} > 0$), whether the kinetics are fast (large α) or slow (small α), whether the phase resetting curve is Type I or Type II, and the size of axonal delays. We shall illustrate this by considering two simple examples with $\widehat{R}(\phi) \approx u_{syn}R(\phi)$ and u_{syn} absorbed into ε: the sign of ε will then determine whether the neurons are excitatory or inhibitory.

Example: Hodgkin-Huxley neuron. In Figure 2 we show a typical phase-resetting curve $R(\phi)$ for the Hodgkin-Huxley model together with the resulting phase interaction function $H(\psi)$ for an α function synapse, calculated using equation (4.15),

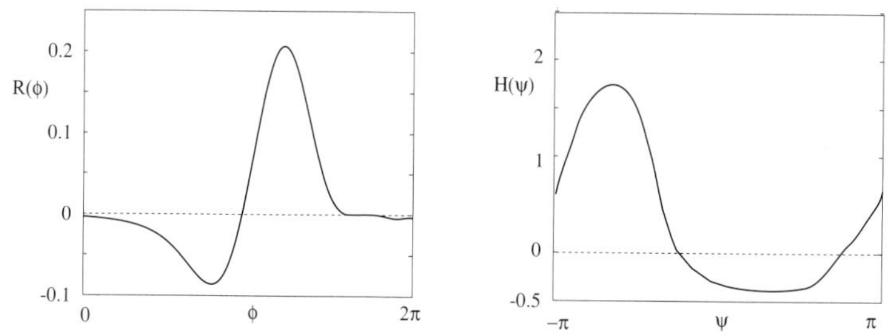

FIGURE 2. Plot of Type II phase resetting curve $R(\phi)$ for the Hodgkin-Huxley model together with effective phase interaction function $H(\psi)$.

and zero axonal delays. If we approximate the PRC by $R(\phi) = -\sin\phi$ then

$$\begin{aligned} H(\psi) &= \frac{1}{2\pi}\int_0^\infty R(\phi-\psi)P(\phi\Delta_0/2\pi)d\phi \\ &= \frac{\alpha^2}{2\pi}\frac{d}{d\alpha}\int_0^\infty \sin(\phi-\psi)\mathrm{e}^{-\alpha\phi\Delta_0/2\pi}d\phi \\ &= \frac{\alpha^2}{2\pi}\mathrm{Im}\frac{d}{d\alpha}\int_0^\infty \mathrm{e}^{i(\phi-\psi)}\mathrm{e}^{-\alpha\phi\Delta_0/2\pi}d\phi. \end{aligned}$$

Now

(4.33) $$\frac{1}{2\pi}\frac{d}{d\alpha}\int_0^\infty \mathrm{e}^{i\phi}\mathrm{e}^{-\alpha\phi\Delta_0/2\pi}d\phi = -\frac{\Delta_0}{(\alpha\Delta_0-2\pi i)^2},$$

and so

(4.34) $$H(\psi) = A\sin\psi - B\cos\psi,$$

where

(4.35) $$A = [(\alpha\Delta_0)^2-(2\pi)^2]K, \quad B = 4\pi\alpha\Delta_0 K, \quad K = \frac{\alpha^2\Delta_0}{[(2\pi)^2+(\alpha\Delta_0)^2]^2}.$$

The odd part is then

(4.36) $$H^-(\psi) = \frac{2\alpha^2\Delta_0[4\pi^2-\alpha^2\Delta_0^2]\sin\psi}{\Delta_0[\alpha^2\Delta_0^2+4\pi^2]^2}.$$

As long as $2\pi/\alpha\Delta_0 \neq 1$ there are phase-locked states at $\psi = 0, \pi$ as expected. The synchronous solution is stable if $\varepsilon H^{-\prime}(0) < 0$. This implies that synchronization will occur for excitatory coupling if $\alpha > 2\pi/\Delta_0$ and for inhibitory coupling if $\alpha < 2\pi/\Delta_0$. Thus, *inhibition* rather than excitation synchronizes the neurons when the synapses have sufficiently slow kinetics. The reverse is true for the anti-phase solution $\psi = \pi$. If higher-order terms in the Fourier series expansion of H are included, $H(\psi) = \sum_n h_n \mathrm{e}^{in\psi}$, then the transition from synchronous to antisynchronous behavior is smoother, that is, the system no longer makes a sudden jump from one to the other at a critical value of α.

Example: Integrate-and-fire model. Substituting into equation (4.14) the PRC of an IF neuron given by equation (1.43), the α-dependence of phase-locked solutions in the case of a symmetric pair of excitatory or inhibitory IF neurons can be determined [**141**]. The results are summarized in Figure 3(a). For excitatory coupling

($\epsilon > 0$) the synchronous state is unstable for all $0 < \alpha < \infty$. On the other hand, the anti-synchronous solution is stable for $\alpha < \alpha_c$ but loses stability when $\alpha > \alpha_c$ with the creation of two stable partially synchronized or asynchronous states. The emergence of these two additional states can be understood in terms of the behavior of the odd part of the interaction function $H^-(\psi)$ as shown in Figure 3(b). In the limit $\alpha \to \infty$ the two asynchronous states asymptotically approach the synchronous state so that the neurons are almost perfectly synchronized. This is consistent with the analysis of Mirollo and Strogatz [**143**] who proved rigorously that globally coupled IF oscillators almost always synchronize in the presence of instantaneous excitatory interactions. The stability properties of all solutions are reversed in the case of inhibitory coupling ($\epsilon < 0$) so that, in particular, the synchronous state is now stable for all α. Introduction of a discrete axonal delay τ_a produces a checkerboard pattern of alternating stable/unstable solution branches that can overlap to produce multistable solutions [**144**].

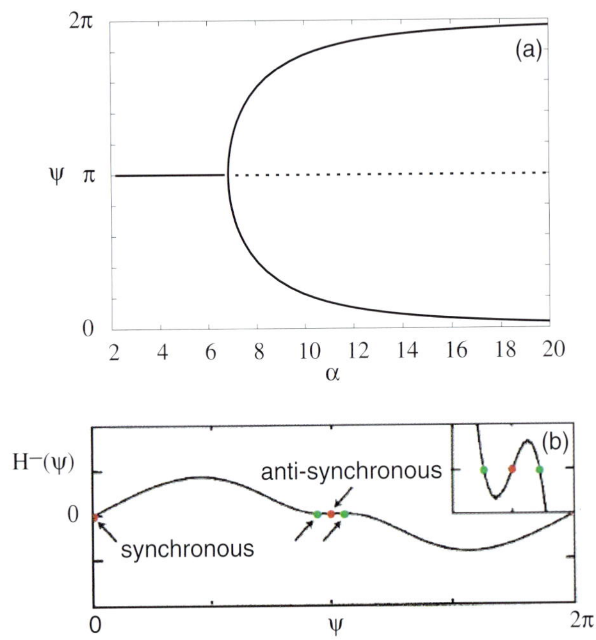

FIGURE 3. (a) Relative phase $\psi = \psi_2 - \psi_1$ for a pair of IF oscillators with symmetric excitatory coupling as a function of α with $I_0 = 2$. In each case the anti-phase state undergoes a bifurcation at a critical value of $\alpha = \alpha_c$ where it becomes unstable and two additional stable solutions ψ, $2\pi - \psi$ are created. The synchronous state remains unstable for all α. (b) Odd part of the phase interaction function for α just above the critical value α_c showing two new phase-locked solutions.

Traveling oscillatory waves on a chain

Both vertebrates and invertebrates have dedicated networks of cells that contain the information necessary to activate different motor neurons in the appropriate sequence and intensity to generate a variety of motor patterns. Such networks are known as central pattern generators (CPGs). The simplest CPGs control reflex actions, swallowing and coughing. At the next level are those that control rhythmic movements such as breathing and locomotion. A well studied example of the latter is swimming locomotion in vertebrates such as the lamprey [**145, 146, 147**]. The basic mechanism for such coordinated movement is alternating left-right activity within each segment of the spinal cord, combined with a phase lag between the activation of consecutive segments along the body (see Figure 4). This generates an undulatory wave that propagates along the body and pushes the animal forwards through the water. An analogous mechanism also occurs in invertebrates such as the leech [**148**].

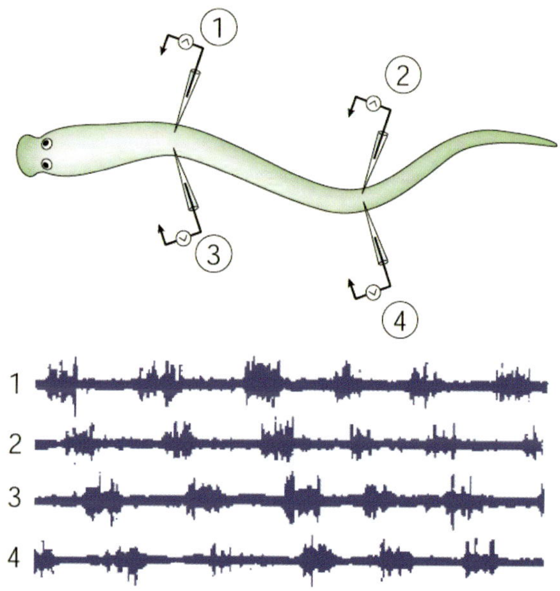

FIGURE 4. Neural activity recorded from four locations on the body of a lamprey during swimming locomotion. The body can be segmented into a bilateral chain of alternating left/right oscillators such as the pairs 1,3 and 2,4. There is also an intersegmental phase-lag between the activity of the pair 1,3 and 2,4.

At the simplest level, the spinal cord may be modeled as a chain N phase oscillators with nearest-neighbor coupling as shown in Figure 5. A traveling wave of activity along the chain then corresponds to a phase-locked solution in which ϕ_n

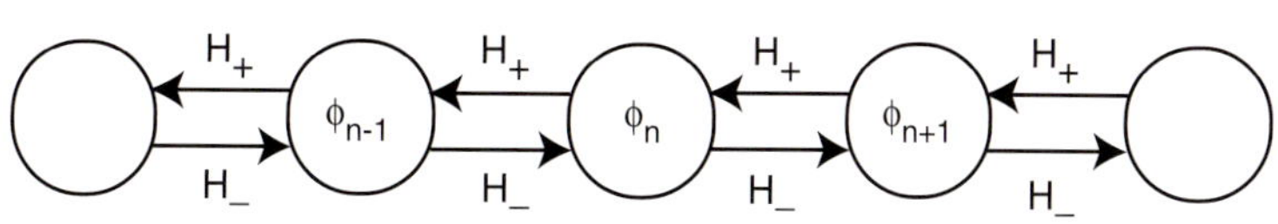

FIGURE 5. A chain of N phase oscillators.

is a monotonically increasing or decreasing function of n. At least two mechanisms have been identified for generating such a phase gradient along the chain [**149, 150**]: a) Heterogeneity in the form of a gradient in intrinsic frequencies of the individual neurons, that is, $\omega_n = \omega_0 + n\omega_1$ where $\omega_{0,1}$ are constants, will induce a gradient in the relative phases of the coupled system. Such a frequency gradient has been found in the leech CPG for swimming [**148**]. The resulting plane wave propagates opposite to the direction of the gradient, that is, from high to low frequencies. Interestingly, a spatial gradient of frequencies has been observed experimentally in slices of molluscan olfactory lobe, with the wave propagating from high to low frequencies as predicted by the theory [**151**].

b) In the lamprey CPG for swimming, there is no evidence for a gradient in frequency. Moreover, unlike the observed wave in the lamprey, waves produced by frequency gradients do not have constant phase lags along the chain, that is, the wave is not of constant speed. However, it is possible to produce approximately constant speed waves in the bulk of an oscillator chain by having anisotropic interactions, $H_+ \neq H_-$, together with the so-called edge property that $H_+(\theta)$ or $H-(\theta)$ vanishes at a non-zero value of θ.
We briefly illustrate the above results using a simple model of a chain with $H_\pm(\phi) = W_\pm H(\phi)$. Such a network is described by the following system of coupled phase equations:

$$\begin{aligned}
\dot{\phi}_1 &= \omega_1 + W_+ H(\phi_2 - \phi_1) \\
\dot{\phi}_i &= \omega_i + W_+ H(\phi_{i+1} - \phi_i) + W_- H(\phi_{i-1} - \phi_i), \quad i = 2, \dots, N-1 \\
\dot{\phi}_N &= \omega_N + W_- H(\phi_{N-1} - \phi_N).
\end{aligned} \tag{4.37}$$

Introducing the intersegmental phase differences $\theta_i = \phi_{i+1} - \phi_i$ leads to the $N-1$-dimensional system of equations

$$\dot{\theta}_i = \Delta\omega_i + W_+[H(\theta_{i+1}) - H(\theta_i)] + W_-[H(-\theta_i) - H(-\theta_{i-1})], \quad i = 1, \dots, N-1, \tag{4.38}$$

with boundary conditions

$$H(-\theta_0) = 0 = H(\theta_N) \tag{4.39}$$

and $\Delta\omega_i = \omega_{i+1} - \omega_i$.

Suppose that there is a gradient of frequencies along the chain, that is, $\Delta\omega_i$ has the same sign for all i. Also take an isotropic, odd interaction function, $W_\pm = 1$ and $H(\theta) = -H(-\theta)$. Then

$$\dot{\theta}_i = \Delta\omega_i + H(\theta_{i+1}) + H(\theta_{i-1}) - 2H(\theta_i), \quad i = 1, \dots, N-1. \tag{4.40}$$

The phase-locked solutions $\Theta = (\theta_1, \dots, \theta_{N-1})$ then satisfy the matrix equation

$$\mathbf{H}(\Theta) = -\mathbf{A}^{-1}\mathbf{D}, \tag{4.41}$$

where $\mathbf{H}(\Theta) = (H(\theta_1), \dots, H(\theta_{N-1}))^T$, $\mathbf{D} = (\Delta\omega_1, \dots \Delta\omega_{N-1})$ and $\mathbf{A}$ is a tridiagonal matrix with elements $A_{ii} = -2$, $A_{i,i+1} = 1 = A_{i+1,i}$. Suppose, for concreteness, that $H(\theta) = \sin\theta$, Then a solution Θ will exist only if every component of $\mathbf{A}^{-1}\mathbf{D}$ lies between ± 1. Let $a_0 = \max\{|\mathbf{A}^{-1}\mathbf{D}_i|\}$. If $a_0 < 1$ then for each $i = 1, \dots, N-1$ there are two distinct solutions $\theta_i^\pm$ in the interval $[0, 2\pi)$ with $H'(\theta_i^-) > 0$ and $H'(\theta_i^+) < 0$. In other words, there are 2^N phase-locked solutions. Linearizing about each phase-locked solution and exploiting the structure of the matrix $\mathbf{A}$, it can be proven that only the solution $\Theta^- = (\theta_1^-, \ldots \theta_{N-1}^-)$ is stable ([**149**]. Assuming that the frequency gradient is monotonic, this solution corresponds to a stable traveling wave. When the frequency gradient becomes too steep to allow phase-locking, that is $a_0 > 1$, two or more pools of oscillators (frequency plateaus) tend to form and oscillate at different frequencies. Waves produced by a frequency gradient do not have a constant speed or, equivalently, constant phase lags along the chain.

Constant speed waves can be generated from equations (4.38) by considering phase-locked solutions defined by $\theta_i = \theta$ for all i with a collective period of oscillation determined using $\dot{\theta}_i = \Omega$ such that

$$\Omega = \omega_1 + W_+ H(\theta), \tag{4.42}$$

such that speed of the wave is Ω/θ. The steady state solutions are then

$$\begin{aligned} \Delta\omega_1 + W_- H(-\theta) &= 0 \\ \Delta\omega_i &= 0, \quad i = 2, \ldots, N-2 \\ \Delta\omega_{N-1} - W_+ H(\theta) &= 0. \end{aligned} \tag{4.43}$$

This implies that all internal oscillators ($i = 2, \ldots, N-2$) have the same intrinsic frequency. If we further assume that $\Delta w_1 = 0 = \Delta\omega_{N-1}$ then a constant speed wave will exist provided that (i) the interactions are unidirectional with $W_- = 0$ say, and (ii) H satisfies the edge property $H(\theta) = 0$ for some $\theta \neq 0$. For example, if $H(\theta) = \sin(\theta + \sigma)$ for a fixed phase-shift σ, then the steady-state phase solution is $\theta = -\sigma$. It also follows that although the speed of the wave can change by changing the intrinsic frequency ω_1, the phase lag θ remains the same. This is consistent with the swimming motion of invertebrates. Finally, note that if coupling is in both directions, as in the lamprey CPG, one finds that for sufficiently long chains and $H^+ \neq H^-$, the system typically behaves like the unidirectional case except that there is a small *boundary layer* close to one end [**150**].

Phase–locking in strongly coupled IF networks

One of the useful features of the IF model is that one can go beyond the weak coupling regime by studying the existence and stability of phase–locked states directly in terms of the firing time map [**152, 153**]. This generalizes the analysis of the firing time map for a single periodically forced IF neuron (see §1.3) to the network level. It can be shown that increasing the strength of the synaptic coupling can lead to new dynamical instabilities not found in the phase–reduced models. Such instabilities generate non–phase–locked states, characterized by periodic or quasiperiodic variations of the inter-spike intervals (ISIs) on closed orbits. The corresponding mean firing rates exhibit complex spatio-temporal patterns including network–induced bursting.

Consider a network of identical leaky IF neurons labeled by the indices $i = 1, \ldots, N$ such that neuron i makes synapses with neuron j:

$$\frac{du_i}{dt} = -u_i + I_0 + \epsilon \sum_{j=1}^{M} w_{ij} \sum_{m \in \mathbf{Z}} g_{\text{syn}}(t - T_j^m) \tag{4.44}$$

where T_j^m is the mth firing time of the jth neuron, ϵw_{ij} is the strength of the synapse between neuron i and j and the function $g_{\text{syn}}(t)$ represents the effects of synaptic filtering. Equation (4.44) is supplemented by the reset condition $\lim_{\delta \to 0^+} u_i(T_i^m +$

$\delta) = 0$. Suppose that an IF neuron operates as an oscillator such that the inter-spike interval $T_j^{m+1} - T_j^m = \Delta$ for all j, m in the absence of synaptic inputs. Conditions for phase-locking can be derived for arbitrary coupling strength ε by solving equation (4.44) under the ansatz that the firing times are of the form $T_j^n = (n - \psi_j/2\pi)\Delta$ for some self-consistent period Δ and constant phases ψ_j [**141, 140**]. Integrating over the interval $t \in (-\Delta\psi_i/2\pi, \Delta - \Delta\psi_i/2\pi)$ and incorporating the reset condition leads to the result

$$1 = (1 - \mathrm{e}^{-\Delta})I_0 + \varepsilon \sum_{j=1}^{N} w_{ij} H(\psi_j - \psi_i) \tag{4.45}$$

with

$$H(\phi) = \mathrm{e}^{-\Delta} \int_0^{\Delta} dt\, \mathrm{e}^{t} \sum_{k \in \mathbf{Z}} g_{\mathrm{syn}}(t + (k + \phi/2\pi)\Delta) \tag{4.46}$$

Note that up to a positive constant factor, $H(\phi)$ reduces to the weak coupling phase interaction function of equation (4.13) for $\Delta \to \Delta_0$ and $\widehat{R}$ taken to be the PRC (1.43). It can be seen that equation (4.45) has an identical structure to that of equation (4.26). However, the former is exact whereas the latter is only strictly valid to $\mathcal{O}(\varepsilon)$, since it is derived under the assumption of weak coupling. Moreover, the collective period of oscillations Δ must be determined self-consistently in equation (4.45), since H depends on Δ rather than the natural period Δ_0. The stability of the phase–locked solutions can be determined by considering perturbations of the firing times, as detailed elsewhere [**152, 153**].

4.3. Oscillations in large homogeneous networks

In §2 we pointed out that neurons often form large homogeneous stochastic networks. This suggests that rather than considering individual neurons as deterministic oscillators, it might be more appropriate to consider collective oscillations at the population level. In this section we show how a fully connected network of IF neurons can exhibit collective oscillations through destabilization of an asynchronous state [**19**]. Consider the following synaptically coupled network of nonlinear IF neurons

$$\frac{du_i}{dt} = F(u_i) + \frac{\varepsilon}{N} \sum_{j=1}^{N} \int_{-\infty}^{\infty} g_{\mathrm{syn}}(t' - \tau_a) \sum_m \delta(t - t' - T_j^m)dt', \tag{4.47}$$

with g_{syn} given by the alpha function (2.3), τ_a a discrete axonal delay and ε determines the strength of coupling. We take a threshold $u_\kappa = 1$ and a reset $u_r = 0$. In the case of a large network, we can replace the sum over delta functions by the population activity $\nu(t)$ so that

$$\frac{du_i}{dt} = F(u_i) + \varepsilon\widehat{\nu}(t), \tag{4.48}$$

where $\widehat{\nu}(t) = \int_{-\infty}^{\infty} g_{\mathrm{syn}}(t' - \tau_a)\nu(t - t')dt'$. Suppose that there exists an asynchronous state $\nu(t) = \nu_0$. Since the α function is normalized to unity it follows that $\widehat{\nu}(t) = \nu_0$ as well. An implicit equation for ν_0 is then obtained by integrating equation (4.48) between successive firing times:

$$\frac{1}{\nu_0} = \int_0^1 \frac{du}{F(u) + \varepsilon\nu_0}. \tag{4.49}$$

We will assume that there exists a unique solution to this equation for given F and ε.

In order to study the stability of the asynchronous state, it is convenient to carry out the change of variables

$$y_i = \nu_0 \int_0^{u_i} \frac{du}{F(u) + \varepsilon\nu_0}, \tag{4.50}$$

with $0 < y_i < 1$ such that equation (4.48) becomes

$$\frac{dy_i}{dt} = \nu_0 + \Gamma(y_i)[\widehat{\nu}(t) - \nu_0] \tag{4.51}$$

and

$$\Gamma(y) = \frac{\nu_0\varepsilon}{F(u) + \nu_0\varepsilon}. \tag{4.52}$$

In order to incorporate the effects of noise, we include an additive white noise term $\xi_i(t)$,

$$\frac{dy_i}{dt} = \nu_0 + \Gamma(y_i)[\widehat{\nu}(t) - \nu_0] + \xi_i(t), \tag{4.53}$$

with

$$\langle \xi_i(t) \rangle = 0, \quad \langle \xi_i(t)\xi_j(t') \rangle = \sigma^2\delta_{ij}\delta(t - t'). \tag{4.54}$$

(Note that diffusive fluctuations of the membrane potential due to stochastic background activity would lead to an additive white noise term in equation (4.48) rather than in equation (4.54). The corresponding stochastic equation for y_i would then involve multiplicative noise, which is much harder to analyze). In the presence of noise the variable y_i can become negative so $-\infty < y_i < 1$. The Langevin equation (4.54) has an associated Fokker–Planck equation

$$\frac{\partial}{\partial t} p(y,t) = -\frac{\partial}{\partial y} J(y,t), \tag{4.55}$$

where $J(y,t)$ is the probability flux

$$J(y,t) = [\nu_0 + \Gamma(y)[\widehat{\nu}(t) - \nu_0]]\, p(y,t) - \frac{\sigma^2}{2}\frac{\partial}{\partial y} p(y,t). \tag{4.56}$$

This is supplemented by the boundary conditions arising from reset (see §3)

$$p(1,t) = 0, \quad J(1,t) = \nu(t), \tag{4.57}$$

$$p(0^+,t) = p(0^-,t), \quad J(0^+,t) - J(0^-,t) = \nu(t), \tag{4.58}$$

We also require $p(-\infty,t) = 0$ and $J(-\infty,t) = 0$. The steady-state solution of the Fokker–Planck equation is $J(y,t) = \nu_0$ and $p(y,t) = p_0(y)$ with

$$p_0(y) = \begin{cases} e^{2\nu_0 y/\sigma^2} - e^{2\nu_0(y-1)/\sigma^2}, & y < 0 \\ 1 - e^{2\nu_0(y-1)/\sigma^2}, & 0 < y < 1 \end{cases}. \tag{4.59}$$

The stability of the steady-state can be determined by setting

$$p(y,t) = p_0(y) + \rho(y)e^{\lambda t}, \quad \nu(t) = \nu_0 + \nu_1 e^{\lambda t} \tag{4.60}$$

and expanding to first order in ρ, ν_1. This gives the eigenvalue equation

$$\lambda\rho(y) = \frac{\sigma^2}{2}\frac{\partial^2}{\partial y^2}\rho(y) - \nu_0\frac{\partial}{\partial y}\rho(y) - \nu_1\widetilde{g}_{\rm syn}(\lambda)\frac{\partial}{\partial y}[\Gamma(y)p_0(y)], \tag{4.61}$$

where $\widetilde{g}_{\rm syn}(\lambda)$ is the Laplace transform

$$\widetilde{g}_{\rm syn}(\lambda) = \int_0^\infty g_{\rm syn}(t-\tau_a)e^{-\lambda t} = \frac{\alpha^2}{(\lambda+\alpha)^2}e^{\tau_a\lambda}. \tag{4.62}$$

Defining the function

$$h(y) = \nu_1\widetilde{g}_{\rm syn}(\lambda)\frac{\partial}{\partial y}[\Gamma(y)p_0(y)], \tag{4.63}$$

we can write equation (4.61) as the inhomogeneous equation

$$[\mathcal{L} - \lambda\mathbf{1}]\rho(y) = h(y;\lambda), \tag{4.64}$$

where

$$\mathcal{L} = \frac{\sigma^2}{2}\frac{\partial^2}{\partial y^2} - \nu_0\frac{\partial}{\partial y}. \tag{4.65}$$

This inhomogeneous equation can be solved in terms of the associated one–dimensional Green's function satisfying $[\mathcal{L} - \lambda\mathbf{1}]G(y,y';\lambda) = \delta(y-y')$ and $G(1,y',\lambda) = 0$:

$$\rho(y) = \int_{-\infty}^1 G(y,y';\lambda)h(y';\lambda)dy' - \nu_1 G(y,0;\lambda) \tag{4.66}$$

with

$$G(y,y';\lambda) = \begin{cases} A\left(e^{\mu_+(\lambda)[y-1]} - e^{-\mu_-(\lambda)[y-1]}\right)e^{\mu_-(\lambda)[y'-1]}, & y'<y<1 \\ A\left(e^{\mu_-(\lambda)[y'-1]} - e^{-\mu_+(\lambda)[y'-1]}\right)e^{\mu_+(\lambda)[y-1]}, & y<y', \end{cases} \tag{4.67}$$

where

$$A = \frac{2}{\sigma^2}\frac{1}{\mu_+ + \mu_-}, \tag{4.68}$$

$$\mu_\pm(\lambda) = \frac{1}{\sigma^2}\left[\sqrt{\nu_0^2 + 2\lambda\sigma^2} \pm \nu_0\right]. \tag{4.69}$$

Note that the term $\nu_1 G(y,0;\lambda)$ ensures that the flux discontinuity at $y=0$ is satisfied. Finally, an implicit equation for the eigenvalues λ can be obtained by substituting equation (4.63) into (4.66) and imposing the boundary condition $J(1,t) = \nu(t)$, which corresponds to the following first–order condition

$$-\frac{\sigma^2}{2}\frac{\partial}{\partial y}\rho(y,t)|_{y=1} = \nu_1. \tag{4.70}$$

The resulting characteristic equation is [**19**]

$$\left(e^{\mu_-(\lambda)} - 1\right) = \mu_-(\lambda)\widetilde{g}_{\rm syn}(\lambda)\int_{-\infty}^1 p_0(y)\Gamma(y)e^{\mu_-(\lambda)y}dy. \tag{4.71}$$

In the zero noise limit $\sigma \to 0$, we have $\mu_-(\lambda) \to \lambda/\nu_0$ and $p_0(y) \to 1$ for $0<y<1$ and is zero otherwise. Thus, equation (4.71) becomes

$$\left(e^{\lambda/\nu_0} - 1\right) = \frac{\lambda}{\nu_0}\widetilde{g}_{\rm syn}(\lambda)\int_0^1 \Gamma(y)e^{\lambda y/\nu_0}dy. \tag{4.72}$$

In the weak coupling regime, solutions of equation (4.72) are of the form $\lambda = 2\pi i n\nu_0 + \Lambda_n$ for integer n with $\Lambda_n = \mathcal{O}(\epsilon)$. The term Λ_n can be calculated by performing a perturbation expansion in the coupling ϵ. The lowest order contribution is simply determined by setting $\lambda = 2\pi i n\nu_0$ on the right-hand side of equation (4.72). In the case of a linear IF model with $F(u) = I_0 - u$, we have $\Gamma(y) = (\varepsilon\nu_0/I_0)e^{y/\nu_0} + \mathcal{O}(\varepsilon^2)$ so that after absorbing a positive factor into ε,

$$\Lambda_n = \epsilon\left(\frac{2\pi i n\nu_0}{1+2\pi i n\nu_0}\right)\widetilde{g}_{\rm syn}(2\pi i n\nu_0) + \mathcal{O}(\epsilon^2). \tag{4.73}$$

We then have the following stability results in the absence of noise [**19, 125**]:
(i) For zero axonal delays ($\tau_a = 0$) and excitatory coupling ($\epsilon > 0$), the asynchronous state is stable with respect to excitation of the nth mode if and only if $\alpha < \alpha_n$ where

$$\alpha_n = -1 + \sqrt{1 + 4n^2\pi^2\nu_0^2}. \tag{4.74}$$

Hence, it is stable for sufficiently slow synapses, that is, $\alpha < \alpha_1$. The asynchronous state is always unstable in the case of inhibitory coupling since the condition for stability with respect to the nth harmonic is now $\alpha > \alpha_n$, which cannot be satisfied for all n.
(ii) The asynchronous state is almost always unstable for non-zero delays (in the noise-free case).
(iii) For large n, $|\Lambda_n| \sim 1/n^2$ so that higher harmonics grow or decay slowly.

Note that although the zero delay case is a singular limit in the absence of noise, it becomes non-singular for arbitrarily small amounts of noise, where instabilities with respect to higher harmonics are suppressed [**19, 125**]. One finds that for sufficiently high noise levels the asynchronous state is always stable. Reducing the noise for fixed delay induces an instability due to excitation of one of the harmonic modes with frequency $\omega \approx \omega_n = 2\pi n\nu_0$. A bifurcation at $\omega \approx \omega_1$ implies that the period of the resulting collective oscillation is identical to the period of the individual oscillators. Higher harmonics correspond to instabilities of the asynchronous state that lead to the formation of cluster states [**125, 154**]: each neuron fires with mean rate ν_0, but the population of neurons splits up into several groups that fire in sequence so that the overall activity exhibits faster oscillations. In Figure 6 we show the stability diagram for the asynchronous state of an IF network calculated within the context of the spike response model [**17**]. In this version of the model, the source of noise is now reset noise rather than membrane fluctuations; each time the neuron fires the membrane potential is reset to a random value u_r generated from a Gaussian distribution of width σ. Nevertheless, the qualitative behavior is very similar. Finally, note that fast oscillations are also found in sparsely connected random networks [**124**].

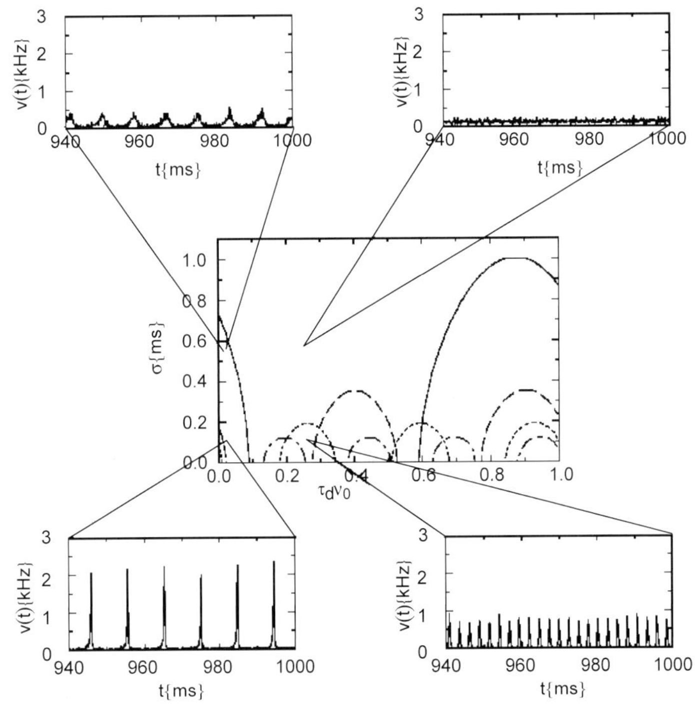

FIGURE 6. Stability diagram (center) for asynchronous state in an IF network as a function of noise amplitude σ and delay τ_a for fixed synaptic time constant $\alpha^{-1} = 4ms$ and membrane time constant $\tau = 10ms$. The input I_0 was chosen so that the population activity of the asynchronous state was $\nu_0 = 0.5\alpha$. The diagram shows the borders of the stability region with respect to the harmonic frequencies ω_1 (solid lines), ω_2 (long–dashed lines), ω_3 (short–dashed lines) and ω_4 (long–short dashed lines). Insets show typical patterns of activity from a simulation of $N = 1000$ neurons. [Reprinted from [**7**], with permission of Cambridge University Press.]

LECTURE 5

Neural pattern formation

Wilson and Cowan [**155, 156**] introduced a rate–based population model of cortical tissue based on the idea that the identity of individual presynaptic neurons is not important, but only the distribution of their level of activity. This leads to a statistical description of cortical activity in which the proportion of active neurons is chosen as a model variable. A justification for a statistical approach can be given in terms of the spatial clustering of neurons with similar response properties in cortex. These cells form vertical columns within the thin convoluted sheet of neural tissue that constitutes the cerebral cortex. For example, Hubel and Wiesel [**157**] reported a columnar-like arrangement of neurons in primary visual cortex that prefer stimuli of similar orientation. Neurons within a cortical column share many inputs and are tightly interconnected, so that it is reasonable to consider the mean activity $\nu(t)$ of the given neural population, rather than keeping track of the spiking of individual cells. However, in order to construct a closed set of equations for population activity, it is necessary that the total input into a population is slowly varying relative to the time-scale of action potential generation. Hence, an implicit assumption in rate–based population models is that neurons within a population have spike trains that are temporally incoherent (see also §3). In this final lecture, we derive integrodifferential equations describing the dynamics of synaptically coupled neuronal populations in spatially structured cortical networks. We then use these equations to investigate a variety of problems regarding large–scale cortical dynamics. First, we consider a Turing–like mechanism for spontaneous pattern formation in visual cortex, and show how this provides a neural mechanism for geometric visual hallucinations. We then show how cortical networks can support persistent spatially coherent states that could provide a basis for working memory. Finally, we study traveling waves in disinhibited cortical slices.

5.1. Reduction to rate models

Suppose that there exist M populations each consisting of N neurons labeled by $i \in \mathcal{I}_p$, $p = 1, \ldots, M$ and $|\mathcal{I}_p| = N$, see Figure 1. Assume that all neurons of a given population are equivalent in the sense that the interaction between neuron $i \in \mathcal{I}_p$ and neuron $j \in \mathcal{I}_q$ only depends on p and q. Each neuron i in population p receives input from all neurons j in population q with strength $w_{ij} = W_{pq}/N$. The total synaptic current to neuron $i \in \mathcal{I}_p$ from the various populations is thus

$$X_p(t) = \sum_{q=1}^{M} W_{pq} \int_0^\infty g_{\rm syn}(\tau)\nu_q^{(N)}(t-\tau)d\tau, \tag{5.1}$$

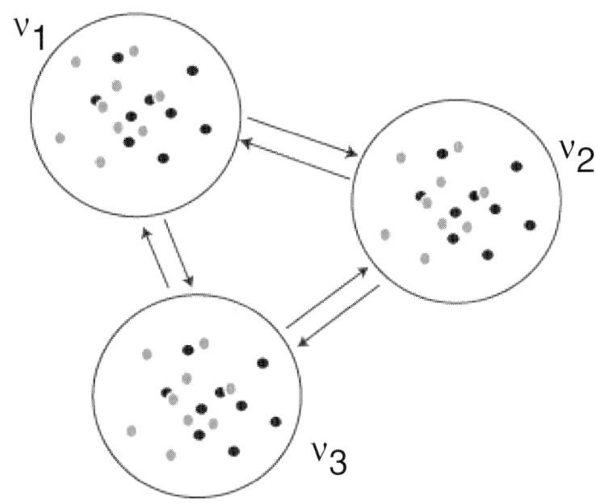

FIGURE 1. Interacting populations of spiking neurons each with its own population activity ν_p, $p = 1, 2, 3$.

where

$$\nu_q^{(N)}(t) = \frac{1}{N} \sum_{j \in \mathcal{I}_q} \sum_{m \in \mathbf{Z}} \delta(t - T_j^m). \tag{5.2}$$

Now take the thermodynamic limit $N \to \infty$ with $\nu_q(t) = \lim_{N\to\infty} \nu_q^{(N)}$. The macroscopic variable $\nu_q(t)$ represents the total activity of the qth population, that is, the fraction of neurons firing at time t. An asynchronous state may then be defined as one with constant activity $\nu_q(t) = \nu_q^{(0)}$ for all $q = 1, \dots, M$. It follows that the total synaptic current is also time-independent with

$$X_p(t) = X_p^{(0)} = \sum_q W_{pq} \nu_q^{(0)}. \tag{5.3}$$

From our previous analysis, we have seen that each homogeneous population's activity is related to the synaptic input (assuming zero external inputs) according to some effective gain function which we denote by f, $\nu_q^{(0)} = f(X_q^{(0)})$. This then yields the steady-state equation

$$X_p^{(0)} = \sum_q W_{pq} f(X_q^{(0)}). \tag{5.4}$$

This analysis can now be extended to the case of time-varying activities provided that each population remains in a state of incoherent firing so that the synaptic input is slowly varying in time. We can then make the approximation $\nu_q(t) \approx f(X_q(t))$ to obtain the closed integral equation

$$X_p(t) = \sum_q W_{pq} \int_0^\infty g_{\text{syn}}(\tau) f(X_q(t-\tau)) d\tau. \tag{5.5}$$

In the case of exponential synapses this reduces to the well–known form

$$\tau_s \frac{dX_p}{dt} = -X_p(t) + \sum_q W_{pq} f(X_q(t)). \tag{5.6}$$

The reduction to a rate equation breaks down, however, when the population activity changes rapidly during one period due to fast transients or the existence of a coherent state exhibiting collective oscillations, such as when the asynchronous state becomes unstable.

Spatial continuum limit

The physical location of a column of neurons in a region of cortex often reflects the task of that population of neurons. For example, in the auditory system, neurons are organized along an axis that reflects the cell's preferred frequency. As one moves along the axis the preferred frequency changes gradually from high to low frequency tones. Thus there is a one–dimensional frequency or tonotopic map in auditory cortex. Similarly, in primary visual cortex (V1) there is an orderly retinotopic mapping of the visual field onto its surface, with left and right halves of the visual field mapped onto right and left V1 respectively. Superimposed upon this are additional two–dimensional maps reflecting the fact that neurons respond preferentially to stimuli with particular features such as orientation [**158**]. This suggests labeling neurons according to their cortical position. We now give a heuristic argument for how such labeling leads to a continuum model of cortex.

Consider a population of neurons distributed along a one–dimensional axis. Suppose that we partition space into segments of length d such that the number of neurons in segment $[nd,(n+1)d]$ is $N=\rho d$ where ρ is the cell density. We treat neurons in that interval as a homogeneous population of cells (cortical column) labeled by the integer n, and assume that synaptic interactions between the nth and mth populations only depend on the discrete locations of the populations on the line. Writing $W_{nm}=\rho d w(nd,md)$ and $X_n(t)=a(nd,t)$, equation (5.5) becomes

$$a(nd,t)=\rho d\sum_m w(nd,md)\int_0^\infty g_{\rm syn}(\tau)f(a(md,t-\tau))d\tau. \tag{5.7}$$

Taking the limit $d\to 0$, the summation on the right–hand side can be replaced by an integral to give

$$a(x,t)=\rho\int w(x,y)\int_0^\infty g_{\rm syn}(\tau)f(a(y,t-\tau))d\tau dy. \tag{5.8}$$

In the case of exponential synapses this reduces to the integro–differential equation

$$\tau\frac{\partial}{\partial t}a(x,t)=-a(x,t)+\rho\int w(x,y)f(a(y,t))dy. \tag{5.9}$$

Equation (5.9) is an example of a neural field equation [**156, 159, 101, 102**], which is widely used to study large–scale spatio–temporal dynamics of cortical tissue. It can easily be extended to higher spatial dimensions as well as to separate populations of excitatory and inhibitory neurons, see below.

5.2. Turing mechanism for cortical pattern formation

Let $a_E(\mathbf{r},t)$ be the activity of excitatory neurons in a given volume element of a slab of neural tissue located at $\mathbf{r}\in\mathbf{R}^2$, and $a_I(\mathbf{r},t)$ be the corresponding activity of inhibitory neurons. Assuming for concreteness a rate-based model with exponential synapses, we have an evolution equation of the form

$$\tau_l\frac{\partial a_l(\mathbf{r},t)}{\partial t}=-a_l(\mathbf{r},t)+\sum_{m=E,I}\int_{\mathbf{R}^2}w_{lm}(\mathbf{r}|\mathbf{r}')f_m[a_m(\mathbf{r}',t)]d\mathbf{r}'+h_l(\mathbf{r},t) \tag{5.10}$$

for $l = E, I$, where $w_{lm}(\mathbf{r}|\mathbf{r}') = w_{lm}(|\mathbf{r}-\mathbf{r}'|)$ gives the weight per unit volume of all synapses to the lth population from neurons of the mth population a distance $|\mathbf{r}-\mathbf{r}'|$ away, f_E and f_I are taken to be smooth output functions

$$f_l(x) = \frac{1}{1+\mathrm{e}^{-\eta_l(x-\kappa_l)}}, \quad l = E, I, \tag{5.11}$$

where η_l determines the slope or sensitivity of the input–output characteristics of the population and κ_l is a threshold, h_E and h_I are external stimuli, and $\tau_{E,I}$ are synaptic time constants. From a mathematical viewpoint, it is often convenient to reduce the above two-population model to an effective one-population model. In particular, suppose that $\tau_I \ll \tau_E$, $w_{II} = 0$ and f_I is a linear function. We can then eliminate a_I in terms of a_E such that

$$\tau\frac{\partial a(\mathbf{r},t)}{\partial t} = -a(\mathbf{r},t) + \int_{\mathbf{R}^2} w(|\mathbf{r}-\mathbf{r}'|)f[a(\mathbf{r}',t)]d\mathbf{r}' + h(\mathbf{r},t), \tag{5.12}$$

where we have dropped the index E and

$$w(|\mathbf{r}-\mathbf{r}'|) = w_{EE}(|\mathbf{r}-\mathbf{r}'|) - \int w_{EI}(|\mathbf{r}-\mathbf{r}''|)w_{IE}(|\mathbf{r}-\mathbf{r}''|)d\mathbf{r}''. \tag{5.13}$$

In the case of a constant external input, $h_l(\mathbf{r}) = \bar{h}_l$, there exists at least one fixed point solution $a_l(\mathbf{r}) = \bar{a}_l$ of equation (5.10), where

$$\bar{a}_l = \sum_{m=E,I} W_{lm} f(\bar{a}_m) + \bar{h}_l \tag{5.14}$$

and $W_{lm} = \int_{\mathbf{R}^2} w_{lm}(\mathbf{r})d\mathbf{r}$. If $\bar{h}_l$ is sufficiently small relative to the threshold κ then this fixed point is unique and stable. Under the change of coordinates $a_l \to a_l - \bar{h}_l$, it can be seen that the effect of $\bar{h}_l$ is to shift the threshold by the amount $-\bar{h}_l$. Thus there are two ways to increase the excitability of the network and thus destabilize the fixed point: either by increasing the external input $\bar{h}_l$ or reducing the threshold κ. The latter can occur through the action of drugs on certain brain stem nuclei which provides a mechanism for generating geometric visual hallucinations [**160, 161**], see below. The local stability of $(\bar{a}_E, \bar{a}_I)$ is found by linearization:

$$\tau_l\frac{\partial b_l(\mathbf{r},t)}{\partial t} = -b_l(\mathbf{r},t) + \mu \sum_{m=E,I}\int_{\mathbf{R}^2} w_{lm}(|\mathbf{r}-\mathbf{r}'|)b_m(\mathbf{r}',t)d\mathbf{r}', \tag{5.15}$$

where $b_l(\mathbf{r},t) = a_l(\mathbf{r},t) - \bar{a}_l$ and we have performed a rescaling of the local weights $f'(\bar{a}_l)w_{lm} \to \mu w_{lm}$ with μ a measure of the degree of network excitability. Assuming solutions of the form $b_l(\mathbf{r},t) = b_l(\mathbf{r})e^{-\lambda t/\tau_l}$ we are left with the spectral problem:

$$\lambda b_l(\mathbf{k}) = -b_l(\mathbf{k}) + \mu\sum_m W_{lm}(|\mathbf{k}|^2)b_m(\mathbf{k}), \tag{5.16}$$

where $b_l(\mathbf{k})$ and $W_{lm}(|\mathbf{k}|^2)$ are, respectively, the Fourier coefficients of $b_l(\mathbf{r})$ and $w_{lm}(\mathbf{r})$. This leads to a matrix *dispersion relation* for λ as a function of $k = |\mathbf{k}|$:

$$\det([\lambda+1]I - \mu\mathbf{W}(k)) = 0, \tag{5.17}$$

where $\mathbf{W}$ is the matrix of Fourier coefficients of the w_{lm}. Similarly, in the corresponding one–population model (5.12), we obtain the dispersion relation

$$\lambda = -1 + \mu W(k) \equiv \lambda(k), \tag{5.18}$$

with $W(k)$ the Fourier transform of $w(\mathbf{r})$. One major difference between the one-population and two-population models is that the latter can exhibit time-periodic solutions [**101**]. We will restrict our discussion to static solutions here.

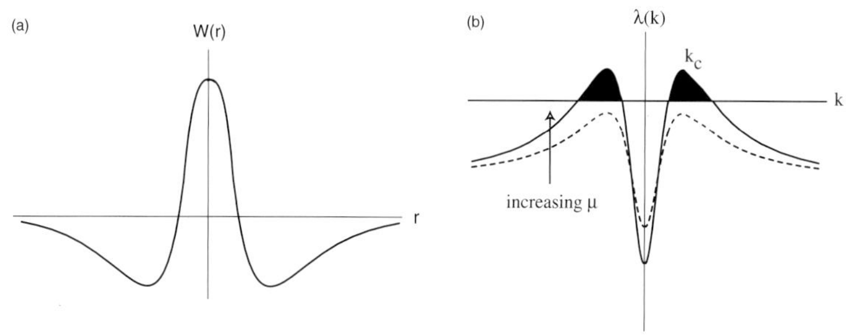

FIGURE 2. Neural basis of the Turing mechanism. (a) Mexican hat interaction function showing short-range excitation and long-range inhibition. (b) Dispersion curves $\lambda(k)$ for Mexican hat function. If the excitability μ of the cortex is increased, the dispersion curve is shifted upwards leading to a Turing instability at a critical parameter $\mu_c = W(k_c)^{-1}$ where $W(k_c) = [\max_k\{W(k)\}]$. For $\mu_c < \mu < \infty$ the homogeneous fixed point is unstable.

We now determine conditions under which the homogeneous state loses stability leading to the formation of spatially periodic patterns. The standard mechanism for such an instability, which is the neural analog of the Turing instability in reaction-diffusion equations, is a combination of short-range excitation and long-range inhibition. In the case of a one-population model this can be represented by the so-called "Mexican Hat" function (see Figure 2(a)):

$$w(|\mathbf{r}|) = (\frac{A_+}{\sigma_+})e^{-r^2/\sigma_+^2} - (\frac{A_-}{\sigma_-})e^{-r^2/\sigma_-^2}, \tag{5.19}$$

the Fourier transform of which is:

$$W(k) = \frac{1}{2}(A_+e^{-\frac{1}{4}\sigma_+^2k^2} - A_-e^{-\frac{1}{4}\sigma_-^2k^2}). \tag{5.20}$$

It is simple to establish that λ passes through zero at the critical value μ_c signalling the growth of spatially periodic patterns with wave number k_c, where $W(k_c) = \max_k\{W(k)\}$, see Figure 2(b). Close to the bifurcation point these patterns can be represented as linear combinations of plane waves

$$b(\mathbf{r}) = \sum_n (c_n e^{i\mathbf{k}_n\cdot\mathbf{r}} + c_n^* e^{-i\mathbf{k}_n\cdot\mathbf{r}}),$$

where the sum is over all wave vectors with $|\mathbf{k}_n| = k_c$. Rotation symmetry implies that the space of such modes is infinite-dimensional. That is, all plane-waves with wave vectors on the critical circle $|\mathbf{k}| = k_c$ are allowed (see Figure 3(a)). However, translation symmetry means that we can restrict the space of solutions to that of doubly-periodic functions corresponding to regular tilings of the plane. That is, $b(\mathbf{r} + \boldsymbol{\ell}) = b(\mathbf{r})$ for all $\boldsymbol{\ell} \in L$ where L is a regular square, rhomboid or hexagonal lattice as illustrated in Figure 3(b). The sum over n is now finite with $\mathrm{N} = 2$ (square, rhomboid) or $\mathrm{N} = 3$ (hexagonal) and, depending on the boundary conditions, various patterns of stripes or spots can be obtained as solutions. Amplitude

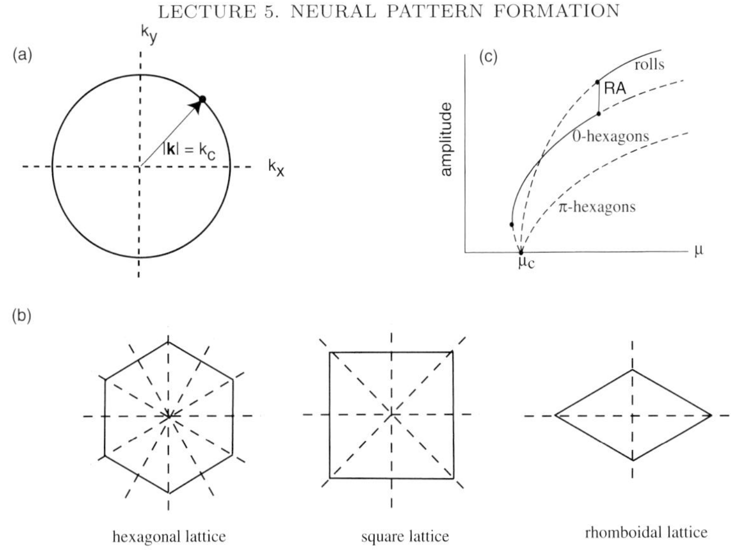

FIGURE 3. (a) Critical circle for Turing instability. (b) Planar lattices: hexagonal, square, rhomboid. (c) Bifurcation diagram showing the variation in amplitude C with parameter μ for patterns on a hexagonal lattice. Solid and dashed curves indicate stable and unstable solutions respectively. The different patterns are distinguished by the coefficients $\mathbf{c} = (c_1, c_2, c_3)$ with $\mathbf{c} = (1, 0, 0)$ for roll or stripe patterns, $\mathbf{c} = (1, 1, 1)$ for 0-hexagons and $\mathbf{c} = (1, 1, -1)$ for π-hexagons. It is also possible for additional patterns to form through secondary bifurcations (such as rectangular (RA) patterns). However, higher-order contributions to the amplitude equation are needed to determine such bifurcations.

equations for the coefficients c_n can then be obtained using perturbation methods [**102**]. However, their basic structure can be determined from the underlying rotation and translation symmetries of the network model. In the case of a square or rhombic lattice, we can take $\mathbf{k}_1 = k_c(1, 0)$ and $\mathbf{k}_2 = k_c(\cos\varphi, \sin\varphi)$ such that (to cubic order)

$$\frac{dc_n}{dt} = c_n \left[\mu - \mu_c - \Gamma_0 |c_n|^2 - 2\Gamma_\varphi \sum_{m \neq n} |c_m|^2 \right], \quad n = 1, 2, \tag{5.21}$$

where Γ_φ depends on the angle φ. In the case of a hexagonal lattice we can take $\mathbf{k}_n = k_c(\cos\varphi_n, \sin\varphi_n)$ with $\varphi_1 = 0, \varphi_2 = 2\pi/3, \varphi_3 = 4\pi/3$ such that

$$\frac{dc_n}{dt} = c_n \left[\mu - \mu_c - \Gamma_0 |c_n|^2 - -\eta c^*_{n-1} c^*_{n+1} \right] - 2\Gamma_{\varphi_2} c_n \left(|c_{n-1}|^2 + |c^2_{n+1}| \right), \tag{5.22}$$

where $n = 1, 2, 3(\mod 3)$. These ordinary differential equations can then be analyzed to determine which particular types of pattern are selected and to calculate their stability [**160, 161, 102**]. The results can be summarized in a bifurcation diagram as illustrated in Figure 3(c) for the hexagonal lattice with $h > 0$ and $2\Gamma_{\varphi_2} > \Gamma_0$.

Geometric visual hallucinations

The primary visual cortex (V1) is the first cortical area to receive visual information from the retina (see Figure 4). The output from the retina is conveyed by ganglion

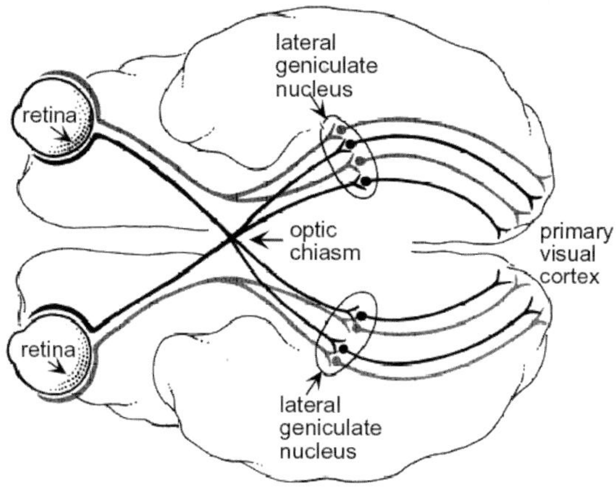

FIGURE 4. Visual pathways from the retina through the lateral geniculate nucleus (LGN) of the thalamus to the primary visual cortex (V1)

cells whose axons form the optic nerve. The optic nerve conducts the output spike trains of the retinal ganglion cells to the lateral geniculate nucleus (LGN) of the thalamus, which acts as a relay station between retina and primary visual cortex (V1). Prior to arriving at the LGN, some ganglion cell axons cross the midline at the optic chiasm. This allows the left and right sides of the visual fields from both eyes to be represented on the right and left sides of the brain, respectively. Note that signals from the left and right eyes are segregated in the LGN and in input layers of V1. This means that the corresponding LGN and cortical neurons are monocular, in the sense that they only respond to stimuli presented to one of the eyes but not the other (ocular dominance).

One of the striking features of the visual system is that the visual world is mapped onto the cortical surface in a topographic manner. This means that neighboring points in a visual image evoke activity in neighboring regions of visual cortex. Moreover, one finds that the central region of the visual field has a larger representation in V1 than the periphery, partly due to a non–uniform distribution of retinal ganglion cells. The retinotopic map is defined as the coordinate transformation from points in the visual world to locations on the cortical surface. In order to describe this map, we first need to specify visual and cortical coordinate systems. Since objects located a fixed distance from one eye lie on a sphere, we can introduce spherical coordinates with the "north pole" of the sphere located at the fixation point, the image point that focuses onto the fovea or center of the retina. In this system of coordinates, the latitude angle is called the eccentricity ϵ and the longitudinal angle measured from the horizontal meridian is called the azimuth φ. In most experiments the image is on a flat screen such that, if we ignore the curvature of the sphere, the pair (ϵ, φ) approximately coincides with polar coordinates on the screen. One can also represent points on the screen using Cartesian coordinates (X, Y). In primary visual cortex the visual world is split in half with the region

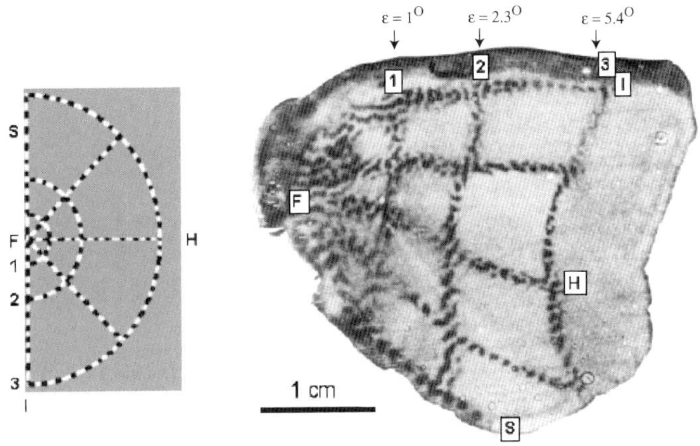

FIGURE 5. An autoradiograph from the primary visual cortex in the left side of a macaque monkey brain. The pattern is a radioactive trace of the activity evoked by the image shown to the left. [Reprinted from [**162**], with permission from J. Neurosci.]

$-90^\circ \leq \varphi \leq 90^\circ$ represented on the left side of the brain, and the reflection of this region represented on the right side brain. Note that the eccentricity ϵ and Cartesian coordinates (X, Y) are all based on measuring distance on the screen. However, it is customary to divide these distances by the distance from the eye to the screen so that they are specified in terms of angles. The structure of the retinotopic map in monkey is shown in Figure 5, which was produced by imaging a radioactive tracer that was taken up by active neurons while the monkey viewed a visual image consisting of concentric circles and radial lines. The fovea is represented by the point F on the left hand side of the cortex, and eccentricity increases to the right. Note that concentric circles are approximately mapped to vertical lines and radial lines to horizontal lines.

Motivated by Figure 5, we assume that eccentricity is mapped on to the horizontal coordinate x of the cortical sheet, and φ is mapped on to its y coordinate. An approximate equation for the retinotopic map can then be obtained for eccentricities greater than 1°:

$$x \approx \lambda \ln(\epsilon/\epsilon_0), \quad y \approx -\frac{\lambda \pi \varphi}{180^\circ}. \tag{5.23}$$

It follows that the retinotopic map can be approximated by a complex logarithm [**163**]. That is, introducing the complex representations $Z = (\epsilon/\epsilon_0)e^{-i\pi\varphi/180^\circ}$ and $z = x + iy$ then $z = \lambda \log Z$. Now suppose that a doubly periodic pattern of activity is formed in cortex. If the cortical patterns are mapped back into visual field coordinates using the inverse of the above retinicortical map, then a number of common visual hallucinations are generated [**160**]. This is illustrated in Figures 6 and 7. A more detailed model of visual hallucinations has recently been developed that takes into account the fact that neurons signal the orientation as well as the location of a visual stimulus [**161, 102**]. This model can reproduce a broader range of hallucinations and also has some interesting symmetries arising from the distribution of long-range connections in cortex.

FIGURE 6. Doubly-periodic activity patterns in cortex

FIGURE 7. Corresponding hallucinatory images generated using the inverse retinocortical map

5.3. Persistent localized states

In addition to spatially periodic patterns, neural networks with Mexican hat interactions also support spatially localized persistent activity states known as bumps. Wilson and Cowan [**156**] showed numerically how solitary stationary bumps can arise in a one-dimensional network of excitatory and inhibitory neurons, with short-range excitation keeping the bump active and longer–range inhibition preventing the bump from spreading. Amari [**159**] subsequently obtained analytical solutions for bump states in a simplified one–population model with a step–function nonlinearity. In contrast to the spatially periodic patterns discussed in §4.2, these bump states are highly nonlinear and are far from any spatially and temporally uniform states. We will present the details of Amari's analytical construction.

Consider the neural field equation

$$\frac{\partial a(x,t)}{\partial t} = -a(x,t) + \int_{-\infty}^{\infty} w(x-x')H[a(x',t)]dx' + h + s(x,t), \tag{5.24}$$

where $w(x)$ is a symmetric integrable function, $H(a)$ is the Heaviside step function

$$H[a] = \begin{cases} 0 & \text{if } a \leq 0 \\ 1 & \text{if } a > 0 \end{cases}. \tag{5.25}$$

h is a homogeneous input and $s(x,t)$ is a spatially patterned input. Following Amari [**159**], we restrict w to be a Mexican hat function. Equilibrium solutions of equation (5.24) with $s(x,t) = 0$ satisfy

$$A(x) = \int_{-\infty}^{\infty} w(x-x')H[A(x')]dx' + h. \tag{5.26}$$

Let $R[A] = \{x|A(x) > 0\}$ be the region over which the field is excited. Equation (5.26) can then be rewritten as

$$A(x) = \int_{R[A]} w(x-x')dx' + h. \tag{5.27}$$

Exploiting the fact that any solution can be arbitrarily translated so that it is centered at the origin, we define a stationary pulse solution of width d to be one that is excited over the interval $(-d, d)$. Let

$$W(x) = \int_0^x w(y)dy \tag{5.28}$$

and

$$W_m = \max_{x>0} W(x), \quad W_\infty = \lim_{x\to\infty} W(x) \tag{5.29}$$

such that $W(0) = 0$ and $W(-x) = -W(x)$. For a bump of width d, equation (5.27) reduces to the form

$$A(x) = h + W(d+x) - W(x-d). \tag{5.30}$$

Since $A(d) = 0$ by definition, we obtain the following necessary condition for the existence of a bump:

$$W(2d) + h = 0. \tag{5.31}$$

This condition is also sufficient for the Mexican hat weight distribution [**159**]. It will be shown below that a bump is stable provided the condition $W'(2d) < 0$ is satisfied. The existence and stability of activity bumps for a given h can thus be determined graphically as illustrated in Figure 8(b). For a certain range of values of $h < 0$ one finds bistability between a stable bump and a rest state for which $R[A] = \emptyset$.

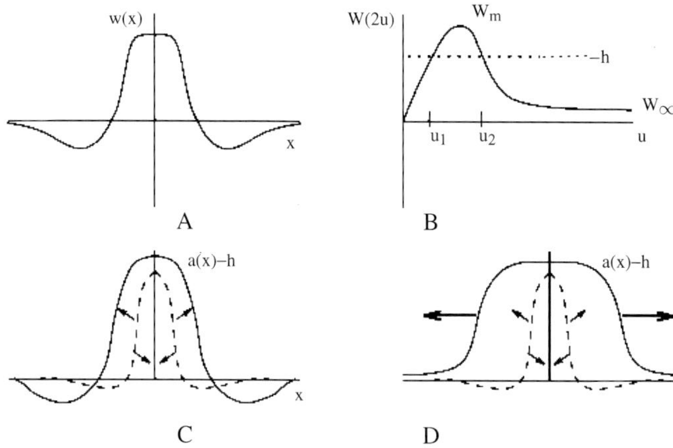

FIGURE 8. Construction of a solitary pulse in the Amari model. (a) The Mexican hat weight distribution w. (b) Integral $W(x)$ of $w(x)$. Horizontal line shows the uniform stimulus $-h$ whose intersections with $W(2d)$ determine the allowed stationary pulse solutions. (c) Unstable pulse (broken) acts as a separatrix between the stable pulse and the rest state. (d) Unstable pulse acts as a separatrix between a wavefront and the rest state.

The linear stability of a stationary pulse can be determined by setting $a(x,t) = A(x) + p(x)e^{\lambda t}$ and expanding to first order in p [**164, 135**]. This leads to the eigenvalue equation

$$(5.32) \qquad (\lambda+1)p(x) = \int_{-\infty}^{\infty} w(x-x')\delta(A(x'))p(x')dx'.$$

Using the identity

$$(5.33) \qquad \delta(A(x)) = \left(\frac{\delta(x-d)}{|A'(d)|} + \frac{\delta(x-+d)}{|A'(-d)|}\right),$$

and setting $|A'(d)| = |A'(-d)| = \gamma^{-1}$, we have

$$(5.34) \qquad (\lambda+1)p(x) = \gamma\left(w(x-d)p(d) + w(x+a)p(-d)\right).$$

There are two types of solution to this equation. The first consist of solutions $p(x)$ that vanish at $x = \pm d$ so that $\lambda = -1$. This defines the essential spectrum and does not contribute to any instabilities. The second class of solution, which generates the discrete spectrum, is obtained by setting $x = \pm d$ in the eigenvalue equation,

$$(5.35) \qquad (\lambda+1)p_+ = \gamma\left(w(0)p_+ + w(2d)p_-\right)$$

$$(5.36) \qquad (\lambda+1)p_- = \gamma\left(w(-2d)p_+ + w(0)p_-\right).$$

This has the solutions $p_- = \pm p_+$ with corresponding eigenvalues

$$(5.37) \qquad \lambda_\pm = -1 + \gamma(w(0) \pm w(2d)).$$

Finally, using the fact that $\gamma^{-1} = w(0) - w(2d)$ we deduce that $\lambda_- = 0$ (reflecting the translation invariance of the system) and $\lambda_+ = \gamma w(2d)$. Thus the bump is stable if $w(2d) = W'(2d) < 0$.

Note that the discrete spectrum is determined completely in terms of the perturbations $p_\pm = p(\pm d)$. This explains why it is also possible to analyze the stability of the bumps by restricting attention to the effects of perturbations at the boundaries of the activity bump as originally formulated by Amari [**159**]. In particular, if $a(x,t) = 0$ at $x = x \pm d + \Delta_\pm(t)$, then

$$\begin{aligned} 0 &= A(\pm d + \Delta_\pm(t)) + p(d + \Delta_\pm(t), t) \\ &= A(\pm d) + A'(\pm d)\Delta_\pm(t) + p(\pm d, t) + \mathcal{O}(\Delta^2), \end{aligned}$$

that is,

$$\Delta_\pm(t) = \pm\gamma p(\pm d, t)$$

since $A(\pm d) = 0$ and $A'(\pm d) = \mp\gamma^{-1}$. It follows that $p_- = p_+$ generates a uniform expansion of the bump ($\Delta_- = -\Delta_+$) and $p_- = -p_+$ generates a shift in the center of the bump ($\Delta_- = \Delta_+$).

There have been various extensions of Amari's original analysis described above. Kishimoto and Amari [**165**]) proved the existence of a solitary pulse for a smooth sigmoidal nonlinearity f rather than a hard threshold using a fixed point theorem. More recent extensions include weight distributions that are not of the lateral inhibition type (for which multiple bump states can arise) [**166, 167**], spiking rather than rate-based models [**168**], two-dimensional bumps [**169, 167, 135, 136**], and weakly interacting bumps [**170**]. (For a recent review see [**171**]). Another interesting issue concerns the effect of inhomogeneous inputs. In the homogeneous case, a stationary pulse is only marginally stable, since arbitrarily small fluctuations can shift its center or peak as a consequence of translation symmetry. Amari [**159**] showed that in the presence of small stationary inputs $s(x)$ the peak moves towards a local maximum of $s(x)$. More recently it has been established that if a spatially localized input is presented to a network with some form of slow negative feedback such as spike frequency adaptation (see §5.4), then a stationary pulse can undergo a Hopf instability leading to the formation of a spatially localized oscillation or *breather* [**135, 136**]. This provides an alternative, recurrent network mechanism for generating stimulus–induced oscillations and synchrony, rather than one based on intrinsic single neuron oscillators (see §4).

Working memory and continuous attractors

One of the reasons why persistent activity bumps are of interest is that they are thought to arise in cortical circuits performing certain spatial working memory tasks. Working memory involves cortical "memory neurons" that establish a representation of a stimulus that persists after the stimulus is removed (see Figure 9(a)). A typical experiment is a delayed response task, in which a primate is required to retain information regarding the location of a sensory cue across a delay period between the stimulus and behavioral response. There are three basic kinds of working memory that can be distinguished according to the type of sensory stimulus [**172, 173**]. *Discrete working memory*: in a delayed match-to-sample experiment the behavioral response depends on the memory of one of two items, for example, the stimulus color red or green. Here working memory circuits store information regarding a discrete set of categorical features such as a face, color or word [**128**]. *Spatial working memory*: neurons display persistent activity that is spatially selective. One example is a delayed oculomotor experiment, in which a saccadic eye movement is guided by the memory of the spatial location of a cue stimulus such as a flashed spot of light. Physiological recordings in prefrontal cortex have shown that spatially localized groups of neurons fire during the recall task and then stop firing once the task has finished [**174**]. The stimulus response of a cell is characterized by a smooth tuning curve that is peaked at a preferred spatial cue and varies from cell to cell. At the network level the memory of cue location is stored as an activity bump. Persistent activity bumps occur in a number of other systems that encode directional or spatial information, including head direction cells in thalamus and basal ganglia [**175**] and place cells in the hippocampus [**103**]. *Parametric working memory*: there is a continuous range of persistent activity that increases monotonically with some stimulus parameter. One example is a delayed somatosensory discrimination task, in which a monkey is trained to compare and discriminate the frequencies of two vibrotactile stimuli separated by a delay period [**176**]. Here the stimulus frequency is encoded by the firing rate of persistent activity. A similar monotonic encoding scheme is used by goldfish oculomotor neurons for memory of the current eye position [**177, 178, 179**]. A change in eye position in one direction is preceded by an increase in the activity of certain ON burst neurons. This transient signal is then transformed into elevated persistent activity in the neurons that encode eye position, supplying motorneurons with the tonic drive necessary to maintain the new eye position. Similarly, OFF burst cells initiate movement in the opposite direction by reducing the level of persistent activity (see Figure 9(b)). An interesting aspect of the oculomotor system, which is shared with the head direction system, is that some form of *neural integrator* is needed in order determine the change in eye (or head) position based on velocity inputs.

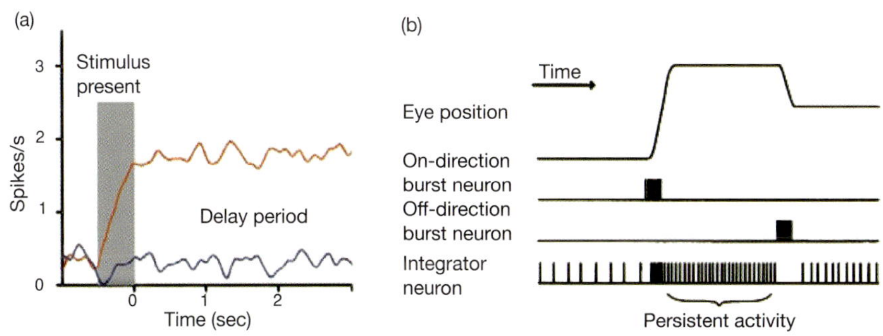

FIGURE 9. (a) An illustration of stimulus–dependent, persistent firing rates in a working memory neuron. (b) Schematic illustration of eye position and neural activity versus time in the oculomotor system. See text for details. [Reprinted from [**179**], Copyright 2000, with permission of Elsevier.]

A characteristic feature of persistent activity in both spatial and parametric working memory is that there exists a continuous manifold of states forming an attractor for the dynamics. Figure 10 shows some examples of such attractors for a simple two–neuron system, in which the firing rates ν_j of the two neurons evolve according to gradient dynamics. That is, there exists a potential function $L(\nu_1, \nu_2)$ such that $\dot{\nu}_j = -\partial L/\partial \nu_j$, $j = 1, 2$. Attracting fixed points of the system then correspond to minima of L. A graded stimulus feature is usually represented by some form of ring attractor in models of spatial working memory, whereas in models of parametric working memory it is represented by a line attractor. If the stimulus feature is represented by the peak location of an activity bump in a spatially-structured network, then the set of allowed features will form a continuous manifold provided that the network is homogeneous. This manifold will reflect the topology and dimensionality of the network. Thus, a periodic stimulus feature can be encoded by a persistent activity bump in a homogeneous ring network. One of

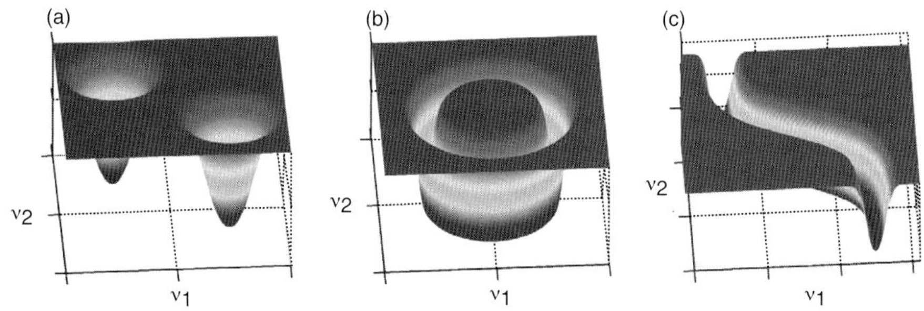

FIGURE 10. Attractor dynamics for a pair of neurons whose firing rates evolve according to a gradient dynamics that minimizes some potential function L. Attracting fixed points correspond to minima of L. (a) A pair of isolated fixed points that could represent discrete categorical features. (b) A ring attractor that could represent head direction or the directional position of some visual cue. (c) A line attractor that could represent eye position or the frequency of a vibrotactile stimulus. [Reprinted from [**173**], Copyright 2003, with permission of Elsevier.]

the major theoretical challenges is to understand how persistent working memory can be implemented robustly in neural circuits. That is, from a dynamical systems perspective, continuous attractors are structurally unstable so that their existence requires a considerable degree of fine tuning of system parameters. For example, it is necessary to impose strong symmetry constraints on the potential function of the two–neuron system shown in Figure 10 in order to generate a ring or line attractor. Similarly, a continuum of activity bumps only occurs in spatially homogeneous networks. A related issue is that a persistent activity state will be marginally stable with respect to perturbations in directions tangential to the attractor manifold. For example, the peak of an activity bump will tend to drift over time in the presence of arbitrarily small levels of noise. This latter effect may actually have a biological interpretation in terms of the degradation in the memory recall performance of a monkey when the delay in the response is increased. One suggested mechanism for making network models more robust to noise and network inhomogeneities is to introduce some form of bistability at the single neuron level. This idea has been applied to bump formation in spatial working memory [**180, 181**], and to the formation of line attractors in parametric working memory [**182**]. Another important issue concerns how to generate persistent states consisting of neurons with low firing rates, given that high levels of recurrent excitation are required in order to maintain persistent activity when a stimulus is removed. (Low firing rates are found in recordings from prefrontal cortex). Large–scale simulations of noisy LIF networks with biophysically realistic synaptic dynamics suggest that robust low activity persistent states can be generated if a significant fraction of the excitatory connections consist of slow NMDA synapses rather than fast AMPA synapses [**183**]. The stability of such states can be analyzed using a mean field reduction of the population dynamics along the lines outlined in §3.4, see [**184**].

5.4. Traveling waves

In the case of excitatory rather than Mexican hat weight functions w, spatially structured rate models can support the propagation of traveling waves of activity. This is consistent with experimental studies that have revealed the propagation of traveling bursts of activity in slices of disinhibited cortical tissue [**185, 186, 187, 188**]. The underlying mechanism for the propagation of such waves appears to be synaptic in origin rather than diffusive, as in the case of action potentials traveling along the axons of individual neurons. Since there is strong vertical coupling between cortical layers, it is possible to treat a thin vertical cortical slice as an effective one–dimensional medium. Mathematically speaking, one can then distinguish between traveling wavefronts and traveling pulses as illustrated in Figure 11. We will restrict our discussion of analytical methods for studying wave propagation to the simpler case of traveling fronts. However, these methods can be extended to the more realistic case of traveling pulses (see below).

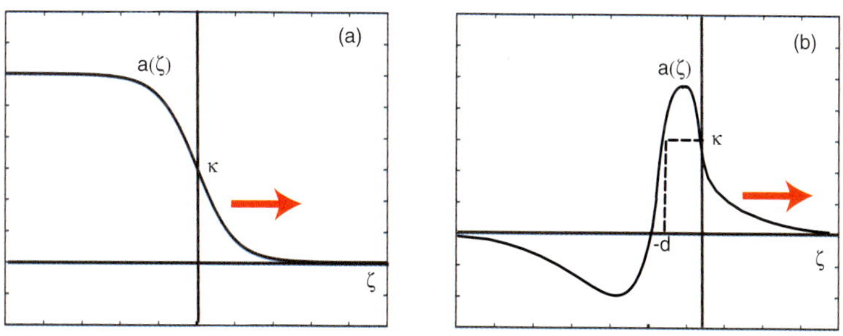

FIGURE 11. (a) Traveling wavefront (b) Traveling pulse of width d as determined by the firing threshold κ

The existence of traveling front solutions in a one–dimensional network has been analyzed by Ermentrout and Mcleod [**189**] for scalar integro–differential equations of the form

$$\frac{\partial a(x,t)}{\partial t} = -a(x,t) + \int_{-\infty}^{\infty} w(x-x')f(a(x',t))dx', \tag{5.38}$$

with

$$f(a) = \frac{1}{1+e^{-\eta(a-\kappa)}}. \tag{5.39}$$

The weight distribution w is assumed to be a positive, even, continuously differentiable function of x with unit normalization $\int_{-\infty}^{\infty} w(y)dy = 1$. Suppose that the function $F(a) = f(a) - a$ has precisely three zeros at $a = A_{\pm}, A_0$ with $A_- < A_0 < A_+$ and $F'(A_{\pm}) < 0$. It can then be shown that (modulo uniform translations) there exists a unique traveling front solution of equation (5.38) such that $a(x,t) = A(\xi)$,

$\xi = x - ct$, with $A(\xi) \to A_\pm$ as $\xi \to \mp\infty$ [**189**]. Moreover, the speed of the wave satisfies

$$c = \frac{\Gamma}{\int_{-\infty}^{\infty} a'^2 f'(a) d\xi}, \tag{5.40}$$

where

$$\Gamma = \int_{A_-}^{A_+} F(a) da. \tag{5.41}$$

Since the denominator of equation (5.40) is positive definite, the sign of c is determined by the sign of the coefficient Γ. In particular, if $\kappa = 0.5$ and $\eta > 4$, there exists a pair of stable homogeneous fixed points with $A_- = -A_+$, which in turn implies that $\Gamma = 0$ and the front solution is stationary, see Figure 12.

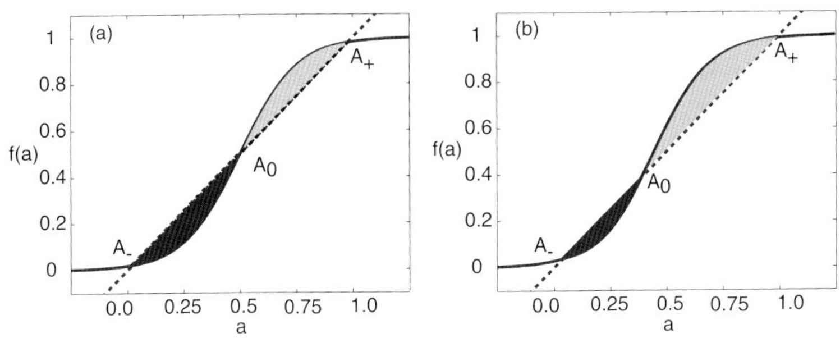

FIGURE 12. Balance condition for the speed of a traveling wave front in a scalar excitatory network with $a(x,t) = A(x - ct)$ such that $A(\mp\infty) = A_\pm$. The solid curve is $f(a) = 1/(1 + e^{-\eta(a-\kappa)})$ with $\eta = 8$ and the dashed line is $g(a) = a$. The wave speed c is positive (negative) if the gray shaded area is larger (smaller) than the black shaded area. (a) $\kappa = 0.5$ such that $c = 0$. (b) $\kappa = 0.4$ such that $c > 0$.

Although the step function $f(a) = H(a - \kappa)$ obtained in the high-gain limit $\eta \to \infty$ is not itself continuously differentiable, the wavefront solution can be calculated explicitly in this case [**189**]. Note that equation (5.38) is then equivalent to equation (5.24). Since equation (5.38) is translation invariant, we are free to impose the condition $A(0) = \kappa$, that is, the wavefront solution passes through threshold at the origin. For concreteness, take $A(\xi)$ to be a monotonically decreasing function of $\xi = x - ct$ such that $A(\xi) < \kappa$ for $\xi > 0$ and $A(\xi) > \kappa$ for $\xi < 0$. Equation (5.38) then reduces to

$$-cA'(\xi) + A(\xi) = \int_{-\infty}^{0} w(\xi - \xi')d\xi' = \int_{\xi}^{\infty} w(x)dx \equiv \widehat{W}(\xi), \tag{5.42}$$

which has the solution

$$A(\xi) = e^{\xi/c}\left[\kappa - \frac{1}{c}\int_0^{\xi} e^{-y/c}\widehat{W}(y)dy\right]. \tag{5.43}$$

Requiring the solution to remain bounded as $\xi \to \infty$ ($\xi \to -\infty$) for $c > 0$ (for $c < 0$) implies that κ must satisfy the condition

$$\kappa = \frac{1}{|c|} \int_0^\infty \mathrm{e}^{-y/|c|} \widehat{W}(\mathrm{sign}(c)y)dy. \tag{5.44}$$

In the case of an exponential weight distribution $w(x) = (2d)^{-1}\mathrm{e}^{-|x|/d}$, with the length–scale fixed by setting $d = 1$, the relationship between wave speed c and threshold κ is

$$\kappa = \frac{1}{2(1+c)} \text{ (for } c > 0), \quad \kappa = \frac{1+2|c|}{2(1+|c|)} \text{ (for } c < 0) \tag{5.45}$$

Wave stability

We now indicate how to determine the stability of a traveling wavefront by constructing the so–called Evans function [**190, 191, 192**]. Consider the evolution of small smooth perturbations $\bar{\varphi}$ of a traveling front solution A of equation (5.38) centered about $\xi = 0$. Linearizing about the wave solution, the perturbations evolve according to

$$\frac{\partial \bar{\varphi}}{\partial t} - c\frac{\partial \bar{\varphi}}{\partial \xi} + \bar{\varphi} = \int_{\mathbb{R}} w(\xi - \eta)H'(A(\eta) - \kappa)\bar{\varphi}(\eta)d\eta. \tag{5.46}$$

Separating variables, $\bar{\varphi}(\xi, t) = \varphi(\xi)e^{\lambda t}$, we find that $\varphi \in \mathcal{C}^1(\mathbb{R}, \mathbb{C})$ satisfies the eigenvalue problem

$$(\mathcal{L} + \mathcal{N}_s)\,\varphi = \lambda\,\varphi \tag{5.47}$$

where

$$\mathcal{L}\varphi = c\frac{\partial \varphi}{\partial \xi} - \varphi, \qquad \mathcal{N}_s\varphi\,(\xi) = \frac{w(\xi)}{|A'(0)|}\,\varphi(0). \tag{5.48}$$

We need to characterize the spectrum of the linear operator $\mathcal{L} + \mathcal{N}_s : \mathcal{C}^1(\mathbb{R}, \mathbb{C}) \longrightarrow \mathcal{C}^0(\mathbb{R}, \mathbb{C})$ in order to determine the linear stability of the traveling pulse. The following definitions concern linear operators $\mathcal{T} : \mathcal{D}(\mathcal{T}) \longrightarrow \mathcal{B}$ where $\mathcal{B}$ is a Banach space and the domain $\mathcal{D}(\mathcal{T})$ of $\mathcal{T}$ is dense in $\mathcal{B}$. In our case $\mathcal{D}(\mathcal{L} + \mathcal{N}_s) = \mathcal{C}^1(\mathbb{R}, \mathbb{C})$ which is dense in $\mathcal{C}^0(\mathbb{R}, \mathbb{C})$. λ is in the resolvent set ρ, if $\lambda \in \mathbb{C}$ is such that $\mathcal{T} - \lambda$ has a range dense in $\mathcal{B}$ and a continuous inverse or resolvent operator $R_\lambda(\mathcal{T}) = (\mathcal{T} - \lambda I)^{-1}$, otherwise λ is in the spectrum σ. We decompose the spectrum into the following disjoint sets. λ is an element of the point or discrete spectrum σ_p, if $R_\lambda(\mathcal{T})$ does not exist; λ is an element of the continuous spectrum σ_c if $R_\lambda(\mathcal{T})$ exists, is defined on a dense subset of $\mathcal{B}$, but is not bounded; λ is an element of the residual spectrum σ_r if $R_\lambda(\mathcal{T})$ exists but its domain is not dense in $\mathcal{B}$. We refer to elements of the discrete spectrum as eigenvalues and the union of the continuous and residual spectra as the essential spectrum. It can be shown that the essential spectrum lies in the left–half complex plane so does not contribute to any instabilities [**190**]. Therefore, we will focus on the discrete spectrum.

Resolvent and the discrete spectrum

We seek to construct a bounded inverse by solving the inhomogeneous equation

$$(\mathcal{L} + \mathcal{N}_s - \lambda)\varphi = -f, \tag{5.49}$$

where $f \in \mathcal{C}^0(\mathbb{R}, \mathbb{C})$, using a variation of parameters approach along the lines of Zhang [**190**]. We write equation (5.49) as

$$\frac{\partial}{\partial \xi}\left(e^{-\left(\frac{1+\lambda}{c}\right)\xi}\,\varphi(\xi)\right) = -\frac{1}{c}\,e^{-\left(\frac{1+\lambda}{c}\right)\xi}\left(\mathcal{N}_s\varphi(\xi) + f(\xi)\right). \tag{5.50}$$

For $(\mathcal{R}e(\lambda)+1)/c > 0$, integrating equation (5.50) over $[\xi, \infty)$ yields

$$\varphi(\xi) - \Lambda_+(\lambda;\xi)\,\varphi(0) \;=\; \mathcal{H}_f(\xi). \tag{5.51}$$

where

$$\Lambda_+(\lambda;\xi) = \frac{1}{c|A'(0)|}\int_\xi^\infty w(\eta)e^{\left(\frac{1+\lambda}{c}\right)(\xi-\eta)}d\eta,$$

$$\mathcal{H}_f(\xi) = \frac{1}{c}\int_\xi^\infty e^{\left(\frac{1+\lambda}{c}\right)(\xi-\eta)}f(\eta)d\eta.$$

Using the Hölder inequality, it can be shown that both $\Lambda_+(\lambda;\xi)$ and $\mathcal{H}_f(\xi)$ are bounded for all $\xi \in \mathbb{R}$ and $f \in \mathcal{C}^0(\mathbb{R}, \mathbb{C})$. It is then seen from equation (5.51) that $\varphi(\xi)$ is determined by its restriction $\varphi(0)$, in which case we obtain

$$\left(1 - \Lambda_+(\lambda;0)\right)\varphi(0) = \frac{1}{c}\int_0^\infty e^{-\left(\frac{1+\lambda}{c}\right)\eta}f(\eta)d\eta.$$

This can be solved for $\varphi(0)$ and, hence for $\varphi(\xi)$, if and only if $1-\Lambda_+(\lambda;0) \neq 0$. This results in a bounded inverse which is defined on all of $\mathcal{C}^0(\mathbb{R}, \mathbb{C})$, and, therefore, all corresponding λ are in the resolvent set. On the other hand, we cannot invert the operator for λ such that $\Lambda_+(\lambda;0) = 1$. In this case, the equation $(\mathcal{L} + \mathcal{N}_s - \lambda)\varphi = 0$ has nontrivial solutions, indicating that λ is in the discrete spectrum. Hence, if we define the function

$$\mathcal{E}_+(\lambda) = 1 - \Lambda_+(\lambda;0), \qquad \frac{\mathcal{R}e(\lambda)+1}{c} > 0,$$

then the eigenvalues correspond to the zeros of $\mathcal{E}_+(\lambda)$. Proceeding along similar lines for $(\mathcal{R}e(\lambda)+1)/c < 0$ by integrating equation (5.50) over $(-\infty, 0]$ shows that the eigenvalues now correspond to the zeros of the function

$$\mathcal{E}_-(\lambda) = 1 - \Lambda_-(\lambda;0),, \qquad \frac{\mathcal{R}e(\lambda)+1}{c} < 0$$

where

$$\Lambda_-(\lambda;\xi) = -\frac{1}{c|A'(0)|}\int_{-\infty}^\xi w(\eta)e^{-\left(\frac{1+\lambda}{c}\right)\eta}d\eta. \tag{5.52}$$

The zeros of the Evans function $\mathcal{E}(\lambda)$, which is defined as the composite of the pair of functions $\mathcal{E}_\pm(\lambda)$, then fully determine the discrete spectrum of the operator obtained by linearizing about the wavefront solution.

Evans function for an exponential weight distribution We now explicitly calculate the zeros of the Evans functions for an exponential weight distribution $w(x) = \mathrm{e}^{-|x|}/2$. The region in the complex plane $\mathbf{D} = \{z \;:\; \mathcal{R}e(z) > -1\}$ is the domain of the Evans function $\mathcal{E}_+$, and we need only consider this region to determine the stability of the wave. For $c > 0$ and $\lambda \in \mathbf{D}$

$$\begin{aligned}\mathcal{E}_+(\lambda) &= 1 - \frac{1}{c|A'(0)|}\int_0^\infty w(\eta)e^{-(\frac{1+\lambda}{c})\eta}d\eta,\\ &= 1 - \frac{1}{2(1+\lambda+c)}\frac{1}{|A'(0)|},\end{aligned}$$

and similarly for $c < 0$ and $\lambda \in \mathbf{D}$

$$\begin{aligned}\mathcal{E}_-(\lambda) &= 1 + \frac{1}{c|A'(0)|}\int_{-\infty}^0 w(\eta)e^{(-\frac{1+\lambda}{c})\eta}d\eta,\\ &= 1 + \frac{1}{2(1+\lambda+c)}\frac{1}{|A'(0)|}.\end{aligned}$$

From this we can directly solve $\mathcal{E}_\pm(\lambda) = 0$ for λ

$$\lambda = -\left(1+|c|\right) + \frac{1}{2|A'(0)|}, \qquad c \in \mathbb{R}, \tag{5.53}$$

with $A'(0)$ determined from equation (5.43), that is, $A'(0) = c^{-1}(\kappa - 1/2)$. It follows that the eigenvalues are given by

$$\lambda = -(1+|c|) + \frac{|c|}{2\left|\kappa - \frac{1}{2}\right|}, \qquad c \in \mathbb{R}. \tag{5.54}$$

Since κ satisfies equation (5.45), we see that the only eigenvalue in $\mathbf{D}$ is the zero eigenvalue $\lambda = 0$. Moreover it can be shown that the eigenvalue is simple [**190**] and, hence, that the natural front is linearly stable modulo uniform translations.

Traveling pulses

Traveling fronts are not particularly realistic representations of waves in cortical slices, since populations of neurons do not stay in the excited state forever. Even in the absence of synaptic inhibition, most neurons posses intrinsic negative feedback mechanisms that slowly bring the cell back to resting voltages after periods of high activity. Hence, rather than a traveling front, propagating activity in the brain is better described as a traveling pulse. We can incorporate a recovery mechanism into the rate equation with the addition of a slow, local negative feedback component q such that equation (5.38) with Heaviside firing rate function becomes [**193**]

$$\begin{aligned}\frac{\partial a(x,t)}{\partial t} &= -a(x,t) + \int_{-\infty}^\infty w(x-x')H(a(x',t)-\kappa)dx' - \beta q(x,t),\\ \frac{1}{\varepsilon}\frac{\partial q(x,t)}{\partial t} &= -q(x,t) + a(x,t).\end{aligned} \tag{5.55}$$

This negative feedback could represent spike frequency adaptation or synaptic depression for example. Introducing the traveling wave coordinates (ξ, t), where $\xi = x - ct$, we define a traveling pulse solution of equation (5.55) to be the pair of functions (A, Q), satisfying the conditions

$$\begin{aligned}&A(\xi_i) = \kappa, \quad i = 1,2; \qquad &&A(\xi) \longrightarrow 0 \quad \text{as } \xi \longrightarrow \pm\infty;\\ &A(\xi) > \kappa, \quad -d < \xi < 0; \qquad &&A(\xi) < \kappa, \quad \text{otherwise},\end{aligned}$$

with $\xi = -d, 0$ defining the points at which the activity A crosses threshold, and d is the pulse width. Taking $a(x,t) = A(x-ct)$ and $q(x,t) = Q(x-ct)$, the profile of the pulse is governed by

$$(5.56) \qquad -c\, A_\xi = -A - \beta Q + \int_{-d}^{0} w(\xi - \eta) d\eta,$$

$$(5.57) \qquad -\frac{c}{\epsilon}\, Q_\xi = -Q\ +\ A.$$

The existence and stability of traveling pulse solutions of equation (5.57) can be analyzed by extending the basic methods used for the scalar equation (5.38) [**190, 191, 192**]. One typically finds two solution branches corresponding to a stable fast wave and an unstable slow wave, see Figure 13.

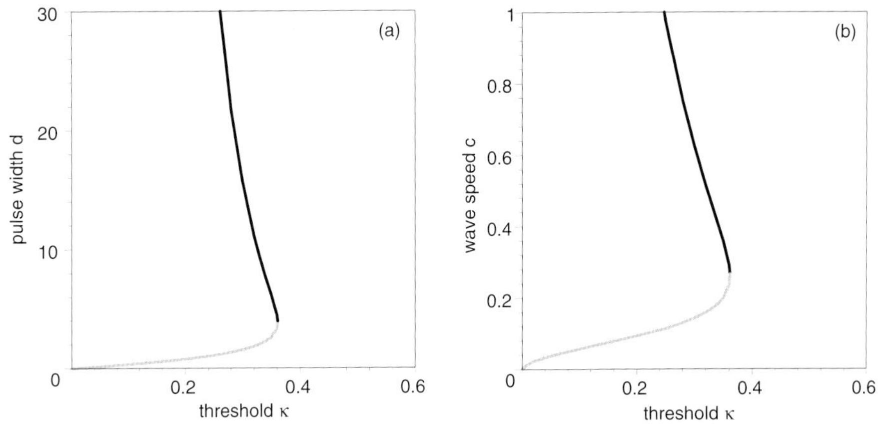

FIGURE 13. Existence of right–moving traveling pulses in the case of the excitatory network (5.55) for an exponential weight distribution with $w(x) = e^{-\sigma|x|}$. Here $\sigma = 1, \varepsilon = 0.01$ and $\beta = 2.5$. (a) Plot of pulse width d against threshold κ. (b) Plot of wave speed c against threshold κ. Stable (unstable) branches indicated by black (gray) curves.

Analysis of the above model provides valuable information regarding how the speed of a traveling wave, which is relatively straightforward to measure experimentally, depends on various features of the underlying neural tissue [**193**]. Indeed, one prediction of the model, concerning how the speed of the wave depends on the firing threshold of the neurons, has recently been confirmed experimentally in disinhibited rat cortical slices [**194**]. External electric fields are used to modulate the threshold and thus control wave propagation, see Figure 14. Finally, note that traveling pulses have also been studied in spiking neuron models, under the simplifying assumption that each neuron only fires once in a propagating sequence [**151, 195**]. This simplification is motivated by simulations of more detailed conductance–models suggesting that the propagation velocity is determined primarily by the response of the postsynaptic neuron to the first one or two presynaptic spikes [**186**].

Thalamic waves

There are significant differences in the mechanisms underlying the spread of activity in cortical and thalamic slices. In computational and experimental studies of disinhibited neocortical slices, one finds that neuronal discharges propagate continuously at a velocity $c \sim 10$ cm/sec [**186**]. On the other hand, in models of thalamic

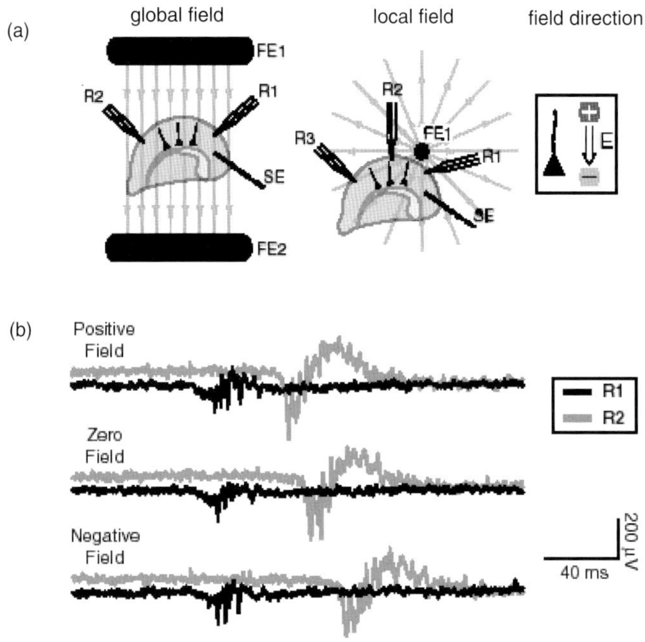

FIGURE 14. (a) Rat cortical slices are bathed in picrotoxin (a $GABA_A$ blocker) and a stimulation electrode (SE) is placed in layers 5-6 to initiate epileptiform bursts. An electric field is applied globally or locally across the slice using Ag/AgCl electrodes (FE1,FE2). Layer 5 neurons have long apical dendrites and are easily polarizable by an electric field, which controls the effective firing threshold of the neuron. (b) The time for an activity pulse to travel between two recording electrodes R1 and R2 depends on the applied electric field, reflecting the dependence of wave speed on the effective firing threshold. [Reprinted from Richardson, Schiff and Gluckman, [**194**], Copyright 2005, with permission from the American Physical Society.]

slices, composed of excitatory thalamocortical neurons and inhibitory reticular thalamic neurons, waves propagate in a lurching manner at a velocity $c \sim 1$ cm/sec [**196**]. This is thought to form the basic mechanism for the generation of 7- to 14-Hz spindle oscillations during the onset of sleep [**28**]. Each recruitment cycle of the lurching waves has two stages (see Figure 15): I. A new group of inhibitory RE cells is excited by synapses from TC cells, and this RE group then inhibits a new group of TC cells. II. The new recruited TC cells rebound from hyperpolarization and fire a burst of spikes, which further recruit more RE cells during next cycle. A detailed analysis of the post inhibitory rebound mechanism underlying the generation of continuous and lurching waves in thalamic networks has recently been carried out using conductance–based [**197**] and integrate–and–fire–or–burst models [**198**].

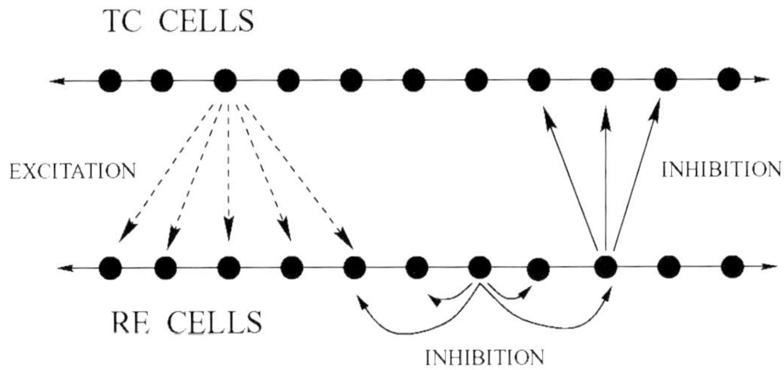

FIGURE 15. Two-layer thalamic model. Inhibitory reticular (RE) cells inhibit excitatory thalamocortical (TC) cells, and TC cells excite RE cells. Mutual inhibition between RE cells has a relatively small effect on discharge propagation.

Bibliography

[1] E Izhikevich. *Dynamical Systems in Neuroscience: The Geometry of Excitability and Bursting.* MIT Press, Cambridge, MA, 2006.

[2] C Koch. *Biophysics of Computation.* Oxford University Press, 1999.

[3] J H Byrne and J L Roberts. *From molecules to networks: an introduction to cellular and molecular neuroscience.* Elsevier, Amsterdam, 2004.

[4] D Terman. An introduction to dynamical systems and neuronal dynamics. In *Tutorials in Mathematical Biosciences I, Lecture Notes in Mathematics, Vol. 1860*, pages 21–68. Springer, Berlin, 2005.

[5] A Borisyuk and J Rinzel. Understanding neuronal dynamics by geometrical dissection of minimal models. In *Methods and Models in Neurophysics, Les Houches Summer School, Session LXXX*, pages 21–68. Elsevier, 2005.

[6] G B Ermentrout. Neural oscillators. In *Tutorials in Mathematical Biosciences I, Lecture Notes in Mathematics, Vol. 1860*, pages 69–106. Springer, Berlin, 2005.

[7] W Gerstner and W Kistler. *Spiking neuron models.* Cambridge University Press, Cambridge, 2002.

[8] A L Hodgkin and A F Huxley. A quantitative description of membrane and its application to conduction and excitation in nerve. *J. Physiol.*, 117:500–544, 1952.

[9] S Coombes and P C Bressloff (eds). *Bursting: The Genesis of Rhythm in the Nervous System.* World Scientific Press, 2005.

[10] A Winfree. *The geometry of biological time.* Springer-Verlag, New York, 1980.

[11] Y Kuramoto. *Chemical Oscillations, Waves and Turbulence.* Springer-Verlag, New-York, 1984.

[12] L Glass and M C Mackey. *From Clocks to Chaos.* Princeton Univ Press, Princeton, 1988.

[13] J Guckenheimer and P J Holmes. *Nonlinear Oscillations, Dynamical Systems and Bifurcations of Vector Fields.* Springer-Verlag, New York, 1983.

[14] C Morris and H Lecar. Voltage oscillations in the barnacle giant muscle fiber. *J. Biophys.*, 35:193–213, 1981.

[15] J P Keener, F C Hoppensteadt, and J Rinzel. Integrate–and–fire models of nerve membrane response to oscillatory input. *SIAM J. Appl. Math.*, 41(3):503–517, 1981.

[16] W Gerstner and J L van Hemmen. Coding and information processing in neural networks. In E Domany, J L van Hemmen, and K Schulten, editors, *Models of Neural Networks II*, pages 1–93. Springer-Verlag, 1994.

[17] W Gerstner and W Kistler. Population dynamics of spiking neurons: fast transients, asynchronous states and locking. *Neural Comput.*, 12:43–89, 2000.

[18] W M Kistler, W Gerstner, and J L van Hemmen. Reduction of the Hodgkin-Huxley equations to a single-variable threshold. *Neural Comput.*, 9:1015–1045, 1997.

[19] L F Abbott and C van Vresswijk. Asynchronous states in networks of pulse–coupled oscillators. *Phys. Rev. E*, 48(2):1483–1490, 1993.

[20] G B Ermentrout and N Kopell. Parabolic bursting in an excitable system coupled with a slow oscillation. *SIAM J. of Appl. Math.*, 46:233–253, 1986.

[21] G B Ermentrout. Type I membranes, phase resetting curves, and synchrony. *Neural Comput.*, 8:979–1001, 1996.

[22] B Gutkin and G B Ermentrout. Dynamics of membrane excitability determine interspike interval variability; a link between spike generation mechanisms and cortical spike train statistics. *Neural Comput.*, 10:1047–1065, 1998.

[23] X-J Wang and J Rinzel. Oscillatory and bursting properties of neurons. In M. Arbib, editor, *Handbook of Brain Theory and Neural Networks.* MIT Press, 1995.

[24] X-J Wang and J Rinzel. Alternating and synchronous rhythms in reciprocally inhibitory model neurons. *Neural Comput.*, 4:84–97, 1992.

[25] G D Smith, C L Cox, S M Sherman, and J Rinzel. Fourier analysis of sinusoidally driven thalamocortical relay neurons and a minimal integrate-and-burst model. *J. Neurophysiol.*, 83:588–610, 2000.

[26] A R R Casti et al. A population study of integrate–and–fire–or burst neurons. *Neural Comput.*, 14:957–986, 2002.

[27] S Coombes, M R Owen, and G D Smith. Mode-locking in a periodically forced integrate-and-fire-or-burst neuron model. *Physical Review E*, 64:041914, 2001.

[28] M Steriade. Coherent oscillations and short-term plasticity in corticothalamic networks. *Trends in Neurosci.*, 2:337–345, 1999.

[29] A Destexhe and T J Sejnowski. *Thalamocortical Assemblies: How Ion Channels, Single Neurons and Large-Scale Networks Organize Sleep Oscillations*. Oxford University Press, Oxford, 2001.

[30] R W Guillery and S M Sherman. Thalamic relay functions and their role in corticocortical communication: generalizations from the visual system. *Neuron*, 33:163–175, 2002.

[31] E F Pace-Schott and J A Hobson. The neurobiology of sleep: genetics, cellular physiology and subcortical networks. *Nature Rev. Neurosci.*, 3:591–605, 2002.

[32] M A Huertas, J R Groff, and G D Smith. Feedback inhibition and throughput properties of an ifb network model of retinogeniculate transmission. *J. Comput. Neurosci.*, 19:147–180, 2005.

[33] B Babadi. Bursting as an effective relay mode in a minimal thalamic model. *J. Comput. Neurosci.*, 18:229–243, 2005.

[34] A Fogelson and R S Zucker. Presynaptic calcium diffusion from various arrays of single channels. implications for transmitter release and synaptic facilitation. *Biophys. J.*, 48:1003–1017, 1985.

[35] R Bertram, G D Smith, and A Sherman. Modeling study of the effects of overlapping calcium microdomains on neurotransmitter release. *Biophys. J.*, 76:735–750, 1999.

[36] E R Kandel and J H Schwartz anf T M Jessel. *Principles of Neural Science*. McGraw Hill, New York, 2000.

[37] A Destexhe, Z F Mainen, and T J Sejnowski. Synthesis of models for excitable membranes synaptic transmission and neuromodulation using a common kinetic formalism. *J. Comput. Neurosci.*, 1:195–231, 1994.

[38] R C Malenka and R A Nicoll. Long-term potentiation–a decade of progress? *Science*, 285:1870–1874, 1999.

[39] H Markram and M Tsodyks. Redistribution of synaptic efficacy between neocortical pyramidal neurons. *Nature*, 382:807–810, 1996.

[40] L F Abbott, J A Varela, K Sen, and S B Nelson. Synaptic depression and cortical gain control. *Science*, 275:220–224, 1997.

[41] L F Abbott and E Marder. Modelling small networks. In C Koch and I Segev, editors, *Methods in Neuronal Modelling*, pages 361–410. MIT Press, 2nd edition, 1998.

[42] R S Zucker. Short term synaptic plasticity. *Ann. Rev. Neurosci.*, 12:13–31, 1989.

[43] G L Millhauser, E E Salpeter, and R E Oswald. Diffusion model of ion–channel gating and the origin of power–law distributions from single–channel recordings. *Proc. Natl. Acad. Sci. USA*, 85:1503–1507, 1988.

[44] S Fusi, P J Drew, and L F Abbott. cascade models of synaptically stored memories. *Neuron*, 45:599–611, 2005.

[45] G Gilboa, R Chen, and N Brenner. History–dependent multiple-time-scale dynamics in a single-neuron model. *J. Neurosci.*, 25:6479–6489, 2005.

[46] W Rall and H A Snir. Cable theory for dendritic neurons. In C Koch and I Segev, editors, *Methods in Neuronal Modelling (2nd ed.)*. MIT Press, Cambridge, 1998.

[47] H C Tuckwell. *Introduction to Theoretical Neurobiology*, volume 1. (Linear cable theory and dendritic structure) of *Cambridge Studies in Mathematical Biology*. Cambridge University Press, 1988.

[48] P C Bressloff and S Coombes. Physics of the extended neuron. *Int. J. Mod. Phys. B*, 11(20):2343–2392, 1997.

[49] G Stuart, N Spruston, and M Hausser. *Dendrites*. Oxford University Press, New York, 1999.

[50] J P Miller, W Rall, and J Rinzel. Synaptic amplification by active membrane in dendritic spines. *Brain Res.*, 325:325–330, 1985.

[51] G M Shepherd, R K Brayton, J P Miller, I Segev, J Rinzel, and W Rall. Signal enhancement in distal cortical dendrites by means of interactions between active dendritic spines. *Proc. Natl. Acad. Sci. USA*, 82:2192–2195, 1985.

[52] G M Shepherd. The dendritic spine: a multifunctional unit. *J. Neurophysiol.*, 75:2197–2210, 1996.

[53] S M Baer and J Rinzel. Propagation of dendritic spikes mediated by excitable spines: a continuum theory. *J. Neurophysiol.*, 65:874–890, 1991.

[54] S Coombes and P C Bressloff. Solitary waves in a model of dendritic cable with active spines. *SIAM J. Appl. Math.*, 61:432–453, 2000.

[55] R Yuste and W Denk. Dendritic spines as a basic functional units of neuronal integration. *Nature*, 375:682–684, 1995.

[56] A Matus. Actin plasticity in dendritic spines. *Science*, 290:754–758, 2000.

[57] R Yuste and T Bonhoeffer. Morphological changes in dendritic spines associated with LTP. *Ann. Rev. Neurosci.*, 24:1071–1089, 2001.

[58] N Mataga, Y Mizuguchi, and T K Hensch. Experience-dependent pruning of dendritic spines in visual cortex by tissue plasminogen activator. *Neuron*, 44:1031–1041, 2004.

[59] S Oray, A Majewska, and M Sur. Dendritic spine dynamics are regulated by monocular deprivation and extracellular matrix degredation. *Neuron*, 44:1021–1030, 2004.

[60] C Koch and A M Zador. The function of dendritic dendritic spines. devices subserving biochemical rather than electrical compartmentalization. *J. Neurosci.*, 13:413–422, 1993.

[61] E A Nimchinsky, B L Sabatini, and K Svoboda. Structure and function of dendritic spines. *Ann. Rev. Physiol.*, 64:313–353, 2002.

[62] K M Franks and T J Sejnowski. Complexity of calcium signaling in synaptic spines. *Bioessays*, 24:1130–1144, 2002.

[63] D Holcman, Z Schuss, and E Korkotian. Calcium dynamics in dendritic spines and spine motility. *Biophys. J.*, 87:81–91, 2004.

[64] D O Hebb. *The organization of behavior.* Wiley, New York, 1949.

[65] T V P Bliss and T Lomo. Long–lasting potentiation of synaptic transmission in the dentate area of unanaeasthetized rabbbit following stimulation of the perforant path. *J. Physiol. (London)*, 232:331–356, 1973.

[66] S M Dudek and M F Bear. Homosynaptic long–term depression in area ca1 of hippocampus and effects of NMDA receptor blockade. *Proc. natl. Acad. Sci.*, 89:4363–4367, 1992.

[67] T V P Bliss and G L Collingridge. A synaptic model of memory: long–term potentiation in the hippocampus. *Nature*, 361:31–39, 1993.

[68] J E Lisman. Long-term potentiation: outstanding questions and attempted synthesis. *Phi. Trans. R. Soc. Lond. B*, 358:829–842, 2003.

[69] R C Malenka and M F Bear. LTP and LTD: an embarrassment of riches. *Neuron*, 44:5–21, 2004.

[70] J C Magee and D Johnston. A synaptically contriolled associative signal for hebbian plasticity in hippocampal neurons. *Science*, 275:209–213, 1997.

[71] H Markram, J Lubke, M Frotscher, and B Sakmann. Regulation of synaptic efficacy by conincidence of postsynaptic aps and epsps. *Science*, 275:213–215, 1997.

[72] G Bi and M Poo. Synaptic modification of correlated activity: Hebb's postulate revisited. *Ann. Rev. Neurosci.*, 24:139–166, 2001.

[73] G Bi and M Poo. Synaptic modification in cultured hippocampal neurons: dependence on spike timing, synaptic strength and postsynaptic cell type. *J. Neurosci.*, 18:10464–10472, 1998.

[74] J E Lisman. Amechanism for the hebb and the anti–hebb processes underlying learning and memory. *Proc. Natl. Acad. Sci*, 86:9574–9578, 1989.

[75] R C Malenka, J A Kauer, R S Zucker, and R A Nicoll. Postsynaptic calcium is sufficient for potentiation of hippocampal synaptic transmission. *Science*, 242:81–84, 1988.

[76] S N Yang, Y G Tang, and R A Nicoll. Selective induction of LTP and LTD by postsynaptic calcium elevation. *J. Neurophysiol.*, 81:781–787, 1999.

[77] H Shouval, M F Bear, and L N Cooper. A unified model of NMDA receptor–dependent bidirectional synaptic plasticity. *Proc. Natl. Acad. Sci.*, 99:10831–10836, 2002.

[78] C Jahr and C Stevens. A quantitative description of NMDA receptor-channel kinetic behavior. *J. Neurosci.*, 10:1830–1837, 1990.

[79] J E Lisman and N Spruston. Postsynaptic depolarization requirements for LTP and LTD: a critique of spike timing–dependent plasticity. *Nat. Neurosci.*, 8:839–841, 2005.

[80] J Rubin, R C Gherkin, G Q Bi, and C C Chow. Calcium time course as a signal for spike–timing–synaptic–plasticity. *J. Neurophsyiol.*, 93:2600–2613, 2005.

[81] G L Collingridge, J T R Isaac, and Y T Wang. Receptor trafficking and synaptic plasticity. *Nature Rev. Neurosci.*, 5:952–962, 2004.

[82] D Choquet and A Trillier. The role of receptor diffusion in the organization of the postsynaptic membrane. *Nat. Rev. Neurosci.*, 4:251–265, 2003.

[83] A Triller and D Choquet. Surface trafficking of receptors between synaptic and extrasynaptic membranes. *Trends Neurosci.*, 28:133–139, 2005.

[84] B Earnshaw and P C Bressloff. A biophysical model of AMPA receptor trafficking and its regulation during ltp/ltd. *J. Neurosci.*, 26:12362–12373, 2006.

[85] J E Lisman and H Schulman anf H Cline. The molecular basis of camkii function in synaptic and behavioral memory. *Nat. Rev. Neurosci.*, 3:175–190, 2002.

[86] A M Zhabotinsky. Bistability in the Ca^{2+}/calmodulin–dependent protein kinase–phosphotase system. *Biophys. J.*, 79:2211–2221, 2000.

[87] J E Lisman and A M Zhabotinsky. A model of synaptic memory: A camkii/pp1 switch that potentiates transmission by organizing an AMPA receptor anchoring system. *Neuron*, 31:191–201, 2001.

[88] P Miller, A M Zhaobotinsky, J E lisman, and X-J Wang. The stability of a stochastic CaMKII switch: dependence on the number of enzyme molecules and protein turnover. *PLoS Biology*, 3:705–717, 2005.

[89] J D Petersen, X Chen, L Vinade, A Dosemeci, and J E Lisman. Distribution of (PSD)–95 and CaMKII to the postsynaptic density. *J. Neurosci.*, 23:11270–11278, 2003.

[90] P N R Rao and T J Sejnowski. Self-organizing neural systems based on predictive learning. *Phil. Trans. R. Soc. Lond.*, 361:1149–1175, 2003.

[91] M R Mehta, M Quirk, and M Wilson. Experience–dependent asymmetric shape of hippocampal receptive fields. *Neuron*, 25:707–715, 2000.

[92] W Schultz. Getting formal with dopamine and reward. *Neuron*, 36:241–263, 2002.

[93] P R Montague, S E Hyman, and J D Cohen. Computational roles for dopamine in behavioural control. *Nature*, 431:760–767, 2004.

[94] F Rieke, D Warland, R van Steveninck, and W Bialek. *Spikes – exploring the neural code*. MIT Press, Cambridge, MA, 1997.

[95] P Dayan and L F Abbott. *Theoretical neuroscience*. MIT, Cambridge, MA, 2001.

[96] R Baddeley, L F Abbott, M J A Booth, F Sengpiel, T Freeman, E A Wakeman, and E T Rolls. Responses of neurons in primary and interior temporal visual cortices to natural scenes. *Proc. Royal Society Lond.*, B264:1775–1783, 1997.

[97] M N Shadlen and W T Newsome. Noise, neural codes and cortical organization. *Curr. Opin. Neurobiology*, 4:569–579, 1994.

[98] S Thorpe, D Fize, and C Marlot. Speed of processing in the human visual system. *Nature*, 381:520–522, 1996.

[99] M Rossumn, G G Turrigiano, and S B Nelson. Fast propagation of firing rates through layered networks of noisy neurons. *J. Neurosci.*, 22:1956–1966, 2002.

[100] M Tsodyks and T J Sejnowski. Rapid switching in balanced cortical network models. *Network Comput. Neural Syst.*, 6:111–124, 1995.

[101] G B Ermentrout. Neural networks as spatio-temporal pattern-forming systems. *Rep. Prog. Phy.*, 61:353–430, 1998.

[102] P C Bressloff. Pattern formation in visual cortex. In C C Chow, B Gutkin, D Hansel, C Meunier, and J Dalibard, editors, *Les Houches 2003: methods and models in neurophysics*, pages 477–574. Elsevier, 2005.

[103] J O'Keefe and M L Recce. Phase relationship between hippocampal place units and the EEG theta rhythm. *Hippocampus*, 3:317–330, 1993.

[104] J E Lisman. Relating hippocampal circuitry to function: recall of memory sequences by reciprocal dentate-ca3 interactions. *Neuron*, 22:233–242, 1999.

[105] G Laurent. Odor encoding as an active, dynamical process: experiments, computation and theory. *Ann. Rev. Neurosci.*, 24:263–297, 2001.

[106] K R Delaney, A Galperin, M S Fee, J A Flores, R Gervais, D W Tank, and D Kleinfeld. Waves and stimulus-modulated dynamics in an oscillating olfactory network. *Proc. Natl. Acad. Sci. USA*, 91:669–673, 1994.

[107] C M Gray. Synchronous oscillations in neuronal systems: mechanisms and functions. *J. Comput. Neurosci.*, 1:11–38, 1994.

[108] Y Dan, J M Alonso snd W M Usrey, and R C Reid. Coding of visual information by precisely correlated spikes in the lateral geniculate nucleus. *Nature Neurosci.*, 1:501–507, 1998.

[109] R C DeCharms and M Merzenich. Primary cortical representation of sounds by the coordination of action potential timing. *Nature*, 381:610–613, 1995.

[110] W Singer and C M Gray. Visual feature integration and the temporal correlation hypothesis. *Ann. Rev. Neurosci.*, 18:555–586, 1995.

[111] M N Shadlen and J A Movshon. Synchrony unbound: a critical evaluation of the temporal binding hypothesis. *Neuron*, 24:67–77, 1999.

[112] C M Gray. The temporal correlation hypothesis of visual feature integration: still alive and well. *Neuron*, 24:31–47, 1999.

[113] E Salinas and T J Sejnowski. Correlated neuronal activity and the flow of neural information. *Nature Rev. Neurosci.*, 4:539–550, 2001.

[114] W Bair, C Koch, W T Newsome, and K H Britten. Power spectrum analysis of bursting cells in area mt in the behaving monkey. *J. Neurosci.*, 14:2870–2893, 1994.

[115] W R Softky and C Koch. Cortical cell should spike regularly but do not. *Neural Comput*, 4:643–646, 1992.

[116] H C Tuckwell. *Introduction to theoretical neurobiology*, volume 2. (Nonlinear and stochastic theories) of *Cambridge studies in Mathematical Biology*. Cambridge University Press, 1988.

[117] van Kampen. *Stochastic processes in physics and chemistry*. North-Holland, Amsterdam, 1992.

[118] C W Gardiner. *Handbook of stochastic methods, 2nd edition*. Springer, Berlin, 1997.

[119] M J E Richardson. Effects of synaptic conductance on the voltage distribution and firing rate of spiking neurons. *Phys. Rev. E*, 69:051918, 2004.

[120] P Konig, A K Engel, and W Singer. Integrator or coincidence detector? the role of the cortical neuron revisited. *Trends Neurosci.*, 19:130–137, 1996.

[121] L Gammaitoni, P Hanggi, P Jung, and F Marchesoni. Stochastic resonance. *Rep. Prog. Phys.*, 70:223–287, 1998.

[122] A Longtin. Stochastic resonance in neuron models. *J. Stat. Phys.*, 70:309–327, 1993.

[123] B Linder, J Garcia-Ojalvo, A Neiman, and L Schimansky-Geier. Effects of noise in excitable systems. *Phys. Rep.*, 392:321–424, 2004.

[124] N Brunel and V Hakim. Fast global oscillations in networks of integrate–and–fire neurons with low firing rates. *Neural Comput.*, 11:1621–1671, 1999.

[125] W Gerstner and J L Van Hemmen. Coherence and incoherence in a globally coupled ensemble of pulse–emitting units. *Phys. Rev. Lett.*, 71(3):312–315, 1993.

[126] C van Vreeswijk and L F Abbott. Self–sustained firing in populations of integrate–and–fire neurons. *SIAM J. Appl. Maths*, 53(1):253–264, 1993.

[127] C van Vreeswijk and H Sompolinsky. Chaotic balanced state in a model of cortical circuits. *Neural Comput.*, 10:1321–1371, 1998.

[128] D J Amit and N Brunel. Model of global spontaneous activity and local structured activity during delay periods in the cerebral cortex. *Cereb Cortex*, 7:237–252, 1997.

[129] N Brunel. Dynamics of sparsely connected networks of excitatory and inhibitory spiking neurons. *J Comput. Neurosci*, 8:183–208, 2000.

[130] A Omurtag, B W Knight, and L Sirovich. On the simulation of large populations of neurons. *J. Comput. Neurosci.*, 8:51–63, 2000.

[131] L Sirovich. Dynamics of neuronal populations: eigenfunction theory; some solvable cases. *Network: Comput. Neural Syst.*, 14:249–272, 2003.

[132] D Cai, L Tao, M Shelley, and D W McLaughlin. An effective kinetic representation of fluctuation–driven neuronal networks with application to simple and complex celss in visual cortex. *Proc. Natl. Acad. Sci. USA*, 101:7757–7562, 2004.

[133] H G Schuster and P Wagner. A model for neuronal oscillations in visual cortex. *Biol. Cybern.*, 64:77–82, 1990.

[134] E R Grannan, D Kleinfeld, and H Sompolinsky. Stimulus dependent synchronization of neuronal assemblies. *Neural Comput.*, 5:550–569, 1993.

[135] S E Folias and P C Bressloff. Breathing pulses in an excitatory neural network. *SIAM J. Appl. Dyn. Syst.*, 3:378–407, 2004.

[136] S E Folias and P C Bressloff. Breathers in two–dimensional neural media. *Phys. Rev. Lett.*, 95:208107, 2005.

[137] G B Ermentrout and N Kopell. Multiple pulse interactions and averaging in systems of coupled neural oscillators. *J. Math Biol.*, 29:195–217, 1991.

[138] F C Hoppensteadt and E Izhekevich. *Weakly Connected Neural Nets.* Springer-Verlag, New York, 1997.

[139] G B Ermentrout. The behaviour of rings of coupled oscillators. *Journal of Mathematical Biology*, 23:55–74, 1985.

[140] P C Bressloff and S Coombes. Symmetry and phase-locking in a ring of pulse-coupled oscillators with distributed delays. *Physica D*, 126:99–122, 1999.

[141] C van Vreeswijk, G B Ermentrout, and L F Abbott. When inhibition not excitation synchronizes neural firing. *J. Comput. Neurosci.*, 1:313–321, 1994.

[142] D Hansel, G Mato, and C Meunier. Synchrony in excitatory neural networks. *Neural Comput.*, 7:2307–2337, 1995.

[143] R E Mirollo and S H Strogatz. Synchronisation of pulse–coupled biological oscillators. *SIAM J. Appl. Math.*, 50(6):1645–1662, 1990.

[144] S Coombes and G J Lord. Desynchronisation of pulse–coupled integrate–and–fire neurons. *Phys. Rev. E*, 55(3):R2104–R2107, 1997.

[145] A H Cohen, G B Ermentrout, T Kiermel, N Kopell, K A Sigvardt, and T L Williams. Modeling of intersegmental coordination in the lamprey central pattern generator for motion. *Trends in Neurosci.*, 15:434–438, 1992.

[146] E Marder and R L Calabrese. Principles of rhythmic motor pattern generation. *Physiol. Rev.*, 76:687–717, 1996.

[147] S Grillner. The motor infrastructure: from ion channels to neuronal networks. *Nat. Rev. Neurosci.*, 4:573–586, 2003.

[148] W O Friesen and R A Pearce. Mechanism of intersegmental coordination in leech locomotion. *Semin. Neurosci.*, 4:41–47, 1993.

[149] G B Ermentrout and N Kopell. Frequency plateaus in a chain of weakly coupled oscillators. *SIAM J. Appl. Math.*, 15:215–237, 1984.

[150] N Kopell and G B Ermentrout. Symmetry and phase-locking in chains of weakly coupled oscillators. *Comm. Pure Appl. Math.*, 39:623–660, 1986.

[151] G B Ermentrout. The analysis of synaptically generated travelling waves. *J. Comput. Neurosc.*, 5:191–208, 1998.

[152] P C Bressloff and S Coombes. Dynamics of strongly coupled spiking neurons. *Neural Comput.*, 12:91–129, 2000.

[153] P C Bressloff and S Coombes. Dynamical theory of spike train dynamics in networks of integrate-and-fire oscillators. *SIAM J. Appl. Math*, 60:828–841, 2000.

[154] D Golomb and J Rinzel. Clustering in globally coupled inhibitory neurons. *Physica D*, 72:259–282, 1994.

[155] H R Wilson and J D Cowan. Excitatory and inhibitory interactions in localized populations of model neurons. *Biophys. J.*, 12:1–23, 1972.

[156] H R Wilson and J D Cowan. A mathematical theory of the functional dynamics of cortical and thalamic nervous tissue. *Kybernetik*, 13:55–80, 1973.

[157] D H Hubel and T N Wiesel. Receptive fields, binocular interaction and functional architecture in the cat's visual cortex. *J. Neurosci.*, 3:1116–1133, 1962.

[158] N V Swindale. The development of topography in the visual–cortex: A review of models. *Network*, 7(2):161–274, 1996.

[159] S Amari. Dynamics of pattern formation in lateral inhibition type neural fields. *Biol. Cybern.*, 27:77–87, 1977.

[160] G B Ermentrout and J Cowan. A mathematical theory of visual hallucination patterns. *Bio. Cybern.*, 34:137–150, 1979.

[161] P C Bressloff J D Cowan, M Golubitsky, P J Thomas, and M Wiener. Geometric Visual Hallucinations, Euclidean Symmetry and the Functional Architecture of Striate Cortex. *Phil. Trans. Roy. Soc. Lond. B*, 356:299–330, 2001.

[162] R. B. H. Tootell, E Switkes, M. S. Silverman, and S.L. Hamilton. Functional anatomy of macaque striate cortex. II. Retinoptic organization. *J. Neurosci.*, 8:1531–1568, 1988.

[163] E Schwartz. Spatial mapping in the primate sensory projection: analytic structure and relevance to projection. *Biol. Cybern.*, 25:181–194, 1977.

[164] D Pinto and G B Ermentrout. Spatially structured activity in synaptically coupled neuronal networks: Ii. lateral inhibition and standing pulses. *SIAM J. Appl. Math*, 62:226–243, 2001.

[165] K Kishimoto and S Amari. Existence and stability of local excitations in homogeneous neural fields. *J. Math. Biol.*, 7:303–318, 1979.

[166] C R Laing, W C Troy, B Gutkin, and G B Ermentrout. Multiple bumps in a neuronal model of working memory. *SIAM J. Appl. Math*, 63:62–97, 2002.

[167] C R Laing and W C Troy. Pde methods for nonlocal models. *SIAM J. Appl. Dyn. Syst.*, 2:487–516, 2003.

[168] C R Laing and C C Chow. Stationary bumps in networks of spiking neurons. *Neural Comput.*, 13:14731494, 2001.

[169] H Werner and T Richter. Circular stationary solutions in two-dimensional neural fields. *Biol. Cybern.*, 85:211–217, 2001.

[170] P C Bressloff. Weakly interacting pulses in synaptically coupled excitable neural media. *SIAM J. Appl. Math.*, 66:57–81, 2005.

[171] S Coombes. Waves, bumps and patterns in neural field theories. *Biol. Cybern.*, 93:91–108, 2005.

[172] X-J Wang. Synaptic reverberation underlying mnemonic persistent activity. *Trends Neurosci.*, 24:455–463, 2001.

[173] C D Brody, R Romo, and A Kepecs. Basic mechanisms for graded persistent activity: discrete attractors, continuous attractors, and dynamic representations. *Curr. Opin. Neuro.*, 13:204–211, 2003.

[174] J M Fuster and G Alexander. Neuron activity related to short-term memory. *Science*, 173:652, 1971.

[175] J S Taube and J P Bassett. Persistent neural activity in head direction cells. *Cereb. Cortex*, 13:1162–1172, 2003.

[176] T Ritz and T J Sejnowski. Neuronal correlates of parametric working memory in the prefrontal cortex. *Nature*, 399:470–473, 1999.

[177] H S Seung. How the brain keeps the eyes still. *Proc. Natl. Acad. Sci. USA*, 93:13339–13344, 1996.

[178] E Aksay, R Baker, H S Seung, and D W Tank. Anatomy and discharge properties of premotor neurons in the goldsh medulla that have eye-position signals during xations. *J. Neurophysiol.*, 84:1035–1049, 2000.

[179] H S Seung, D D Lee, B Y Reis, and D W Tank. Stability of the memory of eye position in a recurrent network of conductance-based model neurons. *Neuron*, 26:1259–271, 2000.

[180] M Camperi and X-J Wang. A model of visuospatial short-term memory in prefrontal cortex: recurrent network and cellular bistability. *J. Comp. Neurosci.*, 5:383–405, 1998.

[181] C P Fall, T J Lewis, and J Rinzel. Background–activity–dependent properties of a network model for working memory that incorporates cellular bistability. *Biol. Cybern.*, 93:109–18, 2004.

[182] A A Koulakov, S Raghavachari, A Kepecs, and J E Lisman. Model for a robust neural integrator. *Nat. Neurosci.*, 5:775–782, 2002.

[183] X-J Wang. Synaptic basis of cortical persistent activity: the importance of nmda receptors to working memory. *J. Neurosci.*, 19:9587–9603, 1999.

[184] A Renart, N Brunel, and X-J Wang. Mean–field theory of irregularly spiking neuronal populations and working memory in recurrent cortical networks. In J Feng, editor, *Computational Neuroscience: a comprehensive approach*, pages 431–490. CRC Press, Boca Raton, Fl, 2004.

[185] R D Chervin, P A Pierce, and B W Connors. Periodicity and directionality in the propagation of epileptiform discharges across neocortex. *J. Neurophysiol.*, 60:1695–1713, 1988.

[186] D Golomb and Y Amitai. Propagating neuronal discharges in neocortical slices: Computational and experimental study. *J. Neurophysiol.*, 78:1199–1211, 1997.

[187] J-Y Wu, L Guan, and Y Tsau. Propagating activation during oscillations and evoked responses in neocortical slices. *J. Neurosci.*, 19:5005–5015, 1999.

[188] D Pinto, S L patrick, W C Huang, and B W Connors. Initiation, propagation, and termination of epileptiform activity in rodent neocortex in vitro involve distinct mechanisms. *J. Neurosci.*, 25:8131–8140, 2005.

[189] G B Ermentrout and J B Mcleod. Existence and uniqueness of travelling waves for a neural network. *Proc. Roy. Soc. Edin. A*, 123:461–478, 1993.

[190] L Zhang. On the stability of traveling wave solutions in synaptically coupled neuronal networks. *Diff. Int. Eq.*, 16:513–536, 2003.

[191] S Coombes and M R Owen. Evans functions for integral neural field equations with heaviside firing rate function. *SIAM J. Appl. Dyn. Syst.*, 4:574–600, 2004.

[192] S E Folias and P C Bressloff. Stimulus–locked traveling pulses and breathers in an excitatory neural network. *SIAM J. Appl. Math.*, 65:2067–2092, 2005.

[193] D Pinto and G B Ermentrout. Spatially structured activity in synaptically coupled neuronal networks: I. traveling fronts and pulses. *SIAM J. Appl. Math*, 62:206–225, 2001.

[194] K A Richardson, S J Schiff, and B J Gluckman. Control of traveling waves in the mammalian cortex. *Phys. Rev. Lett.*, 028103, 2005.

[195] P C Bressloff. Synaptically generated wave propagation in excitable neural media. *Phys. Rev. Lett.*, 82:2979–2982, 1999.

[196] D Golomb, X-J Wang, and J Rinzel. Propagation of spindle waves in a thalamic slice model. *J. Neurophysiol.*, 75:750–769, 1996.

[197] D H Terman, G B Ermentrout, and A C Yew. Propgating activity patterns in thalamic neuronal networks. *SIAM J. Appl. Math*, 61:1578–1604, 2003.

[198] S Coombes. Dynamics of synaptically coupled integrate–and–fire–or–burst neurons. *Phys. Rev. E*, 67:041910, 2003.